WileyPLUS

WileyPLUS is a research-based online environment for effective teaching and learning.

WileyPLUS builds students' confidence because it takes the guesswork out of studying by providing students with a clear roadmap:

- **what to do**
- **how to do it**
- **if they did it right**

It offers interactive resources along with a complete digital textbook that help students learn more. With *WileyPLUS*, students take more initiative so you'll have greater impact on their achievement in the classroom and beyond.

WileyPLUS Tools for Instructors

WileyPLUS enables you to:

- Assign automatically graded homework, practice questions, and quizzes from the end of chapter questions and test bank.

- Track your students' progress in an instructor's grade book.

- Access all teaching and learning resources, including an online version of the text, and student and instructor supplements, in one easy-to-use Web site. These include a **lecturer testbank including over 500 questions**, many of which are algorithmic.

- An all new set of PowerPoint slides, substantially revised to accompany the new edition.

- Create class presentations using Wiley-provided resources, with the ability to customize and add your own materials.

WileyPLUS Resources for Students Within WileyPLUS

- **E-book.** The complete text is available on WileyPLUS with learning links to various features and tools to assist you in learning the subject.

- **Problems in Context** provide students with further help by placing the mathematics into the context of an economics or business topic.

- **Animations.** 16 of the worked examples from the book are animated with a voiceover providing an additional way for students to tackle more difficult material.

- Excel practice exercises and spreadsheets linked to each chapter of the book.

For more information, or to request a test drive, simply visit
www.wileyplus.com

Essential Mathematics for Economics and Business

Fourth Edition

Essential Mathematics for Economics and Business

Fourth Edition

Teresa Bradley

Senan and Ferdia

CONTENTS

Introduction   xiii

CHAPTER 1
Mathematical Preliminaries   1
   1.1   Some Mathematical Preliminaries   2
   1.2   Arithmetic Operations   3
   1.3   Fractions   6
   1.4   Solving Equations   11
   1.5   Currency Conversions   14
   1.6   Simple Inequalities   18
   1.7   Calculating Percentages   21
   1.8   The Calculator. Evaluation and Transposition of Formulae   24
   1.9   Introducing Excel   28

CHAPTER 2
The Straight Line and Applications   37
   2.1   The Straight Line   38
   2.2   Mathematical Modelling   54
   2.3   Applications: Demand, Supply, Cost, Revenue   59
   2.4   More Mathematics on the Straight Line   76
   2.5   Translations of Linear Functions   82
   2.6   Elasticity of Demand, Supply and Income   83
   2.7   Budget and Cost Constraints   91
   2.8   Excel for Linear Functions   92
   2.9   Summary   97

CHAPTER 3
Simultaneous Equations   101
   3.1   Solving Simultaneous Linear Equations   102
   3.2   Equilibrium and Break-even   111
   3.3   Consumer and Producer Surplus   128
   3.4   The National Income Model and the *IS-LM* Model   133
   3.5   Excel for Simultaneous Linear Equations   137
   3.6   Summary   142
       Appendix   143

CHAPTER 4

Non-linear Functions and Applications 147
4.1 Quadratic, Cubic and Other Polynomial Functions 148
4.2 Exponential Functions 170
4.3 Logarithmic Functions 184
4.4 Hyperbolic (Rational) Functions of the Form $a/(bx + c)$ 197
4.5 Excel for Non-linear Functions 202
4.6 Summary 205

CHAPTER 5

Financial Mathematics 209
5.1 Arithmetic and Geometric Sequences and Series 210
5.2 Simple Interest, Compound Interest and Annual Percentage Rates 218
5.3 Depreciation 228
5.4 Net Present Value and Internal Rate of Return 230
5.5 Annuities, Debt Repayments, Sinking Funds 236
5.6 The Relationship between Interest Rates and the Price of Bonds 248
5.7 Excel for Financial Mathematics 251
5.8 Summary 254
     Appendix 256

CHAPTER 6

Differentiation and Applications 259
6.1 Slope of a Curve and Differentiation 260
6.2 Applications of Differentiation, Marginal Functions, Average Functions 270
6.3 Optimisation for Functions of One Variable 286
6.4 Economic Applications of Maximum and Minimum Points 304
6.5 Curvature and Other Applications 320
6.6 Further Differentiation and Applications 334
6.7 Elasticity and the Derivative 347
6.8 Summary 357

CHAPTER 7

Functions of Several Variables 361
7.1 Partial Differentiation 362
7.2 Applications of Partial Differentiation 380
7.3 Unconstrained Optimisation 400
7.4 Constrained Optimisation and Lagrange Multipliers 410
7.5 Summary 422

CHAPTER 8

Integration and Applications 427
8.1 Integration as the Reverse of Differentiation 428
8.2 The Power Rule for Integration 429
8.3 Integration of the Natural Exponential Function 435
8.4 Integration by Algebraic Substitution 436
8.5 The Definite Integral and the Area under a Curve 441

CONTENTS

8.6   Consumer and Producer Surplus   448
8.7   First-order Differential Equations and Applications   456
8.8   Differential Equations for Limited and Unlimited Growth   468
8.9   *Integration by Substitution and Integration by Parts*   website only
8.10   Summary   474

CHAPTER 9
Linear Algebra and Applications   477
9.1   Linear Programming   478
9.2   Matrices   488
9.3   Solution of Equations: Elimination Methods   498
9.4   Determinants   504
9.5   The Inverse Matrix and Input/Output Analysis   518
9.6   Excel for Linear Algebra   531
9.7   Summary   534

CHAPTER 10
Difference Equations   539
10.1   Introduction to Difference Equations   540
10.2   Solution of Difference Equations (First-order)   542
10.3   Applications of Difference Equations (First-order)   554
10.4   Summary   564

Solutions to Progress Exercises   567

Worked Examples   653

Index   659

INTRODUCTION

Many students who embark on the study of economics and/or business are surprised and apprehensive to find that mathematics is a core subject on their course. Yet, to progress beyond a descriptive level in most subjects, an understanding and a certain fluency in basic mathematics is essential. In this text a minimal background in mathematics is assumed: the text starts in Chapter 1 with a review of basic mathematical operations such as multiplying brackets, manipulating fractions, percentages, use of the calculator, evaluating and transposing formulae, the concept of an equation and the solution of simple equations. Throughout the text worked examples demonstrate concepts and mathematical methods with a simple numerical example followed by further worked examples applied to real-world situations. The worked examples are also useful for practice. Start by reading the worked example to make sure you understand the method; then test yourself by attempting the example with a blank sheet of paper! You can always refer back to the detailed worked example if you get stuck. You should then be in a position to attempt the progress exercises. In this new edition, the worked examples in the text are complemented by an extensive question bank in WileyPLUS and MapleTA.

An Approach to Learning

The presentation of content is designed to encourage a metacognitive approach to manage your own learning. This requires a clear understanding of the goals or content of each topic; a plan of action to understand and become competent in the material covered; and to test that this has been achieved.

1. Goals

The learning goals must be clear.

Each chapter is introduced with an overview and chapter objectives. Topics within chapters are divided into sections and subsections to maintain an overview of the chapter's logical development. Key concepts and formulae are highlighted throughout. A summary, to review and consolidate the main ideas, is given within chapters where appropriate, with a final overview and summary at the end.

2. Plan of Action to Understand and Become Competent in the Material Covered

- Understanding the rationale underlying methods is essential. To this end, concepts and methods are reinforced by verbal explanations and then demonstrated in worked examples also interjected with explanations, comments and reminders on basic algebra. More formal mathematical terminologies are introduced, not only for conciseness and to further enhance understanding, but to enable the readers to transfer their mathematical skills to related subjects in economics and business.

- Mathematics is an analytical tool in economics and business. Each mathematical method is followed immediately by one or more applications. For example, demand, supply, cost and revenue functions immediately follow the introduction of the straight line in Chapter 2.

Simultaneous linear equations in Chapters 3 and 9 are followed by equilibrium, taxes and subsides (see Worked Example 3.12 below), break-even analysis, and consumer and producer surplus and input/output analysis.

WORKED EXAMPLE 3.12
TAXES AND THEIR DISTRIBUTION

 Find animated worked examples at **www.wiley.com/college/bradley**

The demand and supply functions for a good are given as

$$\text{Demand function: } P_d = 100 - 0.5Q_d \qquad \textbf{(3.17)}$$

$$\text{Supply function: } P_s = 10 + 0.5Q_s \qquad \textbf{(3.18)}$$

(a) Calculate the equilibrium price and quantity.
(b) Assume that the government imposes a fixed tax of £6 per unit sold.

  (i) Write down the equation of the supply function, adjusted for tax.
  (ii) Find the new equilibrium price and quantity algebraically and graphically.
  (iii) Outline the distribution of the tax, that is, calculate the tax paid by the consumer and the producer.

Further applications such as elasticity, budget and cost constraints, equilibrium in the labour market and the national income model are available on the website **www.wiley.com/college/bradley**. Financial mathematics is an important application of arithmetic and geometric series and also requires the use of rules for indices and logs. Applications and analysis based on calculus in Chapters 6, 7, 8 and 10 are essential for students of economics.

- Graphs help to reinforce and provide a more comprehensive understanding by visualisation. Use Excel to plot graphs since parameters of equations are easily varied and the effect of changes are seen immediately in the graph. Exercises that use Excel are available on the website.
- A key feature of this text is the verbal explanations and interpretation of problems and of their solutions. This is designed to encourage the reader to develop both critical thinking and problem-solving techniques.
- The importance of practice, reinforced by visualisation, is summarised succinctly in the following quote attributed to the ancient Chinese philosopher Lao Tse:

You read and you forget;

You see and you remember;

You do and you learn.

3. Test whether goals were achieved

There are several options:

- Attempt the worked examples without looking at the text.
- Do the progress exercises. Answers and, in some cases, solutions are given at the back of the text.
- In this new edition a large question bank, with questions classified as 'easy', 'average' or 'hard', is provided in WileyPLUS and MapleTA. Many of the questions are designed to reinforce problem-solving techniques and so require several inputs from the reader not just a single final numeric answer.
- A test exercise is given at the end of each chapter. Answers to test exercises are available to lecturers online.

Structure of the Text

Mathematics is a hierarchical subject. The core topics are covered in Chapters 1, 2, 3, 4, 6 and 8. Some flexibility is possible by deciding when to introduce further material based on these core chapters, such as linear algebra, partial differentiation and difference equations, etc., as illustrated in the following chart.

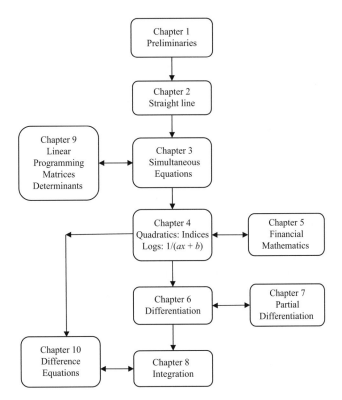

Note: An introductory course may require the earlier sections from the core chapters and a limited number of applications.

WileyPLUS

WileyPLUS is a powerful online tool that provides instructors and students with an integrated suite of teaching and learning resources, including an online version of the text, in one easy-to-use website. To learn more about WileyPLUS, request an instructor test drive or view a demo, please visit www.wileyplus.com.

WileyPLUS Tools for Instructors

WileyPLUS enables you to:

- Assign automatically graded homework, practice, and quizzes from the test bank.
- Track your students' progress in an instructor's grade book.
- Access all teaching and learning resources, including an online version of the text, and student and instructor supplements, in one easy-to-use website. These include an extensive test bank of algorithmic questions; full-colour PowerPoint slides; access to the Instructor's Manual; additional exercises and quizzes with solutions; Excel problems and solutions; and, answers to all the problems in the book.
- Create class presentations using Wiley-provided resources, with the ability to customise and add your own materials.

WileyPLUS Resources for Students within WileyPLUS

In WileyPLUS, students will find various helpful tools, such as an e-book, additional Progress Exercises, Problems in Context, and animated worked examples.

- **e-book** of the complete text is available in WileyPLUS with learning links to various features and tools to assist students in their learning.
- **Additional Progress Exercises** are available affording students the opportunity for further practice of key concepts.
- **Problems in Context** provide students with further exposition on the mathematics elements of a key economics or business topic.
- **Animated Worked Examples** utilising full-colour graphics and audio narration; these animations of key worked examples from the text provide the students with an added way of tackling more difficult material.

Ancillary Teaching and Learning Materials

All materials are housed on the companion website, which you can access at www.wiley.com/ college/bradley.

Students' companion website containing:

- Animations of key worked examples from the text.
- Problems in Context covering key mathematic problems in a business or economics context.
- Excel exercises, which draw upon data within the text.
- Introduction to using Maple, which has been updated for the new edition.
- Additional learning material and exercises, with solutions for practice.

Instructor's companion website containing:

- **Instructor's Manual** containing comments and tips on areas of difficulty; solutions to test exercises; additional exercises and quizzes, with solutions; sample test papers for introductory courses and advanced courses.
- **PowerPoint presentation slides** containing full-colour graphics to help instructors create stimulating lectures.
- **Test bank including algorithmic questions,** which is provided in Microsoft Word format and can be accessed via Maple T.A.
- **Additional exercises and solutions** that are not available in the text.

MATHEMATICAL PRELIMINARIES

1.1 Some Mathematical Preliminaries

1.2 Arithmetic Operations

1.3 Fractions

1.4 Solving Equations

1.5 Currency Conversions

1.6 Simple Inequalities

1.7 Calculating Percentages

1.8 The Calculator. Evaluation and Transposition of Formulae

1.9 Introducing Excel

Chapter Objectives

At the end of this chapter you should be able to:

- Perform basic arithmetic operations and simplify algebraic expressions
- Perform basic arithmetic operations with fractions
- Solve equations in one unknown, including equations involving fractions
- Understand the meaning of no solution and infinitely many solutions
- Convert currency
- Solve simple inequalities
- Calculate percentages.

In addition, you will be introduced to the calculator and a spreadsheet.

1.1 Some Mathematical Preliminaries

Brackets in mathematics are used for grouping and clarity.
Brackets may also be used to indicate multiplication.
Brackets are used in functions to declare the independent variable (see later).
Powers: positive whole numbers such as 2^3, which means $2 \times 2 \times 2 = 8$:

$$(\text{anything})^3 = (\text{anything}) \times (\text{anything}) \times (\text{anything})$$
$$(x)^3 = x \times x \times x$$
$$(x+4)^5 = (x+4)(x+4)(x+4)(x+4)(x+4)$$

Note:
Brackets: $(A)(B)$ or $A \times B$ or AB
all indicate A multiplied by B.

Variables and letters: When we don't know the value of a quantity, we give that quantity a symbol, such as x. We may then make general statements about the unknown quantity or variable, x. For example, 'For the next 15 weeks, if I save x per week I shall have \$4500 to spend on a holiday'. This statement may be expressed as a mathematical equation:

$$15 \times \text{weekly savings} = 4500$$
$$15 \times x = 4500$$

Now that the statement has been reduced to a mathematical equation (see Section 1.4 for more on equations), we may solve the equation for the unknown, x:

$$15x = 4500$$
$$\frac{15x}{15} = \frac{4500}{15} \quad \text{divide both sides of the equation by 15}$$
$$x = 300$$

Algebra: The branch of mathematics that deals with the manipulation of symbols (letters) is called **algebra**. An **algebraic term** consists of letters/symbols: therefore, in the above example, x is referred to as an algebraic term (or simply a term). **An algebraic expression** (or simply an **expression**) is a formula that consists of several algebraic terms (constants may also be included), e.g., $x + 5x + 8$.

Square roots: The square root of a number is the reverse of squaring:

$$(2)^2 = 4 \rightarrow \sqrt{4} = 2$$
$$(2.5)^2 = 6.25 \rightarrow \sqrt{6.25} = 2.5$$

Accuracy: rounding numbers correct to x decimal places

When you use a calculator you will frequently end up with a string of numbers after the decimal point. For example, $15/7 = 2.142\,857\,1\ldots$. For most purposes you do not require all these numbers.

However, if some of the numbers are dropped, subsequent calculations are less accurate. To minimise this loss of accuracy, there are rules for 'rounding' numbers correct to a specified number of decimal places, as illustrated by the following example.

Consider: (a) $15/7 = 2.142\,857\,1$; (b) $6/7 = 0.857\,142\,8$. Assume that three numbers after the decimal point are required. To round correct to three decimal places, denoted as 3D, inspect the number in the fourth decimal place:

- If the number in the fourth decimal place is less than 5, simply retain the first three numbers after the decimal place: (b) $6/7 = 0.857\,142\,8$: use 0.857, when rounded correct to 3D.
- If the number in the fourth decimal place is 5 or greater, then increase the number in the third decimal place by 1, before dropping the remaining numbers: (a) $15/7 = 2.142\,857\,1$, use 2.143, when rounded correct to 3D.

To get some idea of the greater loss of accuracy incurred by truncating (chopping off after a specified number of decimal places) rather than rounding to the same number of decimal places, consider (a) the exact value of 15/7, using the calculator, (b) 15/7 truncated after 3D and (c) 15/7 rounded to 3D. While these errors appear small they can become alarmingly large when propagated by further calculations. A simple example follows in which truncated and rounded values of 15/7 are each raised to the power of 20.

Note: Error = exact value − approximate value.

	(a) Exact value $15/7 = 2.142\,857\,143\ldots$	(b) Truncated to 3D $15/7 = 2.142$	(c) Rounded to 3D $15/7 = 2.143$
Error	0	Truncation error $= 0.000\,857\,143\ldots$	Rounding error $= 0.000\,142\,857\ldots$
Raise to power of 20	$\left(\dfrac{15}{7}\right)^{20} = (2.142\,857\,143\ldots)^{20}$ $= 4\,167\,392$ (integer part of result)	$(2.142)^{20} = 4\,134\,180$ (integer part of result)	$(2.143)^{20} = 4\,172\,952$ (integer part of result)
Error	0	$4\,167\,392 - 4\,134\,180$ $= 33\,213$	$4\,167\,392 - 4\,172\,952$ $= -5560$

1.2 Arithmetic Operations

Addition and subtraction

Adding: If all the signs are the same, simply add all the numbers or terms of the same type and give the answer with the common overall sign.

Subtracting: When subtracting any two numbers or two similar terms, give the answer with the sign of the largest number or term.

If terms are of the same type, e.g., all x-terms, all xy-terms, all x^2-terms, then they may be added or subtracted as shown in the following examples:

Add/subtract with numbers, mostly		Add/subtract with variable terms
$5 + 8 + 3 = 16$	similarity →	$5x + 8x + 3x = 16x$
$5 + 8 + 3 + y = 16 + y$	similarity →	(i) $5x + 8x + 3x + y = 16x + y$
The y-term is different, so it cannot be added to the others		(ii) $5xy + 8xy + 3xy + y = 16xy + y$
		The y-term is different, so it cannot be added to the others
$7 - 10 = -3$	similarity →	(i) $7x - 10x = -3x$
		(ii) $7x^2 - 10x^2 = -3x^2$
$7 - 10 - 10x = -3 - 10x$	similarity →	$7x^2 - 10x^2 - 10x = -3x^2 - 10x$
The x-term is different, so it cannot be subtracted from the others		The x-term is different from the x^2-terms, so it cannot be subtracted from the x^2-terms

WORKED EXAMPLE 1.1
ADDITION AND SUBTRACTION

For each of the following, illustrate the rules for addition and subtraction:

(a) $2 + 3 + 2.5 = (2 + 3 + 2.5) = 7.5$

(b) $2x + 3x + 2.5x = (2 + 3 + 2.5)x = 7.5x$

(c) $-3xy - 2.2xy - 6xy = (-3 - 2.2 - 6)xy = -11.2xy$

(d) $8x + 6xy - 12x + 6 + 2xy = 8x - 12x + 6xy + 2xy + 6 = -4x + 8xy + 6$

(e) $3x^2 + 4x + 7 - 2x^2 - 8x + 2 = 3x^2 - 2x^2 + 4x - 8x + 7 + 2 = x^2 - 4x + 9$

Multiplication and division

Multiplying or dividing two quantities with like signs gives an answer with a positive sign. Multiplying or dividing two quantities with different signs gives an answer with a negative sign.

WORKED EXAMPLE 1.2a
MULTIPLICATION AND DIVISION

Each of the following examples illustrates the rules for multiplication.

(a) $5 \times 7 = 35$ (d) $-5 \times 7 = -35$

(b) $-5 \times -7 = 35$ (e) $7/5 = 1.4$

(c) $5 \times -7 = -35$ (f) $(-7)/(-5) = 1.4$

Remember

It is very useful to remember that a minus sign is a -1,
so -5 is the same as -1×5

(g) $(-7)/5 = -1.4$ (j) $(-5)(-7) = 35$

(h) $7/(-5) = -1.4$ (k) $(-5)y = -5y$

(i) $5(7) = 35$ (l) $(-x)(-y) = xy$

 Remember

$0 \times$ (any real number) $= 0$
$0 \div$ (any real number) $= 0$
But you cannot divide by 0

(m) $2(x + 2) = 2x + 4$ multiply each term inside the bracket by the term outside the bracket

(n) $(x + 4)(x + 2)$ multiply the second bracket by x, then multiply the second bracket by $(+4)$ and add

   $= x(x + 2) + 4(x + 2)$ multiply each bracket by the term outside it; add or
   $= x^2 + 2x + 4x + 8$ subtract similar terms, such as $2x + 4x = 6x$
   $= x^2 + 6x + 8$

(o) $(x + y)^2$ multiply the second bracket by x and then by y; add the
   $= (x + y)(x + y)$ similar terms: $xy + yx = 2xy$
   $= x(x + y) + y(x + y)$
   $= xx + xy + yx + yy$
   $= x^2 + 2xy + y^2$

The following identities are important:

1. $(x + y)^2 = x^2 + 2xy + y^2$
2. $(x - y)^2 = x^2 - 2xy + y^2$
3. $(x + y)(x - y) = (x^2 - y^2)$

 Remember

Brackets are used for grouping terms together to
Enhance clarity.
Indicate the order in which arithmetic operations should be carried out.

The precedence of arithmetic operators is summarised as follows:

 (i) Simplify or evaluate the terms within brackets first.
 (ii) When multiplying/dividing two terms: (1) determine the overall sign first; (2)
       multiply/divide the numbers; (3) finally, multiply/divide the variables (letters).
 (iii) Finally, add and/or subtract as appropriate.

WORKED EXAMPLE 1.2b

The precedence of arithmetic operators is illustrated by the following examples:

(a) $2x - y + 4(x - 2y + 2x) = 2x - y + 4(3x - 2y)$ simplify within brackets first

$$= 2x - y + 12x - 8y \qquad \text{multiply the bracket by 4}$$

$$= 2x + 12x - y - 8y \qquad \text{add/subtract similar terms}$$

$$= 14x - 9y$$

(b) $15 - 5(3y - 2(y + 2)) = 15 - 5(3y - 2y - 4)$ simplify the inner bracket first

$$= 15 - 5(y - 4) \qquad \text{multiply the bracket by} - 5$$

$$= 15 - 5(y) - 5(-4) \qquad \text{be careful with the negatives!}$$

$$= 15 - 5y + 20 = 35 - 5y \qquad \text{add/subtract}$$

1.3   Fractions

Terminology:

$$\text{A fraction} = \frac{\text{numerator}}{\text{denominator}}$$

$\dfrac{3}{7}$ $\leftarrow$ 3 is called the numerator
      $\leftarrow$ 7 is called the denominator

1.3.1   Add/subtract fractions: method

The method for adding or subtracting fractions is:

Step 1: Take a common denominator, that is, a number or term which is divisible by the denominator of each fraction to be added or subtracted. A safe bet is to use the product of all the individual denominators as the common denominator.

Step 2: For each fraction, divide its denominator into the common denominator, then multiply the result by its numerator.

Step 3: Simplify your answer if possible.

WORKED EXAMPLE 1.3
ADD AND SUBTRACT FRACTIONS

Each of the following illustrates the rules for addition and subtraction of fractions.

Numerical example	Same example, but with variables
$\dfrac{1}{7} + \dfrac{2}{3} - \dfrac{4}{5}$	
Step 1: The common denominator is $(7)(3)(5)$	$\dfrac{x}{7} + \dfrac{2x}{3} - \dfrac{4x}{5}$
Step 2: $\dfrac{1}{7} + \dfrac{2}{3} - \dfrac{4}{5} = \dfrac{1(3)(5) + 2(7)(5) - 4(7)(3)}{(7)(3)(5)}$	$= \dfrac{x(3)(5) + 2x(7)(5) - 4x(7)(3)}{(7)(3)(5)}$
	$= \dfrac{15x + 70x - 84x}{105}$
Step 3: $\qquad = \dfrac{15 + 70 - 84}{105} = \dfrac{1}{105}$	$= \dfrac{x}{105}$

Numerical example	Similar example, but with variables
$\dfrac{1}{7} + \dfrac{2}{3}$	$\dfrac{1}{x+4} + \dfrac{2}{x}$
Step 1: The common denominator is $(7)(3)$	
Step 2: $\dfrac{1}{7} + \dfrac{2}{3} = \dfrac{1(3) + 2(7)}{(7)(3)}$	Common denominator $(x+4)(x)$
	$\dfrac{1}{x+4} + \dfrac{2}{x} = \dfrac{1(x) + 2(x+4)}{(x+4)(x)}$
Step 3: $= \dfrac{3 + 14}{21} = \dfrac{17}{21}$	$= \dfrac{x + 2x + 8}{x^2 + 4x}$
	$= \dfrac{3x + 8}{x^2 + 4x}$

1.3.2 Multiplying fractions

In multiplication, write out the fractions, then multiply the numbers across the top lines and multiply the numbers across the bottom lines.

Note: To write whole numbers as fractions put the whole number above the line and 1 below the line.

Terminology: RHS means right-hand side and LHS means left-hand side.

WORKED EXAMPLE 1.4
MULTIPLYING FRACTIONS

(a) $\left(\dfrac{2}{3}\right)\left(\dfrac{5}{7}\right) = \dfrac{(2)(5)}{(3)(7)} = \dfrac{10}{21}$

(b) $\left(-\dfrac{2}{3}\right)\left(\dfrac{7}{5}\right) = \dfrac{(-2)(7)}{(3)(5)} = -\dfrac{14}{15}$

(c) $3 \times \dfrac{2}{5} = \left(\dfrac{3}{1}\right)\left(\dfrac{2}{5}\right) = \dfrac{(3)(2)}{(1)(5)} = \dfrac{6}{5} = 1\dfrac{1}{5}$

The same rules apply for fractions involving variables, x, y, etc.

(d) $\left(\dfrac{3}{x}\right)\dfrac{(x+3)}{(x-5)} = \dfrac{3(x+3)}{x(x-5)} = \dfrac{3x+9}{x^2-5x}$

1.3.3 Dividing by fractions

General rule:

Dividing by a fraction is the same as multiplying by the fraction inverted.

To demonstrate this general rule, consider an example such as dividing 5 by a half; we know the answer is 10 because there are 10 halves in 5: $\dfrac{5}{\left(\dfrac{1}{2}\right)} = 10$.

This answer may also be obtained by applying the rule stated above:

'Dividing by a fraction is the same as multiplying by that fraction inverted'

$$\dfrac{5}{\left(\dfrac{1}{2}\right)} = 5 \times \left(\dfrac{2}{1}\right) = \dfrac{5}{1} \times \left(\dfrac{2}{1}\right) = \dfrac{10}{1} = 10$$

Note: The same rules apply to all fractions, whether the fractions consist of numbers or variables.

WORKED EXAMPLE 1.5

DIVISION BY FRACTIONS

The following examples illustrate how division with fractions operates.

(a) $\dfrac{\left(\dfrac{2}{3}\right)}{\left(\dfrac{5}{11}\right)} = \left(\dfrac{2}{3}\right)\left(\dfrac{11}{5}\right) = \dfrac{22}{15}$

(b) $\dfrac{5}{\left(\dfrac{3}{4}\right)} = 5 \times \dfrac{4}{3} = \dfrac{5}{1} \times \dfrac{4}{3} = \dfrac{20}{3} = 6\dfrac{2}{3}$

(c) $\dfrac{\left(\dfrac{7}{3}\right)}{8} = \dfrac{\left(\dfrac{7}{3}\right)}{\dfrac{8}{1}} = \dfrac{7}{3} \times \dfrac{1}{8} = \dfrac{7}{24}$

(d) $\dfrac{\dfrac{2x}{x+y}}{\dfrac{3x}{2(x-y)}} = \dfrac{2x}{x+y} \times \dfrac{2(x-y)}{3x}$

$\qquad = \dfrac{4x(x-y)}{3x(x+y)} = \dfrac{4(x-y)}{3(x+y)}$

Note: The same rules apply to all fractions, whether the fractions consist of numbers or variables.

Reducing a fraction to its simplest form and equivalent fractions

A fraction that has common factors in the numerator and denominator may be simplified by dividing the common factors into each other giving unity. This is described as the cancellation of common factors and is illustrated in the following examples:

(a) $\dfrac{21}{15} = \dfrac{(3)(7)}{(3)(5)} = \dfrac{(\cancel{3})(7)}{(\cancel{3})(5)} = \dfrac{7}{5}$

Hence $\dfrac{21}{15}$ and $\dfrac{7}{5}$ are described as **equivalent fractions**.

(b) $\dfrac{21(x+2)}{15(x+2)} = \dfrac{(3)(7)(x+2)}{(3)(5)(x+2)} = \dfrac{(\cancel{3})(7)(\cancel{x+2})}{(\cancel{3})(5)(\cancel{x+2})} = \dfrac{7}{5}$

(c) $\dfrac{21x^2}{15x} = \dfrac{(3)(7)(x)(x)}{(3)(5)(x)} = \dfrac{(\cancel{3})(7)(\cancel{x})(x)}{(\cancel{3})(5)(\cancel{x})} = \dfrac{(7)(x)}{(5)} = \dfrac{7x}{5}$

PROGRESS EXERCISES 1.1

Revision on Basics

Show, step by step, how the expression on the left-hand side simplifies to that on the right.

1. $2x + 3x + 5(2x - 3) = 15(x - 1)$

2. $4x^2 + 7x + 2x(4x - 5) = 3x(4x - 1)$

3. $2x(y + 2) - 2y(x + 2) = 4(x - y)$

4. $(x + 2)(x - 4) - 2(x - 4) = x(x - 4)$

5. $(x + 2)(y - 2) + (x - 3)(y + 2)$
   $= 2xy - y - 10$

6. $(x + 2)^2 + (x - 2)^2 = 2(x^2 + 4)$

7. $(x + 2)^2 - (x - 2)^2 = 8x$

8. $(x + 2)^2 - x(x + 2) = 2(x + 2)$

9. $\dfrac{1}{3} + \dfrac{3}{5} + \dfrac{5}{7} = 1\dfrac{68}{105}$

10. $\dfrac{x}{2} - \dfrac{x}{3} = \dfrac{x}{6}$

11. $\dfrac{\left(\dfrac{2}{3}\right)}{\left(\dfrac{1}{5}\right)} = \dfrac{10}{3}$

12. $\dfrac{\left(\dfrac{2}{7}\right)}{3} = \dfrac{2}{21}$

13. $2\left(\dfrac{2}{x} - \dfrac{x}{2}\right) = \dfrac{4 - x^2}{x} = \dfrac{4}{x} - x$

14. $\dfrac{-12}{P}\left(\dfrac{3P}{2} + \dfrac{P}{2}\right) = -24$

15. $\dfrac{\left(\dfrac{3}{x}\right)}{x + 3} = \dfrac{3}{x(x + 3)}$

16. $\dfrac{\left(\dfrac{5Q}{P + 2}\right)}{\left(\dfrac{1}{(P + 2)}\right)} = 5Q$

17. $\dfrac{11}{5} + \dfrac{2}{5} - \dfrac{6}{5} = \dfrac{7}{5}$

18. $3 + 1\dfrac{1}{5} = \dfrac{21}{5}$

19. $\dfrac{7}{5} - 9 + \dfrac{6}{10} = -7$

20. $\dfrac{5}{3}\left(\dfrac{11}{5} + \dfrac{2}{5} - \dfrac{6}{5}\right) = \dfrac{7}{3}$

21. $\dfrac{\dfrac{11}{5} + \dfrac{2}{5} - \dfrac{6}{5}}{3} = \dfrac{7}{15}$

22. $\dfrac{2}{7} + \dfrac{3}{5} - \dfrac{3}{8} = \dfrac{143}{280}$

23. $\dfrac{3}{\left(\dfrac{3}{8}\right)} + \dfrac{3}{2} = \dfrac{19}{2}$

24. $\dfrac{x}{5} + \dfrac{2x}{5} - \dfrac{6}{5} = \dfrac{3x - 6}{5}$

25. $\dfrac{7}{2x} - \dfrac{x}{9} = \dfrac{63 - 2x^2}{18x}$

26. $\dfrac{x - 3}{5} + \dfrac{2}{x} - \dfrac{x}{5} = \dfrac{10 - 3x}{5x}$

27. $\dfrac{192}{54} = \dfrac{32}{9}$

28. $\dfrac{12xy}{4x^2} = \dfrac{3y}{x}$

29. $\dfrac{-(x + 2)^2}{2x(x + 2)} = \dfrac{-(x + 2)}{2x}$

30. $\dfrac{4x^2(5 + x)}{20x + 4x^2} = 2x$

1.4   Solving Equations

An equation is a statement declaring that the expression on the right-hand side (RHS) of the '=' is equal to or equivalent to that on the left-hand side (LHS). Therefore you could visualise an equation as a balance.

$$\frac{\text{LHS} = \text{RHS}}{\triangle}$$

The solution of an equation is simply the value or values of the unknown(s) for which the LHS of the equation is equal to the RHS.

For example, the equation $x + 4 = 10$ has the solution $x = 6$. We say $x = 6$ 'satisfies' the equation. We say this equation has a **unique solution**.

Methods for solving equations

To solve an equation for an unknown, x, rearrange the equation in order to isolate the x on one side. This may be achieved by a variety of techniques such as:

- adding/subtracting a number or term to each side
- multiplying/dividing each side by a number or term (except division by 0!).

It is vitally important to remember that whatever operation is performed on the LHS of the equation must also be performed on the RHS, otherwise the equation is changed.

WORKED EXAMPLE 1.6

SOLVING EQUATIONS

(a) Given the equation $x + 3 = 2x - 6 + 5x$, solve for x.
(b) Given the equation $(x + 3)(x - 6) = 0$, solve for x.
(c) Given the equation $(x + 3)(x - 3) = 0$, solve for x.

SOLUTION

(a)
$$x + 3 = 2x - 6 + 5x$$
$$x + 3 = 7x - 6 \qquad \text{adding the } x\text{-terms on the RHS}$$
$$x + 3 + 6 = 7x - 6 + 6 \quad \text{to cancel the } -6 \text{ on the RHS, add } +6 \text{ to both sides}$$
$$x + 3 + 6 = 7x \qquad \uparrow \text{bringing over } -6 \text{ to the other side}$$
$$x + 9 = 7x$$
$$9 = 7x - x \qquad \text{bringing } x \text{ over to the other side}$$
$$9 = 6x$$
$$\frac{9}{6} = x \qquad \text{dividing both sides by 6}$$
$$1.5 = x$$

> Hence the frequently quoted rule: 'bring the −6 to the other side and change the sign to +6'

(b) $(x + 3)(x - 6) = 0$

The LHS of this equation consists of the product of two terms $(x + 3)$ and $(x - 6)$. A product is equal to 0, if either or both terms in the product $(x + 3)$, $(x + 6)$ are 0.

Here there are two solutions: $(x + 3) = 0$, hence $x = -3$; and $(x - 6) = 0$, hence $x = 6$. Each of these solutions can be confirmed by checking that they satisfy the original equation.

Check the solutions:

Substitute $x = -3$ into $(x + 3)(x - 6) = 0$: $(-3 + 3)(-3 - 6) = 0$: $(0)(-9) = 0$: $0 = 0$. True.

Substitute $x = 6$ into $(x + 3)(x - 6) = 0$: $(6 + 3)(6 - 6) = 0$: $(9)(0) = 0$: $0 = 0$. True.

(c) This is similar to part (b), a product on the LHS, zero on the right, hence the equation $(x - 3)(x + 3) = 0$ has two solutions: $x = 3$ from the first bracket and $x = -3$ from the second bracket.

Alternatively, on multiplying out the brackets, this equation simplifies as follows:

$$(x - 3)(x + 3) = 0$$
$$x(x + 3) - 3(x + 3) = 0$$
$$x(x) + x(3) - 3(x) - 3(3) = 0$$
$$x^2 + 3x - 3x - 9 = 0$$
$$x^2 - 9 = 0$$
$$x^2 = 9 \quad \text{simplified equation}$$
$$x = +\sqrt{9}$$
$$x = 3 \quad \text{or} \quad x = -3$$

The reason for two solutions is that x may be positive or negative; both satisfy the simplified equation $x^2 = 9$.

If $x = 3 \rightarrow x^2 = 9$, so satisfying the equation.

If $x = -3 \rightarrow x^2 = (-3)^2 = 9$, also satisfying the equation.

Therefore, when solving equations of the form $x^2 = $ number, there are always two solutions:

$$\boxed{x = +\sqrt{\text{number}} \quad \text{and} \quad x = -\sqrt{\text{number}}}$$

MATHEMATICAL PRELIMINARIES

Not all equations have solutions. In fact, equations may have no solutions at all or infinitely many solutions. Each of these situations is demonstrated in the following examples.

Case 1: Unique solutions An example of this is given above: $x + 4 = 10$ etc.

Case 2: Infinitely many solutions The equation $x + y = 10$ has solutions $(x = 5, y = 5)$, $(x = 4, y = 6)$, $(x = 3, y = 7)$, etc. In fact, this equation has infinitely many solutions or pairs of values (x, y) which satisfy the formula $x + y = 10$.

Case 3: No solution The equation $0(x) = 5$ has no solution. There is simply no value of x which can be multiplied by 0 to give 5.

WORKED EXAMPLE 1.7
SOLVING A VARIETY OF SIMPLE ALGEBRAIC EQUATIONS

 Find animated worked examples at **www.wiley.com/college/bradley**

In this worked example, try solving the following equations yourself. The answers are given below, followed by the detailed solutions.

(a) $2x + 3 = 5x - 8$ (b) $\dfrac{1}{x} + \dfrac{2}{x} = 5$

(c) $x^2 + 4x - 6 = 2(2x + 5)$ (d) $(x - y) = 4$ (e) $x^3 - 2x = 0$

SOLUTION

Now check your answers:

(a) $x = 11/3$ (b) $x = 0.6$ (c) $x = 4$, $x = -4$
(d) Infinitely many solutions for which $x = y + 4$ (e) $x = 0$, $x = \sqrt{2}$, $x = -\sqrt{2}$

Suggested solutions to Worked Example 1.7

(a) $2x + 3 = 5x - 8$

$3 + 8 = 5x - 2x$

$11 = 3x$

$\dfrac{11}{3} = x$

(b) $\dfrac{1}{x} + \dfrac{2}{x} = 5$

$\dfrac{1 + 2}{x} = 5$

$\dfrac{3}{x} = \dfrac{5}{1}$

$3 = 5x$ multiply each side by x

$5x = 3$ swap sides

$x = \dfrac{3}{5} = 0.6$ dividing each side by 5

(c) $x^2 + 4x - 6 = 2(2x + 5)$

This time, simplify first by multiplying out the brackets and collecting similar terms:

$$x^2 + 4x - 6 = 4x + 10$$

$$x^2 + 4x - 4x = 10 + 6 \quad \text{bring all } x\text{-terms to one side and numbers to the other side}$$

$$x^2 = 16$$

$$x = \pm 4$$

(d) $(x - y) = 4$: Here we have one equation in two unknowns, so it is not possible to find a unique solution. The equation may be rearranged as

$$x = y + 4$$

This equation now states that for any given value of y (there are infinitely many values), x is equal to that value plus 4. So, there are infinitely many solutions.

(e) x is common to both terms, therefore we can separate or factor x from each term as follows:

$$x^3 - 2x = 0$$

$$x(x^2 - 2) = 0$$

$$x = 0 \text{ and/or } (x^2 - 2) = 0 \quad \text{the product } x(x^2 - 2) \text{ is zero if one or both terms are zero}$$

$$x = 0 \text{ and/or } x^2 = 2$$

Solution: $x = 0$, $x = \pm\sqrt{2}$

An exercise for the reader: check that the solutions satisfy the equation.

1.5 Currency Conversions

You may have browsed the internet in order to purchase books, software, music, etc. Frequently prices will be quoted in some currency other than your own. With some simple maths and knowledge of the current rate of exchange, you should have no difficulty in converting the price to your own currency. The following worked examples use the Euro exchange rates in Table 1.1. This table equates each of the currencies listed to 1 Euro on a given day in January 2012.

Table 1.1   Euro exchange rates

Currency	Rate	Currency	Rate
British pound	0.8346	Canadian dollar	1.3164
US dollar	1.3003	Australian dollar	1.2429
Japanese yen	100.89	Polish zloty	4.2984
Danish krone	7.4352	Hungarian forint	302.8
Brazilian real	2.2923	Hong Kong dollar	10.0912
Swiss franc	1.2065	Singaporean dollar	1.6527
Norwegian krone	7.639	South African rand	10.3843
Malaysian ringgit	4.0147	Indian rupee	65.152

WORKED EXAMPLE 1.8
CURRENCY CONVERSIONS

(a) A book is priced at US\$20. Calculate the price of the book in (i) Euros, (ii) British pounds and (iii) Australian dollars correct to four decimal places.

(b) Convert (i) 500 Australian dollars and (ii) \$10 000 Singaporean dollars to pounds sterling.

Give all answers correct to four decimal places.

SOLUTION

The calculations (correct to four decimal places) are carried out as follows:

Step 1: State the appropriate rates from Table 1.1.

Step 2: Set up the identity: 1 unit of *given* currency = y units of *required* currency.

Step 3: Multiply both sides by x: x units of *given* currency = $x \times$ (y units of *required* currency).

(a) (i) The price is *given* in US dollars, the price is *required* (currency) in Euros, hence

Step 1: US\$1.3003 = €1   from Table 1.1

Step 2: $\text{US\$}1 = € \dfrac{1}{1.3003}$ dividing both sides by 1.3003

Step 3: $\text{US\$}20 = € \dfrac{20}{1.3003}$ multiplying both sides by 20

$= €15.3811$

(a) (ii)

Step 1: From Table 1.1 write down the exchange rates for 1 Euro in both British pounds and US dollars:

$$\left. \begin{array}{c} \text{US\$}1.3003 = 1\,\text{Euro} \\ £0.8346 = 1\,\text{Euro} \end{array} \right\} \rightarrow \text{US\$}1.3003 = £0.8346$$

\$1.3003 = £0.8346 since they are each equivalent to 1 Euro

The *given* price is in US dollars, the price is *required* in British pounds

Step 2: $\text{US\$}1 = £1 \times \dfrac{0.8346}{1.3003}$ dividing both sides of the previous equation by 1.3003 to get rate for \$1 in £

Step 3: $\text{US\$}20 = £20 \times \dfrac{0.8346}{1.3003}$ multiplying both sides by 20

$\text{US\$}20 = £12.8370$

(a) (iii)

Step 1: From Table 1.1 write down the exchange rates for 1 Euro in Australian dollars and US dollars:

$$\left.\begin{array}{l} US\$1.3003 = 1\,Euro \\ A\$1.2429 = 1\,Euro \end{array}\right\} \rightarrow US\$1.3003 = A\$1.2429$$

$US\$1.3003 = A\$1.2429$ since they are each equivalent to 1 Euro

The *given* price is US\$, the *required* price (currency) is A\$

Step 2: $US\$1 = A\$1 \times \dfrac{1.2429}{1.3003}$ dividing both sides by 1.3003

Step 3: $US\$20 = A\$20 \times \dfrac{1.2429}{1.3003}$ multiplying both side by 20

$\quad\quad = A\$19.1171$

(b) (i)

Step 1: From Table 1.1 write down the exchange rates for 1 Euro in Australian dollars and pounds sterling:

$$\left.\begin{array}{l} £0.8346 = 1\,Euro \\ A\$1.2429 = 1\,Euro \end{array}\right\} \rightarrow £0.8346 = A\$1.2429$$

Here we are *given* A\$500 and *we require* its equivalent in £ sterling $A\$1.2429 = £0.8346$ since they are each equivalent to 1 Euro

Step 2: $A\$1 = £1 \times \dfrac{0.8346}{1.2429}$ dividing both sides by 1.2429 to get rate for A\$1

Step 3: $A\$500 = £1 \times \dfrac{0.8346}{1.2429} \times 500$ multiplying both sides by 500

$A\$500 = £335.7470$

(b) (ii)

Step 1: From Table 1.1 write down the exchange rates for 1 Euro in Singaporean dollars and British pounds:

$$\left.\begin{array}{l} £0.8346 \ = 1\,Euro \\ S\$1.6527 = 1\,Euro \end{array}\right\} \rightarrow £0.8346 = S\$1.6527$$

$S\$1.6527 = £0.8346$ since they are each equivalent to 1 Euro

Step 2: $S\$1 = £1 \times \dfrac{0.8346}{1.6527}$ dividing both sides by 1.6527

Step 3: $S\$10\,000 = £1 \times \dfrac{0.8346}{1.6527} \times 10\,000$ multiplying both sides by 10 000

$S\$10\,000 = £5049.9183$

PROGRESS EXERCISES 1.2

Use Basics to Solve Equations

Solve the following equations. Remember that equations may have no solution or infinitely many solutions.

1. $2x + 3x + 5(2x - 3) = 30$
2. $4x^2 + 7x - 2x(2x - 5) = 17$
3. $(x - 2)(x + 4) = 0$
4. $(x - 2)(x + 4) = 2x$
5. $(x - 2)(x + 4) = -8$
6. $x(x - 2)(x + 4) = 0$
7. $4x(x - 2)(x - 2) = 0$
8. $2x(y + 2) - 2y(x + 2) = 0$
9. $(x + 2)(y + 2) = 0$
10. $(x + 2)(y + 2) + (x - 3)(y + 2) = 0$
11. $(x - 2)(x + 4) - 2(x - 4) = 0$
12. $(x + 2)^2 + (x - 2)^2 = 0$
13. $(x + 2)^2 - (x - 2)^2 = 0$
14. $x(x^2 + 2) = 0$
15. $\dfrac{x}{3} - \dfrac{x}{2} = \dfrac{2}{3}$
16. $\dfrac{x}{3} = 2x$
17. $\dfrac{2}{x} - \dfrac{3}{2x} = 1$
18. $\dfrac{4x(x - 4)(x + 3.8)}{x^4 - 4x^3 + 7x^2 - 5x + 102} = 0$
19. $\left(\dfrac{-12P}{P}\right)\left(\dfrac{3P}{2} + \dfrac{P}{2}\right) = 480$
20. $\dfrac{2}{Q} = 0$
21. $\dfrac{\left(\dfrac{25}{x}\right)}{x} = 1$
22. $\dfrac{\left(\dfrac{5Q}{P + 2}\right)}{\left(\dfrac{1}{P + 2}\right)} = 20$
23. $10 - \dfrac{1}{10Q} = 0$
24. $4x + 8(x - 2) = 12$
25. $\dfrac{2}{x} = 4$
26. $2 - 0.75x = x - 1.5$
27. $3x(x - 2) = 0$
28. $(x - 2)^2 = 4$
29. $100(1 - x)(1 + x) = 0$
30. $100(1 - x)(1 + x) = 100$
31. $\dfrac{x}{5} + \dfrac{2x}{5} - \dfrac{6}{5} = 3$
32. $\dfrac{7}{2x} - \dfrac{x}{9} = x$
33. $\dfrac{x - 3}{5} + \dfrac{2}{x} - \dfrac{x}{5} = \dfrac{3}{10}$

Refer to exchange rates given in Table 1.1 to answer questions 34 to 40. Give answers correct to two decimal places.

34. An iPad is priced at 400 Canadian dollars. Calculate the price of the iPad in (a) Euros, (b) rupees and (c) ringgits.
35. How many Euros are equivalent to (a) 800 US dollars, (b) 800 zlotys?
36. A flight is priced at 240 South African rand. What is the equivalent price in Brazilian reals?

37. Calculate a table of exchange rates for Hong Kong dollars for the first five currencies in the third column of Table 1.1.
38. A new textbook is priced at £58 sterling. Calculate the equivalent price in Hungarian forints.
39. Convert 500 US dollars to (a) Swiss francs, (b) Australian dollars, (c) Singaporean dollars.
40. You have 400 Polish zlotys. How much would you have left (in Euros) if you bought a book for £35 sterling and a T-shirt for 420 Hong Kong dollars?

1.6 Simple Inequalities

An equation is an equality. It states that the expression on the LHS of the '=' sign is equal to the expression on the RHS. An inequality is a statement in which the expression on the LHS is either greater than (denoted by the symbol >) or less than (denoted by the symbol <) the expression on the RHS. For example, $5 = 5$ or $5x = 5x$ are equations, while

$$5 > 3 \qquad 5x > 3x$$

are inequalities which read, '5 is greater than 3', '5x is greater than 3x' (for any positive value of x).

Note: Inequalities may be read from left to right, as above, or the inequality may be read from right to left, in which case the above inequalities are

$5 > 3$ (5 is greater than 3) is the same as $3 < 5$ ('3 is less than 5')
$5x > 3x$ ('5x is greater than 3x') is the same as $3x < 5x$ ('3x is less than 5x')

Inequality symbols

$>$ greater than $\qquad\qquad$ $<$ less than
$\geq$ greater than or equal to $\qquad$ $\geq$ less than or equal to

The number line

The number line is a horizontal line on which every point represents a real number. The central point is zero, the numbers on the left are negative, numbers on the right are positive, as illustrated for selected numbers in Figure 1.1.

Figure 1.1   Number line, numbers increasing from left to right

Look carefully at the negative numbers; as the numbers increase in value they decrease in magnitude, for example: -1 is a larger number than -2; -0.3 is a larger number than -0.5. Another way of looking at it is to say the numbers become less negative as they increase in value. (Like a bank account, you are better off when you owe £10 (-10) than when you owe £1000 (-1000).)

An inequality statement, such as $x > 2$, means all numbers greater than, but not including, 2. This statement is represented graphically as every point on the number line to the right of the number 2, as shown in Figure 1.2, with an open circle at the point 2 to show that it is not included in the interval (a solid circle at 2 would indicate that 2 is included in the interval).

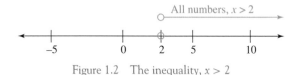

Figure 1.2 The inequality, $x > 2$

In economics, it is meaningful to talk about positive prices and quantities. In this text, we shall assume that the variable $x \geq 0$ when solving inequalities.

Intervals defined by inequality statements

When an application uses values within a certain range only, then inequality signs are often used to define this interval precisely. For example, suppose a tax is imposed on all incomes, (£Y), between £10 000 and £15 000 inclusive; we say the tax is imposed on all salaries within the interval $10\,000 \leq Y \leq 15\,000$.

A certain bus fare applies to all children of ages (x) 4 but less than 16; we say the fare applies to those whose ages are in the interval $4 \leq x < 16$. The age 4 is included, and all ages up to, but not including, 16.

Manipulating inequalities

Inequalities may be treated as equations for many arithmetic operations. The inequality remains true when constants are added to or subtracted from both sides of the inequality sign, or when both sides of the inequality are multiplied or divided by positive numbers or variables. For example, the equalities above are still true when 8 or –8 is added to both sides,

$$5 + 8 > 3 + 8, \quad \text{that is,} \quad 13 > 11, \quad \text{similarly} \quad 5x + 8 > 3x + 8$$

$$5 - 8 > 3 - 8, \quad \text{that is} \quad -3 > -5. \quad \text{Remember} -5 \text{ is less than} -3. \quad \text{See Figure 1.1.}$$

However, when both sides of the inequality are multiplied or divided by negative numbers or variables, then the direction of the inequality changes: $>$ becomes $<$ and vice versa. For example, multiply both sides of the inequality $5 > 3$ by –2:

$$5(-2) > 3(-2), \quad \text{or} \quad -10 > -6 \quad \text{is not true}$$

$$5(-2) < 3(-2), \quad \text{or} \quad -10 < -6 \quad \text{is true}$$

Solving inequalities

The solution of an equation is the value(s) for which the equation statement is true. For example, $x + 4 = 10$ is true when $x = 6$ only. On the other hand, the solution of an inequality is a range of values for which the inequality statement is true; for example, $x + 4 > 10$ is true when $x > 6$.

WORKED EXAMPLE 1.9
SOLVING SIMPLE INEQUALITIES

Find the range of values for which the following inequalities are true, assuming that $x > 0$. State the solution in words and indicate the solution on the number line.

(a) $10 < x - 12$ (b) $\dfrac{-75}{x} > 15$ (c) $2x - 6 \leq 12 - 4x$

SOLUTION

(a) $10 < x - 12 \rightarrow 10 + 12 < x \rightarrow 22 < x \,(\text{or } x > 22)$

The solution states: 22 is less than x or x is greater than 22. Hence the solution is represented by all points on the number line to the right of, but not including, 22, as shown in Figure 1.3.

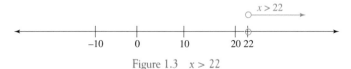

Figure 1.3 $x > 22$

(b) $\dfrac{-75}{x} > 15$

Multiply both sides of the inequality by x. Since $x > 0$, the direction of the inequality sign does not change.

$$-75 > 15x$$
$$-5 > x \qquad \text{dividing both sides by 15}$$

The solution states: -5 is greater than x or x is less than -5. But we assumed $x > 0$. x cannot be less than -5 and greater than 0 at the same time. Hence there is no solution, as shown in Figure 1.4.

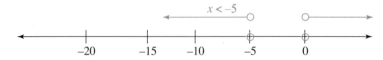

Figure 1.4 $x < -5$ and $x > 0$ is not possible

(c) $2x - 6 \leq 12 - 4x$

$2x + 4x - 6 \leq 12$ add $4x$ to both sides

$6x \leq 12 + 6$ add 6 to both sides

$6x \leq 18$

$x \leq 3$ dividing both sides by 6

But we assume $x > 0$, hence the solution is $0 < x \leq 3$.

This is represented by all points on the number line to the right of 0 and up to and including 3, as shown in Figure 1.5.

Figure 1.5 $0 < x \leq 3$

1.7 Calculating Percentages

• When we speak of 5% of a number we mean

$$\frac{5}{100} \times \text{number}$$

• When we say a number increases by x%, then the increase in the number

$$\text{increase} = \frac{x}{100} \times \frac{\text{number}}{1} = \frac{x \times \text{number}}{100}$$

• The increased number

$$= \text{number} + \text{increase} = \text{number} + \frac{x \times \text{number}}{100} = \text{number} \left(1 + \frac{x}{100}\right)$$

$$= \text{number} \times \left(\frac{100 + x}{100}\right)$$

You should notice that, in calculations, percentages are always written as fractions: quote percentage/100. These definitions and other calculations with percentages are best illustrated using worked examples.

WORKED EXAMPLE 1.10

CALCULATIONS WITH PERCENTAGES

(a) Calculate   (i) 23% of 1534   (ii) 100% of 1534
(b) A salary of £55 240 is to be increased by 12%. Calculate (i) the increase, (ii) the new salary.
(c) In 2013, a holiday apartment is valued at £63 600. This is a drop of 40% on the price paid for the apartment in 2007. Calculate the price paid in 2007.

SOLUTION

(a) In calculations, quoted percentages are always written as the fraction, quoted percentage/100.

(i) $23\% \text{ of } 1534 = \dfrac{23}{100}\dfrac{1534}{1} = \dfrac{(23)(1534)}{100} = \dfrac{35\,282}{100} = 352.82$

(ii) $100\% \text{ of } 1534 = \dfrac{100}{100}\dfrac{1534}{1} = \dfrac{(100)(1534)}{100} = 1534$

So, 100% of (anything) = (anything)

(b) (i) $12\% \text{ of } 55\,240 = \dfrac{12}{100}\dfrac{55\,240}{1} = \dfrac{12(5524)}{10} = \dfrac{66\,288}{10} = 6628.8 = \text{increase}$

So the increase in salary is £6628.8
(ii) The new salary is £55 240 + £6628.8 = £61 868.8

Alternatively, the new salary is 112% of the previous salary and may be calculated as

$$55\,240 \times \left(\dfrac{112}{100}\right) = £61\,868.8$$

(c) Let the 2007 price be the basic price. The price in 2013 is 60% of the 2007 price, i.e., 2013 price = 60% × basic price.

So £63 400 = 60% of the basic price and we want to find 100% of the basic price.

Method

$$63\,600 = \dfrac{60}{100} \times \text{basic price} \qquad 60\% \times \text{basic price}$$

$$\dfrac{63\,600}{60} = \dfrac{1}{100} \times \text{basic price} \qquad \text{find 1\% of the basic price}$$

$$\dfrac{63\,600}{60} \times \dfrac{100}{1} = \dfrac{100}{100} \times \text{basic price} \qquad \text{100\% of the basic price}$$

$$106\,000 = \text{basic price}$$

So, in 2007, the apartment cost £106 000.

PROGRESS EXERCISES 1.3

Percentages and Inequalities

1. Graph the intervals given by the following inequalities on the number line:

    (a) $x > 2$ (b) $x < 25$ (c) $x > -4$ (d) $x \geq -1.5$ (e) $-4 \geq x$ (f) $60 < x$

2. Solve the following inequalities, stating the solution in words. Graph the inequality on the number line.

    (a) $x - 25 > 7$ (b) $5 < 2x + 15$ (c) $\dfrac{25}{x} < 10$ (d) $\dfrac{x}{2} + \dfrac{x}{3} \geq \dfrac{17}{6}$ (e) $3x - 29 \leq 7x + 11$

3. Calculate: (a) 12% of 5432.7; (b) 85% of 23.65; (c) 11.5% of 6.5.

4. A fast-food chain proposes to increase the basic hourly rate of pay by 14%. If the present rate is £5.65 per hour, calculate: (a) the increase in the hourly rate; (b) the new hourly rate.

5. The 2016 price of a basic computer will be 35% lower than the 2013 price. If the 2016 price is £910, calculate the price in 2013.

6. A company which produces printers proposes to increase its output by 6% each week. Calculate the company's projected output for the next three weeks (to the nearest whole printer) if the present output is 720 per week.

7. A company plans to phase out a particular model of car by reducing the output by 20% each week. If the present output is 400 cars per week, calculate the number of cars to be produced per week over the next six weeks.

8. The price of a new washing machine is £485. The price includes a value added tax (VAT) of 21% of the selling price (VAT is explained on page 35). Calculate the price without VAT.

9. A retailer sells a TV for £658. If the cost price was £480, calculate his profit as a percentage of the cost price. (**Note:** profit = selling price − cost price.)

10. A retailer sells a video recorder for £880. The price includes VAT at 21% and a profit of 34%. Calculate the cost price of the recorder.

11. A retailer buys TVs for £425. He must then pay VAT at 21%. What price should he charge if he is to make a profit of 25% of the cost price?

12. 154 students attend a maths lecture. If 22 students are absent, calculate the percentage of students absent.

13. A house was valued at €365 000 in 2003 and €332 150 in 2013.
    (a) Calculate the change in the value between 2003 and 2013 as a percentage of the 2003 valuation.
    (b) Calculate the amount of stamp duty that would be due if the house had been sold in 2003 when stamp duty was charged at 6.0% on houses valued at €381 000 or less.
    (c) If house prices are predicted to rise by 2% each year between 2013 and 2016, calculate the projected value of the house in 2014, 2015 and 2016.

14. The number of first-year science students registered in a college in 2006 is 348 females and 676 males.
    (a) What proportion of students are (i) male and (ii) female?
    (b) Calculate the percentage of (i) male students and (ii) female students.
    (c) If the gender balance had been 40% male and 60% female, calculate the numbers of male and female students.

15. (a) A large consignment of components from supplier A are known to contain 4% defectives.
If a batch of 8500 components is received, calculate:
  (i) The proportion of defectives.
  (ii) The number of defectives.
(b) Components from supplier B contain 2.5% defectives. If a batch of 12 400 components are received, calculate:
  (i) The proportion of defectives.
  (ii) The number of defectives.
(c) The batches of components from both suppliers are combined and sent to the production department in an assembly plant.
  (i) Calculate the number of defective components in the combined batch.
  (ii) Calculate the percentage of defective components in the combined batch.

1.8 The Calculator. Evaluation and Transposition of Formulae

1.8.1 The Calculator

You will require a scientific calculator for the remainder of the text. The keys which will be required most frequently are:

Mathematical operators and functions add: [+]   subtract: [−]   multiply: [×]   divide: [÷]   change of sign: [+/−]   squares: [x^2]   square roots: [√]   powers: [x^y]   roots: [$x^{1/y}$]   logs to base 10: [log] antilog to base 10 [10^x]   logs to base e: [ln]   antilog to base e: [e^x]

The memory keys: clear memory; put into memory; add to the contents of memory; subtract from the contents of memory.

The symbols representing the above functions may vary from one brand of calculator to another, so check the instruction booklet which is supplied with your calculator.

The calculator is an extremely useful aid, but it can only produce the correct answer if the data is keyed in correctly and in the correct order.

For example, 200(34/690) + 124 − 80/5 is evaluated by keying in the numbers in the following order:

$$200 \ [\times] \ 34 \ [\div] \ 690 \ [+] \ 124 \ [−] \ 80 \ [\div] \ 5 \ [=] \ 117.855$$

Also check that the result produced by the calculator (117.855) is about the right order of magnitude. For example, 200(34/690) = 6800/690, which is approximately 10; 124; is 124: 80/5 is 16. So the result should be approximately 10 + 124 − 16 = 118.

1.8.2 Evaluating formulae using the calculator

The formula (or equation) $P_t = P_0(1 + i \times t)$ gives the value P_t that has accrued from an investment P_0 made t years ago when simple interest at a rate of i% pa was used (see Chapter 5).

Note: The subscripts on P refer to the number of years that the principal or investment is on deposit: P_t means that the investment has been on deposit for t years, while P_0 means the investment was deposited at $t = 0$.

The single symbol, P_t, on the left-hand side of the equation is called the 'subject' of the formula. P_t can be calculated when the values of the other three variables, P_0, i and t, are given. Unless these are small integer values then one would normally use the calculator to evaluate P_t. For example, suppose $P_0 = £100$, $i = 0.05$ and $t = 10$ years, then $P_t = P_0(1 + i \times t) \rightarrow P_t = 100(1 + 0.05(10)) = 100(1 + 0.5) = 100(1.5) = 150$.

So, at the end of 10 years the investment is worth £150.

One must be careful to observe the following when evaluating formulae, in particular:

1. The precedence for arithmetic operations, summarised in Section 1.2.
2. Be careful when substituting negative values, particularly when negative signs are also involved.
3. Fractions: when evaluating fractions which have several terms in the numerator and/or denominator, evaluate the numerator and denominator separately, then divide the number in the numerator by the number in the denominator.

These points will be illustrated in the following worked example.

WORKED EXAMPLE 1.11
EVALUATION OF FORMULAE

(a) The sum of the first n terms of an arithmetic series is calculated by the formula

$$S_n = \frac{n}{2}(2a + (n-1)d)$$

Calculate the value of S_n when $n = 35$, $a = 200$ and $d = -2.5$.

Note: The subscript in S_n simply indicates that S is the sum of n terms: the sum of 35 terms is written symbolically as S_{35}.

(b) In statistics, the formula for the intercept, a of a least-squares line is given by the formula

$$a = \frac{\sum y - b \sum x}{n}$$

Evaluate a when $n = 9$, $\sum y = 3$, $b = -2.35$ and $\sum x = -21$.

SOLUTION

(a) Substitute the values given into the formula

$S_n = \frac{n}{2}(2a + (n-1)d)$ evaluate the part of the formula within the bracket first

$S_{35} = \frac{35}{2}(2 \times 200 + (35 - 1) \times (-2.5))$

use a bracket when substituting the negative value, -2.5

$S_{35} = \frac{35}{2}(400 + (34) \times (-2.5))$

$$S_{35} = \frac{35}{2}(400 - (34) \times (2.5)) \quad \text{multiplying two unlike signs results in a negative value}$$

$$S_{35} = \frac{35}{2}(400 - (85))$$

$$S_{35} = \frac{35}{2}(315) \qquad\qquad \text{subtract the two numbers within the bracket}$$

$$S_{35} = 17.5(315) = 5512.5 \qquad \text{multiplying by the 35/2 outside the bracket}$$

(b) Substitute $n = 9$, $\sum y = 3$, $b = -2.35$ and $\sum x = -21$ into the formula

$$a = \frac{\sum y - b \sum x}{n} \qquad\qquad \text{using brackets when substituting negative quantities}$$

$$a = \frac{3 - (-2.35)(-21)}{9} \qquad\qquad \text{evaluate the top line (numerator) to a single figure}$$

$$a = \frac{3 - (49.35)}{9} \qquad\qquad \text{multiplying two like signs gives a positive number}$$

$$a = -\frac{(46.35)}{9} \qquad\qquad \text{divide the numerator by the denominator}$$

$$a = -5.15$$

1.8.3 Transposition of formulae (Making a variable the subject of a formula)

In the previous section, the single variable on the LHS of a formula (called the subject of the formula) was evaluated when the values of the other variables were known. Any variable in a formula may be evaluated provided the values of all the others are given. If the variable to be evaluated is not already the subject of the formula then the formula must be rearranged to make it the subject: such a rearrangement is called the transposition of the formula.

For example, consider the simple interest formula: $P_t = P_0(1 + i \times t)$. If an investment, P_0, had been on deposit for $t = 10$ years at a simple interest rate of 5% pa ($i = 5/100 = 0.05$) and it is now worth £5400, we would be interested in calculating P_0, the initial deposit 10 years ago. Therefore, start by rearranging the formula $P_t = P_0(1 + i \times t)$ to make P_0 the subject:

$$P_t = P_0(1 + i \times t) \dots \text{divide both sides by } (1 + i \times t)$$

$$\frac{P_t}{(1 + i \times t)} = \frac{P_0\cancel{(1 + i \times t)}}{\cancel{(1 + i \times t)}} = P_0$$

Hence

$$P_0 = \frac{P_t}{(1 + i \times t)} \quad \text{and } P_0 \text{ is the subject of the formula}$$

Evaluate P_0 by substituting the values given for the other variables into the transposed formula

$$P_0 = \frac{5400}{(1 + 0.05 \times 10)} = \frac{5400}{1.5} = 3600$$

Note: Making a variable the subject of a formula is the same as solving the equation for that variable. These problems are not difficult but you need to solve for the required subject of the formula step by step.

WORKED EXAMPLE 1.12

TRANSPOSITION AND EVALUATION OF FORMULAE

(a) The following equation gives the price ($€P$) of a concert ticket when there are Q tickets demanded:

$$P = 12\,000 - 4Q$$

(i) Make Q the subject of the formula.
(ii) Evaluate (1) P when $Q = 2980$ and (2) Q when the price per ticket $P = €40$.
(b) The sum of the first n terms of an arithmetic series is calculated by the formula

$$S_n = \frac{n}{2}(2a + (n-1)d)$$

Note: In S_n the subscript simply indicates that S is the sum of n terms.

(i) Make d the subject of the formula.
(ii) Calculate the value of d when $n = 35$, $a = 200$ and the sum of the first 35 terms, $S_n \to S_{35} = 5512.5$.

SOLUTION

(a)(i) To make Q the subject of the formula, solve for Q:

$$P = 12\,000 - 4Q$$

$$4Q = 12\,000 - P$$

$$Q = 3000 - P/4 \quad \text{divide both sides by 4}$$

(a)(ii) (1) To evaluate P substitute $Q = 2980$ into $\quad P = 12\,000 - 4Q$
Hence $\qquad\qquad\qquad\qquad\qquad\qquad\qquad\quad P = 12\,000 - 4(2980) = 80$
(2) To evaluate Q substitute $P = €40$ into $\quad Q = 3000 - P/4$
Hence $\qquad\qquad\qquad\qquad\qquad\qquad\qquad\quad Q = 3000 - 40/4 = 2990$

(b)(i) Transpose the formula $S = \dfrac{n}{2}(2a + (n-1)d)$ to make d the subject of the formula:

$$S_n = \frac{n}{2}(2a + (n-1)d)$$

$$S_n \times 2 = \cancel{2} \times \frac{n}{\cancel{2}}(2a + (n-1)d) \quad \text{multiply both sides by 2: the twos on the RHS}$$
$$\text{cancel, hence}$$

$$2S_n = n(2a + (n-1)d)$$

$$\frac{2S_n}{n} = \frac{\cancel{n}(2a + (n-1)d)}{\cancel{n}} \qquad \text{divide both sides by } n \text{ and simplify the RHS}$$

$$\frac{2S_n}{n} = 2a + (n-1)d \qquad\qquad \text{now subtract } 2a \text{ from each side}$$

$$\frac{2S_n}{n} - 2a = (n-1)d \qquad\qquad \text{bring the } 2a \text{ to the other side}$$

$$\left(\frac{2S_n}{n} - 2a\right) \times \frac{1}{n-1} = (n-1)d \times \frac{1}{n-1} \qquad \text{multiplying both sides by } \times \frac{1}{n-1}$$

$$\left(\frac{2S_n}{n} - 2a\right) \times \frac{1}{n-1} = d$$

That is,

$$d = \left(\frac{2S_n}{n} - 2a\right) \times \frac{1}{n-1}$$

Hence d is the subject of the formula.

One would normally simplify this expression if possible. In this case we could add the terms inside the bracket and factor out the '2'

$$d = \left(\frac{2S_n}{n} - \frac{2a}{1}\right)\frac{1}{(n-1)} = \left(\frac{2S_n - 2an}{n}\right)\frac{1}{(n-1)} = \frac{2(S_n - an)}{n}\frac{1}{(n-1)}$$

$$= \frac{2(S_n - an)}{n(n-1)}$$

Hence, in its most simplified form, the formula for d is

$$d = \frac{2(S_n - an)}{n(n-1)}$$

(ii)　　Next, substitute $n = 35$, $a = 200$ and $S_n = S_{35} = 5512.5$:

$$d = \frac{2(5512.5 - 200 \times 35)}{35(35-1)} = \frac{2(5512.5 - 7000)}{35(34)} = \frac{2(-1487.5)}{1190} = -\frac{2975}{1190} = -2.5$$

If you look back at Worked Example 1.11, where S was evaluated, you will see that this problem is the same except d is evaluated given the other three variables $n = 35$, $a = 200$ and $S_n = 5512.5$.

1.9　Introducing Excel

A spreadsheet is another aid to mathematical and statistical problem solving. An electronic spreadsheet is a software package that accepts data in the rows and columns of its worksheet as shown in Figure 1.6.

One single location on the spreadsheet is called a 'cell'. Cells A4, C2, C7, C9, E2 and H8 are highlighted in Figure 1.6. A cell is referenced by the column letter followed by the row number, for example, cell C7 is located in column C, row 7. When the data is entered, the user may then perform calculations, such as:

- Sum the data in specified rows and/or columns of the spreadsheet.

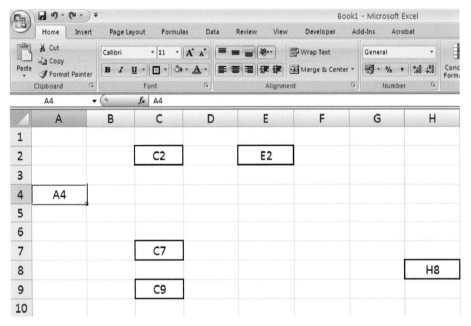

Figure 1.6 Cell reference on a spreadsheet

- Enter formula which calculates results from data in the spreadsheet.
- Plot graphs from data in the spreadsheet.

Examples will be given in Excel where appropriate throughout the text. Further material on Excel is available online.

A variety of different spreadsheet packages are available. We shall use Excel, which is available in Microsoft Office.

WORKED EXAMPLE 1.13
USING EXCEL TO PERFORM CALCULATIONS AND PLOT GRAPHS

Part-time staff are paid on an hourly basis. The number of hours worked per week with the hourly rate of pay for seven staff are as follows:

Name	J.M	P.M	D.H	K.C	J.McM	A.B	C.McK
Hours	6	12.6	34	23	45.8	18	12.6
Rate	£27.5	£27.5	£16.6	£19.2	£50.5	£27.5	£27.5

(a) Enter the data onto a spreadsheet.
(b) Enter a formula to calculate the total pay for each member of staff.
(c) Plot a bar chart, showing the total weekly pay received by each member of staff.

SOLUTION

(a) Enter the data onto the spreadsheet as shown in Figure 1.7. You may start entering the data anywhere in the spreadsheet. In this example the data was entered starting at cell A2.

(b) Next, enter a formula to calculate total pay for each individual in the table.

$$\text{Total pay} = \text{number of hours} \times \text{hourly rate}$$

To indicate to the computer that a formula is being entered into a given cell, the first character typed must be $=$, followed by the formula:

- Position the cursor in cell B5 (so that hours worked, the hourly rate and total pay for J.M are all in the same column).
- Now, from the keyboard, enter the formula to calculate total pay by typing $= \text{B3*B4}$. See Figure 1.8b.
- Repeat this process for each cell in row 5: '$= \text{C3 * C4}$', '$= \text{D3 * D4}$', etc.

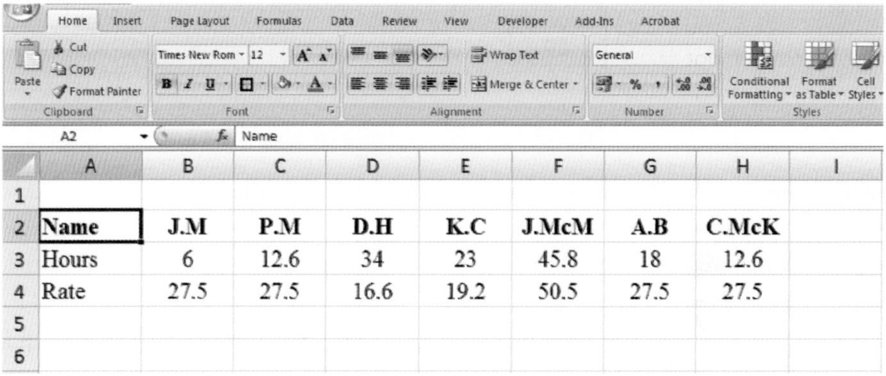

	A	B	C	D	E	F	G	H	I
1									
2	Name	J.M	P.M	D.H	K.C	J.McM	A.B	C.McK	
3	Hours	6	12.6	34	23	45.8	18	12.6	
4	Rate	27.5	27.5	16.6	19.2	50.5	27.5	27.5	
5									
6									

Figure 1.7   Data for Worked Example 1.13 entered on a spreadsheet

The results of the calculations are given in row 5, Figure 1.8a.

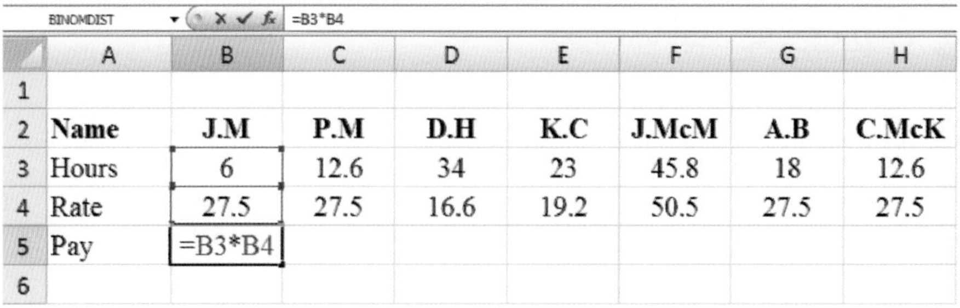

	A	B	C	D	E	F	G	H
1								
2	Name	J.M	P.M	D.H	K.C	J.McM	A.B	C.McK
3	Hours	6	12.6	34	23	45.8	18	12.6
4	Rate	27.5	27.5	16.6	19.2	50.5	27.5	27.5
5	Pay	=B3*B4						
6								

Figure 1.8 a Type the formula $= \text{B3*B4}$ into cell B5 to calculate pay $= 165$ for J.M.

Note: The advantage of entering the cell names instead of the actual values is that the result is automatically recalculated if the numbers in these cells are changed. Try it!

Alternatively, 'drag and drop' the formula in cell B5 across row 5, instead of keying the formula into each individual cell. This is accomplished by clicking on cell B5. It is now the active cell with a black box around it. Point the mouse to the bottom right-hand corner of the black box until a solid black cross appears, see Figure 1.8b.

2	Name	J.M	P.M	D.H	K.C	J.McM	A.B	C.McK
3	Hours	6	12.6	34	23	45.8	18	12.6
4	Rate	27.5	27.5	16.6	19.2	50.5	27.5	27.5
5	Pay	165.0						
6								

Figure 1.8 b Preparing to drag and drop the formula in cell B5 across row 5

Hold down the left button on the mouse and drag across cells C5 to H5. (Use the HELP button or consult some of the many reference books on Excel.)

(c) To plot the total pay for each individual select (by holding down the left mouse button) the names in row 2; then hold down the control (ctrl) button on the keyboard and select row 5, which contains the pay for each individual. See Figure 1.9.

To plot the graph, click on Insert in the menu bar, followed by 'column' and '2-D Column', also illustrated in Figure 1.9.

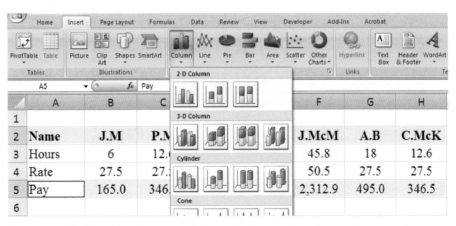

Figure 1.9   Highlight the data required for graph plotting, then click on 'Insert' on the main menu bar: select 'Column' to plot a bar or column graph

The basic bar graph is given in Figure 1.10.

Format the graph

Adding the overall title, axis titles, grid lines, etc., is called 'formatting the graph':

To add titles, first click anywhere on the graph to access 'Chart tools' from the main menu bar. Click on 'Layout'; from the 'Layout' menu you may select 'Chart Title' and 'Axis Titles' to enter these on your graph. See Figure 1.11.

You should try (trial and error!) the various options offered on the 'Chart tools' menu to produce a clear, well-labelled graph. Figure 1.12 is an example of a very basic finished graph.

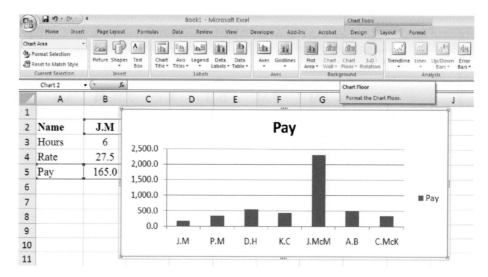

Figure 1.10   Basic bar chart without titles

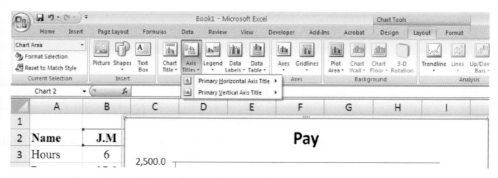

Figure 1.11   Chart tools: use these to add titles and format your graph

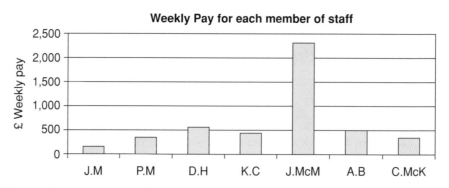

Figure 1.12   A very basic plot of weekly pay for seven members of staff

MATHEMATICAL PRELIMINARIES

PROGRESS EXERCISES 1.4

Electronic Aids: the Calculator and Spreadsheets

Evaluate the following expressions using a calculator.

1. (a) $(23.6)\dfrac{23}{108} - 56 + \dfrac{90}{108}$ (b) $\sqrt{\dfrac{128}{20.4}}$ (c) $\sqrt{\dfrac{24.88}{92.5} + 256}$

2. (a) $\dfrac{23(1.90) - (-39)(1.8)}{12(1.55) + 18.6}$ (b) $\dfrac{23}{1.56} - \dfrac{14.57}{2.55}$ (c) $\sqrt{12 + \dfrac{89}{0.045}}$

3. (a) Calculate the value of $v = 0.8(m^2 - n^2)$ when (i) $m = 3.6$ and $n = 2.1$, (ii) $m = 2.3$ and $n = -4.8$.

   (b) The price (in pence) of a carton of orange juice in a school canteen is given by the formula $P = 1639 - 25x + 15y$ where x and y are the number of cartons of orange juice and milk sold. Find the price of a carton of orange juice when: (i) $x = 125$ and $y = 100$; (ii) $x = 173$ and $y = 182$.

   (c) The slope of a line is given by the equation $m = \dfrac{y_2 - y_1}{x_2 - x_1}$. Calculate the value of m when: (i) $x_1 = 4$, $x_2 = 2$, $y_1 = 3$ and $y_2 = 1.8$; (ii) $x_1 = -0.4$, $x_2 = 2.0$, $y_1 = -16.2$ and $y_2 = -11.4$.

   (d) The formula for the standard error of estimate in regression analysis may be expressed as $s_e = \sqrt{\dfrac{n}{n-2} \times (\sigma_y^2 - b^2 \sigma_x^2)}$. Evaluate s_e when $n = 18$, $\sigma_y^2 = 3.215$, $b = 0.87$ and $\sigma_x^2 = 2.85$.

   (e) In statistics, the formula for the slope, b, of a least-squares line is given by the formula
   $$b = \dfrac{n \sum xy - \sum x \sum y}{n \sum x^2 - \left(\sum x\right)^2}$$
   Evaluate b correct to two decimal places when $n = 9$, $\sum x = -21$, $\sum y = 3$, $\sum xy = -256$, $\sum x^2 = 154.96$.

4. (a) The price (in pence) of a carton of orange juice in a school canteen is given by the formula $P = 1639 - 25x + 15y$ where x and y are the number of cartons of orange juice and milk sold.

   (i) Make y the subject of the formula. Hence evaluate y when $x = 125$ and $P = 50$p.

   (ii) Make x the subject of the formula. Hence evaluate x when $y = 180$ and $P = £1.5$.

   (b) Consider the formula $v = 0.8(m^2 - n^2)$: (i) make m the subject of the formula. (ii) make n the subject of the formula.

(c) The formula for an average cost (AC) function for 18 inch flat-screen TVs is given as $AC = 1500 + \dfrac{2.55}{Q}$, where Q is the quantity of TVs produced. (i) Make Q the subject of the formula; (ii) Evaluate Q when $AC = 1680$.

(d) The formula for the intercept a of a regression line is given as $a = \dfrac{\sum y - b \sum x}{n}$. (i) Make b the subject of the formula. (ii) Evaluate b when $n = 9$, $\sum x = -21$, $\sum y = 3$, $a = -2.5$.

(e) The formula for the standard error of estimate in regression analysis may be expressed as $s_e = \sqrt{\dfrac{n}{n-2} \times (\sigma_y^2 - b^2 \sigma_x^2)}$. Make n the subject of the formula. Hence evaluate n when $s_e = 1.4$, $\sigma_y^2 = 3.215$, $b = 0.9$ and $\sigma_x^2 = 2.85$.

5. The hours worked, expenses due and hourly rate of pay for a group of staff are given as:

Initials	J.C	J.M	S.T	R.G	P.M	K.McK
Hours	16	34	64	12.5	15.5	14.5
Rate	£12.6	£38.8	£102.5	£15.5	£32.5	£45
Expenses	£120	£55	0	£25	£12	£155

Use a spreadsheet to calculate:
(a) The total pay
(b) The net pay when a flat 20% tax is deducted from each employee
(c) The net pay, plus expenses
(d) Plot a bar chart and a pie chart of
   (i) Hours worked
   (ii) Net pay plus expenses
   (iii) Expenses

www.wiley.com/college/bradley

Go to the website for Problems in Context

TEST EXERCISES 1

1. Simplify
   (a) $2x(4) - 3(x + 1)$ (b) $(x - 1)(x + 2)$ (c) $\dfrac{1}{2x} - 4x$

2. Solve the equations
   (a) $x(x + 1) = 0$ (b) $\dfrac{1}{x} + 2 = 5$ (c) $\dfrac{x - 1}{x^2 + 5x + 11} = 0$

3. Determine the range of values for which the following inequalities are true:

(a) $x + 2 > 10$ (b) $x < 2x - 12$ (c) $\dfrac{x - 1}{x^2 + 5x + 11} = 0$

4. A motorised bike is priced at £3750.
   (a) If the price includes the retailer's profit of 20%, calculate the cost price of the bike.
   (b) If the customer must pay 21% tax on the retail price, calculate the total paid by the customer.

5. Solve the equations

(a) $(x + 2)(x - 1) = 0$ (b) $(x + 2)(x - 1) = x^2$

6. Determine whether the expression on the LHS simplifies to the expression on the RHS.

(a) $\dfrac{x}{2} + \dfrac{5x}{4} = \dfrac{7x}{4}$ (b) $\dfrac{4x}{1/3} = 12x$

7. Theatre tickets are classified A (£35), B (£27), C (£17.50), D (£12.50). The numbers of each type sold are 180, 260, 450, 240, respectively.
   (a) Set up a table to calculate the revenue from each type of ticket.
   (b) Plot a bar chart of: (i) the numbers of each class of ticket sold; and (ii) the revenue received from each class of ticket.

8. A company's imports of tea are given by country of origin and weight: 400 tons from India, 580 tons from China, 250 tons from Sri Lanka, 120 tons from Burma.
   (a) Plot the weight of tea imported on: (i) a bar chart; and (ii) a pie chart.
   (b) Calculate the percentage of the company's imports from each country.

9. Solve the equations

(a) $20x - 3x^2 = 10(2x - 3)$ (b) $\dfrac{2}{x} = \dfrac{x}{2x} + 1$

10. Solve the equations

(a) $2x(x - 2) = 2x(x + 2)$ (b) $3\dfrac{x}{5} - 4\dfrac{x}{20} = 2$

11. The exchange rates for US dollars are 1.32 British pounds and 1.23 Euros.
    (a) A notepad costs US$850. Calculate the price in: (i) British pound; and (ii) Euros.
    (b) Convert 400 Euros to: (i) US dollars; and (ii) British pounds.

Value added tax: Value added tax (VAT) is a tax levied at each stage of the production and distribution of certain goods and services. VAT is called an *ad valorem* tax, in that it is set at a fixed percentage of the value (or price) of the good or service.

THE STRAIGHT LINE AND APPLICATIONS

2

2.1   The Straight Line

2.2   Mathematical Modelling

2.3   Applications: Demand, Supply, Cost, Revenue

2.4   More Mathematics on the Straight Line

2.5   Translations of Linear Functions (on the website)

2.6   Elasticity of Demand, Supply and Income

2.7   Budget and Cost Constraints (on the website)

2.8   Excel for Linear Functions

2.9   Summary

Chapter Objectives

The straight line is the simplest mathematical function yet it is used to model a very wide range of concepts in economics and business. At the end of this chapter you should be able to:

- Plot the straight line when given the value of slope and intercept
- Write down the equation of the straight line when given (i) the value of slope and intercept (ii) any two points on the line (iii) the slope of the line and a point on the line
- Plot a straight line when given its equation
- Plot linear demand, supply, cost and revenue functions
- Verbally describe linear demand, supply, cost and revenue functions
- Write down the equations of and plot cost and budget constraints (website)
- Translate linear functions horizontally and vertically and write down the equation of the translated function (website)
- Calculate the price elasticity of demand and supply for linear demand and supply functions
- Plot any linear function in Excel
- Use Excel to illustrate the rules for vertical and horizontal translations.

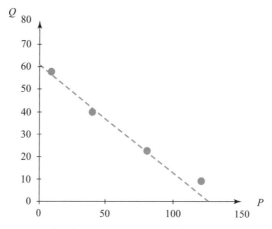

Figure 2.1   Graph of quantity demanded (Q) at set prices (P)

This chapter begins to establish a formal relationship between mathematics and its use in business and economics. Economics studies the relationship between different parts of an economic system; for example, the relationship between the price of a good and quantity demanded, between personal income and consumption, between production and costs. These relationships may be described by both qualitative and quantitative means. Qualitatively, the relationship between price and quantity demanded may be described by the statement 'as price increases, quantity demanded decreases'. This verbal statement, however, is not precise as it does not tell us by how much quantity demanded decreases when price increases. It is therefore necessary to establish a quantitative relationship between price and quantity demanded. The first step could be to make several observations of the quantity demanded (Q) at different prices (P) (see Table 2.1). The results are plotted on a graph (see Figure 2.1).

From the graph, the mathematical form of the relationship, a formula relating Q to P, can usually be deduced. In the example given, the relationship is quite closely modelled by a straight line, or we say the relationship is linear. This formula, or equation, relating Q to P is called a mathematical model. Mathematical models are essential if a system is to be analysed and the exact relationships between economic quantities identified and understood. The simplest quantitative relationship is the straight line or linear function. Due to its simplicity, the linear function serves as an introductory mathematical model for a variety of business and economic applications. This chapter introduces some of these applications. Initially, however, it is necessary to study the straight line.

2.1   The Straight Line

At the end of this section you should be able to:

- Define the slope and intercept of a straight line and plot its graph
- Understand what 'the equation of a line' means
- Plot any straight line given the equation in the form $y = mx + c$ or $ax + by + d = 0$.

Table 2.1   Price and quantity observations

Price	Quantity demanded
$P = 10$	$Q = 58$
$P = 40$	$Q = 40$
$P = 80$	$Q = 23$
$P = 120$	$Q = 10$

2.1.1   The straight line: slope, intercept and graph

The straight line is one of the simplest mathematical functions: it is easy to manipulate and to graph. It is, therefore, essential to fully understand the straight line, what each part of the equation means and the effect of changes in each part on the graph of the line.

Introductory background on graphs

Figure 2.2 shows the horizontal axis x and the vertical axis y, which intersect at the origin where $x = 0$ and $y = 0$. Any point can be plotted on this graph if the coordinates, (x, y), are known; that is, the x-coordinate (with sign) is stated first and measured along the horizontal axis from the origin, followed by its y-coordinate (with sign) measured along the vertical axis. The points $(2, 2)$, $(1, 1)$, $(2, -2)$, $(-4, -3)$, $(-3, 2)$ and $(0, 0)$ are plotted in Figure 2.2.

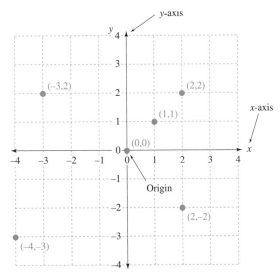

Figure 2.2   Plotting points on a graph

How to define a straight line

A straight line may be defined by two properties:

> • **Slope**, usually represented by the symbol m
> • **Vertical intercept**, the point at which the line crosses the y-axis, usually represented by the symbol c.

Note: The vertical or y-intercept is often simply called 'the intercept'. This is the convention adopted throughout this text.

Slope

The slope of a line is simply the '**slant**' of the line. The slope is negative if the line is falling from left to right and the slope is positive if the line is rising from left to right (see Figure 2.3). In Figure 2.3, each line has a different slope and a different intercept. Note the horizontal line which has a zero slope and the vertical line which has an infinite slope.

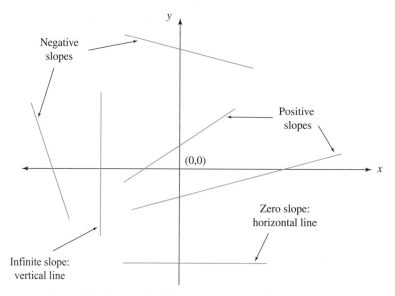

Figure 2.3   Lines with different slopes and different intercepts

Measuring the slope of a line

Figure 2.4 shows how the slope of a line is measured.

• The line through the points A and B has a positive slope since it is rising from left to right. The slope is measured as follows:

$$\text{Slope} = \frac{\text{change in height}}{\text{change in distance}} = \frac{2}{4} = \frac{1}{2} = \frac{0.5}{1} = 0.5$$

That is, the height increases by 0.5 units when the horizontal distance increases by 1 unit.

THE STRAIGHT LINE AND APPLICATIONS

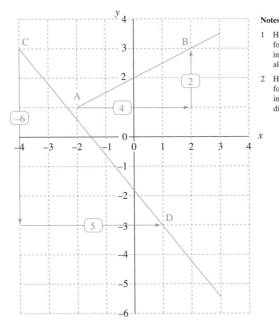

Notes

1 Height increases by 0.5 for every one unit increase in distance along the line AB

2 Height decreases by 1.2 for every one unit increase in horizontal distance along CD

Figure 2.4   Measuring slope

Notes:

1. For the line AB, height increases by 0.5 for every one unit increase in horizontal distance.
2. For the line CD, height decreases by 1.2 for every one unit increase in horizontal distance.

• The line through the points C and D has a negative slope since the line is falling from left to right. A decrease in height is indicated by a minus sign. The slope is measured as follows:

$$\text{Slope} = \frac{\text{change in height}}{\text{change in distance}} = \frac{-6}{5} = \frac{-1.2}{1} = -1.2$$

That is, height decreases by 1.2 units when the horizontal distance increases 1 unit.

Note: The change in height and change in distance are generally referred to as Δy and Δx, respectively, where the symbol Δ means change.

The slope of a line may be described in several ways:

1. The change in height (Δy) divided by the corresponding increase in horizontal distance (Δx):

$$\text{slope}, m = \frac{\Delta y}{\Delta x}$$

2. The change in y per unit increase in x.
3. The number of units by which y changes when x increases by 1 unit.

• Lines with the same slope, but different intercepts are different lines: they are parallel lines.
• Lines with the same intercept, but different slopes are different lines.
• Two lines are identical only if their slopes and intercepts are identical.

WORKED EXAMPLE 2.1
PLOTTING LINES GIVEN SLOPE AND INTERCEPT

Plot the lines given the following information on slope and intercept:

(a) Slope $= 1$, intercept $= 0$ ($m = 1, c = 0$)   (b) Slope $= 1$, intercept $= 2$ ($m = 1, c = 2$)

SOLUTION

(a) Locate the vertical intercept at $y = 0$ on the y-axis. This is the point $(0,0)$ which is the origin. The slope of the line is unity: $m = 1$ which means $\Delta y / \Delta x = 1$; hence $\Delta y = \Delta x$. To draw the line start at the intercept $(0, 0)$, move forward a horizontal distance of $\Delta x = 1$ unit, then move vertically up by a height $\Delta y = 1$ unit; see Figure 2.5. This gives the point $(1, 1)$. Join the points $(0, 0)$ and $(1, 1)$ and continue the straight line until it is the required length. This line is often referred to as the 45° line, the line through the origin which makes an angle of 45° with the horizontal axis. Any change in the horizontal distance is accompanied by an equal change in the vertical height.

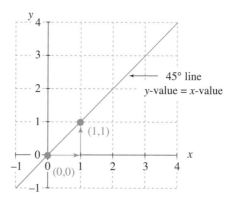

Figure 2.5   45° line through the origin

(b) Locate the vertical intercept, $y = 2$. At $y = 2$, draw a line with a slope of 1, by moving forward a horizontal distance of 1 unit and then vertically up by 1 unit. This gives the point $(1, 3)$. Join the points $(0, 2)$ and $(1, 3)$; continue the straight line until it is the required length. This line is illustrated in Figure 2.6.

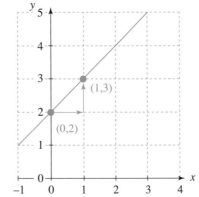

Figure 2.6   Line with slope = 1, intercept = 2

PROGRESS EXERCISES 2.1

Points, Slopes and Intercepts for Linear Functions

1. Plot the points $(-2,0)$, $(0,2)$, $(2,4)$, $(4,6)$ and $(5,7)$. The plotted points lie on a straight line. Measure the slope and vertical intercept.
2. Plot the points $(-1,7)$, $(0,6)$, $(1,5)$, $(3,3)$, $(5,1)$ and $(7,-1)$. The plotted points lie on a straight line. Measure the slope and vertical intercept.

   Draw the following lines given in questions 3 to 6. Measure the point where each line cuts the horizontal axis.
3. Slope = 2, y-intercept = 0
4. Slope = 1, y-intercept = -2
5. Slope = -2, y-intercept = 0
6. Slope = -1, y-intercept = 2.

2.1.2   The equation of a line

Worked Example 2.1 demonstrated how a straight line is determined uniquely in terms of its slope and intercept. It is not, however, convenient for mathematical manipulations and other applications to work with a line defined in terms of slope m and intercept c separately. It will be shown in Worked Example 2.2 that a line may be defined by a formula or equation that incorporates slope m and intercept c as follows:

$$y = mx + c \qquad\qquad (2.1)$$

The formula, $y = mx + c$, which calculates the y-coordinate for any given value of x on the line, is called **the equation of the line**.

WORKED EXAMPLE 2.2

DETERMINE THE EQUATION OF A LINE GIVEN SLOPE AND INTERCEPT

(a) Draw the line with slope $m = 3$ and intercept $c = 1$. State the values of the y-coordinate for the points whose x-coordinate is $x = 0, 1, 2$ and 3. Hence, use these points to demonstrate that the equation $y = 1 + 3x$ calculates the value of the y-coordinate from the value of the x-coordinate.

(b) Write down the equations of the lines with slope and intercept given in Worked Example 2.1:

(i) $m = 1, c = 0$ (ii) $m = 1, c = 2$.

SOLUTION

(a) In Figure 2.7, the line is drawn, starting at the intercept $c = 1$ on the y-axis, i.e., at the point $x = 0, y = 1$. Then, since slope $m = 3$, each unit increase in x is followed by a 3 unit increase in y.

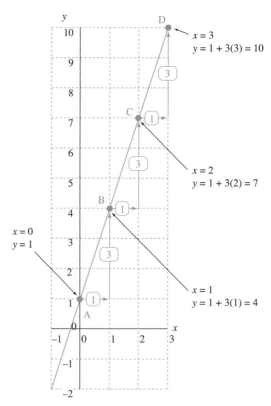

Figure 2.7   The formula $y = 1 + 3x$ calculates the y-coordinate for any given value of x on the line; $y = 1 + 3x$ is called **the equation of the line**

Remember

Slope is the change in y per unit increase in x.

To demonstrate that the equation $y = 1 + 3x$ calculates the value of the y-coordinate from the value of the x-coordinate, observe the pattern relating the x- and y-values for selected points on the line.

x-value	y-value	Calculation of y-value
$x = 0$	$y = 1$	this is the intercept
$x = 1$	$y = 4$	$y = 1 + 3 \times 1$
$x = 2$	$y = 7$	$y = 1 + 3 \times 2$
$x = 3$	$y = 10$	$y = 1 + 3 \times 3$
any x-value	$y = 1 + 3x$	$y = 1 + 3 \times (x\text{-value})$

Hence the equation $y = 1 + 3x$ describes the relationship between the x- and y-coordinates for any point on the line:

$$y = 1 + 3x \text{ is called the equation of the line}$$

In general, $y = mx + c$ is the equation of any line, slope $= m$ and intercept $= c$. See Figure 2.7.
In the example above, $y = 1 + 3x$ is the equation of a line with slope $m = 3$ and intercept $c = 1$.

(b) (i) Figure 2.8 is simply Figure 2.5 redrawn with additional points labelled. The equation of the line may be determined by substituting $m = 1$, $c = 0$ into the general formula $y = mx + c$:

$$y = 1x + 0$$
$$y = x$$

(2.2)

This is the equation of the line.

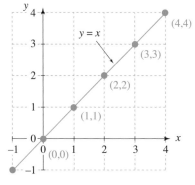

Figure 2.8   Graph of the line $y = x$

Remember

$y = x$ is the equation of the 45° line through the origin.

Observe that, for each point on the line, the y-coordinate is equal to the x-coordinate. The equation of the line is $y = x$.

(ii) The equation of the line with $m = 1, c = 2$ is determined by substituting $m = 1, c = 2$ into the general formula:

$$y = mx + c$$
$$y = 1x + 2 \qquad \qquad \textbf{(2.3)}$$
$$y = x + 2$$

See Figure 2.9 for the graph of this line.

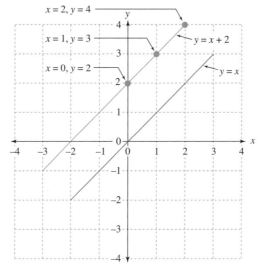

Figure 2.9   Comparing the lines $y = x$ and $y = x + 2$

Notes:

(i) Observe that for each point on the line $y = x + 2$, the value of the y-coordinate is equal to the value of the x-coordinate plus 2; hence its equation.

(ii) The lines $y = x$ and $y = x + 2$ have the same slope, $m = 1$, with different intercepts; they are parallel lines.

Horizontal intercepts

The above analysis has been concerned with the slope of a line and the vertical intercept only insofar as these two properties may be used to define a straight line. However, there are times when the calculation of the horizontal intercept is required, that is, the point where the line crosses the horizontal axis. This calculation uses the fact that every point on the horizontal axis has a y-coordinate: $y = 0$. Since y is known, substitute $y = 0$ into the equation of the line, $y = mx + c$, and solve for x: this value of x ($x = -c/m$) is the horizontal intercept. See the following worked example.

WORKED EXAMPLE 2.3
CALCULATION OF HORIZONTAL AND VERTICAL INTERCEPTS

Calculate the horizontal and vertical intercepts for each of the following lines:

(i) $y = mx + c$ (ii) $y = x + 2$ (iii) $y = 1 + 3x$.

Method	(i) $y = mx + c$	(ii) $y = x + 2$ (Figure 2.9)	(iii) $y = 1 + 3x$ (Figure 2.7)
Equation of line	$y = mx + c$	$y = x + 2$	$y = 1 + 3x$
(i) Horizontal or x-intercept			
Line cuts x-axis at $y = 0 \rightarrow$	$0 = mx + c$	$0 = x + 2$	$0 = 1 + 3x$
Solve for $x \rightarrow$	$-c = mx$	$-2 = x$	$-1 = 3x$
	$-c/m = x$		$-1/3 = x$
The horizontal intercept is	$x = -\dfrac{c}{m}, y = 0$	$x = -2, y = 0$	$x = -\dfrac{1}{3}, y = 0$
(ii) Vertical or y-intercept[*]	$y = mx + c$	$y = x + 2$	$y = 1 + 3x$
Line cuts y-axis at $x = 0 \rightarrow$	$y = m(0) + c$	$y = 0 + 2$	$y = 1 + 3(0)$
Substitute $x = 0$, solve for y	$y = c$	$y = 2$	$y = 1$
The vertical intercept is	$x = 0, y = c$	$x = 0, y = 2$	$x = 0, y = 1$

[*]Alternatively, read off the vertical intercept, $y = c$, from the equation $y = mx + c$

Note: The horizontal intercept may be denoted by $x(0)$; this simply means the 'value of x when $y = 0$' or x (when $y = 0$), abbreviated to $x(0)$. Similarly, the vertical intercept, which is the value of y when x is zero, may be denoted by $y(0)$.

Equation of horizontal and vertical lines

Figure 2.10 illustrates that the equation of a horizontal line simply takes the value of the coordinate where the line crosses the y-axis, and the equation of a vertical line simply takes the value of the coordinate where the line crosses the x-axis.

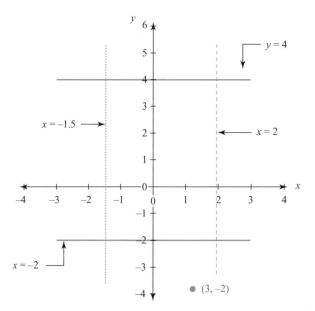

Figure 2.10   Equation of horizontal and vertical lines

For any straight line

The equation is the formula which relates the *x*- and *y*-coordinates for every point on the line (or curve, as we shall see later).

$$y = \text{slope (change in } y \text{ per unit increase in } x \times \text{(units moved from } x = 0) + \text{intercept}$$

$$y = \qquad m \qquad \times \qquad x \qquad + \qquad c$$

2.1.3   To graph a straight line from its equation

To graph a straight line when its equation is given in the form $y = mx + c$

- Earlier in this chapter lines were plotted, starting at the intercept (first point), then plotting further points using the fact that the slope is the change in *y* per unit increase in *x*.
- In this section a straight line is graphed from its equation by calculating at least two points, then drawing a line through them. To calculate the required points, use either of the following methods.

Method A
Calculate a table of points (at least two points).

Step 1: Rearrange the equation into the form $y = mx + c$, if it is not already in this form.
Step 2: Choose any *x*-values (choose convenient values of *x*, such as $x = 0$, $x = 10$, $x = -2$); calculate the corresponding *y*-values from the rearranged formula.
Step 3: Plot the points and draw the line through these points.

Method B

In economics, graphs are frequently required in the region where x-values and y-values are positive. To plot a line in this region, calculate the horizontal and vertical intercepts; draw a line through these points.

Step 1: Rearrange the equation into the form $y = mx + c$ if necessary.
Step 2: The x-intercept is at $(x = -c/m, y = 0)$; the y-intercept is at $(x = 0, y = c)$.
Step 3: Draw the line through the points.

Note: While a minimum of two points are required to plot a line, it is a good idea to calculate at least three points, just in case of mistakes in calculations. All points should lie perfectly on the line.

WORKED EXAMPLE 2.4

TO GRAPH A STRAIGHT LINE FROM ITS EQUATION

$y = mx + c$

A line has the equation (or formula) $y = 2x - 1$
Plot the line over the interval $x = -2$ to $x = 3$.

SOLUTION

Method A

Step 1: $y = 2x - 1$. No rearranging necessary.
Step 2: Since the graph is required for values of x from $x = -2$ to $x = 3$, calculate y for several x-values within this interval. You may choose any x-values you wish, for example, $x = -2, -1, 0, 1, 2, 3$. Calculate the corresponding y-coordinates by simply substituting the chosen x-value into the equation (Table 2.2).

Table 2.2   Calculating the y-coordinate given the x-coordinate

x	$y = 2x - 1$	Points (x, y)
−2	$y = 2(-2) - 1 = (-4) - 1 = -5$	−2, −5
−1	$y = 2(-1) - 1 = (-2) - 1 = -3$	−1, −3
0	$y = 2(0) - 1 = (0) - 1 = -1$	0, −1
1	$y = 2(1) - 1 = (2) - 1 = 1$	1, 1
2	$y = 2(2) - 1 = (4) - 1 = 3$	2, 3
3	$y = 2(3) - 1 = (6) - 1 = 5$	3, 5

Step 3: Plot the points, draw the line; see Figure 2.11.

Method B

Step 1: $y = 2x - 1$. No rearranging necessary.

Step 2: **Vertical intercept:** the line cuts the vertical axis at $y = c \rightarrow y = -1$. Hence the vertical intercept is $(0, -1)$.

Horizontal intercept: the line cuts the horizontal axis at $x = -c/m \rightarrow x = 0.5$. Hence the horizontal intercept is $(0.5, 0)$.

Note: Rather than memorising formulae, such as $x = -c/m$, $y = 0$ for the horizontal intercept, you may prefer to solve for the x-intercept directly, as follows. Since the line cuts the x-axis at $y = 0$, substitute $y = 0$ into the equation and solve for x:

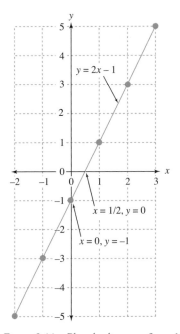

Figure 2.11   Plot the line $y = 2x - 1$

$$\left.\begin{array}{l} y = 2x - 1 \\ 0 = 2x - 1 \\ 1 = 2x \\ \dfrac{1}{2} = x \end{array}\right\} \rightarrow \text{hence horizontal intercept at } x = 0.5, \ y = 0$$

Step 3: Plot the vertical and horizontal intercept on the axis, then draw the line through these points. See Figure 2.11.

To graph a straight line given its equation in the form $ax + by + d = 0$

Any equation of the form $ax + by + d = 0$, where a, b and d are constants, is the equation of a straight line. When the equation is rearranged into the form $y = mx + c$, the slope and intercept on the y-axis can then be written down immediately. It is helpful to have a rough idea of the graph before you start plotting. Therefore, by writing down m and c you know where the line crosses the vertical axis and whether the slope is positive or negative:

$$
\begin{aligned}
ax + by + d &= 0 \\
ax + by &= -d && \text{bringing across the } d \\
by &= -ax - d && \text{bringing across the } ax \\
\frac{1}{b}by &= \frac{1}{b}(-ax - d) && \text{dividing each side by } b \\
y &= \frac{1}{b}\frac{(-ax)}{1} + \frac{1}{b}\frac{(-d)}{1} && \text{multiplying each term inside the bracket by } \frac{1}{b} \\
y &= -\frac{a}{b}x - \frac{d}{b} && \text{simplifying the fractions}
\end{aligned}
$$

This is a straight line that cuts the vertical axis at $y = b$ with a slope $-a/b$. The slope $-a/b$ will be positive or negative depending on the signs of a and b.

Note: It is not necessary to memorise these formulae for intercept and slope. In numerical examples, simply rearrange the equation with y on the LHS (this is called making **y the subject of the formula**), then read off the value of the intercept and slope, as shown in the following worked example.

WORKED EXAMPLE 2.5
PLOT THE LINE $ax + by + d = 0$

The equation of a line is given as $4x + 2y - 8 = 0$.

(a) Write the equation of the line in the form $y = mx + c$, hence write down the slope and y-intercept. Give a verbal description of slope and intercept.
(b) Which variable should be plotted on the vertical axis and which on the horizontal axis when the equation of the line is written in the form $y = mx + c$?
(c) Plot the line by (i) calculating a table of points (**method A**) and (ii) calculating the intercepts (**method B**).

SOLUTION

(a) Rearrange the equation of the line with y as the only variable on the LHS:

$$
\begin{aligned}
4x + 2y - 8 &= 0 \\
2y &= 8 - 4x && \text{bringing all non-}y\text{-terms to the RHS} \\
\frac{2y}{2} &= \frac{8}{2} - \frac{4x}{2} && \text{dividing everything on both sides by 2} \\
y &= 4 - 2x && \text{simplifying}
\end{aligned}
$$

From the rearranged equation we can read off useful information before going ahead to plot the line. Intercept: the line cuts the y-axis at $y = 4$. So when $x = 0$, $y = 4$. Slope: the line has slope $m = -2$, i.e., the line is falling from left to right, dropping by 2 units for each horizontal increase of 1 unit.

(b) When the equation is written in the form $y = mx + c$ with the variable by itself on one side of the equation, y is usually plotted on the vertical axis and the other variable, x, is plotted on the horizontal axis.

(c)

(i) Calculate a table of points	(ii) Calculate the intercepts
Step 1: $y = -2x + 4$: see part (a)	**Step 1:** $y = -2x + 4$: see part (a)
Step 2: From the equation of the line $y = -2x + 4$, calculate y for several x-values, such as: $x = 0$, $y = -2(0) + 4 = 4$: the point is $(0, 4)$ $x = 1$, $y = -2(1) + 4 = 2$: the point is $(1, 2)$ $x = 2$, $y = -2(2) + 4 = 0$: the point is $(2, 0)$ $x = 3$, $y = -2(3) + 4 = -2$: the point is $(3, -2)$	**Step 2:** The equation is $y = -2x + 4$ The line cuts the y-axis at $y = 4$ Hence the vertical intercept is $(0, 4)$ The line cuts the x-axis at $x = -c/m = -4/(-2) = 2$ Hence the horizontal intercept is $(2, 0)^*$
Step 3: Plot these points and draw the line through them (Figure 2.12)	**Step 3:** Plot the intercepts, join them and you have the graph of the line (Figure 2.12)

*Alternatively, solve for the x-intercept by using the fact that $y = 0$ at the point where the line cuts the x-axis

$y = -2x + 4$
$0 = -2x + 4$, since $y = 0$ on the x-axis $\Big\}$ hence the horizontal intercept is $x = 2$, $y = 0$
$2x = 4$, i.e., $x = 2$

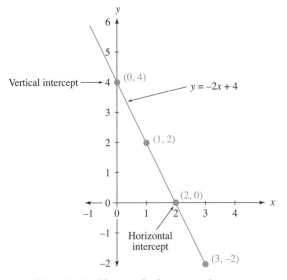

Figure 2.12   Plotting the line $y = -2x + 4$

Some mathematical notations

Take the line $y = mx + c$. In general, when y is written in terms of x we say that '**y is a function of** x'. A function of x may be described as a rule for 'operating' on x. In the case of $y = mx + c$, the rule is: multiply x by the constant, m, then add the constant, c.
x is called the independent variable (when plotting a line from a given equation, any convenient x-values are chosen, as in Worked Example 2.4).

The independent variable, x, is plotted on the horizontal axis.

---

y is called the dependent variable: since the y-values are calculated from the independently chosen x-values. Usually the dependent variable, y, is the variable which is written by itself on the LHS of the equation.

The dependent variable, y, is plotted on the vertical axis.

General notations

'y is a function of x' may be denoted in several ways: $y = f(x); y = g(x); y = h(x)$, etc., where the letters f, g and h are used to distinguish different formulae in x. (The brackets here do not mean multiplication, they mean '**y depends on** x'.) For example:

(a) $y = -20x + 41$ or $f(x) = -20x + 41$
(b) $y = 11x - 12$ or $g(x) = 11x - 12$

Inverse function: When equations are used in applications later, there are times when we write y in terms of x, denoted as $y = f(x)$, but in other applications we require x in terms of y, denoted as $x = f^{-1}(y)$. This latter function, $x = f^{-1}(y)$, is called the inverse function. For example:

$$y = 11x - 12 \qquad \text{written symbolically as } y = f(x) = 11x - 12$$

$$x = \frac{y + 12}{11} \qquad \text{rearranging, writing } x \text{ as the variable on the LHS}$$

$$x = \frac{y}{11} + 1.09$$

This is the inverse function written symbolically as

$$x = f^{-1}(y) = \frac{y}{11} + 1.09$$

PROGRESS EXERCISES 2.2

Straight Lines: Equations, Slopes and Intercepts

1. Deduce the equations of the lines in Progress Exercises 2.1, questions 3, 4, 5, 6.
   (3) Slope = 2, y-intercept = 0. (4) Slope = 1, y-intercept = −2.
   (5) Slope = −2, y-intercept = 0. (6) Slope = −1, y-intercept = 2.
2. Given the equations of the lines:
   (a) $y = x + 2$ (b) $y = -4x + 3$ (c) $y = 0.5x - 2$ (d) $2y = 6x + 4$
   For each line (a), (b) and (c)
   (i) Write down the slope and intercepts (horizontal and vertical).
   (ii) Calculate the values of y when $x = -2, 0, 2, 4, 6$.
   (iii) Plot each line over the interval $x = -2$ to $x = 6$.
3. Write the equation for each of the following lines in the form $y = f(x)$, hence write down the slope, intercepts (if any):
   (a) $y = 2$ (b) $x = -2$ (c) $5x + y + 4 = 0$ (d) $y = x$ (e) $x - y + 5 = 0$
   Plot the intercepts, hence plot the line by joining the intercepts.
4. Given the equations of the following lines:
   (a) $2y - 5x + 10 = 0$ (b) $x = 10 - 2y$ (c) $y + 5x = 15$
   (i) Write each of the equations in the form $y = f(x)$.
   (ii) Calculate the intercepts. Plot each line by joining the intercepts.
   (iii) From the graph, show that the magnitude of slope is the ratio: vertical intercept/horizontal intercept.
   From the graph, how would you determine whether the slope is positive or negative?
5. (See question 4) For each line (a), (b), (c)
   (i) Write the equations in the form $x = g(y)$.
   (ii) Plot each line as $x = g(y)$, showing both intercepts.
   Compare the graphs, particularly intercepts and slopes, in questions 4 and 5.
6. Determine which of the following points lie on the given line.
   (a) Points (x, y): A = (1, 3), B = (−1, −1), C = (0, 1) and line, $y = 2x + 1$.
   (b) Points (P, Q): A = (90, 5), B = (8, 10), C = (70, 15) and line, $Q = 50 - 0.5P$.
   (c) Points (Q, TC): A = (2, 14), B = (14, 18), C = (6, 22) and line, $TC = 10 + 2Q$.
7. (a) Plot the points (4, 4), (2, 4), (12, 4) and (−2, 4).
   (b) Plot the points (2, 10), (2, 5), (2, 1) and (2, −2).
   The points plotted in parts (a) and (b) each lie on a straight line. Deduce the equation of the line for the points in (a) and (b).
8. On the same diagram plot the following lines in the form $y = f(x)$:
   (a) $y = -2x + 5$ (b) $y + 2x + 5 = 0$ (c) $0.2y + 0.4x = 2$
   (i) State any properties which the lines (a), (b) and (c) have in common.
   (ii) Do these common properties change when each equation (a), (b) and (c) is written in the form $x = g(y)$?

2.2 Mathematical Modelling

At the end of this section you should:

• Understand what the term 'mathematical modelling' means

- Understand the importance of economic models
- Understand the interactions between the various parts of an economy: circular flow model.

2.2.1 Mathematical modelling

The idea of a 'model' is used in many areas of life to represent aspects of reality. In the broadest sense, models may be classified as physical or abstract.

- **Physical models (static):** such as fashion models, miniature replicates of buildings, cars.
- **Physical models (dynamic):** such as the aerodynamics of a car in a wind tunnel.
- **Abstract models (static):** such as a chart showing the interrelationships within a large organisation; the design of a car engine; the age/sex structure of a population; the relationship between the price of a good and the quantity demanded, etc. These are called static models, since the model is concerned with reproducing the state of the system.
- **Abstract models (dynamic):** such as the flow of information within an organisation; how the car engine works when switched on; how the population changes in time.

Models may be further classified as deterministic or stochastic.

- **A deterministic model:** the outcome can be determined exactly when all the inputs are given. The models which follow are all deterministic. For example, if P is the price of a lunch and Q is the number of lunches served, then the equation $Q = 20 - 2P$ describes how the number of lunches (quantity) demanded depends on the price. The variables are P and Q. According to this equation or model, when any price, P, is specified, then Q, the number of lunches demanded, may be calculated exactly.
- **A stochastic model:** the outcome is calculated with a certain probability. For example, when a fair die is tossed, the possible values which turn up are the numbers 1, 2, 3, 4, 5, 6. The probability of getting any one number is 1/6, so these numbers are called random variables (not variables). Equations may be written, based on probability, not to calculate any outcome exactly but rather to calculate the probability of any outcome occurring, such as two sixes on two tosses, etc. In this text, we are not concerned with stochastic models.

Figure 2.13 outlines the place of mathematical modelling in the overall scheme of modelling.

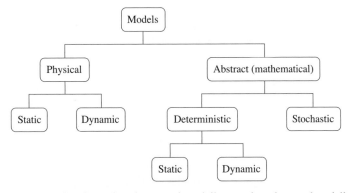

Figure 2.13   The place of mathematical modelling in the scheme of modelling

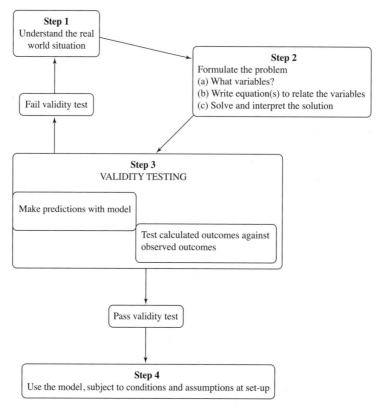

Figure 2.14   Steps in mathematical modelling

Suggested steps in the construction of a mathematical model

The suggested steps in mathematical modelling are now outlined very briefly, and illustrated graphically in Figure 2.14. Abstract models are usually a mathematical equation or system of equations which attempt to represent or 'model' a real-life situation or system. It is extremely difficult to represent real systems completely and the mathematics required is usually very complex; therefore mathematical models usually start off by representing some aspects of the real system. This requires a good deal of intuition.

Step 1:   Understand the real situation in order to ask the pertinent question(s). For example, 'Does price determine the number of units (quantity) of a good demanded?' such as, say, the number of lunches demanded daily in the Hot Pot restaurant. There are several other possible factors that affect demand: taste; alternatives; income of consumers; substitutes, etc. However, for the present, the effect of these factors will be considered to be less significant than the price.

Step 2:   Formulate the mathematical model. Ascertain that the question(s) posed by the model can be treated mathematically.

(a) Decide exactly which variables are important and which have little or no bearing on the situation. In this example, the variables are P and Q, both of which are easily measured.

(b) Describe the problem in terms of mathematical equations. For a normal good, the quantity demanded decreases as the price increases. The simplest mathematical equation is the linear function or straight line, therefore, the proposed equation is of the form: $P = a - bQ$. In the lunches example, suppose it is observed that 50 lunches are demanded when the price is £1, but only 20 lunches when the price is £4. Therefore, the equation of the demand function is: $Q = 60 - 10P$. (Use your knowledge of the straight line to confirm that the observed data is described by this equation.)

(c) Solve the equations and interpret the solution(s). In this example, the solution and interpretation are straightforward. However, in more complex situations, the solution may require complex mathematical methods.

Step 3: Check the validity of the model. Models are generally used to describe the workings of a system and/or to make predictions. Therefore, test the model with real data and check that the outcome calculated by the model compares reasonably with the actual outcome. In the lunches example, check that this equation does predict the quantity demanded when different prices are charged. Remember that you are assuming that the other factors mentioned earlier do not change. For instance, if a competing restaurant closes down, then the 'alternative' assumption changes, so the model for the demand at Hot Pot is no longer valid.

Step 4: If the model does not pass the validity test, then start again, as indicated in Figure 2.14. If the model is satisfactory, then it may be used, subject to the conditions under which it was set up remaining constant.

When you have studied the mathematical methods which follow in this text or others, you may be interested in a more in-depth treatment of mathematical modelling.

2.2.2 Economic models

Economics is a social science which studies how individuals within an economy make economic decisions on the allocation, distribution and utilisation of resources in order to satisfy their needs and wants. As a modern economy is very complex with individuals making millions of possible economic decisions, how can one model such an economic system? (See the circular flow of economic activity, next.)

Simplified abstractions of reality are used in the form of mathematical models. In this text, we shall refer to mathematical models as economic models since they are used in the context of economics. So, the economic system is studied through the use of economic models.

Economic models fall into two categories:

- **Microeconomics:** this studies the economic decisions of **individual** households and firms. It also analyses how governments can affect these decisions through the use of taxes and how individual markets operate.
  Examples: Individual household demand and individual firm supply of good X.
- **Macroeconomics:** this studies the economy as a whole. It studies the **aggregate** of all economic decisions.
  Examples: Total planned consumption, total planned savings and total planned investment in the economy.

Circular flow of economic activity

A modern economic system is complex, with consumers, firms and the government making millions of output and input decisions. The circular flow model, as described in Figure 2.15, is a simplified representation of this economic system (it is an abstraction of reality). It illustrates that economic activity in the economy is circular.

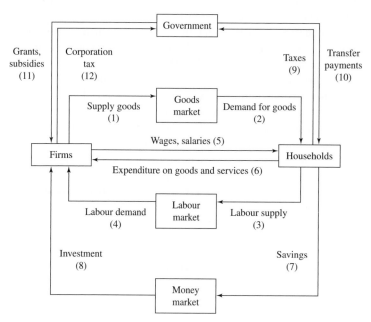

Figure 2.15   The circular flow model

The circular flow model assumes that the economy is composed of the following:

- Three economic agents: households (consumers), firms (producers) and the government.
- Three markets: the goods market, the labour market and the money market.
- The model shows that both households and firms operate simultaneously in the goods and labour market. Firms are suppliers of goods to the goods market (flow 1) and demanders of labour from the labour market (flow 4). On the other hand, households are demanders of goods from the goods market (flow 2) and suppliers of labour to the labour market (flow 3). Therefore, both households and firms have a dual role in the economy.
- Flows 1–4 illustrate the physical flows of goods and labour in the economy. These physical flows have monetary counterparts. Firms pay wages and salaries to households for supplying their labour services (flow 5). In turn, households use this income for expenditure on the goods and services produced by firms (flow 6).
- Households do not spend all their income; part of it may be saved in financial institutions (flow 7). In turn, these savings may be used to provide funds to firms for investment purposes (flow 8).
- The government has also an important role to play in the economy. Again, the government has a dual role in the model. It takes taxes from both households and firms (flows 9 and 12). These are known as **withdrawals** from the circular flow. In addition, the government puts money into the circular flow through transfer payments (social welfare) and grants/subsidies (flows 10 and 11). These are known as **injections** into the circular flow.

The economic models that follow in the remainder of this text are mathematical in nature. The simplest mathematical function or equation is the straight line which is used to model demand, supply, revenue and cost in Section 2.3.

2.3 Applications: Demand, Supply, Cost, Revenue

At the end of this section you should be familiar with how to model (i) graphically and (ii) algebraically (writing down their equations) the following basic functions:

- Demand and supply
- Cost
- Revenue
- Profit.

2.3.1 Demand and supply

Demand and supply decisions by consumers, firms and the government determine the level of economic activity within an economy. As these decisions play a vital role in business and consumer activity, it is important to possess mathematical tools that enable us to analyse them.

The demand function

There are several variables that influence the demand for a good X. These may be expressed by the general demand function

$$Q = f(P, Y, P_S, P_C, Ta, A, \ldots)$$

We say Q is a function of the independent variables P, Y, P_S, P_C, Ta, A, etc.

where

Q is the quantity demanded of good X
P is the price of good X
Y is the income of the consumer
P_S is the price of substitute goods
P_C is the price of complementary goods
Ta is the taste or fashion of the consumer
A is the level of advertising

Definitions

A substitute good is one that can be used instead of another good, e.g., trains and buses.
A complementary good is one that is consumed in conjunction with another, e.g., petrol/cars.

The simplest model for the demand function is written as

$$Q = f(P)$$

We say, Q is a function of P

Therefore, quantity demanded depends on price only, so long as the other variables upon which demand depends remain constant; that is, Y, P_S, P_C, Ta, A, ... are constant.

For example, the demand for good X may be given by the linear equation, $Q = 200 - 2P$. This equation describes the *law of demand*, a basic economic hypothesis which states that there is a negative relationship between quantity demanded and price; that is, when the price of a good increases, the quantity demanded will decrease, all other variables remaining constant. The demand function $Q = 200 - 2P$ is graphed in Figure 2.16(a) with Q, the variable by itself on the LHS of the equation, plotted on the vertical axis and P plotted on the horizontal one. In most applications in economics, however, Q is plotted on the horizontal axis and P on the vertical one. In order to do so, the inverse demand function, $P = f^{-1}(Q) = g(Q)$, is required. For example, the inverse demand function of $Q = 200 - 2P$ is given as

$$Q = 200 - 2P$$

$$2P = 200 - Q$$

$$P = 100 - 0.5Q$$

The inverse demand function, $P = 100 - 0.5Q$, is plotted in Figure 2.16(b). Note that P is written by itself on the LHS of this equation: $P = f(Q)$. Most introductory economics texts plot P and Q this way.

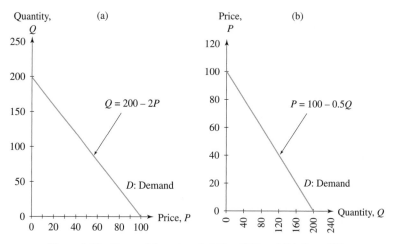

Figure 2.16   Demand functions (a) $Q = f(P)$ and (b) $P = g(Q)$

Note: For convenience, we do not distinguish between the terms 'demand function' and 'inverse demand function' throughout the remainder of this text. However, it is understood that a function written in the form $Q = f(P)$ refers to the demand function while a function in the form $P = g(Q)$ refers to the inverse demand function.

The equation of the demand function

The demand function, $P = g(Q)$, can be modelled by the simple linear equation

$$P = a - bQ \qquad\qquad (2.4)$$

where a and b are constants. This is the equation of a straight line. The general format for the equation of a straight line is $y = mx + c$ (or $y = c + mx$); therefore the following can be deduced:

$$P = a + (-b)Q$$
$$\vdots \quad \vdots \qquad \vdots \quad \vdots$$
$$y = c + \quad (m)\, x$$

$a > 0$: the **vertical intercept** of the demand function is positive ($a > 0$ denotes 'a greater than zero'). When $Q = 0$ (the quantity demanded is zero) the price per unit is $P = a$. In other words, when the price per unit $P = a$ there is no demand for the good. The value of a depends on the size of the terms on the RHS of the general demand function, except the price of the good itself.
$(-b) < 0$: the **slope** of the demand function is negative ($-b < 0$ denotes '$-b$ less than zero').
The negative slope indicates that price drops by b units for each unit increase in quantity. The demand function, $P = a - bQ$, is represented graphically in Figure 2.17.

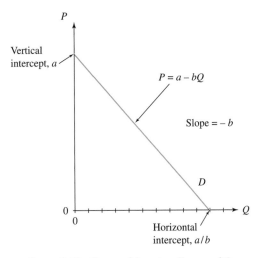

Figure 2.17   Demand function $P = a - bQ$

Note: $P = a - bQ$ intersects the horizontal axis at the point where $P = 0$. Substituting $P = 0$ into the demand function gives

$$0 = a - bQ \;\rightarrow\; bQ = a \;\rightarrow\; Q = \frac{a}{b}$$

Therefore, the horizontal intercept occurs at ($Q = a/b$, $P = 0$).

WORKED EXAMPLE 2.6
LINEAR DEMAND FUNCTION

The demand function is given by the equation $P = 100 - 0.5Q$.

(a) State and give a verbal description of the slope and intercepts.
(b) What is the quantity demanded when $P = 5$?
(c) Plot the demand function $P = 100 - 0.5Q$ for $0 < Q < 200$.
(d) Find an expression for the demand function in the form $Q = f(P)$ and graph it.

SOLUTION

(a) The slope and vertical intercept may be deduced as follows:

$$P = 100 + (-0.5)\,Q$$
$$\vdots \quad \vdots \qquad \vdots \quad \vdots$$
$$P = a + (-b)\,Q$$
$$\vdots \quad \vdots \qquad \vdots \quad \vdots$$
$$y = c + m\,x$$

The vertical intercept is 100. This means that $P = 100$ when $Q = 0$.
Alternatively when the price $P = 100$ there is no demand for the good. The slope $\Delta P/\Delta Q = -0.5$. This indicates that the price drops by 0.5 units for each successive unit increase in quantity demanded.
The horizontal intercept is calculated by substituting $P = 0$ into the equation of the demand function, that is,

$$P = 100 - 0.5Q \;\rightarrow\; 0 = 100 = 0.5Q \;\rightarrow\; 0.5Q = 100 \;\rightarrow\; Q = \frac{100}{0.5} = 200$$

Therefore, the horizontal intercept is at $(Q = 200,\ P = 0)$. We could describe this situation by saying that the quantity demanded $Q = 200$ when $P = 0$, so $Q = 200$ when the good is free!

(b) The quantity demanded when $P = 5$ is calculated by substituting $P = 5$ into the demand function

$$P = 100 - 0.5Q$$
$$5 = 100 - 0.5Q$$
$$0.5Q = 100 - 5$$
$$0.5Q = 95$$
$$Q = \frac{95}{0.5} = 190$$

(c) **Method A:** To plot the demand function over the range $0 \leq Q \leq 200$, choose various quantity values within this range. In Table 2.3 equal intervals of 40 units of quantity are used. Substitute these quantity values into the equation of the demand function to derive corresponding values for P. Plot these points and graph the demand function as in Figure 2.18. Table 2.3 and Figure 2.18 are easily set up using Excel.

Table 2.3   Demand schedule

Quantity	Price
0	100
40	80
80	60
120	40
160	20
200	0

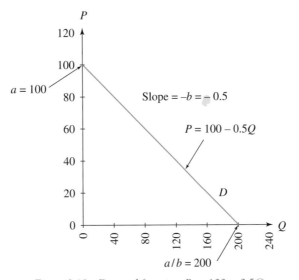

Figure 2.18   Demand function $P = 100 - 0.5Q$

Method B: Plot the horizontal and vertical intercepts calculated in part (a). Draw a line through these points.

(d) The demand function $Q = f(P)$ is graphed in Figure 2.19. Its equation is derived as

$$P = 100 - 0.5Q \quad \text{in general form as} \quad P = a - bQ$$
$$0.5Q = 100 - P \qquad\qquad\qquad\qquad bQ = a - P$$
$$Q = 200 - 2P \qquad\qquad\qquad\qquad Q = \frac{a}{b} - \frac{1}{b}P$$

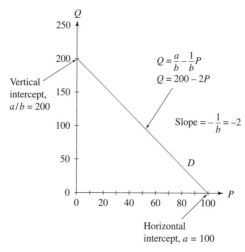

Figure 2.19   Demand function $Q = 200 - 2P$

The supply function

There are several variables that influence the quantity supplied of good X. These may be expressed by the general supply function

$$Q = f(P, C, P_0, Te, N, O, \ldots)$$

> We say, Q is a function of the independent variables P, C, P_0, Te, N, O, etc.

where

Q	is the quantity of good X supplied
P	is the price of the good itself
C	is the cost of production
P_0	is the price of other goods
Te	is the available technology
N	is the number of producers in the market
O	is other factors, e.g., tax/subsidies

The simplest model for the supply function is written as $Q = f(P)$, i.e., quantity supplied depends on price only, so long as the other variables upon which supply depends remain constant. For example, in Worked Example 2.7(a) the supply is given by the linear equation, $Q = 2P - 10$. This describes **the law of supply**, a basic economic hypothesis which states that there is a positive relationship between quantity supplied and price; that is, when the price of a good increases, the quantity supplied will also increase, all other variables remaining constant.

The equation of the supply function

The supply function $P = h(Q)$ can be modelled by the simple linear equation

> We say P is a function of Q

$$P = c + dQ \tag{2.5}$$

where c and d are constants.

$c > 0$: the vertical intercept is positive. Its value depends on the size of the terms on the RHS of the general supply function except the price of the good itself.

$d > 0$: the slope of the supply function is positive. Price increases by d units for every unit increase in quantity supplied.

The supply function is graphically represented in Figure 2.20.

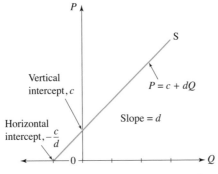

Figure 2.20   Supply function $P = c + dQ$

Note: $P = c + dQ$ intersects the horizontal axis at the point where $P = 0$. Substitute $P = 0$ into the supply function and solve for Q: the value of Q at $P = 0$ is the horizontal intercept:

$$P = c + dQ$$
$$0 = c + dQ$$
$$-c = dQ$$
$$-\frac{c}{d} = Q$$

The horizontal intercept occurs at $(Q = -c/d, P = 0)$. A minus quantity is not economically meaningful; therefore, many economic textbooks illustrate the supply function in the positive quadrant only.

WORKED EXAMPLE 2.7
ANALYSIS OF THE LINEAR SUPPLY FUNCTION

 Find animated worked examples at **www.wiley.com/college/bradley**

A supplier will only start to supply T-shirts when the price per unit exceeds £5. He or she will then increase output (Q) by 2 units (2 T-shirts) for every unit increase in price.

(a) Plot the supply function in the form $Q = f(P)$.
(b) Write down the equation of the supply function.
(c) Find the value of Q when $P = 15$ from the graph. Confirm your answer from the equation.
(d) Write the equation of the supply function in the form $P = h(Q)$, i.e., write P in terms of Q. Plot the graph of P in terms of Q.

SOLUTION

(a) Since a graph of the form $Q = f(P)$ is required, plot Q on the vertical axis and P on the horizontal axis. In this case the graph crosses the horizontal axis at $P = 5$; that is, the supplier will supply no goods ($Q = 0$) (T-shirts) at prices of $P = £5$ per unit or lower. Plot the point $P = 5, Q = 0$. From this point, draw a line with a slope of 2. Recall: a slope $m = 2$ means that Q increases by 2 when P increases by 1: it follows that Q increases by 10 when P increases by 5, etc. In Figure 2.21 you can see that this line cuts the vertical axis at $Q = -10$.

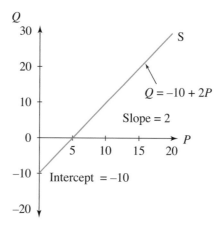

Figure 2.21   Supply function $Q = -10 + 2P$

(b) The general format for the equation of a line is $y = mx + c$. In this case, $y \equiv Q$ and $x \equiv P$. From Figure 2.21, slope $m = 2$ and vertical intercept $c = -10$. The required equation, therefore, for the supply function, is $Q = 2P - 10$ or $Q = -10 + 2P$.

(c) From the graph, when $P = 15$, $Q = 20$. Using the equation, substitute $P = 15$, then evaluate Q as follows:

$$Q = -10 + 2P \rightarrow Q = -10 + 2(15) \rightarrow Q = 20$$

(d) The equation of the supply function in the form $P = h(Q)$ is derived as follows:

$$Q = -10 + 2P$$
$$-2P = -10 - Q$$
$$P = 5 + 0.5Q \quad \text{dividing across by } -2$$

The graph $P = h(Q)$ (hence P on the vertical axis and Q on the horizontal) is given in Figure 2.22. Show how the graph was plotted.

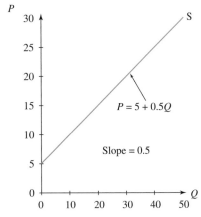

Figure 2.22   Supply function $P = 5 + 0.5Q$

WORKED EXAMPLE 2.8
LINEAR SUPPLY FUNCTION 2

The supply function is given by $P = 10 + 0.5Q$.

(a) State and verbally describe the slope and intercepts.
(b) Plot the supply function, $P = 10 + 0.5Q$, for $0 \leq Q \leq 100$.

SOLUTION

(a) The vertical intercept is $c = 10$. This means that the firm will supply no units at $P \leq 10$. The slope $\Delta P/\Delta Q = 0.5$. This means that the price increases by 0.5 units for each successive unit increase in quantity supplied.
The horizontal intercept is calculated by substituting $P = 0$ into the equation of the supply function, then solving for the corresponding value of Q

$$P = 10 + 0.5Q$$
$$0 = 10 + 0.5Q$$
$$-10 = 0.5Q$$
$$-\frac{10}{0.5} = Q \rightarrow Q = -20$$

Therefore, the horizontal intercept is at ($Q = -20$, $P = 0$); however, $Q = -20$ is not economically meaningful.
(b) **Method A:** To plot the supply function over the range, $0 \leq Q \leq 100$, choose various quantity values within this range. In Table 2.4 $Q = 0, 20, 40, 60, 80$ and 100. Substitute each value of Q into the equation of the supply function $P = 10 + 0.5Q$ to derive corresponding values for P. Plot these points and graph the supply function as in Figure 2.23.

Table 2.4   Supply schedule

Quantity	Price
0	10
20	20
40	30
60	40
80	50
100	60

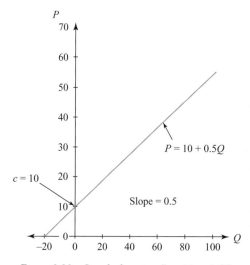

Figure 2.23   Supply function $P = 10 + 0.5Q$

Method B: Plot the horizontal and vertical intercepts calculated in part (a). Draw the supply function by joining these points.

PROGRESS EXERCISES 2.3

Linear Functions: Supply and Demand

1. For each of the following lines:
   (i) $y = 10 - 2.5x$ (ii) $y = -2.5 + 10x$ (iii) $4y + 2x = 60$
   (a) Calculate the slopes and intercepts.
   (b) Write down the number of units by which y changes when x increases by 1 unit.
   (c) Plot the lines.
2. The equation of a demand function is given by $Q = 64 - 4P$ where Q is the number of helicopter flights demanded daily; P is the price per flight, in £00s.
   (a) Plot the demand function with Q on the vertical axis.
   (b) What is the change in demand (Q) when price (P) increases by 1 unit?

(c) What is the demand when $P = 0$?

(d) What is the price when $Q = 0$?

Describe the answers to (b), (c) and (d) verbally.

3. The equation of the demand function is given by $P = 65$, where P is the price of bread (per loaf) in pence.

   (a) What is the slope of this demand function? Describe what this means.

   (b) Plot the graph of this demand function.

4. Given the supply function $P = 500 + 2Q$, where P is the price of a bottle of cognac in Euros and Q is the number of litres supplied.

   (a) Graph the supply function, $P = f(Q)$. What is the meaning of the value at the vertical intercept?

   (b) What is the value of Q when $P = 600$ Euros?

   (c) What is the value of P when $Q = 20$?

   Describe the answers to (a), (b) and (c) verbally.

5. (See question 4)

   (a) Express the supply function given in question 4 in the form $Q = f(P)$.

   (b) Hence, determine the increase in the number of litres supplied for each unit increase in price.

6. The equation of the supply function is given as $P = 50$.

   (a) Write down the slope of the supply function. What is the change in price when the quantity supplied increases by 10 units?

   (b) Plot the graph of this supply function.

7. The equation of the supply function is given as $Q = 1200$, where Q is the number of hospital dinners supplied daily.

   (a) What is the slope of this supply function?

   (b) Plot the graph of this supply function. Describe the situation modelled by the equation verbally.

   In the following exercises give a verbal description of the answers above whenever possible.

8. The equation of the demand function for sparkling wine is given by $Q = 1520 - 5P$ where Q is the number of bottles demanded weekly and P is the price per bottle in €.

   (a) Calculate P when $Q = 0$.

   (b) What is the value of Q when $P = 0$?

   (c) Plot the demand function with Q on the vertical axis.

   (d) What is the change in demand (Q) when price (P) increases by €1?

   (e) What is the value of Q when $P = €8.50$?

   (Give a verbal description of answers whenever possible.)

9. The demand equation for football tickets is given by the equation $P = 8000 - 0.625Q$ where P is the price in £ sterling and Q is the number of tickets demanded at the given price.

   (a) Write the equation of the demand function in the form $Q = f(P)$. Hence plot the demand function with Q on the vertical axis.

   (b) Give a verbal description of the slope of this function.

   (c) Calculate the values of the intercepts. Give a verbal description of the intercepts.

   (d) Calculate the value of Q when $P = £20$.

10. (a) The demand equation for new potatoes is given by the equation

$$Q = 210 - 3.5P$$

where P is the price in £ and Q is the quantity (numbers of crates).
   (i) Calculate P when $Q = 0$.
   (ii) What is the value of Q when $P = 0$?
   (iii) Plot the demand function with P on the vertical axis.
(b) The supply function for new potatoes is given by the equation

$$P = 0.25Q + 22.5$$

where P is the price in £ and Q is the quantity in crates.
   (i) Calculate P when $Q = 0$.
   (ii) What is the value of Q when $P = 0$?
   (iii) On the same diagram as (a)(iii), plot the supply function with P on the vertical axis.
   State and interpret the point where the demand function and supply function meet.

11. The demand and supply functions for hotel accommodation per room during the low season are given by the equations $P = 96 - 0.8Q$ and $P = 40 + 0.4Q$ where Q is the number of bookings and P is the price in €.
(a) Graph the demand and supply functions on the same diagram.
(b) For each function (demand and supply):
   (i) Calculate the value of Q when $P = 0$.
   (ii) Calculate the price when $Q = 0$.
   (iii) How many bookings will be demanded and supplied when the price is €60 per room?

2.3.2 Cost

Firms incur costs when they employ inputs, such as capital and labour, in order to produce goods for sale in the market. The total cost of producing a good will normally consist of:

 (i) Fixed costs, FC: costs that are fixed irrespective of the level of output, e.g., rent on premises.
(ii) Variable costs, VC: costs which vary with the level of output, e.g., each extra unit of a good produced will require additional units of raw materials, labour, etc.

Total cost, TC, is therefore the sum of fixed costs and variable costs:

$$TC = FC + VC \tag{2.6}$$

These costs can be modelled by a linear cost function, such as

$$TC = 20 + 4Q$$

where $FC = 20$ (the vertical intercept) and $VC = 4Q$, where $4 =$ slope of the line.

WORKED EXAMPLE 2.9
LINEAR TOTAL COST FUNCTION

A pumpkin grower has fixed production costs of £10 for the rent of a stall at a farmers' market and variable production costs of £2 per pumpkin

(a) Write down the equation for the total cost function.
(b) Graph the total cost function.

SOLUTION

(a) $FC = £10$; while $VC = £2 \times Q$, since to produce

$$
\begin{array}{ll}
1 \text{ unit} & VC = 2(1) \\
2 \text{ units} & VC = 2(2) \\
3 \text{ units} & VC = 2(3) \\
Q \text{ units} & VC = 2Q
\end{array}
$$

That is, the total variable cost incurred in producing Q pumpkins is £2Q. With $TC = FC + VC$, the total cost incurred in producing Q pumpkins is

$$TC = 10 + 2Q$$

Note: This is the same as the line $y = 10 + 2x$, where $y \equiv TC$ and $x \equiv Q$.

(b) Graphically, the TC function is a straight line, with costs measured on the vertical axis and units of the good produced on the horizontal axis. Since the vertical intercept = 10 (level of FC) and the slope = 2 are known, the total cost function may be plotted. Alternatively, total costs may be plotted by calculating at least two points on the line and joining these points. A number of points are calculated in Table 2.5 and then graphed in Figure 2.24.

Table 2.5   Fixed, variable and total costs

Q	Variable costs (at £2 per unit)	$TC = FC + VC$ ($TC = 10 + 2Q$)	Point (Q, TC) (x, y)
0	0	$TC = 10 + 0 = 10$	(0, 10)
1	$1(2) = 2$	$TC = 10 + 2 = 12$	(1, 12)
2	$2(2) = 4$	$TC = 10 + 4 = 14$	(2, 14)
3	$3(2) = 6$	$TC = 10 + 6 = 16$	(3, 16)
4	$4(2) = 8$	$TC = 10 + 8 = 18$	(4, 18)
5	$5(2) = 10$	$TC = 10 + 10 = 20$	(5, 20)
6	$6(2) = 12$	$TC = 10 + 12 = 22$	(6, 22)

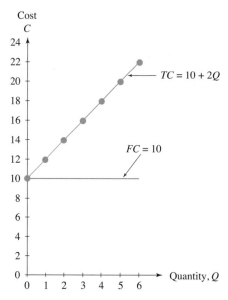

Figure 2.24   Linear total cost function

2.3.3   Revenue

A firm receives revenue when it sells output. The total revenue, TR, received is the price of the good, P, multiplied by the number of units sold, Q, that is,

$$TR = P \cdot Q \tag{2.7}$$

Total revenue may be modelled by a straight line if the price of each unit sold is the same, that is, each firm is a price-taker, denoted by a horizontal demand function. A price-taking firm is known as a perfectly competitive firm. In this case, the total revenue function would take the form

$$TR = P_0 Q$$

where P_0 is the constant price per unit of the good and is represented by the vertical intercept of a horizontal demand function. For example, if $P_0 = 10$ (a horizontal demand function), then the total revenue function is linear

$$TR = 0 + 10Q = 10Q$$
$$\vdots \quad \vdots \quad \vdots \quad \vdots$$
$$y = c + m \ \ x = mx$$

This is a line through the origin $(0, 0)$ since the vertical intercept $c = 0$ and slope $m = 10$.

WORKED EXAMPLE 2.10a
LINEAR TOTAL REVENUE FUNCTION

Suppose that each chicken snack box is sold for £3.50 irrespective of the number of units sold.

(a) Write down the equation of the total revenue function.
(b) Graph the total revenue function.

SOLUTION

(a) Total revenue is price multiplied by the number of units sold, that is,

$$TR = 3.5Q$$

Note that price is constant at £3.50 irrespective of the value of Q.

(b) Total revenue is represented graphically by a straight line, with slope $= 3.5$ and intercept $= 0$. It is graphed by calculating values of TR for any values of Q, for example, $Q = 0, 2, 4, 6$. These points are outlined in Table 2.6 and then plotted in Figure 2.25.

Table 2.6   Total revenue

Q	$TR = PQ = 3.5Q$	Point (Q, TR)
0	$TR = 3.5(0) = 0$	$(0, 0)$
2	$TR = 3.5(2) = 7$	$(2, 7)$
4	$TR = 3.5(4) = 14$	$(4, 14)$
6	$TR = 3.5(6) = 21$	$(6, 21)$

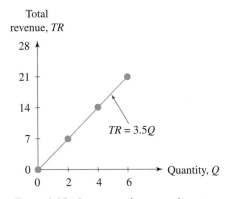

Figure 2.25   Linear total revenue function

2.3.4   Profit

Profit is the total revenue received less the total cost. Profit is represented by the Greek letter π, since the letter P is reserved to represent price.

Hence, profit $\pi = TR - TC$.

A company or firm is said to 'break even' when total revenue is the same as total cost and hence profit is zero.

At break-even, $\pi = 0$; hence $TR = TC$.

When: (i) $TR > TC$ the company makes a profit; (ii) $TR < TC$ the company makes a loss.

WORKED EXAMPLE 2.10b
LINEAR PROFIT FUNCTION

The company that sells chicken snack boxes for £3.50 each has fixed costs of £800 per day and variable costs of £1.50 for each chicken snack box.

(a) Write down the equation for (i) total revenue, (ii) total costs, (iii) profit.
(b) Calculate the number of chicken snack boxes that must be produced and sold if the company is to break even.
(c) Graph the profit function. From the graph estimate the break-even quantity.

SOLUTION

(a) (i) $TR = 3.5Q$, (ii) $TC = 800 + 1.5Q$, (iii) $\pi = 3.5Q - (800 + 1.5Q) = 2Q - 800$.
(b) The firm breaks even when $TR = TC$:

$$3.5Q = 800 + 1.5Q$$

$$2Q = 800$$

$$Q = 400$$

Alternatively, the firm breaks even when profit is zero: $\pi = 2Q - 800 = 0$; $Q = 400$.
(c) Profit is represented graphically by a straight line, $\pi = 2Q - 800$. It may be graphed by calculating values of π for any values of Q, e.g., $Q = 0, 200, 400, 800$. These points are calculated in Table 2.7 and then plotted in Figure 2.26.

Table 2.7

Q (compare to x)	$\pi = 2Q - 800$ ($y = mx + c$)	Point (Q, π)
0	$\pi = 2(0) - 800 = -800$	$(0, -800)$
200	$\pi = 2(200) - 800 = -400$	$(200, -400)$
400	$\pi = 2(400) - 800 = 0$	$(400, 0)$
800	$\pi = 2(800) - 800 = 800$	$(800, 800)$

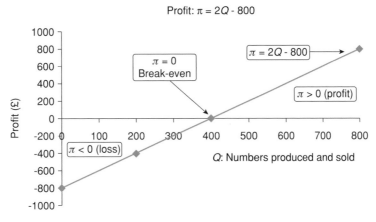

Figure 2.26   Linear profit function

PROGRESS EXERCISES 2.4

Linear Functions: Cost, Revenue

1. A firm faces the following cost function, $TC = 5Q$.
   (a) What is the value of fixed costs?
   (b) Graph the total cost function.
   (c) What is the total cost when $Q = 10$?
   (d) What is the cost of producing each additional unit of this good?
2. A firm sells its product in a perfectly competitive output market (i.e., the price is the same for each unit sold). The total revenue function is given as $TR = 10Q$.
   (a) What is the price per unit charged by the perfectly competitive firm?
   (b) Plot the total revenue function.
3. The canoeing club provides swimming lessons to boost club funds. The club has fixed costs of £250 daily (insurance) when offering these lessons. The variable cost is £25 for each lesson given.
   (a) Write down the equation for total cost and plot its graph for $Q = 0$ to 60.
   (b) Calculate the cost of providing 28 lessons.
   (c) Calculate the number of lessons provided when total costs are £1400.
   (d) Confirm the answers in (b) and (c) graphically.
4. (See question 3)
   If each student is charged £32 for a canoeing lesson,
   (a) Write down the equation for total revenue.
   (b) Calculate the number of students when the total revenue is £1024.
   (c) If the club has 44 students on a given day, determine whether the revenue exceeds costs.
5. A firm producing lamps has fixed costs of €1000 per week. A lamp costs €15 to produce.
   (a) Write down the equation for total cost.
   (b) Graph the total cost function for $Q = 0$ to 100.

(c) Calculate the cost of producing 25 lamps.

(d) How many lamps are produced when total costs are €7000?

6. (See question 5)

If the lamps are sold for €35 each,

(a) Write down the equation for total revenue.

(b) Graph the total revenue function for $Q = 0$ to 100.

(c) How many lamps are sold when total revenue is €1750?

(d) Does the firm's revenue exceed costs when 80 lamps are sold? Calculate the profit/loss when 80 lamps are sold.

7. A trader, who sells watches at the Saturday market, has fixed costs of £90 and buys watches wholesale at £3 each. He sells the watches for £6 each.

(a) Write down equations for:

(i) fixed costs

(ii) variable costs

(iii) total costs

(iv) total revenue.

(b) Plot the total cost equation for $Q = 0$ to 40.

Calculate the total cost when the trader buys 100 watches for the Saturday market

(c) Graph the total revenue function.

How many watches are sold when total revenue is £330?

8. A Canteen has fixed costs of £1500 per week. Meals cost £5 each and are sold at a fixed price of £9 each.

(a) Write down the equation for total costs and plot its graph for $Q = 0$ to 160.

(b) Calculate the cost of producing 100 meals per week.

(c) Calculate the number of meals produced when total costs are £2300.

(d) Write down the equation for the total revenue.

(e) Calculate the number of meals sold when the total revenue is £3780.

(f) Calculate the profit when 500 meals are sold per week.

2.4 More Mathematics on the Straight Line

At the end of this section you should be able to:

- Calculate the slope of a line given two points on the line
- Determine the equation of a line given the slope and any point on the line
- Determine the equation of a line given any two points on the line.

2.4.1 Calculating the slope of a line given two points on the line

Section 2.1 introduced how the slope, m, of a line is measured. This section repeats that exercise in more detail. If two points on a line are known, then the slope is defined as the change in y divided by the corresponding change in, given as

$$\text{Slope} = \frac{\text{Change in } y}{\text{Change in } x} = \frac{\Delta y}{\Delta x} = \frac{y_2 - y_1}{x_2 - x_1}$$

(2.8)

WORKED EXAMPLE 2.11
CALCULATING THE SLOPE GIVEN TWO POINTS ON THE LINE

Calculate the slope of a line given the following two points on the line,

$$(x_1, y_1) \quad \text{and} \quad (x_2, y_2)$$
$$(2, 1) \quad \text{and} \quad (-4, -2)$$

SOLUTION
The slope, m, is calculated using equation (2.8):

$$m = \frac{\Delta y}{\Delta x} = \frac{y_2 - y_1}{x_2 - x_1} = \frac{-2 - (1)}{-4 - (2)} = \frac{-3}{-6} = \frac{1}{2} = 0.5$$

This is exactly the same result that you would obtain if you plotted these points on a graph and measured the slope directly from the graph as height, 3, divided by distance, 6, as illustrated in Figure 2.27. Furthermore, the slope of a straight line is the **same** when measured between any two points, that is, the slope of a straight line is constant.

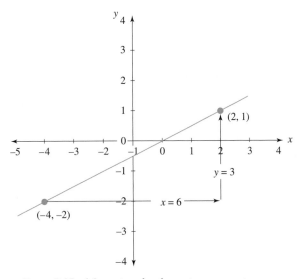

Figure 2.27   Measuring the slope given two points

2.4.2 The equation of a line given the slope and any point on the line

Since the slope of a line is the same when measured between any two points, such as the points (x_1, y_1) and (x, y), we may write

$$\frac{\Delta y}{\Delta x} = m = \frac{y - y_1}{x - x_1}$$

therefore, $m(x - x_1) = y - y_1$. This equation is usually written as

$$y - y_1 = m(x - x_1) \tag{2.9}$$

the equation of a straight line with slope, m, which passes through the specific point (x_1, y_1).

WORKED EXAMPLE 2.12
EQUATION OF A LINE GIVEN THE SLOPE AND A POINT ON THE LINE

(a) Deduce the equation of a line whose slope is of 1.7 and which passes through the point $(2, 5)$.
(b) Plot the graph of the line.

SOLUTION

(a) Using equation (2.9), the equation of the line is deduced as

$$y - y_1 = m(x - x_1)$$
$$y - 5 = 1.7(x - 2)$$
$$y - 5 = 1.7x - 3.4$$
$$y = 1.6 + 1.7x$$

This line has a slope of 1.7 and it intersects the y-axis at $y = 1.6$.
(b) To plot the line it is necessary to calculate at least two points on the line; however, it is a good idea to calculate three points in case of error. All three points should lie exactly on the line. Using **method A**, take suitable x-values and calculate the corresponding y-coordinates from the equation of the line as shown in Table 2.8. These points are plotted in Figure 2.28.

Table 2.8   Calculating points on the line, $y = 1.6 + 1.7x$

x	$y = 1.6 + 1.7x$	Point (x, y)
-2	$y = 1.6 + 1.7(-2) = -1.8$	$(-2, -1.8)$
0	$y = 1.6 + 1.7(0) = 1.6$	$(0, 1.6)$
2	$y = 1.6 + 1.7(2) = 5$	$(2, 5)$

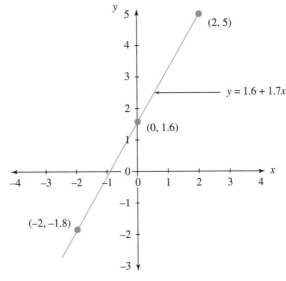

Figure 2.28   Graph of line $y = 1.6 + 1.7x$

2.4.3   The equation of a line given two points

There are two steps involved when deducing the equation of a line given two points, say points (x_1, y_1) and (x_1, y_2).

Step 1: Calculate the slope of the line using equation (2.8).
Step 2: Determine the equation of the line by substituting the slope (as calculated in step 1), and either point, into equation (2.9).

WORKED EXAMPLE 2.13
EQUATION OF A LINE GIVEN TWO POINTS ON THE LINE

(a) A line passes through the points (2, 4) and (6, 1). Deduce the equation of the line.
(b) Plot the graph of the line.

SOLUTION

(a) **Step 1:** The slope of the line is calculated by substituting the points into equation (2.8):

$$m = \frac{y_2 - y_1}{x_2 - x_1} = \frac{1 - 4}{6 - 2} = \frac{-3}{4} = -0.75$$

Step 2: The equation of the line is deduced by substituting this slope, and either point, into equation (2.9):

$$y - y_1 = m(x - x_1)$$
$$y - 4 = -0.75(x - 2) \quad \text{using point } (2, 4)$$
$$y - 4 = -0.75x + 1.5$$
$$y = 5.5 - 0.75x$$

The equation of the line has a negative slope of -0.75 and y-intercept of 5.5.

(b) Following **method A**, use the equation of the line to calculate any three points as in Table 2.9. These points are then plotted in Figure 2.29.

Table 2.9   Calculating points on the line $y = 5.5 - 0.75x$

x	$y = 5.5 - 0.75x$	Point (x, y)
0	$y = 5.5 - 0.75(0) = 5.5$	$(0, 5.5)$
2	$y = 5.5 - 0.75(2) = 4$	$(2, 4)$
4	$y = 5.5 - 0.75(4) = 2.5$	$(4, 2.5)$

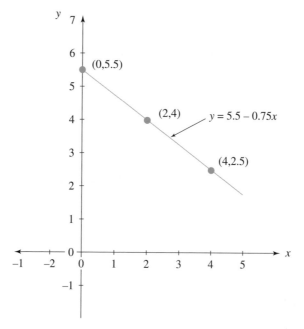

Figure 2.29   Graph of line $y = 5.5 - 0.75x$

Summary

Given one point and slope	Given two points on the line
$$y - y_1 = m(x - x_1)$$ may be used to determine the equation of a straight line when the slope of the line and a point on the line are given.	Given any two points (x_1, y_1), (x_2, y_2) (1) Calculate the slope: $$m = (y_2 - y_1)/(x_2 - x_1)$$ (2) Use this slope with one of the points to determine the equation of the line using $$y - y_1 = m(x - x_1)$$

PROGRESS EXERCISES 2.5

Determine Equation of a Line from Two Points and Applications

1. (a) Plot the points
   (i) (0, 0) and (2, 1)   (ii) (2, 3) and (−1, −3).
   (b) Measure the slope of the lines formed by joining the points in (i) and (ii).
2. Find the equations of the following lines and, hence, plot their graphs.
   (a) A straight line passes through the point (2, 4) and has a slope of 1.
   (b) A straight line passes through the points (2, 4) and (8, 16).
   (c) A straight line intersects the y-axis at $y = 4$ and the x-axis at $x = 8$.
3. The variable cost of a product increases by £1.50 for each unit produced while fixed costs are £55.
   (a) Write down the equation for and graph the total cost function.
   (b) Determine the total cost, both algebraically and graphically, when 8 units are produced.
4. It is known that the number of lunches demanded is 80 units when the price is £5 and 45 when the price is £12.
   (a) Determine the equation of the demand function in the form $Q = f(P)$. Plot the graph of the demand function.
   (b) Use the equation of the demand function to calculate the change in demand when the price (i) increases by £3, (ii) decreases by £2.
   (c) Estimate the decrease in price for each lunch when the number of lunches demanded (quantity demanded) increases by 15.
5. A supplier supplies 50 football scarves when the price is £6 each and 90 units when the price is £11 each.
   (a) Determine the equation of the supply function in the form $P = h(Q)$.
   (b) How many additional scarves are supplied for each successive £1 increase in price?
   (c) Calculate the quantity supplied when the price is £8.5 per scarf.

(d) Calculate the price when 120 scarves are supplied.

(e) What is the price below which no scarves are supplied?

6. (a) Calculate the equation of a line having a slope of 1.5 and which passes through the point (2, 4).

   (b) Plot the graph of the line.

7. (a) A line passes through the points (4, 2) and (2, 6). Calculate the equation of the line.

   (b) Graph the line.

8. (a) Calculate the slope of a line which passes through the points (3, 8) and (1, 4).

   (b) Hence find the equation of the line.

9. The variable cost of a product increases by €4 for each unit produced and the fixed costs are €64.

   (a) Write down the equation for the total cost function.

   (b) Graph the equation.

   (c) What is the slope of the line?

   (d) Determine the total cost both algebraically and graphically when 10 units are produced.

10. A distributor supplies 100 DVDs when the price is €15 and 125 when the price is €20 each.

   (a) Determine the equation of the supply function in the form $P = f(Q)$. Plot the graph of the supply function.

   (b) How many additional units are supplied for each successive €1 increase in price?

   (c) Calculate the price when 150 DVDs are supplied.

   (d) Calculate the quantity supplied when the price is €30 per DVD.

11. It is known that the number of concert tickets demanded is 150 when the price is €40 and 100 when the price is €60.

   (a) Determine the equation of the demand function in the form $Q = f(P)$ and plot the graph of the function.

   (b) Use the equation to calculate the change in demand when the price (i) increases by €10, (ii) decreases by €8.

2.5   Translations of Linear Functions

www.wiley.com/college/bradley

Go to the website for the following additional material that accompanies Chapter 2:

2.5.1   Meaning of translation

2.5.2   Equation of translated lines

• Vertical translations

   Worked Example 2.14 Vertical translations of linear functions

   Figure 2.30 Vertical translations of the line $y = 1 + 2x$

• Calculate the size of the horizontal shift that occurs in conjunction with a vertical translation

• Horizontal translations

   Worked Example 2.15 Horizontal translations of linear functions

   Figure 2.31 Horizontal translations of the line $y = 1 + 2x$

- Horizontal and vertical translations together
  Worked Example 2.16 Horizontal and vertical translations of linear functions
  Figure 2.32 Vertical and horizontal translations of the line $y = 1 + 2x$
  2.5.3   Some applications of translations
- Excise tax
  Worked Example 2.17 The effect of an excise tax on a supply function
  Figure 2.33 Effect of an excise tax (10 Euros per unit) on the supply function
- Value added tax (VAT)
- Translating cost functions
  Worked Example 2.18 Translating linear cost functions
  Figure 2.34 The effect of (i) a subsidy and (ii) changes in fixed costs on a total cost function
  Translations of linear functions and applications
  Solutions to Progress Exercises 2.6

2.6   Elasticity of Demand, Supply and Income

At the end of this section you should be able to:
- Calculate the coefficient of price elasticity of demand at a point on the demand function
- Illustrate how the coefficient of point elasticity varies along a linear demand function (website)
- Illustrate that the coefficient of price elasticity of demand depends on price and vertical intercept and not on the slope of the linear demand function
- Calculate the coefficient of price elasticity of demand averaged along an arc or interval on the demand function
- Calculate the coefficient of price elasticity of supply
- Calculate the coefficient of income elasticity of demand (website).

'The prices of cars increase by 5%'
'The prices of computers decrease by 8%'

Significant as they may seem, the above statements alone are of little value to firms. From the demand function, firms can predict that as price increases, quantity demanded decreases and vice versa. However, this information is not sufficient. What firms require is a numerical value that outlines the responsiveness of quantity demanded to such price changes at various price levels. For example, when the price of a car is £9000, will a 5% increase in price result in a 5% decrease in demand, or a larger percentage decrease in demand? When the price of a car is £19 200, will a 5% increase in price result in a smaller or larger percentage decrease in demand?

This information is invaluable to firms as their total revenue depends on both prices and quantity sold. Will the additional revenue generated by the price increase be offset by the loss in revenue due to the decrease in units sold?

The concept of **elasticity**, as used in economics, is the ratio of the percentage change in the quantity demanded (or supplied) to a percentage change in an economic variable, such as price, income, etc. Elasticity may be used to predict the responsiveness of demand (and supply) to changes in such economic variables.

There are various symbols used to denote elasticity. In this text it is denoted by the symbol ε, pronounced 'epsilon'. Three types of elasticity are introduced: price elasticity of demand, ε_d, price elasticity of supply, ε_s, and income elasticity of demand, ε_y.

2.6.1 Price elasticity of demand

Price elasticity of demand measures the responsiveness (sensitivity) of quantity demanded to changes in the good's own price. It is also referred to as own-price elasticity. It is represented by the general elasticity formula

$$\varepsilon_d = \frac{\text{Percentage change in quantity demanded}}{\text{Percentage change in price}} = \frac{\%\Delta Q_d}{\%\Delta P} = \frac{\dfrac{\Delta Q}{Q} \cdot 100}{\dfrac{\Delta P}{P} \cdot 100} \qquad (2.10)$$

which simplifies to:
$$\varepsilon_d = \frac{\Delta Q}{\Delta P} \cdot \frac{P}{Q} \qquad (2.11)$$

Remember

Dividing by a fraction is the same as multiplying by the fraction inverted.

The numerical value or coefficient for price elasticity of demand is normally negative since $\Delta Q/\Delta P$ is negative. That is, for the linear demand function, $P = a - bQ$, slope

$$\frac{\Delta P}{\Delta Q} = -\frac{b}{1}$$

Hence, inverting both sides gives

$$\frac{\Delta Q}{\Delta P} = -\frac{1}{b}$$

which is negative.

Note:

- The negative sign associated with ε_d indicates the direction and magnitude of the responsiveness of one variable w.r.t. another. A negative sign indicates that an increase in one variable is accompanied by a decrease in the other or vice versa. A positive sign indicates that an increase (decrease) in one is accompanied by an increase (decrease) in the other.
- There is a convention in some economic textbooks to give the numerical value of elasticity without the sign. This is known as the absolute value or magnitude of elasticity, $|\varepsilon|$, which only indicates the magnitude of the responsiveness of one variable to a change in another and not the direction of the responsiveness.

In this text, both conventions will be used as appropriate:
There are two approaches to measuring price elasticity: point elasticity and arc elasticity. Point elasticity measures elasticity at a point on the demand function. Arc elasticity measures elasticity over an interval on the demand function.

Point elasticity of demand

Given the linear demand function, $P = a - bQ$, the formula for point elasticity of demand at any point (P_0, Q_0) is

$$\varepsilon_d = \frac{\Delta Q}{\Delta P}\frac{P}{Q}$$

or

$$\varepsilon_d = -\frac{1}{b}\frac{P_0}{Q_0} \quad \text{since slope } \frac{\Delta Q}{\Delta P} = -\frac{1}{b} \tag{2.12}$$

WORKED EXAMPLE 2.19
DETERMINING THE COEFFICIENT OF POINT ELASTICITY OF DEMAND

Given the demand function for computers as $P = 2400 - 0.5Q$.

(a) Determine the coefficient of point elasticity of demand when (i) $P = 1800$, (ii) $P = 1200$, and (iii) $P = 600$. Give a verbal description of each result.
(b) If the price of computers increases by 12%, calculate the percentage change in the quantity demanded at $P = 1800$, $P = 1200$ and $P = 600$.
(i) First use the definition of elasticity:

$$\varepsilon_d = \frac{\text{percentage change in quantity demanded}}{\text{percentage change in price}} = \frac{\Delta Q_d(\%)}{\Delta P(\%)}$$

(ii) Then calculate the exact percentage changes and compare them with the answers in (i).
(c) Graph the demand function, indicating where demand is elastic, unit elastic and inelastic.

SOLUTION

(a) (i) At $P = 1800$ the quantity of computers demanded, Q, is calculated by substituting $P = 1800$ into the demand function:

$$P = 2400 - 0.5Q \rightarrow 1800 = 2400 - 0.5Q \rightarrow 0.5Q = 600 \rightarrow Q = 1200$$

The value of point elasticity of demand at $(P = 1800, Q = 1200)$ is calculated by substituting these values along with $b = 0.5$ into formula (2.12):

$$\varepsilon_d = -\frac{1}{b}\frac{P_0}{Q_0}$$

$$= -\frac{1}{0.5}\frac{1800}{1200} = -\frac{1800}{600} = -\frac{3}{1} = -3$$

The coefficient of point elasticity of demand is $\varepsilon_d = -3$, which indicates that at the price $P = 1800$ a 1% increase (decrease) in price will cause a 3% decrease (increase) in the quantity of computers demanded. Demand is elastic, $|\varepsilon_d| > 1$.

(ii) and (iii) The calculations for price elasticity of demand for $P = 1200$ and $P = 600$ are carried out in the same way as (i) and are summarised below.

P_0	1800	1200	600
Calculate Q_0	$P_0 = 2400 - 0.5Q_0$	$P_0 = 2400 - 0.5Q_0$	$P_0 = 2400 - 0.5Q_0$
	$1800 = 2400 - 0.5Q_0$	$1200 = 2400 - 0.5Q_0$	$600 = 2400 - 0.5Q_0$
	$0.5Q_0 = 600$	$0.5Q_0 = 1200$	$0.5Q_0 = 1800$
	$Q_0 = 1200$	$Q_0 = 2400$	$Q_0 = 3600$
	$\varepsilon_d = -\frac{1}{b}\frac{P_0}{Q_0}$	$\varepsilon_d = -\frac{1}{b}\frac{P_0}{Q_0}$	$\varepsilon_d = -\frac{1}{b}\frac{P_0}{Q_0}$
ε_d for $P = 2400 - 0.5Q$	$= -\frac{1}{0.5}\frac{1800}{1200} = -3$	$= -\frac{1}{0.5}\frac{1200}{2400}$	$= -\frac{1}{0.5}\frac{600}{3600}$
		$= -\frac{1200}{1200}$	$= -\frac{1}{3}$
		$= -1$ (see notes)	≈ -0.33 (notes)
Demand	Demand is elastic $|\varepsilon_d| > 1$	Demand is unit elastic $|\varepsilon_d| = 1$	Demand is inelastic $|\varepsilon_d| < 1$

Notes:

$\varepsilon_d = -1$ indicates that at $P = 1200$, a 1% increase (decrease) in price will cause a 1% decrease (increase) in the quantity of computers demanded.

$\varepsilon_d = -0.33$ (correct to 2 decimal places) indicates that at $P = 600$, a 1% increase (decrease) in price will cause a 0.33% decrease (increase) in the quantity of computers demanded.

(b) (i) The definition

$$\varepsilon_d = \frac{\text{percentage change in quantity demanded}}{\text{percentage change in price}} = \frac{\Delta Q_d(\%)}{\Delta P(\%)}$$

may be rearranged as

$$\Delta Q(\%) = \Delta P(\%) \times \varepsilon_d \tag{2.13}$$

- At $P = 1800$ we have calculated that $\varepsilon_d = -3$. To calculate the percentage change in Q, substitute $\varepsilon_d = -3$ and $\Delta P = 12\%$ into equation (2.13):

$$\Delta Q = 12\% \times -3 = -36\%$$

The quantity demanded decreases by 36%, a larger percentage decrease than the 12% price increase. Demand is strongly responsive to price change and is described as elastic demand.
- $P = 1200$: $\varepsilon_d = -1$, $\Delta P = 12\%$. Substitute into equation (2.13):

$$\Delta Q = 12\% \times -1 = -12\%$$

The quantity demanded decreases by 12%; the percentage increase in price is 12%. This is described as unit elastic demand.
- $P = 600$: $\varepsilon_d = -0.33$, $\Delta P = 12\%$. Substitute into equation (2.13):

$$\Delta Q = 12\% \times -0.33 = -3.96\%$$

The quantity demanded decreases by 3.96%, a smaller percentage decrease than the 12% price increase. So demand is weakly responsive to price, hence it is described as inelastic demand.

(ii) The calculation of the exact percentage change in Q requires basic arithmetic. Start by calculating the price which results from an increase of 12% on the initial price.

$$\text{Increase } P = 1800 \text{ by } 12\% \rightarrow P_{new} = \frac{112}{100}P = 1.12(1800) = 2016$$

Then calculate the corresponding values of Q from the equation of the demand function. When $P = 1800$, $Q = 1200$ and when $P = 2016$, $Q = 768$.

Initial P, Q	New P, Q	Percentage change in Q by direct calculation	ε_d	$\Delta Q(\%) = \Delta P(\%) \times \varepsilon_d$ by formula (2.13)
1800, 1200	2016, 768	$\frac{768 - 1200}{1200} \times 100 = -36\%$	-3	$\Delta Q = -3 \times 12 = -36\%$
1200, 2400	1344, 2122	$\frac{2112 - 2400}{2400} \times 100 = -12\%$	-1	$\Delta Q = -1 \times 12 = -12\%$
600, 3600	672, 3456	$\frac{3456 - 3600}{3600} \times 100 = -4\%$	$-\frac{1}{3} \approx -0.33$	$\Delta Q = -\frac{1}{3} \times 12 \approx -4\%$

$\approx$ means approximately equal to

Comment: The exact value of elasticity, $-1/3$, not the rounded value, -0.33, was used for accurate results when $P = 600$. The rounded value of elasticity, -0.33, would give approximately the same answer, -3.96%, as obtained by direct calculation. At prices $P = 1800$ and $P = 1200$, no rounding of ε_d was required; the percentage change in demand was exactly the same, whether calculated directly or calculated by equation (2.13). In general, the computer firm is not only interested in the percentage change in the price of computers, but also in the percentage change in demand which results from the price changes.

(c) The graph of the demand function is given in Figure 2.35.

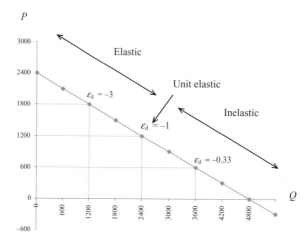

Figure 2.35 Variation of price elasticity of demand with price along the demand function $P = 2400 - 0.5Q$

Coefficient of price elasticity of demand

There are three categories of price elasticity: **elastic, inelastic** and **unit elastic.**

Coefficient of price elasticity of demand	Description
$-\infty < \varepsilon_d < -1$ Coefficient of price elasticity of demand is greater than minus infinity and less than -1. **Note:** when $\varepsilon_d < -1$, $|\varepsilon_d| > 1$	Elastic demand: demand is strongly responsive to changes in price; that is, the percentage change in demand is greater than the percentage change in price. When $\varepsilon_d \to -\infty$, demand for the good is perfectly elastic. The demand function is horizontal with slope $= 0$; that is, $$\frac{\Delta P}{\Delta Q} = \frac{0}{\Delta Q} \to \frac{\Delta Q}{\Delta P} = \frac{\Delta Q}{0} = \infty \quad (P \neq 0)$$
$-1 < \varepsilon_d < 0$ Coefficient of price elasticity of demand is greater than -1 but less than or equal to zero. **Note:** $|\varepsilon_d| < 1$	Inelastic demand: demand is weakly responsive to changes in price; that is, the percentage change in demand is less than the percentage change in price. When $\varepsilon_d = 0$, demand for the good is perfectly inelastic. The demand function is vertical with slope $= \infty$; that is, $$\frac{\Delta P}{\Delta Q} = \frac{\infty}{1} \to \frac{\Delta Q}{\Delta P} = \frac{1}{\infty} = 0 \quad (Q \neq 0)$$
Coefficient of price elasticity of demand is equal to -1.	Unit elastic demand: the percentage change in demand is equal to the percentage change in price

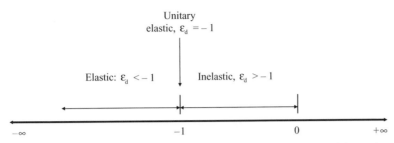

Figure 2.36 Numerical scale for the coefficient of price elasticity of demand

The numerical scale for the coefficient of price elasticity of demand is shown in Figure 2.36.

Note: ∞ denotes infinity.

www.wiley.com/college/bradley

Go to the website for the following additional material that accompanies Chapter 2:

• Point elasticity of demand varies along a demand function
  Worked Example 2.20 Calculating the coefficient of point elasticity at different prices
  Table 2.10 Coefficient of point elasticity of demand for silicon chips
  Figure 2.37 Elasticity varies along the demand of function

Point elasticity of demand depends on price and vertical intercept only

Is a relatively flat demand function more elastic than a steeper demand function? Not necessarily. If two demand functions are drawn to the same scale, then

• The point elasticity of demand is the same at any given price for two demand functions having the same vertical intercept, but different slopes.
• The point elasticity of demand is different at any given price for two demand functions having different vertical intercepts (slopes may be the same or different).

To prove the above statements proceed as follows.
Write the price elasticity formula (2.11) for the demand function $P = a - bQ$, in terms of P only:

$$\varepsilon_d = \frac{\Delta Q}{\Delta P}\frac{P}{Q} = \frac{1}{-b}\frac{P}{Q} = \frac{P}{P - a} \tag{2.14}$$

since $P = a - bQ \rightarrow -bQ = P - a$. This equation states that the coefficient of point elasticity of demand is dependent on price, P, vertical intercept, a, and independent of the slope, b, of the linear demand function (no b appears in formula 2.14).

Arc price elasticity of demand

Arc price elasticity of demand measures the elasticity of demand over an interval on the demand function. Instead of using the price and quantity at a point, as in point elasticity, arc elasticity uses the average of the prices and quantities at the beginning and end of the stated interval:

$$\varepsilon_d = \frac{\Delta Q}{\Delta P} \cdot \frac{\frac{1}{2}[P_1 + P_2]}{\frac{1}{2}[Q_1 + Q_2]} = \frac{\Delta Q}{\Delta P} \cdot \frac{P_1 + P_2}{Q_1 + Q_2} \tag{2.15}$$

Formula (2.15) is often referred to as the **midpoints elasticity formula** or **arc price elasticity formula**.

Note: For linear functions, the arc price elasticity of demand formula can also be written as

$$\varepsilon_d = \frac{\Delta Q}{\Delta P} \cdot \frac{P_1 + P_2}{Q_1 + Q_2} = -\frac{1}{b} \cdot \frac{P_1 + P_2}{Q_1 + Q_2}$$

Price elasticity of supply

The same ideas as developed with respect to the price elasticity of demand can be applied to the analysis of the price elasticity of supply at a point on the supply function or averaged along an arc on the supply function. As the supply function is positively sloped, the coefficient of price elasticity of supply will be positive.

The coefficient of price elasticity of supply is given by the formula

$$\varepsilon_s = \frac{\text{Percentage change in quantity supplied}}{\text{Percentage change in price}} = \frac{\%\Delta Q_s}{\%\Delta P} = \frac{\Delta Q}{\Delta P} \cdot \frac{P}{Q} \tag{2.16}$$

Given the supply function, $P = c + dQ$, slope $= d = \Delta P/\Delta Q$. Then, $1/d = \Delta Q/\Delta P$; hence, the point elasticity of supply formula is

$$\varepsilon_s = \frac{1}{d} \cdot \frac{P}{Q}$$

Note: $\varepsilon_s > 1$: supply is elastic; $\varepsilon_s < 1$: supply is inelastic; $\varepsilon_s = 1$: supply is unit elastic.

 www.wiley.com/college/bradley

Go to the website for the following additional material that accompanies Chapter 2:

2.6.2   The coefficient of income elasticity of demand
Figure 2.38 Numerical scale for the coefficient of income elasticity of demand
Worked Example 2.21 Calculating the coefficient of income elasticity of demand

PROGRESS EXERCISES 2.7

Elasticity for Linear Functions

1. (a) Why is the price elasticity of demand negative?
   (b) When the price of a good is £20, the price elasticity of demand is −0.7. Calculate the percentage change in demand Q when
      (i) The price increases by 5%.
      (ii) The price decreases by 8%.
2. Given the demand function, $Q = 250 − 5P$, where Q is the number of children's watches demanded at $£P$ each:
   (a) Derive an expression for the point elasticity of demand in terms of P only.
   (b) Calculate the point elasticity at each of the following prices, $P = 20; 25; 30$.
   Describe the effect of price changes (expressed as % changes) on demand at each of these prices.
3. Given the demand function, $P = 60 − 0.2Q$, calculate the arc price elasticity of demand when:
   (a) The price decreases from £50 to £40 and (b) the price decreases from £20 to £10.
4. For the demand functions, (i) $P = a − bQ$, (ii) $Q = c − dP$:
   (a) Derive expressions for the price elasticity of demand in terms of Q only.
   (b) Calculate ε_d for each of the following demand functions at $Q = 100; 500; 900$.
      (i) $Q = 1800 − 0.0P$,   (ii) $P = 60 − 0.5Q$.
5. $P = 90 − 0.05Q$ is the demand function for graphics calculators in an engineering college.
   (a) Derive expressions for ε_d in terms of (i) P only, (ii) Q only.
   (b) Calculate the value of ε_d when the calculators are priced at $P = £20; £30; £70$.
   (c) Determine the number of calculators demanded when $\varepsilon_d = −1: \varepsilon_d = 0$.
6. (See question 5)
   (a) Explain the difference between the slope of a demand function and the price elasticity of demand.
   (b) Use ε_d to calculate the response (% change in Q) to a 10% increase in the price of calculators at each of the following prices: $P = £20; £30; £45; £70; £90$. Comment on the results.
7. Given the supply function, $P = 20 + 0.5Q$,
   (a) Calculate the arc price elasticity of supply when the price increases from £40 to £60. Interpret your result.
   (b) Calculate the percentage change in quantity supplied in response to a price increase of 10% when $P = £40$:
      (i) By using the definition of elasticity in equation (2.16).
      (ii) Exactly by using ordinary arithmetic.
   Comment on and explain the difference between (i) and (ii).

2.7 Budget and Cost Constraints

At the end of this section you should be familiar with:

- Budget constraints: effect of price and income changes
- Cost constraints: effect of price changes and expenditure limit.

Go to the website for the following additional material that accompanies Chapter 2:

2.7.1   Budget constraints
• Equation of budget constraint
  Worked Example 2.22 Graphing and interpreting the budget constraint
  Table 2.11 Consumption choices along the budget constraint
  Figure 2.39 Budget constraint or budget line
  Table 2.12 Effect of changes in prices and income on the budget constraint
  Worked Example 2.23 Budget constraint, changes in price and income
  Figure 2.40 ΔP_X and its effect on the budget constraint
  Figure 2.41 ΔP_Y and its effect on the budget constraint
  Figure 2.42 ΔM and its effect on the budget constraint
2.7.2   Cost constraints
• Equation of the isocost line
  Figure 2.43 Isocost line
  Progress Exercises 2.8 Linear Budget and Cost Constraints
  Solutions to Progress Exercises 2.8

2.8 Excel for Linear Functions

You should find Excel extremely useful for carrying out the tedious task of calculating the table of points from which various graphs are plotted. For example, to plot $y = 5x + 2$ from $x = -2$ to $x = 10$, proceed as follows:

• **Set up the table**

1. Use the required values of the independent variable (x), spaced at equal intervals, for the first row of the table, so we shall use $x = -2, 0, 2, 4, 6, 8, 10$ in row 1 as shown in Figure 2.44.
2. In the second row of the table calculate the corresponding y-values using the formula facility in Excel. Point to the cell B2 and type: '$= 5*B1 + 2$'. The '$=$' indicates that a formula follows; the formula states: 'multiply the value in B1 by 5, then add 2'. You could enter the actual value in

	A	B	C	D	E	F	G	H
1	x	−2	0	2	4	6	8	10
2	y	−8	2	12	22	32	42	52

First, enter the formula, by typing: $= 5*B1 + 2$

Next: copy the formula by holding down the left mouse button at the +, then dragging across the row as far as cell H2, then drop by releasing the mouse button

Figure 2.44   Excel sheet

B1 ($x = -2$), but if you then decide to change the x-values in row 1, then the formula must be changed as well. With the reference to B1, the formula automatically recalculates y. The second reason for using the reference to the cell B1 is that the first formula may be copied along the row to automatically calculate the other y-values as indicated in Figure 2.44.

3. When several graphs are required on the same diagram, set up additional y-rows (with the appropriate formulae) to calculate the y-coordinates for each graph.

• **Plot the graph(s)**

1. Select the table of data with the x-values in row 1 (or column 1 if the data is in columns).
2. Follow the steps for graph plotting given in Chapter 1, Worked Example 1.13.
3. The finished graph is very basic. However, the graph is easily formatted with titles, labels, etc., as briefly introduced in Chapter 1. Use the HELP facility in Excel or consult the reference manual.

WORKED EXAMPLE 2.24
USE EXCEL TO SHOW THE EFFECT OF PRICE AND INCOME CHANGES ON A BUDGET CONSTRAINT

Pocket money, £15, may be spent on either pool (60p per hour) or skating (£1.20 per hour).

(a) Write down the equation of the budget constraint. Graph the constraint.
(b) Show by calculation and graphically how the budget constraint changes when:
    (i) The price of pool increases to 75p, the other variables do not change.
    (ii) The price of skating drops to 90p, the other variables do not change.
    (iii) Pocket money decreases to £10, the other variables do not change.

SOLUTION

(a) Let x = number of hours of pool, $P_X = 60$; y = number of hours skating, $P_Y = 120$; $M = 1500$, the income limit. The budget constraint is given by the equation

$$x P_X + y P_Y = M \rightarrow 60x + 120y = 1500$$

> **Notation**
> Let X refer to the name of a good/service
> Then P_X = price per unit of good
> x = number of units of good X

Write the budget constraint in the form $y = c + mx \rightarrow y = 12.5 - 0.5x$.
To plot this line in Excel, determine the range of values of x for which x and y are both positive. What we actually require is the x-intercept to plot the graph for values of x between $x = 0$ and $x = x$-intercept.
The budget constraint or budget line cuts the x-axis at $y = 0$, hence

$$0 = 12.5 - 0.5x \rightarrow x = 25$$

Therefore, set up a table in Excel to calculate y (use the formula command) for values of x (at equal intervals), over the range $x = 0$ to 25. See Figure 2.45.

	A	B	C	D	E	F	G
1	x (pool)	0	5	10	15	20	25
2	y (skating)	12.5	10	7.5	5	2.5	0

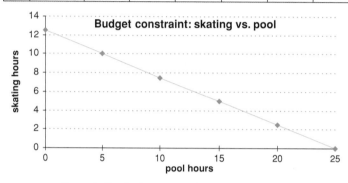

Figure 2.45   Budget constraint: skating versus pool

(b) (i) Pool increases to 75p per hour, so fewer hours of pool are affordable. This will cause the x-intercept to decrease towards the origin or, in other words, the budget line pivots downwards from the unchanged vertical intercept. The new budget line is given by the equation

$$x\,P_X + y\,P_Y = M \rightarrow 75x + 120y_{bi} = 1500$$

$$\text{or} \quad y_{bi} = 12.5 - 0.625x$$

Again, use Excel to calculate a table of points and plot this budget line along with the original budget line. See Figure 2.46.

	A	B	C	D	E	F	G
20	x (pool)	0	5	10	15	20	25
21	y (skating)	12.5	10	7.5	5	2.5	0
22	y_{bi}	12.5	9.375	6.25	3.125	0	−3.125

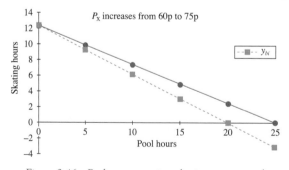

Figure 2.46   Budget constraint: skating versus pool

Note: The subscript '*bi*' (part (b), section (i) of question) attached to the variable y is used only for referencing the new budget line in the figure. The same applies to '*bii*' and '*biii*' below.

(ii) Skating decreases from 120p to 90p per hour, so more hours of skating are affordable. This will cause the y-intercept to increase away from the origin, that is, the budget line pivots upwards from the unchanged horizontal intercept. The new budget line is given by the equation

$$x P_X + y P_Y = M \rightarrow 60x + 90y_{bii} = 1500$$
$$\text{or} \quad y_{biii} = 16.67 - 0.67x$$

Again, use Excel to calculate a table of points and plot this budget line along with the original budget line. See Figure 2.47.

		A	B	C	D	E	F	G
40	x (pool)	0	5	10	15	20	25	
41	y (skating)	12.5	10	7.5	5	2.5	0	
42	y_{bii}	16.67	13.32	9.97	6.62	3.27	0	

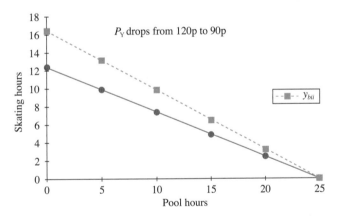

Figure 2.47   Budget constraints and the decrease in the price of skating

(iii) This time income decreases from £15 to £10, so less of everything is affordable. The new budget constraint drops, but runs parallel to the original. The new budget constraint is given by the equation

$$x P_X + y P_Y = M \rightarrow 60x + 120y_{bii} = 1000$$
$$\text{or} \quad y_{biii} = 8.33 - 0.5x$$

Again, use Excel to calculate a table of points and plot this budget line along with the original budget line. See Figure 2.48.

	A	B	C	D	E	F	G
60	x (pool)	0	5	10	15	20	25
61	y (skating)	12.5	10	7.5	5	2.5	0
62	y_{biii}	8.33	5.83	3.33	0.83	−1.67	−4.17

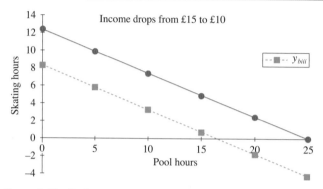

Figure 2.48   Budget constraints and the decrease in pocket money

PROGRESS EXERCISES 2.9

Excel or Otherwise

1. A line cuts the vertical axis 2 units above the origin and has a slope of 1.
   (a) Set up a table which gives five points on this line.
   (b) Plot the line.
   (c) Write down the equation of the line in the form (i) $y = f(x)$ and (ii) $P = h(Q)$.
2. It is known that when the price of a good is £2, 50 units are demanded, but when the price increases to £4 the quantity demanded drops to 45 units (that is, a line contains the points (2, 50) and (4, 45)).
   (a) Calculate the change in demand that results from a 1 unit increase in price.
   (b) Calculate the quantity demanded at prices of £5, £8 and £10 per unit.
   (c) Graph the demand function.
   (d) Write down the equation of the demand function in the form,
       (i) $P = g(Q)$ and   (ii) $Q = f(P)$.
3. Given the line, $y = 65 - 2x$:
   (a) Set up a table of points, hence plot the line over the interval $0 \le x < 34$.
   (b) Write down values of and describe verbally the slope and intercepts.
4. A demand function is given by the equation $P = 80 - 4Q$.
   (a) Set up a demand schedule for values of Q in the first quadrant (when P and Q are positive).
   (b) Plot the demand function.
   (c) Write down the values of the slope and intercepts.
   (d) Write the demand function in the form $Q = f(P)$ and plot the graph for values of P in the first quadrant.
5. Repeat question 4 for the supply function $P = 4 + 2.5Q$.

6. Given the following demand and supply functions

$$P = 60 - 3.5Q \quad Q = 40 - 0.5P \quad P = 5 + 1.2Q$$

State which of these functions are demand functions and which are supply functions.
(a) Describe how to set up a table to calculate the values for price elasticity of demand or supply (point).
(b) Set up a table to calculate the values for the arc elasticity for each 5 unit increase in P from $P = 0$ to $P = 20$.

7. Pocket money of £5 may be spent on either ice-cream or soft drinks. Ice-cream costs 12p per unit while drinks cost 20p per unit.
(a) Write down the equation of the budget constraint. Graph the constraint.
(b) Show by calculation, and graphically, how the budget constraint changes if the price of ice-cream drops to 9p, while pocket money and the price of drinks do not change.
(c) Show by calculation, and graphically, how the budget constraint changes if the price of soft drinks increase to 25p, while pocket money and the price of ice-cream do not change.
(d) If pocket money increases to £7.50, and the prices of ice-cream and drinks remain the same, how does the budget constraint alter? Graph the new budget constraint.

8. The demand function for good X is given by the equation

$$Q_X = 120 - 2P_X + 5P_Y + 0.3Y$$

Calculate the arc elasticity of demand when P_Y increases from
(i) £10 to £20, (ii) £20 to £30, (iii) £30 to £40, (iv) £40 to £50, (v) £50 to £65 while the price of good X and income (Y) remain fixed at £25 and £80 respectively.

2.9 Summary

Mathematics

1. A straight line is uniquely defined by stating the slope (m) and vertical intercept (c).
2. The slope of a line is the 'slant'.
   (a) The slope may be described as the rate of change in y per unit change in x: $m = \Delta y / \Delta x$.
   (b) The slope may be measured between any two points on the line: (x_1, y_1), (x_2, y_2):

$$m = \frac{\Delta y}{\Delta x} = \frac{y_2 - y_1}{x_2 - x_1}$$

3. The equation of a straight line defines the relationship that exists between the x- and y-coordinates for every point on the line. It may be written in several forms:
   (a) $y = mx + c$: vertical intercept $= c$; slope $= m$; horizontal intercept $= (-c)/m$.
   (b) $ax + by + d = 0$ which may be rearranged so that the slope and vertical intercept may be read off:

$$y = -\frac{d}{b} + \left(\frac{-a}{b}\right) x : \quad \text{intercept, } c = -\frac{d}{b}, \quad \text{slope, } m = \frac{-a}{b}$$

4. The equation of a line may be calculated when certain information is given:
   (a) Given slope $= m$ and vertical intercept $= c$, the equation is $y = mx + c$.
   (b) Given slope $= m$ and one point on the line: (x_1, y_1), the equation is $y - y_1 = m(x - x_1)$.
   (c) Given two points on the line: (x_1, y_2), (x_2, y_2), then calculate the slope,

$$m = \frac{y_2 - y_1}{x_2 - x_1}$$

   The equation is $y - y_i = m(x - x_i)$, where $i = 1$ or 2.
5. Translations of $y = f(x)$:

Vertical translations. $\uparrow$ by c units,   replace y by $(y - c)$
                           $\downarrow$ by c units,   replace y by $(y + c)$

                           corresponding horizontal shift $\Delta x = \dfrac{\Delta y}{m}$

Horizontal translations. $\rightarrow$ by c units,   replace every x by $(x - c)$
                           $\leftarrow$ by c units,   replace every x by $(x + c)$

Applications

Demand function: $P = a - bQ$, negative relationship between price per unit and quantity.
Supply function: $P = c + dQ$, positive relationship between price per unit and quantity.
Total revenue: $TR = P \times Q$, price per unit multiplied by quantity sold.
Total cost: $TC = FC + VC = FC + kQ$, where k is the cost of producing each unit.
Supply function when a tax of t per unit is imposed, $P - t = c + dQ$.
Total cost function (linear) when a tax of t per unit is imposed, $TC = FC + (k + t)Q$.
Elasticity of demand:

$$\varepsilon_d = \frac{\% \text{ change in quantity demanded}}{\% \text{ change in price per unit}} = \frac{\Delta Q}{\Delta P}\frac{P}{Q}$$

Point elasticity of demand for the demand function, $P = a - bQ$, is

$$\varepsilon_d = -\frac{1}{b}\frac{P}{Q} = \frac{P}{P - a}$$

Arc elasticity of demand between price and quantity (P_1, Q_1), (P_2, Q_2) is

$$\varepsilon_d = \frac{\Delta Q}{\Delta P}\frac{P_1 + P_2}{Q_1 + Q_2}$$

Income elasticity of demand:

$$\varepsilon_y = \frac{\Delta Q}{\Delta Y}\frac{Y_1 + Y_2}{Q_1 + Q_2}$$

Budget constraint or budget line: $xP_X + yP_Y = M$
Cost constraint: $wL + rK = C$

Excel: Useful for calculating tables and graph plotting, hence visually a good sense of slope, i.e., positive, negative, steep, flat, can be gained. Also very useful for plotting and comparing several functions when one or more of the components change, such as the prices in the budget line. All these tasks can be carried out by hand, but are slow, time consuming and open to human error.

www.wiley.com/college/bradley

Go to the website for Problems in Context

TEST EXERCISES 2

1. A straight line has a slope of 2 and intersects the y-axis at $y = -2$. Deduce the equation of the line and plot it.
2. Determine the slope and y-intercept (if any) for each of the following lines:
   (a) $x + 2y - 4 = 0$ (b) $2Q = P + 8$ (c) $P - 2Q = 4 + 3Q$
   (d) $5Q + P = P - 5$ (e) $y = 1$ (f) $x = 4$ (g) $y = x$
3. Deduce the equation of the following lines given slope and/or a point(s):
   (a) $m = 1.4, (1, 3)$ (b) $m = -0.5, (-2, 5)$ (c) $(3, 1), (5, 3)$
4. Given the following demand functions

$$Q = 25 - 5P \quad \text{Demand function (1)}$$
$$Q = 80 - 2.5P \quad \text{Demand function (2)}$$

   (a) Plot the graphs of the demand functions.
   (b) Calculate P when $Q = 15$ from the graph. Confirm your answer algebraically.
   (c) Calculate the price elasticity of demand at $Q = 15$.
   (d) Write the functions in the form $P = f(Q)$ and plot the graph.
5. (a) The quantity demanded of a product is 102 units when $P = 0$. If the demand, Q, drops by 5 units for every unit increase in price, write down the equation of the demand function in the form $Q = f(P)$ and plot its graph.
   (b) From the graph in (a), calculate the arc price elasticity of demand when the quantity demanded increases from 55 to 65. Confirm your answer algebraically.
6. A supplier will start producing a product when a price of £24 per unit is available and output, Q, will increase by 5 units for each unit increase in price.
   (a) Plot the graph of the supply function, plotting Q on the horizontal axis.
   (b) Deduce the equation of the supply function.
   (c) From the graph calculate Q when $P = 45$. Confirm your answer algebraically.
7. A supply function is given by the equation $20P = 80 + 5Q$. Deduce the equation of the supply function if a tax of £1.50 is imposed on the price of each unit produced.
8. A student has £140 per month to spend on football and on the cinema. If each attendance at a football match costs £12, while an outing to the cinema costs £8:
   (a) Write down the equation of the budget constraint. Graph the constraint.
   (b) Analyse the effect on the constraint if the price of a football outing increases to £20.

SIMULTANEOUS EQUATIONS

3.1   Solving Simultaneous Linear Equations
3.2   Equilibrium and Break-even
3.3   Consumer and Producer Surplus
3.4   The National Income Model and the *IS-LM* Model
3.5   Excel for Simultaneous Linear Equations
3.6   Summary
      Appendix

Chapter Objectives

In this chapter methods for solving simultaneous equations analytically and graphically are introduced. These methods are then applied to the analysis of equilibrium in the goods, labour and money markets and consumer and producer surplus. At the end of this chapter, using linear functions, you should be able to:

- Solve two equations in two unknowns and illustrate the solution graphically
- Distinguish between unique solutions, no solutions and infinitely many solutions
- Solve three equations in three unknowns
- Calculate the equilibrium price and quantity in the goods market and illustrate the solution graphically
- Analyse and illustrate graphically the effect of intervention in the goods market (price ceilings and price floors)
- Analyse and illustrate graphically the effect of taxes and subsidies in the goods market
- Calculate and illustrate graphically break-even, profit and loss
- Calculate consumer and producer surplus
- Calculate the equilibrium conditions for the national income model and illustrate equilibrium graphically
- Calculate and illustrate graphically the equilibrium values of national income and interest rates based in the *IS-LM* model
- Use Excel to plot all of the above.

3.1 Solving Simultaneous Linear Equations

At the end of this section you should be familiar with:

- How to solve two simultaneous equations in two unknowns using:
  (a) Algebra
  (b) Graphical methods.
- Determining when two equations in two unknowns have:
  (a) A unique solution
  (b) No solution
  (c) Infinitely many solutions.
- Solving three simultaneous equations in three unknowns.

In many applications, both practical and theoretical, there will be several equations with several variables or unknowns. These are referred to generally as simultaneous equations. This section introduces some methods to solve simultaneous linear equations.

Reminder

A solution of an equation in an unknown, say x, is simply the value for x for which the left-hand side (LHS) of the equation is equal to the right-hand side (RHS).

For example, consider the equation $x + 4 = 6$.
The solution of this equation is $x = 2$.
$x = 2$ is the only value of x for which the LHS = RHS.

The statement '$x = 2$ satisfies the equation' is another way of saying that $x = 2$ is a solution.

The solution, therefore, of a set of simultaneous equations is a set of values for the variables which satisfy all the equations.

3.1.1 Two equations in two unknowns

A standard method for solving two linear equations in two unknowns is outlined in Worked Examples 3.1 to 3.3.

WORKED EXAMPLE 3.1
SOLVING SIMULTANEOUS EQUATIONS 1

Given the simultaneous equations

$$x + 3y = 4 \qquad\qquad (1)$$

$$-x + 2y = 6 \qquad\qquad (2)$$

(a) Solve for x and y algebraically.
(b) Solve for x and y graphically.

SOLUTION

(a) **Method:** Eliminate x from the system of equations by adding equations (1) and (2). The two equations reduce to a single equation in which the only unknown is y. Solve for y, then substitute the value of y into either of the original equations and solve for x:

Step 1: Adding equations (1) and (2):

$$x + 3y = 4$$

$$\underline{-x + 2y = 6}$$

$$0 + 5y = 10 \quad \text{adding}$$

Each equation states that the LHS = RHS. Therefore, for two (or more) equations the 'sum of the LHSs = sum of RHSs'

(1)

(2)

Solving for y:

$$y = \frac{10}{5} = 2$$

Step 2: Solve for x by substituting $y = 2$ into either equation (1) or equation (2):

$$-x + 2(2) = 6 \qquad \text{substituting } y = 2 \text{ into equation (2)}$$

$$-x = 6 - 4 \qquad \text{hence} - x = 2 \text{ or } x = -2$$

$$x = -2$$

Step 3: Check the solution $x = -2$, $y = 2$ by substituting these values into equations (1) and (2) and confirm that both equations balance.

- Substitute $x = -2$ and $y = 2$ into equation (1):

$$x + 3y = 4$$

$$(-2) + 3(2) = 4 \quad \text{substituting in } x = -2 \text{ and } y = 2$$

$$-2 + 6 = 4$$

$$4 = 4 \quad \text{so equation (1) balances and } x = -2, \ y = 2 \text{ is a solution}$$

- Substitute $x = -2$ and $y = 2$ into equation (2):

$$-x + 2y = 6$$

$$-(-2) + 2(2) = 6 \quad \text{substituting in } x = -2 \text{ and } y = 2$$

$$2 + 4 = 6$$

$$6 = 6 \quad \text{so equation (2) balances and } x = -2, \ y = 2 \text{ is a solution}$$

Since the point $(-2, 2)$ satisfies equations (1) and (2), then this point is at the point of intersection of the lines represented by equations (1) and (2) as shown in Figure 3.1.

(b) The two lines are plotted in Figure 3.1. The point of intersection is the solution. The coordinates of this point are $x = -2$ and $y = 2$. In this case, it is a **unique solution**: that is, the lines intersect at only one point. This point is on the first line, so it satisfies equation (1), and also on the second line, so it satisfies equation (2).

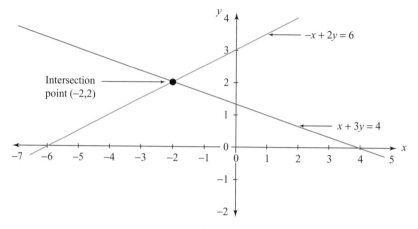

Figure 3.1   Unique solution

WORKED EXAMPLE 3.2
SOLVING SIMULTANEOUS EQUATIONS 2

Given the simultaneous equations

$$2x + 3y = 12.5 \tag{1}$$

$$-x + 2y = 6 \tag{2}$$

(a) Solve for x and y algebraically.
(b) Solve for x and y graphically.

SOLUTION

(a) In this example, neither the x- nor the y-terms are the same. However, all terms on both sides of any equation may be multiplied by a constant without affecting the solution of the equation. So, if equation (2) is multiplied by 2, the x-terms in both equations will be the same with opposite signs. Then, proceed as in Worked Example 3.1.

Step 1: Eliminate x from the system of equations by multiplying equation (2) by 2 and then add the equations; $2x$ and $-2x$ cancel to leave a single equation in one unknown, y:

$$
\begin{array}{ll}
2x + 3y = 12.5 & \text{(1) as given} \\
\underline{-2x + 4y = 12} & \text{(2)} \times 2 \\
0 + 7y = 24.5 & \text{adding the two equations}
\end{array}
$$

$$y = \frac{24.5}{7} = 3.5 \qquad \text{solving for } y$$

Step 2: Solve for the value of x by substituting $y = 3.5$ into either equation (1) or equation (2):

$$-x + 2(3.5) = 6 \quad \text{substituting } y = 3.5 \text{ into (2) to find the value of } x$$

$$-x = 6 - 7$$

$$-x = -1 \rightarrow x = 1$$

Step 3: Check the solution $x = 1$, $y = 3.5$, by substituting these values into equations (1) and (2) and confirm that both equations balance.

- Substitute $x = 1$ and $y = 3.5$ into equation (1):

$$2x + 3y = 12.5$$
$$2(1) + 3(3.5) = 12.5 \quad \text{substituting in } x = 1 \text{ and } y = 3.5$$
$$2 + 10.5 = 12.5$$
$$12.5 = 12.5 \quad \text{so (1) balances and } x = 1, \ y = 3.5 \text{ is a solution}$$

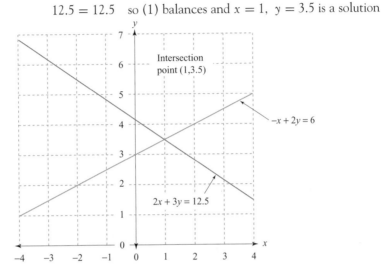

Figure 3.2   Unique solution

- Substitute $x = 1$ and $y = 3.5$ into equation (2):

$$-x + 2y = 6$$
$$-(1) + 2(3.5) = 6 \quad \text{substituting in } x = 1 \text{ and } y = 3.5$$
$$-1 + 7 = 6$$
$$6 = 6 \quad \text{so (2) balances and } x = 1, \ y = 3.5 \text{ is a solution}$$

Therefore, the solution of equations (1) and (2) is at the point of intersection of the lines represented by equations (1) and (2), as shown in Figure 3.2.

(b) The two lines are plotted in Figure 3.2. The point of intersection is the solution. The coordinates of this point are $x = 1$ and $y = 3.5$. In this case it is a **unique solution**: that is, the lines intersect at only one point. This point is on the first line, so it satisfies equation (1), and also on the second line, so it satisfies equation (2).

3.1.2 Solve simultaneous equations by methods of elimination and substitution

The method of elimination

In Worked Examples 3.1 and 3.2 the simultaneous equations were solved by adding a multiple of one equation to the other in order to eliminate one of the variables. Hence the method is called the 'method of elimination'.

The method of substitution

Alternatively, from either equation derive an expression for one variable in terms of the other: *substitute* this expression into the other equation and solve. This is illustrated for the equations given in Worked Example 3.1 as follows:

$$2x + 3y = 12.5 \tag{1}$$

$$-x + 2y = 6 \tag{2}$$

From (2), $x = 2y - 6$ is the expression for x in terms of y (x is the subject of the formula).

Substitute $2y - 6$ for x in equation (1)

$$2x + 3y = 12.5 \rightarrow 2\,(2y - 6) + 3y = 12.5$$
$$4y - 12 + 3y = 12.5$$
$$7y = 24.5$$
$$y = 3.5$$

Substitute $y = 3.5$ into any of the equations to solve for x. For example,

$$x = 2y - 6 \rightarrow x = 2(3.5) - 6 = 1$$

Note: The method of substitution is particularly suitable for solving simultaneous equations where one equation is linear and the other non-linear. See Chapter 4.

WORKED EXAMPLE 3.3
SOLVING SIMULTANEOUS EQUATIONS 3

Given the simultaneous equations

$$2x + 3y = 0.75 \tag{1}$$
$$5x + 2y = 6 \tag{2}$$

Solve for x and y algebraically.

SOLUTION

In these two equations, neither the x- nor the y-terms are the same. If equation (1) is multiplied by 2 and equation (2) is multiplied by -3, the y-terms in both equations will be the same with opposite signs. Then proceed as in Worked Example 3.1 above.

Step 1: Eliminate y-terms from the system of equations:

$$
\begin{array}{ll}
4x + 6y = 1.5 & (1) \times 2 \\
-15x - 6y = 18 & (2) \times -3 \\
\hline
-11x = -16.5 & \text{adding}
\end{array}
$$

$$x = \frac{-16.5}{-11} = 1.5 \quad \text{solving for } x$$

Step 2: Solve for y by substituting $x = 1.5$ into either equation (1) or equation (2):

$$2(1.5) + 3y = 0.75 \qquad \text{substituting } x = 1.5 \text{ into equation (1)}$$

$$3y = 0.75 - 3$$

$$y = \frac{-2.25}{3}$$

Step 3: Checking the solution: $x = 1.5$, $y = -\frac{2.25}{3}$ is left as an exercise for the reader.

3.1.3 Unique, infinitely many and no solutions of simultaneous equations

A set of simultaneous equations may have:

- A unique solution.
- No solution.
- Infinitely many solutions.

Unique solution

This occurs when a set of equations has one set of values which satisfy all equations. See Worked Examples 3.1 to 3.3.

No solution

This occurs when a set of equations has no set of values which satisfy all equations.

WORKED EXAMPLE 3.4
SIMULTANEOUS EQUATIONS WITH NO SOLUTION

Given the simultaneous equations

$$y = 1 + x \qquad \textbf{(1)}$$
$$y = 2 + x \qquad \textbf{(2)}$$

(a) Solve for x and y algebraically.
(b) Solve for x and y graphically.

SOLUTION

(a)

$$y = 1 + x \quad (1)$$
$$y = 2 + x \quad (2)$$
$$0 = -1 \quad (1) - (2)$$

$0 = -1$ is not possible, therefore, there is no solution. Even from a purely practical point of view, you can see that there is no way that these equations can both be true. How can y be equal to $(1 + x)$ and $(2 + x)$ at the same time?

Note: A false statement (or a contradiction) like $0 = -1$ indicates a set of equations with no solution.

(b) The two equations are plotted in Figure 3.3. The lines will never meet since they are parallel and thus will never have a point (solution) in common.

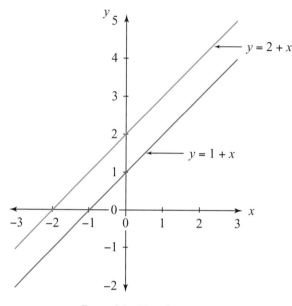

Figure 3.3   No solution

Infinitely many solutions

A set of equations has infinitely many solutions when there is an infinite number of sets of values that satisfy all equations.

WORKED EXAMPLE 3.5
SIMULTANEOUS EQUATIONS WITH INFINITELY MANY SOLUTIONS

Given the simultaneous equations

$$y = 2 - x \tag{1}$$

$$2y = 4 - 2x \tag{2}$$

(a) Solve for x and y algebraically.
(b) Solve for x and y graphically.

SOLUTION

(a) When equation (2) is divided by 2, the result is exactly the same as equation (1), since

$$\frac{2y}{2} = \frac{4}{2} - \frac{2x}{2} \rightarrow y = 2 - x$$

So, equations (1) and (2) are the same! There is only one equation in two unknowns. If x is given any value, the corresponding y-value can be calculated.

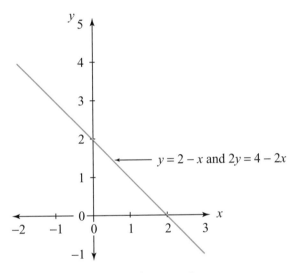

Figure 3.4   Infinitely many solutions

For example, when

$$x = 2 \rightarrow y = 2 - 2 = 0$$
$$x = 3 \rightarrow y = 2 - 3 = -1$$
$$x = 5 \rightarrow y = 2 - 5 = -3 \text{ etc.}$$

There is an infinite number of (x, y) pairs which satisfy equations (1) and (2).

(b) Equations (1) and (2) are plotted in Figure 3.4. Note that these equations represent coincident lines, therefore every point on one line is also a point on the other line. Since a line has infinitely many points, there are infinitely many solutions or points in common.

3.1.4   Three simultaneous equations in three unknowns

The methods used above to solve two equations in two unknowns may be extended to three equations in three unknowns, four equations in four unknowns, etc. The strategy is to eliminate one of the variables first by adding multiples of equations to other equations and hence reducing the problem to two equations in two unknowns. This is best demonstrated by Worked Example 3.6.

WORKED EXAMPLE 3.6
SOLVE THREE EQUATIONS IN THREE UNKNOWNS

Solve the equations

$$2x + y - z = 4 \tag{1}$$
$$x + y - z = 3 \tag{2}$$
$$2x + 2y + z = 12 \tag{3}$$

SOLUTION

The simplest approach is to add equation (3) to equation (1), and hence eliminate z, giving an equation in x and y. Then add equation (3) to equation (2), eliminating z again, giving another equation in x and y.

$$
\begin{array}{ll}
2x + y - z = 4 & (1) \\
2x + 2y + z = 12 & (3) \\
\hline
4x + 3y + 0 = 16 & \text{(4) adding equations (1) and (2)}
\end{array}
$$

$$
\begin{array}{ll}
x + y - z = 3 & (2) \\
2x + 2y + z = 12 & (3) \\
\hline
3x + 3y + 0 = 15 & \text{(5) adding equations (2) and (3)}
\end{array}
$$

Equations (4) and (5) are the usual two equations in two unknowns, so solve for x and y. Then solve for z later.

$$
\begin{array}{ll}
4x + 3y = 16 & (4) \\
-3x - 3y = -15 & \text{(6) equation (5) multiplied by } -1 \\
\hline
x = 1 & \text{adding equations (4) and (6)}
\end{array}
$$

So, $x = 1$. Substitute $x = 1$ into equations (4), (5) or (6) to solve for y.
  Substituting $x = 1$ into equation (4) gives $4(1) + 3y = 16 \rightarrow y = 4$.
  Finally, find z by substituting $x = 1$, $y = 4$ into any of the equations (1), (2) or (3). For example, substituting into equation (2),

$$1 + 4 - z = 3 \rightarrow z = 2$$

Therefore, the values which satisfy all three equations (1), (2) and (3) are $x = 1$, $y = 4$, $z = 2$.

PROGRESS EXERCISES 3.1

Simultaneous Equations: Solving for Two and Three Unknowns

Solve the following simultaneous equations:

1. $y = x$
$y = 3 - x$

2. $x + y = 10$
$x - y = 4$

3. $x + y = 19$
$x - 8y = 10$

4. $3y + 2x = 5$
$4y - x = 3$

5. $5x - 2y = 11$
$3x + 3y = 15$

6. $y = 2x + 3$
$y = 7 - 2x$

7. $2y + 4x - 1 = 7$
$3y - 2x = 12$

8. $x + 2y = 10$
$y = -0.5x + 5$

9. $4x - y = 12$
$2y - 3x = 11.2$

10. $5x - 2y = 15$
$15x - 45 = 6y$

11. $2x - 5y = 7$
$2 = 3x - 2.5y$

12. $4P - 3Q = 4$
$2Q + 1.5P = 20$

13. $5 + 2P = 6Q$
$5P + 8Q = 25$

14. $x - y + z = 0$
$2y - 2z = 2$
$-x + 2y + 2z = 29$

15. $P_1 - 3P_2 = 0$
$5P_2 - P_3 = 10$
$P_1 + P_2 + P_3 = 8$

16. $3x - 2y + 1 = 0$
$0.5x + 2.5y - 14 = 0$

17. $4x - 3y + 1 = 13$
$0.5x + y - 3 = 4$

18. $\dfrac{5}{2}q - 3p = \dfrac{7}{2}$
$3p = 3(q - 3)$

19. $P = \dfrac{18 - 10Q}{5}$

$2 = \dfrac{3Q + 5P}{2}$

20. $3(x - 16) + y = 0$
$\dfrac{5y + 10x}{5} = 20$

21. $2x + y + 2z = 4$
$3x + z = 2 + y$
$x + 2y + 4z + 1 = 0$

22. In questions 16 to 20, plot the graphs to confirm your answer.

3.2 Equilibrium and Break-even

At the end of this section you should be familiar with:

- Equilibrium in the goods market and labour market
- Price controls and government intervention in various markets
- Market equilibrium for substitute and complementary goods
- Taxes, subsidies and their distribution between producer and consumer
- Break-even analysis.

The method of simultaneous equations is now applied to the determination of equilibrium conditions in various markets; for example, the goods, labour and money markets. In addition, situations are considered that prevent the occurrence of equilibrium. Furthermore, the analysis also considers factors that change the state of equilibrium from one position to another.

Note: A state of equilibrium within a model is a situation that is characterised by a lack of tendency to change.

3.2.1 Equilibrium in the goods and labour markets

Goods market equilibrium

Goods market equilibrium (market equilibrium) occurs when the quantity demanded (Q_d) by consumers and the quantity supplied (Q_s) by producers of a good or service are equal. Equivalently, market equilibrium occurs when the price that a consumer is willing to pay (P_d) is equal to the price that a producer is willing to accept (P_s). The equilibrium condition, therefore, is expressed as

$$Q_d = Q_s \quad \text{and} \quad P_d = P_s \tag{3.1}$$

Note: In equilibrium problems, once the equilibrium condition is stated, Q and P are used to refer to the equilibrium quantity and price respectively.

WORKED EXAMPLE 3.7
GOODS MARKET EQUILIBRIUM

The demand and supply functions for a good are given as

$$\text{Demand function: } P_d = 100 - 0.5Q_d \tag{3.2}$$

$$\text{Supply function: } P_s = 10 + 0.5Q_s \tag{3.3}$$

Calculate the equilibrium price and quantity algebraically and graphically.

SOLUTION

Market equilibrium occurs when $Q_d = Q_s$ and $P_d = P_s$. Since the functions are written in the form $P = f(Q)$ with P as the only variable on the LHS of each equation, it is easier to equate prices, thereby reducing the system to an equation in Q only; hence, solve for Q:

$$P_d = P_s$$

$$100 - 0.5Q = 10 + 0.5Q \quad \text{equating the RHS of equations (3.2) and (3.3)}$$

$$100 - 10 = 0.5Q + 0.5Q$$

$$90 = Q \quad \text{equilibrium quantity}$$

Now solve for the equilibrium price by substituting $Q = 90$ into either equation (3.2) or (3.3):

$$P = 100 - 0.5(90) \quad \text{substituting } Q = 90 \text{ into equation (3.2)}$$

$$P = 55 \qquad\qquad\quad \text{equilibrium price}$$

Check the solution, $Q = 90$, $P = 55$, by substituting these values into either equation (3.2) or (3.3). This exercise is left to the reader.

Figure 3.5 illustrates market equilibrium at point E_0 with equilibrium quantity, 90, and equilibrium price, £55. The consumer pays £55 for the good which is also the price that the producer receives for the good. There are no taxes (what a wonderful thought!).

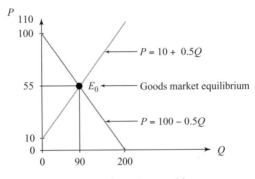

Figure 3.5   Goods market equilibrium

Labour market equilibrium

Labour market equilibrium occurs when the labour demanded (L_d) by firms is equal to the labour supplied (L_s) by workers or, equivalently, when the wage that a firm is willing to offer (w_s) is equal to the wage that workers are willing to accept (w_d). Labour market equilibrium, therefore, is expressed as

$$L_d = L_s \quad \text{and} \quad w_d = w_s \tag{3.4}$$

Again, in solving for labour market equilibrium, once the equilibrium condition is stated, L and w refer to the equilibrium number of labour units and the equilibrium wage, respectively.

WORKED EXAMPLE 3.8
LABOUR MARKET EQUILIBRIUM

The labour demand and supply functions are given as

$$\text{Labour demand function: } w_d = 9 - 0.6L_d \tag{3.5}$$

$$\text{Labour supply function: } w_s = 2 + 0.4L_s \tag{3.6}$$

Calculate the equilibrium wage and equilibrium number of workers algebraically and graphically. (In this example 1 worker $\equiv$ 1 unit of labour.)

SOLUTION

Labour market equilibrium occurs when $L_d = L_s$ and $w_d = w_s$. Since the functions are written in the form $w = f(L)$, equate wages, thereby reducing the system to an equation in L only;

hence, solve for L:

$$w_d = w_s$$

$$9 - 0.6L = 2 + 0.4L \quad \text{equating equations (3.5) and (3.6)}$$

$$9 - 2 = L$$

$$7 = L \qquad \text{equilibrium number of workers}$$

Now solve for w by substituting $L = 7$ into either equation (3.5) or (3.6):

$$w = 9 - 0.6(7) \qquad \text{substituting } L = 7 \text{ into equation (3.5)}$$

$$w = 4.8 \qquad \text{equilibrium wage}$$

Figure 3.6 illustrates labour market equilibrium at point E_0 with equilibrium number of workers, 7, and equilibrium wage, £4.80. Each worker receives £4.80 per hour for his or her labour services, which is also the wage that the firm is willing to pay.

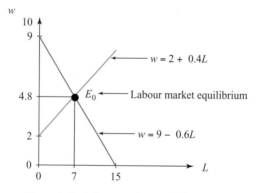

Figure 3.6   Labour market equilibrium

3.2.2   Price controls and government intervention in various markets

In reality, markets may fail to achieve market equilibrium due to a number of factors: for example, the intervention of governments or the existence of firms with monopoly power. Government intervention in the market through the use of price controls is now analysed.

Price ceilings

Price ceilings are used by governments in cases where they believe that the equilibrium price is too high for the consumer to pay. Thus, price ceilings operate **below market equilibrium** and are aimed at protecting consumers. Price ceilings are also known as maximum price controls, where the price is not allowed to go above the maximum or 'ceiling' price (for example, rent controls or maximum price orders).

WORKED EXAMPLE 3.9
GOODS MARKET EQUILIBRIUM AND PRICE CEILINGS

The demand and supply functions for a good are given by

$$\text{Demand function: } P_d = 100 - 0.5Q_d \tag{3.7}$$

$$\text{Supply function: } P_s = 10 + 0.5Q_s \tag{3.8}$$

(a) Analyse the effect of the introduction of a price ceiling of £40 in this market.
(b) Calculate the profit made by black marketeers if a black market operated in this market.

SOLUTION

(a) The demand and supply functions are the same as those in Worked Example 3.7 where the equilibrium price and quantity were £55 and 90 units, respectively. The price ceiling of £40 is below the equilibrium price of £55. Its effect is analysed by comparing the levels of quantity demanded and supplied at $P = £40$.

The quantity demanded at $P = 40$ is	The quantity supplied at $P = 40$ is
$P_d = 100 - 0.5Q_d$ equation (3.7)	$P_s = 10 + 0.5Q_s$ equation (3.8)
$40 = 100 - 0.5Q_d$ substituting $P = 40$	$40 = 10 + 0.5Q_s$ substituting $P = 40$
$0.5Q_d = 100 - 40$	$-0.5Q_s = 10 - 40$
$Q_d = 120$	$-0.5Q_s = -30$
	$Q_s = 60$

Since the quantity demanded ($Q_d = 120$) is greater than the quantity supplied ($Q_s = 60$), there is an excess demand (XD) of: $XD = Q_d - Q_s = 120 - 60 = 60$. This is also referred to as a shortage in the market. It is illustrated in Figure 3.7.

(b) The existence of price ceilings often leads to the establishment of black markets where goods are sold illegally at prices above the legal maximum. Black marketeers would buy the 60 units supplied at the controlled price of £40 per unit. However, as there is a shortage of goods, consumers are willing to pay a higher price for these 60 units. The price that consumers are willing to pay is calculated from the demand function for $Q = 60$. Substitute $Q = 60$ into the demand function:

$$P_d = 100 - 0.5Q_d$$

$$P_d = 100 - 0.5(60) = 100 - 30 = 70$$

So, $P_d = 70$ is the price consumers are willing to pay.

Therefore, black marketeers buy the 60 units at the maximum price of £40 per unit, costing them $60 \times £40 = £2400$, and then sell these 60 units at £70 per unit, generating revenue of $60 \times £70 = £4200$. Their profit (π) is the difference between revenue and costs:

$$\pi = TR - TC$$

$$= (70 \times 60) - (40 \times 60)$$

$$= 4200 - 2400$$

$$= 1800$$

This is illustrated as the shaded area in Figure 3.7.

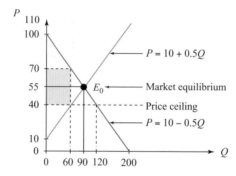

Figure 3.7   Price ceiling and black market

Price floors

Price floors are used by governments in cases where they believe that the equilibrium price is too low for the producer to receive. Thus, price floors operate **above market equilibrium** and are aimed at protecting producers. Price floors are also known as minimum prices, where the price is not allowed to go below the minimum or 'floor' price (for example, the Common Agricultural Policy (CAP) in the European Union and minimum wage laws).

WORKED EXAMPLE 3.10
LABOUR MARKET EQUILIBRIUM AND PRICE FLOORS

Given the labour demand and supply functions as

$$\text{Labour demand function: } w_d = 9 - 0.6L_d \qquad \textbf{(3.9)}$$

$$\text{Labour supply function: } w_s = 2 + 0.4L_s \qquad \textbf{(3.10)}$$

analyse the effect on the labour market if the government introduces a minimum wage law of £6 per hour.

SOLUTION

The labour demand and supply functions are the same as those in Worked Example 3.8 where the equilibrium wage and units of labour were £4.80 per hour and 7 labour units, respectively. The minimum wage law (price floor) of £6 is above market equilibrium. Its effect is analysed by comparing the levels of labour demanded and supplied at $w = 6$.

Labour demanded at $w = 6$ is	Labour supplied at $w = 6$ is
$w_d = 9 - 0.6L_d$ equation (3.9) $6 = 9 - 0.6L_d$ substituting $w = 6$ $0.6L_d = 3$ $L_d = 5$	$w_s = 2 + 0.4L_s$ equation (3.10) $6 = 2 + 0.4L_s$ substituting $w = 6$ $4 = 0.4L_s$ $10 = L_s$

Since labour supplied ($L_s = 10$) is greater than labour demanded ($L_d = 5$), there is an excess supply of labour: $XS = L_s - L_d = 10 - 5 = 5$. This is also referred to as a surplus, that is, there is unemployment in the labour market. The graphical illustration of this result is left to the reader.

PROGRESS EXERCISES 3.2

Determination of Equilibrium for Linear Functions

1. (a) Determine the equation of the line which has a slope $m = 1.5$, and which passes through the point $x = 4, y = 12$.
   (b) A supplier is known to supply a quantity of goods $Q = 30$ when the market price P is 25 per unit. If the quantity supplied increases by 3 for each unit increase in price, determine the equation of the supply function in the form $Q = f(P)$.
2. The demand and supply functions for fashion rings are given by the equations

$$\text{Demand function: } P_d = 800 - 2Q$$

$$\text{Supply function: } P_s = -40 + 8Q$$

   (a) Calculate the equilibrium quantity and price.
   (b) Calculate the level of excess supply ($Q_s - Q_d$) for rings when $P = 720$.
   (c) Find the level of excess demand ($Q_d - Q_s$) when $P = 560$.
3. The demand and supply functions for a product are given by the equations

$$\text{Demand function: } P_d = 400 - 5Q$$

$$\text{Supply function: } P_s = 3Q + 24$$

   (a) Calculate the equilibrium price and quantity.
   (b) Plot the graphs of the demand and supply functions, hence confirm the answer in (a).

4. A company hires out cabin cruisers on a daily basis. The demand and supply functions are given by the equations:

$$Q_d = 920 - 8P \quad \text{and} \quad Q_s = -120 + 2P$$

   (a) Calculate the equilibrium price and quantity algebraically and graphically.
   (b) Calculate the level of excess demand $(Q_d - Q_s)$ when $P = 90$.

5. The demand and supply functions for a good (jeans) are given by:

$$\text{Demand function: } P_d = 50 - 3Q$$

$$\text{Supply function: } P_s = 14 + 1.5Q$$

   where P is the price of a pair of jeans; Q is the number of pairs of jeans.
   (a) Calculate the equilibrium price and quantity.
   (b) Calculate the level of excess supply $(Q_s - Q_d)$ when $P = 38$.

6. (See question 5)
   (a) Calculate the level of excess demand $(Q_d - Q_s)$ when $P = 20$.
   (b) Calculate the profit made on the black market if a maximum price of £20 per pair of jeans is imposed.

7. The demand and supply functions for labour are given by:

$$\text{Labour demand function: } w_d = 70 - 4L$$

$$\text{Labour supply function: } w_s = 10 + 2L$$

   (a) Calculate the equilibrium number of workers employed and the equilibrium wage per hour.
   (b) Calculate the excess demand for labour $(L_d - L_s)$ when $w = 20$.
   (c) Calculate the excess supply for labour $(L_s - L_d)$ when $w = 40$.

3.2.3 Market equilibrium for substitute and complementary goods

Complementary goods are goods that are consumed together (for example, cars and petrol; computer hardware and computer software). One good cannot function without the other. On the other hand, substitute goods are consumed instead of each other (for example, coffee versus tea; bus versus train on given routes).

The general demand function is now written as

$$Q = f(P, P_s, P_c)$$

that is, the quantity demanded of a good is a function of the price of the good itself and the prices of those goods that are substitutes and complements to it.

Note: In this case, P_s refers to the price of substitute goods, not to be confused with P_s which is used to refer to the supply price of a good.

Consider two goods, X and Y. The demand function for good X is written differently depending on whether good Y is a substitute to X or a complement to X.

X and Y are substitutes	**X and Y are complements**
$$Q_X = a - bP_X + dP_Y$$	$$Q_X = a - bP_X - dP_Y$$
Note: the positive sign before dP_Y	**Note:** the negative sign before dP_Y
There is a positive relationship between the quantity demanded of good X and the price of good Y, since	There is a negative relationship between the quantity demanded of good X and the price of good Y, since
$$Q_X = (a + dP_Y) - bP_X$$	$$Q_X = (a - dP_Y) - bP_X$$
Therefore, as P_Y increases, so does Q_X. Similarly, the demand function for good Y is	Therefore, as P_Y increases, Q_X decreases. Similarly, the demand fucntion for good Y is
$$Q_Y = \alpha + \beta P_Y - \delta P_X$$	$$Q_Y = \alpha - \beta P_Y - \delta P_X$$
Example: if train fares increase, individuals will reduce their demand for train journeys and increase their demand for bus journeys.	**Example:** if car prices increase, individuals will reduce their demand for cars and consequently the demand for petrol decreases.

WORKED EXAMPLE 3.11
EQUILIBRIUM FOR TWO SUBSTITUTE GOODS

Find the equilibrium price and quantity for two substitute goods X and Y given their respective demand and supply equations as,

$$Q_{dX} = 82 - 3P_X + P_Y \tag{3.11}$$
$$Q_{sX} = -5 + 15P_X \tag{3.12}$$
$$Q_{dY} = 92 + 2P_X - 4P_Y \tag{3.13}$$
$$Q_{sY} = -6 + 32P_Y \tag{3.14}$$

SOLUTION

The equilibrium condition for this two-goods market is

$$Q_{dX} = Q_{sX} \text{ and } Q_{dY} = Q_{sY}$$

Therefore, the equilibrium prices and quantities are calculated as follows:

$$82 - 3P_X + P_Y = -5 + 15P_X \quad \text{equating equations (3.11) and (3.12)}$$
$$-18P_X + P_Y = -87 \quad \text{simplifying} \tag{3.15}$$

and

$$92 + 2P_X - 4P_Y = -6 + 32P_Y \quad \text{equating equations (3.13) and (3.14)}$$
$$2P_X - 36P_Y = -98 \quad \text{simplifying} \tag{3.16}$$

Equations (3.15) and (3.16) are two equations in two unknowns, P_X and P_Y.

Therefore, solve these simultaneous equations for the equilibrium prices, P_X and P_Y

$$-18P_X + P_Y = -87 \qquad \text{equation (3.15)}$$

$$\frac{18P_X - 324P_Y = -882}{-323P_Y = -969} \qquad \text{equation (3.16) multiplied by 9}$$

$$P_Y = 3$$

Solve for P_X by substituting $P_Y = 3$ into either equation (3.15) or equation (3.16):

$$-18P_X + 3 = -87 \quad \text{substituting } P_Y = 3 \text{ into equation (3.15)}$$
$$-18P_X = -90$$
$$P_X = 5$$

Now, solve for Q_X and Q_Y

Solve for Q_X by substituting $P_X = 5$ and $P_Y = 3$ into either equation (3.11) or equation (3.12) as appropriate:

$$Q_X = -5 + 15P_X \quad \text{using equation (3.12)}$$
$$Q_X = -5 + 15(5) \quad \text{substituting } P_X = 5$$
$$Q_X = 70$$

Solve for Q_Y by substituting $P_Y = 3$ and $P_X = 5$ into either equation (3.13) or equation (3.14) as appropriate:

$$Q_Y = -6 + 32P_Y \quad \text{using equation (3.14)}$$
$$Q_Y = -6 + 32(3) \quad \text{substituting } P_Y = 3$$
$$Q_Y = 90$$

The equilibrium prices and quantities in this two-goods market are

$$P_X = 5, \quad Q_X = 70, \quad P_Y = 3, \quad Q_Y = 90$$

3.2.4 Taxes, subsidies and their distribution

Taxes and subsidies are another example of government intervention in the market. A tax on a good is known as an indirect tax. Indirect taxes may be:

- A fixed amount per unit of output (excise tax); for example, the tax imposed on petrol and alcohol. This will translate the supply function vertically upwards by the amount of the tax.
- A percentage of the price of the good; for example, value added tax. This will change the slope of the supply function. The slope will become steeper since a given percentage tax will be a larger absolute amount the higher the price.

Fixed tax per unit of output

When a tax is imposed on a good, two issues of concern arise:

- How does the imposition of the tax affect the equilibrium price and quantity of the good?
- What is the distribution (incidence) of the tax; that is, what percentage of the tax is paid by consumers and producers, respectively?

In these calculations:

- The consumer always pays the equilibrium price.
- The supplier receives the equilibrium price minus the tax.

WORKED EXAMPLE 3.12
TAXES AND THEIR DISTRIBUTION

 Find animated worked examples at **www.wiley.com/college/bradley**

The demand and supply functions for a good are given as

$$\text{Demand function: } P_d = 100 - 0.5Q_d \qquad \textbf{(3.17)}$$

$$\text{Supply function: } P_s = 10 + 0.5Q_s \qquad \textbf{(3.18)}$$

(a) Calculate the equilibrium price and quantity.
(b) Assume that the government imposes a fixed tax of £6 per unit sold.

    (i) Write down the equation of the supply function, adjusted for tax.
   (ii) Find the new equilibrium price and quantity algebraically and graphically.
  (iii) Outline the distribution of the tax, that is, calculate the tax paid by the consumer and the producer.

SOLUTION
(a) The equilibrium quantity and price are 90 units and £55, respectively.

 Remember

> The equilibrium price of £55 (with no taxes) means that the price the consumer pays is equal to the price that the producer receives.

(b) The tax of £6 per unit sold means that the effective price received by the producer is $(P_s - 6)$. The equation of the supply function adjusted for tax is

$$P_s - 6 = 10 + 0.5Q$$

$$P_s = 16 + 0.5Q \qquad \textbf{(3.19)}$$

The supply function is translated vertically upwards by 6 units (with a corresponding horizontal leftward shift). This is illustrated in Figure 3.8 as a line parallel to the original supply function.

Remember

(i) Translations, Chapter 2.

(ii) The new equilibrium price and quantity are calculated by equating the original demand function, equation (3.17), and the supply function adjusted for tax, equation (3.19):

$$P_d = P_s$$

$$100 - 0.5Q = 16 + 0.5Q \quad \text{equating equations (3.17) and (3.19)}$$

$$Q = 84$$

Substitute the new equilibrium quantity, $Q = 84$, into either equation (3.17) or equation (3.19) and solve for the new equilibrium price:

$$P = 100 - 0.5(84) \quad \text{substituting } Q = 84 \text{ into equation (3.17)}$$
$$P = 58$$

The point (84, 58) is shown as point E_1 in Figure 3.8.

(iii) The consumer always pays the equilibrium price, therefore the consumer pays £58, an increase of £3 on the original equilibrium price with no tax, which was £55. This means that the consumer pays 50% of the tax. The producer receives the new equilibrium price, minus the tax, so the producer receives £58 – £6 = £52, a reduction of £3 on the original equilibrium price of £55. This also means that the producer pays 50% of the tax.

In this example, the tax is evenly distributed between the consumer and producer. The reason for the 50:50 distribution is due to the fact that the slope of the demand function is equal to the slope of the supply function (ignoring signs). This suggests that changes in the slope of either the demand or supply functions will alter this distribution.

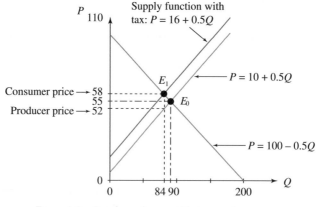

Figure 3.8 Goods market equilibrium and taxes

Subsidies and their distribution

Similar ideas may be analysed with respect to subsidies and their distribution. In the case of subsidies, one would be interested in analysing how the **benefit** of the subsidy is distributed between the producer and consumer.

In the analysis of subsidies, a number of important points need to be highlighted:

- A subsidy per unit sold will translate the supply function vertically downwards, that is, the price received by the producer is $(P + \text{subsidy})$.
- The equilibrium price will decrease (the consumer pays the new lower equilibrium price).
- The price that the producer receives is the new equilibrium price **plus** the subsidy.
- The equilibrium quantity increases.

WORKED EXAMPLE 3.13
SUBSIDIES AND THEIR DISTRIBUTION

The demand and supply functions for a good ($£P$ per ton of potatoes) are given as

$$\text{Demand function: } P_d = 450 - 2Q_d \qquad (3.20)$$

$$\text{Supply function: } P_s = 100 + 5Q_s \qquad (3.21)$$

(a) Calculate the equilibrium price and quantity.
(b) The government provides a subsidy of £70 per unit (ton) sold:

    (i) Write down the equation of the supply function, adjusted for the subsidy.
    (ii) Find the new equilibrium price and quantity algebraically and graphically.
    (iii) Outline the distribution of the subsidy, that is, calculate how much of the subsidy is received by the consumer and the supplier.

SOLUTION

(a) The solution to this part is given over to the reader. Show that the equilibrium quantity and price are 50 units and £350, respectively.
(b) (i) With a subsidy of £70 per unit sold, the producer receives $(P_s + 70)$. The equation of the supply function adjusted for subsidy is

$$P_s + 70 = 100 + 5Q$$
$$P_s = 30 + 5Q \qquad (3.22)$$

The supply function is translated vertically downwards by 70 units. This is illustrated in Figure 3.9 as a line parallel to the original supply function.

(ii) The new equilibrium price and quantity are calculated by equating the original demand function, equation (3.20), and the supply function adjusted for the subsidy, equation (3.22):

$$P_d = (P_s + \text{subsidy})$$
$$450 - 2Q = 30 + 5Q \quad \text{equating equations (3.20) and (3.22)}$$
$$Q = 60$$

Substitute the new equilibrium quantity $Q = 60$ into either equation (3.20) or equation (3.22) and solve for the new equilibrium price:

$$P = 450 - 2Q$$
$$P = 450 - 2(60) \quad \text{substituting } Q = 60 \text{ into equation (3.20)}$$
$$P = 330$$

The point $(P = 330, Q = 60)$ is shown as point E_1 in Figure 3.9.

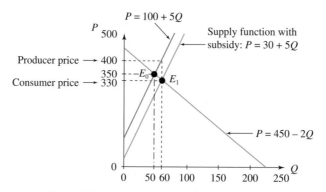

Figure 3.9   Goods market equilibrium and subsidies

(iii) The consumer always pays the equilibrium price, therefore, the consumer pays £330, a decrease of £20 on the equilibrium price with no subsidy (£350). This means that the consumer receives 20/70 of the subsidy. The producer receives the equilibrium price, plus the subsidy, so the producer receives £330 + £70 = £400, an increase of £50 on the original price of £350. The producer receives 50/70 of the subsidy.

In this case, the subsidy is not evenly distributed between the consumer and producer; the producer receives a greater fraction of the subsidy than the consumer. The reason? The slope of the supply function is greater than the slope of the demand function (ignoring signs).

Distribution of taxes/subsidies

The fraction of the tax/subsidy that the consumer pays/receives is given by the equation

$$\frac{|m_d|}{|m_d| + |m_s|}$$

The fraction of the tax/subsidy that the producer pays/receives is given by the equation

$$\frac{|m_s|}{|m_s| + |m_d|}$$

where m_d and m_s are the slopes of the demand and supply functions, respectively. See Appendix to this chapter for proof of these formulae.

3.2.5 Break-even analysis

The break-even point for a good occurs when total revenue is equal to total cost.

Break-even point: total revenue = total cost

WORKED EXAMPLE 3.14
CALCULATING THE BREAK-EVEN POINT

The total revenue and total cost functions are given as follows:

$$TR = 3Q \qquad \text{(3.23)}$$

$$TC = 10 + 2Q \qquad \text{(3.24)}$$

(a) Calculate the equilibrium quantity algebraically and graphically at the break-even point.
(b) Calculate the value of total revenue and total cost at the break-even point.

SOLUTION

(a) The break-even point is algebraically solved by equating total revenue, equation (3.23), and total cost, equation (3.24):

$$3Q = 10 + 2Q$$
$$Q = 10$$

The equilibrium quantity at the break-even point is $Q = 10$. This is illustrated in Figure 3.10.

(b) The value of total revenue and total cost at the break-even point is calculated by substituting $Q = 10$ into the respective revenue and cost functions:

$$TR = 3Q = 3(10) = 30$$
$$TC = 10 + 2Q = 10 + 2(10) = 30$$

At $Q = 10$, $TR = TC = 30$.

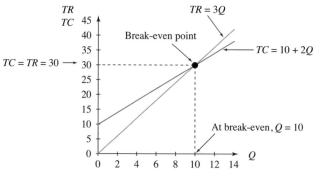

Figure 3.10   Break-even point

PROGRESS EXERCISES 3.3

Equilibrium, Break-even

1. The following demand and supply functions for a safari holiday package are

$$\text{Demand function: } Q = 81 - 0.05P$$

$$\text{Supply function: } Q = -24 + 0.025P$$

   (a) Calculate the equilibrium price and quantity, algebraically and graphically.
   (b) Graph the supply and demand function, showing the equilibrium.
2. A perfectly competitive firm producing lamps has fixed costs of €1000 per week; each lamp costs €15 to produce and is sold at €35.
   (a) Calculate the break-even quantity.
   (b) Does the firm make a profit or loss when: (i) 500 lamps; (ii) 1000 lamps are produced and sold?
   (c) Confirm the answers to (a) and (b) graphically.
3. A craftsman has fixed costs of €90 and a cost of €3 for each bracelet he produces. Bracelets are sold at €6 each.
   (a) Calculate the break-even quantity.
   (b) Confirm your answers graphically.
4. A canteen has fixed costs of £1500 per week. Meals cost £5 each and are sold at a fixed price of £9 each.
   (a) Calculate the break-even quantity.
   (b) Confirm your answers graphically.
5. (See question 1)
   The government imposes a tax of £120 on each safari holiday.
   (a) Write down the equation of the supply function adjusted for tax, hence graph it on the diagram in 1(b).
   (b) Calculate the equilibrium price and quantity when the tax is imposed.
   (c) Outline the distribution of the tax, that is, calculate the tax paid by the consumer and by the travel agent.

6. The demand and supply functions for two complementary products X and Y, pitching wedges and putters (for pitch and putt), respectively, are given as:

$$Q_{dX} = 190 - 2P_X - 2P_Y \qquad Q_{dY} = 240 - 2P_X - 4P_Y$$
$$Q_{sX} = -10 + 2P_X \qquad Q_{sY} = -40 + P_Y$$

Find the equilibrium price and quantity for each good.

7. The demand and supply functions for golf lessons at Greens Club are

$$\text{Demand function: } P = 200 - 5Q$$
$$\text{Supply function: } P = 92 + 4Q$$

(a) Calculate the equilibrium price and quantity algebraically and graphically.
(b) The government imposes a tax of £9 per lesson:
   (i) Write down the equation of the supply function adjusted for tax, hence graph it on the same diagram as in part (a).
   (ii) Calculate the equilibrium price and quantity when the tax is imposed.
   (iii) Outline the distribution of the tax, that is, how much of the tax is paid by the customer and the club (supplier).

8. The demand function for a perfectly competitive firm (same price charged for each good) is given as $P = £30$. The firm has fixed costs of £200 and variable costs of £5 per unit sold.
(a) Calculate the equilibrium quantity at the break-even point.
(b) Calculate the value of total revenue and total cost at the break-even point.

9. The demand and supply functions for free-range Christmas turkeys are given by the equations:

$$P_d = 80 - 0.4Q_d \quad \text{and} \quad P_s = 20 + 0.4Q_s$$

(a) Calculate the equilibrium price and quantity.
(b) If the government provides a subsidy of £4 per bird:
   (i) Rewrite the equation of the supply function to include the subsidy.
   (ii) Calculate the new equilibrium price and quantity.
   (iii) Outline the distribution of the subsidy, that is, how much of the subsidy is received by the customer and by the supplier.

10. A firm which makes travel alarm clocks has a total cost function $TC = 800 + 0.2Q$.
(a) If the price of the clock is £6.6, write down the equation of the total revenue function. Calculate the number of clocks which must be made and sold to break even.
(b) When the firm charges a price P for each alarm clock the break-even point is $Q = 160$. Write down the equation for break-even, hence calculate the price charged per clock.
(c) Graph the total revenue functions (a) and (b) with the total cost on the same diagram, showing each break-even point.

11. The supply and demand functions for complementary goods (jeans and shirts) are given by the equations:

$$P_{dX} = 100 - 5Q_X - Q_Y \qquad P_{dY} = 240 - 10Q_X - 8Q_Y$$
$$P_{sX} = 50 + Q_X \qquad P_{sY} = 40 + 2Q_Y$$

Find the equilibrium price and quantity for each good.

3.3 Consumer and Producer Surplus

At the end of this section you should be familiar with:

- The meaning of consumer and producer surplus
- How to measure consumer and producer surplus.

3.3.1 Consumer and producer surplus

Consumer surplus (CS)

This is the difference between the expenditure a consumer is willing to make on successive units of a good from $Q = 0$ to $Q = Q_0$ and the actual amount spent on Q_0 units of the good at the market price of P_0 per unit. To explain how consumer surplus is calculated geometrically consider the demand function, $P = 100 - 0.5Q$, which is graphed in Figure 3.11.

To calculate consumer surplus when the market price is 55, proceed as follows. When the market price per unit is 55 the consumer will purchase 90 units, since, according to the demand function, $55 = 100 - 0.5Q \rightarrow 0.5Q = 100 - 55 = 45 \rightarrow Q = 90$. The consumer, therefore, spends a total of $P \times Q = (55)(90) = £4950$. This is equivalent to the area of the rectangle $0P_0E_0Q_0$ (since area = length × breadth = $P \times Q$).

The consumer pays the same price, 55, for each of the 90 units purchased; however, he or she is willing to pay more than 55 for each of the units preceding the 90th when the good was scarcer, $Q < 90$. (These higher prices, which the consumer is willing to pay, may be calculated from the demand function.) The total amount which the consumer is willing to pay for the first 90 units is given by the area under the demand function between $Q = 0$ and $Q = 90$, that is, area $0AE_0Q_0$.

Since consumer surplus is the difference between the amount that the consumer is willing to pay and the amount that the consumer actually pays, then

$$CS = 0AE_0Q_0 - 0P_0E_0Q_0 = AP_0E_0$$

$$= 0.5(90)(55) = 2475 \quad \text{(See } \textit{Area of triangles}, \text{ below)}$$

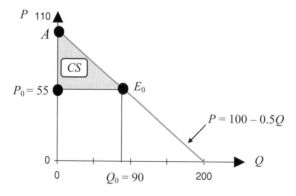

Figure 3.11   Consumer surplus

Area AP_0E_0 is a benefit to the consumer as the amount which the consumer is willing to pay exceeds the amount which is actually paid.

Producer surplus (PS)

This is the difference between the revenue the producer receives for Q_0 units of a good when the market price is P_0 per unit and the revenue that the producer was willing to accept for successive units of the good from $Q = 0$ to $Q = Q_0$. To explain how producer surplus is calculated geometrically, consider the supply function, $P = 10 + 0.5Q$, which is graphed in Figure 3.12.

To calculate producer surplus when the market price is 55 per unit, proceed as follows. When the market price per unit is 55 the producer will supply 90 units since

$$P = 10 + 0.5Q \; \rightarrow \; 55 = 10 + 0.5Q \; \rightarrow \; 45 = 0.5Q \; \rightarrow \; 90 = Q$$

So, the producer receives 55 for each of the 90 units, giving a total revenue of $P \times Q = (55)(90) = £4950$. This is equivalent to the area of rectangle $0P_0E_0Q_0$.

The producer, however, is willing to supply each unit, up to the 90th unit, at prices less than 55. (These lower prices may be calculated from the equation of the supply function.) The revenue the producer is willing to accept for units below the 90th unit is given by the area under the supply function between $Q = 0$ and $Q = 90$, that is, area $0BE_0Q_0$.

Since producer surplus is the difference between the revenue the producer receives at $Q = 90$ and the revenue that the producer was willing to accept for units supplied up to the 90th, then

$$PS = 0P_0E_0Q_0 - 0BE_0Q_0 = BP_0E_0$$
$$= 0.5(90)(45) = 2025 \quad (\text{See } \textit{Area of triangles, } \text{below})$$

Therefore, area BP_0E_0 is a benefit to the producer.

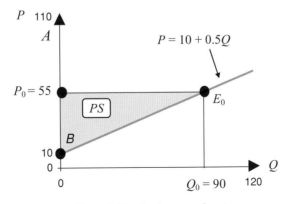

Figure 3.12   Producer surplus

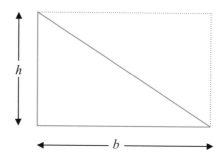

Figure 3.13   Area of triangle $= 0.5 \times$ area of rectangle $= 0.5 \times (b \times h)$

Total surplus (TS)

This is the sum of consumer and producer surplus.

Area of triangles: The consumer and producer surplus are each represented by the area of a triangle. The area of a triangle is half the area of the rectangle formed by the two sides which meet at right angles. This is usually referred to as 'half the length of the base (b) multiplied by the length of the perpendicular height (h)', as illustrated in Figure 3.13.

WORKED EXAMPLE 3.15
CONSUMER AND PRODUCER SURPLUS AT MARKET EQUILIBRIUM

The demand and supply functions of a good (shirts) are given as

$$\text{Demand function: } P = 60 - 0.6Q$$
$$\text{Supply function: } P = 20 + 0.2Q$$

(a) Calculate the equilibrium price and quantity for shirts algebraically and graphically.
(b) Calculate the values of consumer and producer surplus at market equilibrium. Illustrate CS and PS on the graph in (a).
(c) What is the value of total surplus?

SOLUTION

(a) The algebraic solution to this part is given over to the reader. Show that the equilibrium quantity and price of shirts are 50 units and £30, respectively. The graphical solution is the point E_0 illustrated in Figure 3.14.

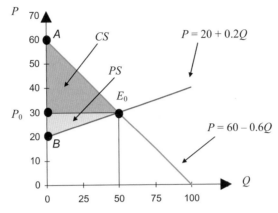

Figure 3.14   Consumer and producer surplus

(b) Consumer and producer surplus at market equilibrium are calculated as follows:

$$\text{At } P = 30, \; Q = 50 \qquad CS = \text{triangle } AP_0E_0 = 0.5 \times 50 \times 30 = 750$$
$$\text{At } P = 30, \; Q = 50 \qquad PS = \text{triangle } BP_0E_0 = 0.5 \times 50 \times 10 = 250$$

(c) Total surplus is the sum of consumer and producer surplus; therefore

$$TS = CS + PS = 750 + 250 = 1000$$

PROGRESS EXERCISES 3.4

CS/PS for Linear Functions

1. (a) Define (i) consumer surplus, (ii) producer surplus, (iii) total surplus, at the equilibrium price P_0. Illustrate all three surpluses graphically.
   (b) Define (i) consumer surplus, (ii) producer surplus, (iii) total surplus, at price P_A, which is below the equilibrium price P_0. Illustrate all three surpluses graphically.
2. The demand and supply functions for seats on a certain weekend bus route are given by

$$\text{Demand function: } P = 58 - 0.2Q$$

$$\text{Supply function: } P = 4 + 0.1Q$$

   (a) Calculate the equilibrium price and quantity. Plot the demand and supply functions and illustrate consumer and producer surplus at equilibrium.
   (b) Calculate:
      (i) The amount consumers pay for bus journeys at equilibrium.
      (ii) The amount consumers are willing to pay for bus journeys up to equilibrium.
      (iii) The consumer surplus (CS); hence, show that the CS = (ii) − (i).
   (c) Calculate:
      (i) The amount the producer (bus company) receives for bus journeys at equilibrium.
      (ii) The amount the producer is willing to accept for bus journeys up to equilibrium.
      (iii) The producer surplus (PS); hence, show that the PS = (i) − (ii).

3. The demand and supply functions for a product (helicopter rides) are given by

$$\text{Demand function: } Q = 50 - 0.1P$$

$$\text{Supply function: } Q = -10 + 0.1P$$

(a) Calculate the equilibrium price and quantity. Plot the demand and supply functions in the form $P = g(Q)$. Illustrate graphically the consumer and producer surplus at equilibrium.

(b) Calculate the consumer surplus at equilibrium.

(c) Calculate the producer surplus at equilibrium.

(d) Calculate the total surplus at equilibrium.

4. (See question 3) The price per helicopter ride decreases to £250.

(a) Calculate the number of helicopter rides demanded at the reduced price £250.

(b) Calculate the consumer surplus at £250, graphically illustrating your answer. What is the change in CS as a result of the price reduction?

5. (See question 3) The price per helicopter ride decreases to £250.

(a) Calculate the number of helicopter rides supplied at the reduced price £250.

(b) Calculate the producer surplus at £250, graphically illustrating your answer. What is the change in PS as a result of the price reduction?

6. The demand and supply functions for a product are

$$\text{Demand function: } P_d = 200 - 4Q$$

$$\text{Supply function: } P_s = 50 + Q$$

(a) Find the equilibrium price and quantity.

(b) Plot the demand and supply functions. Illustrate graphically the consumer (CS) and producer surplus (PS) at equilibrium.

(c) Calculate the consumer surplus at equilibrium.

(d) Calculate the producer surplus at equilibrium.

(e) Calculate the total surplus at equilibrium.

7. The demand and supply functions for a product are

$$P_d = 255 - 4Q \quad P_s = 25 + 7.5Q$$

(a) Confirm that the equilibrium point is at $Q = 20$, $P = 175$.

(b) Calculate the consumer surplus at equilibrium.

(c) Calculate the producer surplus at equilibrium.

(d) Illustrate the CS and the PS graphically.

8. The demand and supply functions for a product are

$$\text{Demand function: } P = 120 - 3Q$$

$$\text{Supply function: } P = 24 + 5Q$$

(a) Calculate the equilibrium price and quantity.

(b) Graph the demand and supply functions, illustrating the equilibrium point.

(c) Calculate the consumer and producer surplus at equilibrium.

(d) What is the value of the total surplus?

3.4 The National Income Model and the *IS-LM* Model

At the end of this section you should be familiar with:

- The national income model: national income equilibrium and expenditure multipliers
- The *IS-LM* model: determination of equilibrium national income and interest rates.

3.4.1 National income model

National income, Y, is the total income generated within an economy from all productive activity over a given period of time, usually one year. Equilibrium national income occurs when aggregate national income, Y, is equal to aggregate planned expenditure, E, that is,

$$Y = E \tag{3.25}$$

Note: In the discussion which follows it is assumed that all expenditure is planned expenditure.

Aggregate expenditure, E, is the sum of households' consumption expenditure, C; firms' investment expenditure, I; government expenditure, G; foreign expenditure on domestic exports, X; **minus** domestic expenditure on imports, M, that is,

$$E = C + I + G + X - M \tag{3.26}$$

Note: Expenditure on imports is income lost to the economy, hence the minus sign.

Therefore, substituting equation (3.26) into equation (3.25) gives the equation for equilibrium national income:

$$Y = C + I + G + X - M \tag{3.27}$$

That is, equilibrium national income exists when total income is equal to total expenditure.

Note: Aggregate expenditure on goods and services (E) is one method of measuring national income. Alternatively, national income may be measured by aggregating total income received by firms and individuals (total income) or aggregating total production (total output). For the purposes of this text, these differences in measuring national income may be ignored.

Steps for deriving the equilibrium level of national income

Step 1: Express expenditure in terms of income, Y: $E = f(Y)$.
Step 2: Substitute expenditure, expressed as a function of Y, into the RHS of the equilibrium condition, $Y = E$. Solve the equilibrium equation for the equilibrium level of national income, Y_e.

Graphical solution: The point of intersection of the equilibrium condition, $Y = E$ (the 45° line through the origin), and the expenditure equation, $E = C + I + G + X - M$, gives the equilibrium level of national income.

Note: When graphing the national income equations, Y is plotted on the horizontal axis. As a reminder: 'E on the vertical, Y on the horizontal', all equations will be written in the form $E = f(Y)$.

Equilibrium level of national income when $E = C + I$

Initially, the model assumes the existence of only two economic agents, households and firms, operating in a closed economy (no foreign sector) with no government sector and no inflation. Households' consumption expenditure, C, is modelled by the equation $C = C_0 + bY$, where C_0 is autonomous consumption, that is, consumption which does not depend on income. b $(0 < b < 1)$ is called the marginal propensity to consume. $b = MPC = \Delta C / \Delta Y$ measures the change in consumption per unit increase in income. A firm's investment expenditure is autonomous, $I = I_0$.

WORKED EXAMPLE 3.16
EQUILIBRIUM NATIONAL INCOME WHEN $E = C + I$

In a two-sector economy, autonomous consumption expenditure, $C_0 = £50m$, autonomous investment expenditure, $I_0 = £100m$, and $b = 0.5$.

(a) Determine (i) the equilibrium level of national income, Y_e, and (ii) the equilibrium level of consumption, C_e, algebraically.
(b) Plot the consumption function, $C = C_0 + bY$, the expenditure function, $E = C + I_0$, and the equilibrium condition, $Y = E$, on the same diagram.
    Hence, determine the equilibrium level of national income, Y_e, and the equilibrium level of consumption, C_e.
(c) Given that $Y = C + S$, determine the equilibrium level of savings. Plot the savings function. Plot the investment function on the same diagram. Comment.

SOLUTION

(a) (i) **Step 1:** Households' consumption expenditure and firms' investment expenditure are the only components of aggregate expenditure, therefore

$$E = C + I = C_0 + bY + I_0 = 50 + 0.5Y + 100 = 150 + 0.5Y$$

Step 2: At equilibrium, $Y = E$ (equation 3.25), therefore:

Example:	**In general:**
$Y = 150 + 0.5Y$	$Y = (C_0 + bY) + I_0$
$Y - 0.5Y = 150$	$Y - bY = C_0 + I_0$
$0.5Y = 150$	$Y(1 - b) = C_0 + I_0$
$Y = \dfrac{150}{0.5}$	$Y = \dfrac{C_0 + I_0}{1 - b}$
$Y_e = \dfrac{1}{0.5} 150 = 300$	$Y_e = \dfrac{1}{1 - b}(C_0 + I_0)$ **(3.28)**

The equilibrium level of national income $Y_e = 300$.

(ii) When the equilibrium level of income has been found, the equilibrium level of consumption is calculated directly from the consumption function,

$$C_e = C_0 + bY_e$$
$$= 50 + 0.5(300) = 50 + 150 = 200$$

(3.29)

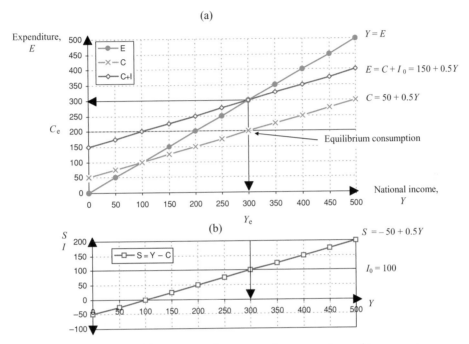

Figure 3.15   Equilibrium national income with consumption and investment

(b) The consumption function $C = 50 + 0.5Y$, the expenditure function $E = 150 + 0.5Y$ and the equilibrium condition $Y = E$ are plotted in Figure 3.15, with E plotted vertically and Y plotted horizontally. The equilibrium condition, $Y = E$, is represented by a 45° line from the origin, provided the number scale is the same on both axes. Graphically, equilibrium national income Y_e is illustrated in Figure 3.15(a) with the equilibrium point occurring at the point of intersection of the expenditure equation and the line $Y = E$. The point of intersection is at $Y = 300 = E$. Graphically, the equilibrium level of consumption, C_e, is at the point of intersection of the consumption function and the vertical line $Y_e = 300$.

(c) Since $C_e = 200$, the equilibrium level of savings $S_e = Y_e - C_e = 300 - 200 = 100$. The savings function $S = Y - (C_0 + bY) = -C_0 + (1 - b)Y = -50 + 0.5Y$ is plotted in Figure 3.15(b). Graphically, investment expenditure is illustrated by a horizontal line in Figure 3.15(b). Notice that, at equilibrium national income, savings is equal to investment,

$$S_e = I$$

Therefore, in this example, one can also say that equilibrium national income occurs when savings (leakages) are equal to investment (injections).

PROGRESS EXERCISES 3.5

The National Income Model

1. Given the condition for equilibrium national income $Y = E$, and the expenditure equation $E = C$, where $C = C_0 + bY$:
   (a) Describe the constants C_0 and b. Are there any restrictions on the range of values which b can assume?
   (b) Find an expression for the equilibrium level of income (the reduced form).
   (c) Deduce how the equilibrium level of income changes as:
      (i) b increases.
      (ii) b decreases.

2. Given the national income model $Y = E$; $E = C + I$ where $C = 280 + 0.6Y$, $I_0 = 80$:
   (a) Write down the value of the intercept and slope of the expenditure equation.
   (b) Graph the equilibrium equation and the expenditure equation on the same diagram; hence, determine the equilibrium level of income Y_e (i) graphically, (ii) algebraically.
   (c) How will the equilibrium level of income change if the marginal propensity to consume increases to 0.9?

3. Given the national income model $Y = E$; $E = C + I$; where $C = 280 + 0.6Y_d$, $I_0 = 80$, $T = 0.2Y$ (that is, $t = 0.2$):
   (a) Write down the value of the intercept and slope of the expenditure equation.
   (b) Graph the equilibrium equation and the expenditure equation on the same diagram; hence determine the equilibrium level of income Y_e (i) graphically, (ii) algebraically.
   (c) If the marginal propensity to consume increases to 0.9, how will:
      (i) The expenditure equation change?
      (ii) The equilibrium level of income change?

4. Assuming that the initial level of national income is £800m, calculate the new level of national income when $b = 0.6$ and $\Delta I = $ £500m.

www.wiley.com/college/bradley

Go to the website for the following additional material that accompanies Chapter 3:

- Expenditure multiplier
  Table 3.1 Relationship between the expenditure multiplier and MPC, b
  Worked Example 3.17 Effect of changes in MPC and I_0 on Y_e
- Government expenditure and taxation: $E = C + I + G$
- Expenditure multiplier with taxes
  Worked Example 3.18 Equilibrium national income and effect of taxation
  Figure 3.16 Equilibrium national income and effect of taxation
- Foreign trade
  Worked Example 3.19 Expenditure multiplier with imports and trade balance
  Table 3.2 Summary of national income model

Progress Exercises 3.5, questions 5, 6, 7
Solutions to questions 5, 6, 7
3.4.2 IS-LM model: determination of equilibrium national income and interest rates
- IS schedule
- LM schedule
- Equilibrium national income and equilibrium interest rates
  Worked Example 3.20 IS-LM analysis
  Figure 3.17 Equilibrium income and interest rates
  Progress Exercises 3.6 IS-LM analysis
  Solutions to Progress Exercises 3.6

3.5   Excel for Simultaneous Linear Equations

If you found Excel useful for plotting graphs in Chapter 2, you should find it equally useful in this chapter as the same skills are required in graph plotting. However, since this chapter is mainly concerned with finding solutions to simultaneous equations, you will need to show the solution, that is, the point of intersection of the lines on the graph. To ensure that the solution is shown on the graph, solve the equations algebraically first. Once the solution is known, the axis may then be scaled by choosing a range of x-values which includes the x-value of the solution.

Alternatively, if the equations are not easy to solve algebraically then a table of y-values calculated for each of the simultaneous equations $y_1 = f(x)$, $y_2 = g(x)$ may be used to detect whether the solution $(y_1 = y_2)$ is included in the table. If $y_1 > y_2$ at the start of the table, the first function is greater than the second; if $y_1 < y_2$ at the end of the table, the first function is less than the second, or vice versa, then the solution, $y_1 = y_2$, is included.

WORKED EXAMPLE 3.21
COST, REVENUE, BREAK-EVEN, PER UNIT TAX WITH EXCEL

A firm receives £2.5 per unit for a particular good. The fixed costs incurred are £44 while each unit produced costs £1.4.

(a) Write down the equations for (i) total revenue, and (ii) total cost.
(b) Calculate the break-even point algebraically.
(c) If the government imposes a tax of £0.70 per unit, recalculate the break-even point. Show the graphical solutions to parts (b) and (c) on the same diagram (using Excel).

SOLUTION

(a)   (i) $TR = P \times Q = 2.5Q$
      (ii) $TC = FC + VC = 44 + 1.4Q$

(b) Break-even occurs when $TR = TC$:

- Algebraically:

$$2.5Q = 44 + 1.4Q$$

$$1.1Q = 44$$

$$Q = 40$$

When $Q = 40$, then $TR = TC = 100$.

(c) If a tax per unit is imposed, either the total revenue function or the total cost function may be adjusted for the tax as follows:

- Algebraically:

The net revenue per unit is (price – tax): $TR = (2.5 - 0.7)Q = 1.8Q$
Break-even is at $TR = TC \rightarrow 1.8Q = 44 + 1.4Q \rightarrow Q = 110$

- Graphically:

To show the break-even points on a graph, choose values of Q such as $Q = 0$ to $Q = 160$. Set up the table of points in Excel and plot the graph as shown in Figure 3.18. (Since the graph is a straight line, a minimum of two points is required.)

	A	B	C	D	E	F
1	Q	0	40	80	120	160
2	TR (no tax)	0	100	200	300	400
3	TR (taxed)	0	72	144	216	288
4	TC	44	100	156	212	268

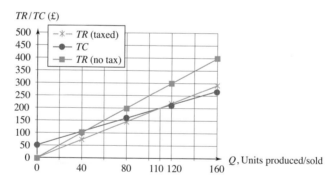

Figure 3.18   Break-even with tax and no tax

WORKED EXAMPLE 3.22

DISTRIBUTION OF TAX WITH EXCEL

(a) Show graphically that the equilibrium price and quantity for the demand and supply functions given by the pairs of equations (i) and (ii) is the same.

(i) $\begin{aligned} P_d &= 120 - 2Q \\ P_s &= 10 + 2Q \end{aligned}$ and   (ii) $\begin{aligned} P_d &= 120 - 2Q \\ P_s &= 37.5 + Q \end{aligned}$

(b) If a tax of 20 is imposed on each unit produced, recalculate the equilibrium price for (i) and (ii) above, hence determine the distribution of the tax. Show the distribution of tax graphically.

(c) Can you deduce a general rule describing how the distribution of the tax changes according to the function with the flatter slope?

SOLUTION

(a) The table of points and the graphs of equations (i) and (ii) are shown in Figure 3.19. The equilibrium point for each pair is the same, $Q = 27.5, P = 65$.

	A	B	C	D	E	F
1	Q	0	10	20	30	40
2	P_d	120	100	80	60	40
3	P_{s1}	10	30	50	70	90
4	P_{s2}	37.5	47.5	57.5	67.5	77.5

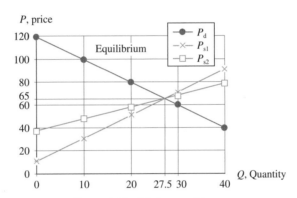

Figure 3.19   Market equilibrium

(b) When a tax of 20 is imposed, the price the supplier receives is the (original price – tax); therefore, replace P in the supply functions by $(P - 20)$.

  (i) With the tax, the set of equations is now

$$P_d = 120 - 2Q, \quad \text{stays the same}$$

$$P_s - 20 = 10 + 2Q \rightarrow P_s = 30 + 2Q$$

Solve for equilibrium at $Q = 22.5, P = 75$.

 The consumer (who always pays the equilibrium price) pays $(P_{e(tax)} - P_e) = 75 - 65 = 10$ more than before the tax. The producer receives $(P_{e(tax)} - tax) = 75 - 20 = 55$. This is 10 units less than before the tax was imposed. So when the slopes of the demand and supply functions are equal the distribution of tax is 50:50.

 A table of values is set up to plot this pair of graphs with the untaxed supply function. The range of Q-values is selected so that the graph focuses on the original and the new equilibrium points as shown in Figure 3.20

(ii) With the tax, the set of equations is now

$$P_d = 120 - 2Q$$

$$P_s - 20 = 37.5 + Q \rightarrow P_s = 57.5 + Q$$

Solve for equilibrium at $Q = 20.83$, $P = 78.33$.

	A	B	C	D	E	F
22	Q	20	22	24	26	28
23	P_d	80	76	72	68	64
24	P_{s1tax}	70	74	78	82	86
25	P_{s1}	50	54	58	62	66

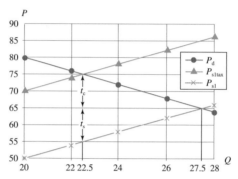

Figure 3.20   Distribution of tax for equations (i)

The consumer pays $(P_{e(tax)} - P_e) = 78.33 - 65 = 13.33$ more than before the tax. The producer receives $(P_{e(tax)} - tax) = 78.33 - 20 = 58.33$. This is 6.67 units less than before the tax was imposed. As expected, the function whose slope is greater in magnitude pays the greater share of the tax.

A table of values is set up to plot this pair of graphs with the untaxed supply function. The range of Q-values is selected so that the graph focuses on the original and the new equilibrium points as shown in Figure 3.21.

	A	B	C	D	E	F
50	Q	20	22	24	26	28
51	P_d	80	76	72	68	64
52	P_{s2tax}	77.5	79.5	81.5	83.5	85.5
53	P_{s2}	57.5	59.5	61.5	63.5	65.5

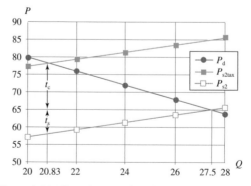

Figure 3.21   Distribution of tax for equations (ii)

(c)

> ## Remember
>
> The distribution of tax for linear functions is given as
>
> $$\text{Consumer pays } \frac{|m_d|}{|m_d| + |m_s|} \times \text{tax} \qquad \text{Producer pays } \frac{|m_s|}{|m_s| + |m_d|} \times \text{tax}$$
>
> See Appendix to chapter. The distribution of tax is graphically illustrated in Figures 3.20 and 3.21.

PROGRESS EXERCISES 3.7

Excel or Otherwise

1. Solve the following simultaneous equations (i) algebraically, (ii) graphically with Excel.

   (a) $x + y - 3 = 0$
   $2x - y = 15$

   (b) $3.8P - 0.75Q = 12$
   $P = 5Q - 6$

2. Set up a table in Excel to find the approximate solution to the following equations:

$$P - 8Q = 120$$
$$3Q - 1.5P + 270 = 0$$

   Hence, determine the solution:
   (a) algebraically, (b) graphically.

3. Given the demand and supply functions,
   $P = 124 - 4.5Q$ and $Q = -16.5 + 0.5P$, respectively.
   (a) Find the equilibrium point (i) graphically, (ii) algebraically.
   (b) If a tax of 30 per unit is imposed, calculate:
      (i) the equilibrium point, (ii) the distribution of tax.
   Show the distribution of tax graphically.

4. On the same diagram plot:
   (a) (i) the 45° line $Y = E$, (ii) $E = C + I$, where $C = 125 + 0.65Y$ and $I = 20$. (b) (i) the 45° line $Y = E$, (ii) $E = C + I$, where $C = 125 + 0.65 Y_d$, $I = 20$, $T = 0.2Y$. Hence, determine the equilibrium level of income (i) graphically, (ii) algebraically.

3.6 Summary

Mathematics

The solution of a set of simultaneous equations is the values of x and y which satisfy all equations.

(a) To solve the equations algebraically, eliminate all but one variable, solve for this one variable, then solve for the other(s).
(b) To solve the equations graphically, plot the graphs. The solution is given by the coordinates of the point of intersection of the graphs.
Simultaneous equations may have:
  (i) A unique solution.
  (ii) No solution.
  (iii) Infinitely many solutions.

Applications

- **Goods market equilibrium:** $Q_d = Q_s$ and $P_d = P_s$
- **Labour demand:** $w_d = a - bL$: a negative relationship between the number of labour units and the wage rate (price per unit).
- **Labour supply:** $w_s = c + dL$: a positive relationship between the number of labour units and the wage rate (price per unit).
- **Labour market equilibrium:** $L_d = L_s$ and $w_d = w_s$
Revise price ceilings, black market profits, price floors.
- **Equilibrium for complementary and substitute goods:**
when $Q_{dX} = Q_{sX}$ and $Q_{dY} = Q_{sY}$: $P_{dX} = P_{sX}$ and $P_{dY} = P_{sY}$: X and Y are substitutes $\rightarrow Q_X = a - bP_X + dP_Y$; X and Y are complements $\rightarrow Q_X = a - bP_X - dP_Y$.
- **Tax and its distribution between consumer and producer:**
When a tax per unit is imposed the price to the supplier is $(P - tax)$

$$P_s = c + dQ \rightarrow P_s - tax = c + dQ$$

The fraction of tax paid by the consumer and producer, respectively:

$$\frac{|m_d|}{|m_d| + |m_s|} \text{ (consumer)}, \quad \frac{|m_s|}{|m_s| + |m_d|} \text{ (producer)}$$

- **Break-even:** $TR = TC$.
- **Consumer surplus:** (CS) is the difference between the expenditure a consumer is willing to make on successive units of a good from $Q = 0$ to $Q = Q_0$ and the actual amount spent on Q_0 units of the good at the market price P_0 per unit.
- **Producer surplus:** (PS) is the difference between the revenue the producer receives for Q_0 units of a good when the market price is P_0 per unit, and the revenue that the producer was willing to accept for successive units of the good from $Q = 0$ to $Q = Q_0$.
Revise the effect of price increases and decreases on CS and PS.
- **National income model:** Equilibrium exists when income (Y) = expenditure (E).

Expenditure may consist of (i) consumption: $C = C_0 + bY$, less tax, (ii) investment, I_0, (iii) government expenditure, G_0, (iv) exports, X_0, (v) less imports, $M = M_0 + mY$, hence $E = C + I + G + X - M$.

To find the level of income (Y_e) at which equilibrium exists, solve the equation $Y = E$ for Y. The solution is Y_e. Hence, equilibrium consumption and taxation: $C_e = C_0 + bY_e$: $T_e = tY_e$.

Revise the reduced expressions for Y_e, with multipliers. These are formulae from which T_e may be calculated directly for various standard national income models.

- **IS-LM model:** Equilibrium in the goods market when $Y = E$ (national income), but consider investment as a function of interest rate: $I = I_0 - dr$, thus

$$Y = \frac{1}{1 - b(1 - t)} \cdot (C_0 + I_0 - dr + G_0)$$

hence the equation: $r = f(Y)$. This is the IS schedule.

Equilibrium in the money market when money supply = money demand.

Money demand: $M_d = M_d^T + M_d^P + M_d^S = L_1 + L_2 = kY + (a - hr)$

Money supply: $M_s = M_0$

Equilibrium, $M_s = M_d \to M_0 = kY + a - hr$

This equation may also be written as $r = g(Y)$. This is the LM schedule. The goods and money markets are simultaneously in equilibrium for the values of r and Y which satisfy the simultaneous IS and LM equations: for example,

$$IS\,\text{schedule:}\, r = 32 - 0.014\,Y$$

$$LM\,\text{schedule:}\, r = -2.0 + 0.0025\,Y$$

- **Excel:** Useful as described in Chapter 2. In this chapter, points of intersection and equilibrium points may be viewed. It should prove helpful for problems on the national income model, to view the effect of various sectors on the equilibrium level of income.

www.wiley.com/college/bradley

Go to the website for Problems in Context: 'Goods Market Equilibrium'

Appendix

- **Distribution of tax paid by the consumer and producer**

See Figure 3.22 where $|m_d|$ and $|m_s|$ represent the magnitudes of the slopes of the demand and supply functions respectively.

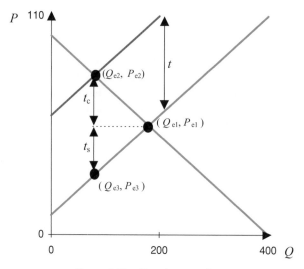

Figure 3.22   Distribution of tax

Consumer pays $(P_{e2} - P_{e1})$ of the tax: t_c

Supplier (producer) pays $(P_{e1} - P)$ of the tax: t_s

$$\left.\begin{array}{l} |m_d| = \dfrac{t_c}{x} \rightarrow x = \dfrac{t_c}{|m_d|} \\[2mm] |m_s| = \dfrac{t_s}{x} \rightarrow x = \dfrac{t_s}{|m_s|} \end{array}\right\} \quad \dfrac{t_c}{|m_d|} = \dfrac{t_s}{|m_s|}$$

Therefore

$$t_c\,|m_s| = t_s\,|m_d|$$

$$t_c\,|m_s| = (t - t_c)\,|m_d| \ \text{ where } t_c + t_s = t$$

$$t_c\,(|m_s| + |m_d|) = t\,|m_d|$$

$$t_c = t\,\frac{|m_d|}{|m_s| + |m_d|} \qquad t_s = t\,\frac{|m_s|}{|m_s| + |m_d|}$$

TEST EXERCISES 3

1. Solve the following simultaneous equations (i) algebraically, (ii) graphically.

(a) $2x + y = 12$ (b) $3.5P - Q = 12$
   $x - y = 15$ $P = 5Q - 6$

2. Given the demand and supply functions, $P = 50 - 1.5Q$ and $Q = -11 + 0.5P$, respectively:
   (a) Find the equilibrium point (i) graphically, (ii) algebraically.
   (b) If a tax of 15.05 per unit is imposed calculate:
      (i) the equilibrium point, (ii) the distribution of tax.
   Show the distribution of tax graphically.

3. On the same diagram plot:
   (a) the $45°$ line $Y = E$, $E = C + I$, where $C = 100 + 0.8Y$ and $I = 50$;
   (b) the $45°$ line $Y = E$, $E = C + I$, where $C = 100 + 0.8Y_d$, $I = 50$, $T = 0.2Y$.
   Determine the equilibrium level of income for (a) and (b): (i) graphically, (ii) algebraically.
   Comment on the effect of taxation.
4. (See question 3) Determine the equilibrium levels of consumption and taxation (i) graphically,
   (ii) algebraically.
5. The demand and supply functions for a good are given by the equations

$$P = 80 - 2Q \text{ and } P = 20 + 4Q$$

   respectively.
   (a) Calculate the equilibrium price and quantity.
   (b) Calculate the consumer and producer surplus at equilibrium.
6. (See question 5) The supplier pays an excise tax of 12 per unit sold.
   (a) Write down the equation for the supply function when the excise tax is imposed.
   (b) Calculate the equilibrium price and quantity.
   (c) Calculate the tax paid by (i) the consumer, (ii) the producer.
   (d) Calculate the consumer and producer surplus at equilibrium.
7. The demand and supply functions for complementary goods X and Y are given by the equations:

$$Q_{d,X} = 200 - 4P_X - 4P_Y \quad Q_{s,X} = -65 + 6P_X$$
$$Q_{d,Y} = 80 - P_X - P_Y \quad Q_{s,Y} = -20 + 6P_Y$$

   respectively. Calculate the equilibrium price and quantity for each good.
8. Solve the simultaneous equations

$$x + y + z = 9 \quad 2x - y + 4z = 19 \quad 3x + 6y - z = 16$$

NON-LINEAR FUNCTIONS AND APPLICATIONS

4

4.1   Quadratic, Cubic and Other Polynomial Functions

4.2   Exponential Functions

4.3   Logarithmic Functions

4.4   Hyperbolic (Rational) Functions of the Form $a/(bx + c)$

4.5   Excel for Non-linear Functions

4.6   Summary

Chapter Objectives

This chapter introduces the student to several types of non-linear functions which are frequently used in economics, management and business studies. At the end of this chapter you should be able to:

- Recognise the general form of the equation representing a non-linear function as well as the main characteristics of the graph representing the function
- Manipulate non-linear functions algebraically, particularly in economic applications such as demand, supply, revenue, cost and profit
- Use exponentials and logs in a range of applications, such as production and consumption
- Plot quadratic, cubic and other functions using Excel.

4.1 Quadratic, Cubic and Other Polynomial Functions

At the end of this section you should be able to:

- Solve quadratic equations
- Sketch any quadratic function
- Use quadratics in economics, for example, total revenue and profit
- Recognise any cubic or other polynomial function
- Use cubic equations in economics, for example, cost and break-even.

Linear functions, covered in the previous three chapters, are very good introductory models for demand, supply, cost, etc.; however, these models provide limited representations of real-life situations, hence the need for more versatile non-linear models. For example, the total revenue function in Worked Example 2.10 was given as $TR = 3.5Q$. This models a linear total revenue function representative of a perfectly competitive firm whose demand function is given as $P = 3.5$. On the other hand, if the firm is a monopolist, and assuming that the demand function for the monopolist is given as $P = a - bQ$, for example, $P = 50 - 2Q$, then the total revenue is

$$TR = P \times Q$$
$$TR = (50 - 2Q)Q$$
$$TR = 50Q - 2Q^2 \tag{4.1}$$

Equation (4.1) is a quadratic function where TR is represented by an inverted U-shaped curve (see Figure 4.1).

Another example which demonstrates the limitations of linear models is the total cost function, $TC = 10 + 2Q$, as given in Worked Example 2.9. It is not reasonable to assume that costs rise by the same amount (2 units) for each extra unit of output produced, whether the extra unit is the 2nd or the 200th. It is more realistic to assume that after the initial outlay, the cost of producing an extra unit will eventually decrease; that is, total costs rise at a decreasing rate, and possibly at an increasing rate further on, when further outlay for expansion is required. This situation is best described by a cubic function, $TC = aQ^3 - bQ^2 + cQ + d$ (a, b, c and d are constants), as illustrated in Figure 4.2.

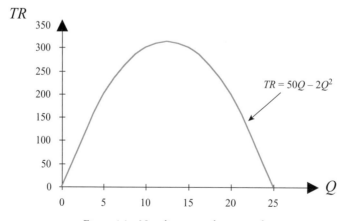

Figure 4.1 Non-linear total revenue function

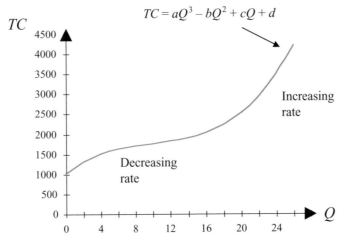

Figure 4.2 A cubic total cost function

4.1.1 Solving a quadratic equation

A quadratic equation has the general form

$$ax^2 + bx + c = 0$$

where a, b and c are constants. The solution of this equation may be found by using the quadratic formula (the '**minus b**' formula),

$$x = \frac{-(b) \pm \sqrt{(b)^2 - 4(a)(c)}}{2(a)} \tag{4.2}$$

Notice that a, b and c are all placed inside brackets, so that the values of a, b and c, with signs included, are substituted into the brackets. This should help avoid mistakes with negative signs.

The roots of quadratic equations: an overview

An equation of the form

$$ax^2 + bx + c = 0 \tag{4.3}$$

is defined as a quadratic equation. This is the most general form of a quadratic equation and is solved by means of the 'minus b' formula; see formula (4.2). However, the only condition for equation (4.3) to be classified as a quadratic is that $a \neq 0$. Equations (4.4), (4.5) and (4.6) are also quadratic equations:

$$ax^2 = 0 \qquad \text{quadratic with } b = 0 \text{ and } c = 0 \tag{4.4}$$

$$ax^2 + bx = 0 \qquad \text{quadratic with } c = 0 \tag{4.5}$$

$$ax^2 + c = 0 \qquad \text{quadratic with } b = 0 \tag{4.6}$$

These latter quadratic equations may be solved without recourse to the 'minus b' formula, as demonstrated in Worked Example 4.1.

WORKED EXAMPLE 4.1
SOLVING LESS GENERAL QUADRATIC EQUATIONS

Solve the following equations:

(a) $5x^2 = 0$ (b) $2x^2 - 32 = 0$ (c) $2x^2 + 32 = 0$ (d) $2x^2 - 32x = 0$

SOLUTION

(a)
$$5x^2 = 0$$
$$x^2 = \frac{0}{5}$$
$$x^2 = 0$$
$$x = \pm 0$$

A repeated real root (or solution) $x = 0$, repeated

(b)
$$2x^2 - 32 = 0$$
$$2x^2 = 32$$
$$x^2 = 16$$
$$x = \pm 4$$

Two real roots (or solutions) $x = -4$, $x = 4$

(c)
$$2x^2 + 32 = 0$$
$$2x^2 = -32$$
$$x^2 = -16$$
$$x = \pm\sqrt{-16}$$
$$* \quad x = \pm 4i$$

Two imaginary roots (or solutions) $x = 4i$, $x = -4i$

(d)
$$2x^2 - 32x = 0$$
$$x\underbrace{(2x - 32)} = 0$$
$$\downarrow$$
$$x = 0$$
or
$$(2x - 32) = 0$$
$$x = \frac{32}{2} = 16$$

Two real roots (or solutions) $x = 0$ and $x = 16$

*We define the imaginary number i, such that

$$(i)^2 = -1 \rightarrow \sqrt{(i)^2} = \sqrt{-1} \rightarrow i = \sqrt{-1}$$

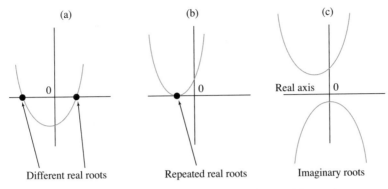

Figure 4.3   Quadratics: (a) real roots, (b) repeated roots, (c) complex roots

Notice that there are always two solutions of a quadratic equation, where the solutions may be:

1. Two different solutions (real roots).
2. Two identical solutions (real roots), that is, a repeated solution (repeated real root).
3. Two imaginary or complex roots (no real roots – these are of no interest in this text).

Reasons for three different types of solutions (roots)

The value of the term inside the square root sign ($\sqrt{\;}$) in the 'minus b' formula gives rise to different types of roots:

$$x = \frac{-(b) \pm \sqrt{(b)^2 - 4(a)(c)}}{2(a)}$$

- $b^2 - 4ac > 0$, then the term under the square root is positive, leading to two different solutions (real roots). The graph will cut the x-axis at two different points as shown in Figure 4.3(a).
- $b^2 - 4ac = 0$, then the term under the square root is zero, leading to two identical solutions (two identical real roots). The graph will touch the x-axis at one point only as shown in Figure 4.3(b).
- $b^2 - 4ac < 0$, then the term under the square root is negative, leading to two different complex roots. The graph will never cut or touch the x-axis as shown in Figure 4.3(c).

Quadratic equations with each type of root are solved in Worked Example 4.2.

WORKED EXAMPLE 4.2
SOLVING QUADRATIC EQUATIONS

Solve each of the following equations, classifying the type of solutions (roots) found as real and different, real and repeated, or complex:

(a) $x^2 + 6x + 5 = 0$ (b) $x^2 + 6x + 9 = 0$ (c) $x^2 + 6x + 10 = 0$

SOLUTION

Use equation (4.2), the 'minus b' formula, to solve for the values of x.

(a)

$$x^2 + 6x + 5 = 0$$

$$x = \frac{-(b) \pm \sqrt{(b)^2 - 4(a)(c)}}{2(a)}$$

$a = 1,\ b = 6,\ c = 5$

$$x = \frac{-6 \pm \sqrt{(6)^2 - 4(1)(5)}}{2(1)}$$

$$x = \frac{-6 \pm \sqrt{36 - 20}}{2}$$

$$x = \frac{-6 \pm \sqrt{16}}{2}$$

$$x = \frac{-6 \pm 4}{2}$$

root $1 = \dfrac{-6 + 4}{2} = \dfrac{-2}{2} = -1$

root $2 = \dfrac{-6 - 4}{2} = \dfrac{-10}{2} = -5$

Different solutions (real roots)

(b)

$$x^2 + 6x + 9 = 0$$

$$x = \frac{-(b) \pm \sqrt{(b)^2 - 4(a)(c)}}{2(a)}$$

$a = 1,\ b = 6,\ c = 9$

$$x = \frac{-(6) \pm \sqrt{(6)^2 - 4(1)(9)}}{2(1)}$$

$$x = \frac{-6 \pm \sqrt{36 - 36}}{2}$$

$$x = \frac{-6 \pm \sqrt{0}}{2}$$

$$x = \frac{-6 \pm 0}{2}$$

root $1 = \dfrac{-6 + 0}{2} = \dfrac{-6}{2} = -3$

root $2 = \dfrac{-6 - 0}{2} = \dfrac{-6}{2} = -3$

Repeated solutions (real roots)

(c)

$$x^2 + 6x + 10 = 0$$

$$x = \frac{-(b) \pm \sqrt{(b)^2 - 4(a)(c)}}{2(a)}$$

$a = 1,\ b = 6,\ c = 10$

$$x = \frac{-(6) \pm \sqrt{(6)^2 - 4(1)(10)}}{2(1)}$$

$$x = \frac{-6 \pm \sqrt{36 - 40}}{2}$$

$$x = \frac{-6 \pm \sqrt{-4}}{2}$$

$$x = \frac{-6 \pm 2i}{2}$$

root $1 = \dfrac{-6 + 2i}{2} = -3 + i$

root $2 = \dfrac{-6 - 2i}{2} = -3 - i$

Different solutions (complex roots)

PROGRESS EXERCISES 4.1

Solve Quadratic Equations

Solve the following quadratic equations:

1. $x^2 - 6x + 5 = 0$
2. $2Q^2 - 7Q + 5 = 0$
3. $-Q^2 + 6Q - 5 = 0$
4. $Q^2 + 6Q + 5 = 0$
5. $P^2 - 7 = 0$
6. $Q^2 - 6Q + 9 = 0$
7. $Q^2 - 6Q - 9 = 0$
8. $Q^2 = 6Q$
9. $x^2 - 6x = 7 + 3x$
10. $P^2 + 12 = 3$
11. $P + 10 = 11P^2 - P + 1$
12. $Q^2 - 8Q = Q^2 - 2$
13. $12 = P^2 - 2P + 12$
14. $5 + P = 4P^2 - 4 + P$
15. $-4P^2 - 2P + 1 = P$
16. $x(x - 3)(x + 3) = 0$
17. $x^2 - 9x + 8 = 9x + 12$
18. $\dfrac{Q}{2} + \dfrac{12}{Q} = 5$
19. $\dfrac{2}{P + 1} = \dfrac{P - 1}{6}$
20. $Q(2Q - 9) = 4(Q + 3)$

21. The total revenue (TR) for an accounting journal is given by the equation

$$TR = 1800Q - Q^2 - 44\,375 \text{ where } Q \text{ is the number of journals sold.}$$

(a) How many journals are sold when TR is zero?
(b) How many journals are sold when TR is €765 625?

4.1.2 Properties and graphs of quadratic functions: $f(x) = ax^2 + bx + c$

Many of the properties of functions may be deduced from their graphs. The properties of quadratics are illustrated in the sketches of quadratic functions in Worked Examples 4.3, 4.4 and 4.5.

To plot the graph of a quadratic function:

- Calculate a table of x, y values (remember, y is the short way of indicating $f(x)$). Be extremely careful with the negative values. Remember, a minus sign is a (-1).
- Plot the calculated points and join the points by a smooth curve.
- The points of intersection with the x-axis are the roots of the quadratic equation $f(x) = 0$

Since the graph crosses the x-axis when $y = 0$, that is, at $0 = ax^2 + bx + c$, then the solution of this equation, $0 = ax^2 + bx + c$, is the point(s) of intersection with the x-axis, i.e., the roots of $f(x) = 0$.

Graphical representation of the roots of a quadratic

- Distinct real roots: the graph cuts the x-axis in two distinct points. See Figure 4.3(a).
- Repeated roots: the graph touches the x-axis at one point. See Figure 4.3(b).
- Complex roots: the graph never cuts the x-axis. See Figure 4.3(c).

WORKED EXAMPLE 4.3
SKETCHING A QUADRATIC FUNCTION $f(x) = \pm x^2$

(a) Plot the graphs of $y = x^2$ and $y = -x^2$ on the same diagram.
(b) Comment on the general properties of the graphs.

What is the effect on the graph of multiplying the right-hand side of each equation by (-1)?

SOLUTION

(a) Calculate a table of (x, y) points as in Table 4.1. Then graph these points as in Figure 4.4.
(b) From Figure 4.4, note the following. The graphs are symmetrical about the x-axis; that is, the graph of $y = -x^2$ is simply the graph of $y = x^2$ inverted. Therefore, multiplying the RHS of an equation $y = f(x)$ by (-1) simply reflects its graph through the x-axis.

Table 4.1   Calculation of points for $y = x^2$ and $y = -x^2$

x	$f(x) = x^2$	$f(x) = (-1)x^2$
-3	$(-3)^2 = 9$	$(-1)(-3)^2 = -9$
-2	$(-2)^2 = 4$	$(-1)(-2)^2 = -4$
-1	$(-1)^1 = 1$	$(-1)(-1)^2 = -1$
0	$(0)^2 = 0$	$(-1)(0)^2 = 0$
1	$(1)^2 = 1$	$(-1)(1)^2 = -1$
2	$(2)^2 = 4$	$(-1)(2)^2 = -4$
3	$(3)^2 = 9$	$(-1)(3)^2 = -9$

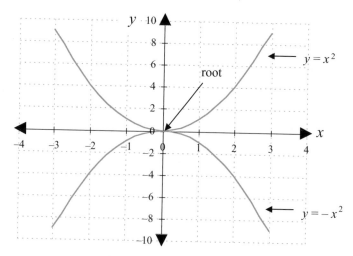

Figure 4.4 $y = x^2$ and $y = -x^2$

www.wiley.com/college/bradley

Go to the website for the following additional material that accompanies Chapter 4:

Worked Example 4.4 Comparing graphs of quadratic functions
Table 4.2
Figure 4.5

Graphs and equations of translated quadratics

In Section 2.5 (on the website), the rules for writing down the equations for translated linear functions were given. These same rules apply to the translation of any mathematical function and are illustrated for quadratic functions as follows:

- General rule for vertical translations
  If $y = f(x)$ is the original graph then:

    $y = f(x) + c$ is the equation of the original graph translated vertically up by c units.
    $y = f(x) - d$ is the equation of the original graph translated vertically down by d units.

- General rule for horizontal translations
  If $y = f(x)$ is the original graph then:

    $y = f(x - c)$ is the equation of the original graph translated horizontally to the right by c units.
    $y = f(x + d)$ is the equation of the original graph translated horizontally back to the left by d units.

WORKED EXAMPLE 4.5
VERTICAL AND HORIZONTAL TRANSLATIONS
OF QUADRATIC FUNCTIONS

(a) Sketch the graph of (i) $y = x^2$, (ii) $y = x^2$ when translated vertically up the y-axis by 4 units, (iii) $y = x^2$ when translated down the y-axis by 2 units.

Write down the equation of the translated graphs.

(b) Sketch the graph of (i) $y = x^2$, (ii) $y = x^2$ when translated horizontally forward (to the right) along the x-axis by 1 unit (iii) $y = x^2$ when translated horizontally backwards (to the left) along the x-axis by 2 units.

In all cases, write down the equation of the translated graphs.

SOLUTION

(a) The graphs are given in Figure 4.6a.
   (i) When the quadratic, $y = x^2$, is translated up the y-axis by 4 units, then the equation of the translated graph is $y = x^2 + 4$.
   (ii) When the quadratic, $y = x^2$, is translated down the y-axis by 2 units, then the equation of the translated graph is $y = x^2 - 2$.
(b) The graphs are given in Figure 4.6b.
   (i) When the quadratic, $y = x^2$, is translated forward (to the right) along the x-axis by 1 unit, then the equation of the translated graph is $y = (x - 1)^2$.
   (ii) When the quadratic, $y = x^2$, is translated horizontally backwards (to the left) along the x-axis by 2 units, then the equation of the translated graph is $y = (x + 2)^2$.

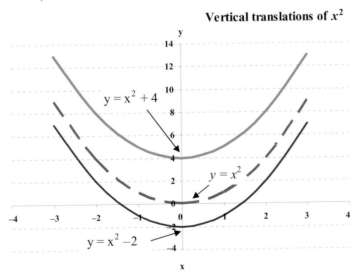

Vertical translations of x^2

Figure 4.6a   Vertical translations of $y = x^2$

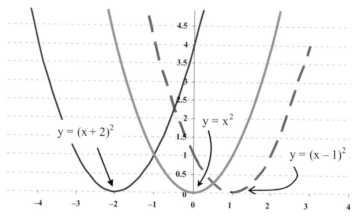

Figure 4.6b   Horizontal translations of $y = x^2$

To sketch any quadratic $y = ax^2 + bx + c$

In the following worked examples, quadratic curves are sketched by calculating a table of points for the range of x-values given.

WORKED EXAMPLE 4.6

SKETCHING ANY QUADRATIC EQUATION

(a) Sketch the graph of $y = 2x^2 - 7x - 9$ for values of x from −2 to 6.
(b) Calculate the points of intersection with the axes; hence, find the roots and mark the roots of $2x^2 - 7x - 9$ on the diagram.
(c) Estimate the coordinates of the turning point from the graph.
(d) Measure the difference between the x-coordinate of the turning point and each root.

SOLUTION

(a) Calculate the y-values given x, ranging from $x = -2$ to 6. The results are given in Table 4.3 and sketched in Figure 4.7.

Table 4.3   Calculation of points for $y = 2x^2 - 7x - 9$

x	y
−2	13
−1	0
0	−9
1	−14
2	−15
3	−12
4	−5
5	6
6	21

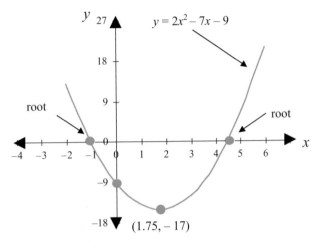

Figure 4.7   Graph for Worked Example 4.5

(b) The points of intersection with the axes are calculated as follows:
   • The graph cuts the y-axis when $x = 0$, that is, at $y = -9$.
   • The graph cuts the x-axis at $y = 0$, that is, at $2x^2 - 7x - 9 = 0$.
   See the quadratic equation whose roots are $x = -1$ and $x = 4.5$.
   See Figure 4.6.
(c) From the graph the minimum point is at $x = 1.75$, $y = -17$ approximately. These points can be found exactly using differentiation in Chapter 6.
(d) The two roots are an equal distance, 2.75, on either side of the vertical line drawn through the minimum point.

Summary to date

Quadratics of the form: $y = ax^2 + bx + c$:

• Are continuous curves (no breaks).
• Have only one turning point. If a is positive, the turning point is a minimum; if a is negative, the turning point is a maximum (turning points are estimated graphically in this chapter, but maximum and minimum points will be calculated in Chapter 6).
• Are symmetrical about the vertical line drawn through the turning point.
• Have y-intercept $= c$.
• Are wider than the $y = x^2$ graph if $a < 1$, and narrower than the $y = x^2$ graph if $a > 1$.
• Have roots of the quadratic that are the solutions of $ax^2 + bx + c = 0$; graphically the roots are the points where the graph crosses the x-axis. The roots are symmetrical about the vertical line drawn through the turning point.

PROGRESS EXERCISES 4.2

Graphs and Solutions of Quadratic Equations

1. Graph each of the following functions on the same diagram over the range -2 to 2.
   (a) $y = x^2$ (b) $y = -3x^2$ (c) $y = 0.5x^2$

2. Graph each of the following functions on the same diagram.
   (a) $P = (Q + 2)^2$ (b) $P = (Q - 1)^2$ (c) $P = Q^2 + 1$
   What is the relationship between each graph (a), (b) and (c) to $P = Q^2$?

3. The following functions were given in Worked Example 4.2.
   (a) $y = x^2 + 6x + 5$ (b) $y = x^2 + 6x + 9$ (c) $y = x^2 + 6x + 10$
   (i) Plot each graph for $-8 < x < 2$, indicating
   - The points of intersection with the axes.
   - The roots of the quadratic.
   - Turning points.
   (ii) Calculate the points of intersection with the axes algebraically.
   (iii) What is the relationship between graphs (a), (b) and (c)?

4. Plot the graph of $P = Q^2$.
   (a) Sketch the graph of $P = Q^2$, when translated vertically downwards by 9 units.
   (b) Write down the equation of the translated graph. Calculate the points of intersection with the axes.

5. (a) Plot the graph of the quadratic $y = x(10 - x)$. Mark in the points of intersection with the axes and estimate the turning points.
   (b) Calculate the points of intersection with the axes.

6. Given the following quadratic functions,
   (a) $P = -Q^2$ (b) $P = -Q^2 + 4$ (c) $P = -(Q - 3)^2 + 4$
   (i) Plot each graph for $-3 < x < 6$, indicating
   - The points of intersection with the axes.
   - The roots of the quadratic.
   - Turning points.
   (ii) Calculate the points of intersection with the axes algebraically.
   (iii) What is the relationship between graphs (a), (b) and (c)?

7. (a) Sketch three possible graphs of a quadratic function which has roots at $x = 2$ and $x = 6$.
   (b) Write down an expression for any quadratic equation which has roots at $x = 2$ and $x = 6$. What characteristic of the quadratic equation determines whether the graph is a minimum type or a maximum type?

4.1.3 Quadratic functions in economics

Non-linear supply and demand functions

In Chapter 3, market equilibrium was calculated for linear demand and supply functions. Now, market equilibrium is calculated for non-linear demand and supply functions. This is demonstrated in Worked Example 4.7.

WORKED EXAMPLE 4.7
NON-LINEAR DEMAND AND SUPPLY FUNCTIONS

The supply and demand functions for a particular market are given by the equations:

$$P_s = Q^2 + 6Q + 9 \quad \text{and} \quad P_d = Q^2 - 10Q + 25$$

(a) Sketch the graph of each function over the interval $Q = 0$ to $Q = 5$.
(b) Find the equilibrium price and quantity graphically and algebraically.

SOLUTION

(a) Calculate the P values for $Q = 0$ to 5. The results are given in Table 4.4 and sketched in Figure 4.8. Note that the vertical and horizontal intercepts of the demand function are $P = 25$ (when $Q = 0$) and $Q = 5$ (when $P = 0$), respectively. The vertical intercept of the supply function is $P = 9$ (when $Q = 0$).

Table 4.4 Points for $P_s = Q^2 + 6Q + 9$ and $P_d = Q^2 - 10Q + 25$

Q	$P_s = Q^2 + 6Q + 9$	$P_d = Q^2 - 10Q + 25$
0	9	25
1	16	16
2	25	9
3	36	4
4	49	1
5	65	0

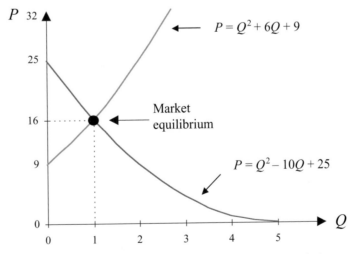

Figure 4.8   Market equilibrium with non-linear demand and supply functions

(b) From the graph, market equilibrium is at the point of intersection of the two functions, estimated at $Q = 1, P = 16$. This solution may be found exactly using algebra. Algebraically, market equilibrium exists when $P_d = P_s$ and $Q_d = Q_s$. In this case, it is easiest to equate the prices, so at equilibrium,

$$P_s = P_d$$
$$Q^2 + 6Q + 9 = Q^2 - 10Q + 25$$
$$Q^2 + 6Q - Q^2 + 10Q = 25 - 9$$
$$16Q = 16$$
$$Q = 1$$
$$\text{when } Q = 1, \quad P = 16 \quad \text{Check!}$$

Total revenue for a profit-maximising monopolist

Total revenue functions are frequently represented by maximum type quadratics which pass through the origin.

Remember

Equation (4.1) outlined a non-linear total revenue function as $TR = 50Q - 2Q^2$.

WORKED EXAMPLE 4.8
NON-LINEAR TOTAL REVENUE FUNCTION

The demand function for a monopolist is given by the equation $P = 50 - 2Q$.

(a) Write down the equation of the total revenue function.
(b) Graph the total revenue function for $0 \leq Q \leq 30$.
(c) Estimate the value of Q at which total revenue is a maximum and estimate the value of maximum total revenue.

SOLUTION

(a) Since $P = 50 - 2Q$, and $TR = P \times Q$, then $TR = (50 - 2Q)Q = 50Q - 2Q^2$.
(b) Calculate a table of values for $0 \leq Q \leq 30$ such as those in Table 4.5. The graph is plotted in Figure 4.9.

Table 4.5   Points for
$TR = 50Q - 2Q^2$

Q	TR
0	0
5	200
10	300
15	300
20	200
25	0
30	−300

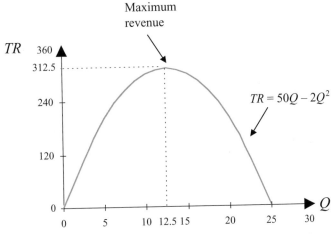

Figure 4.9   Non-linear total revenue function

(c) A property of quadratic functions is that the turning point (in this case a maximum) lies halfway between the roots (solutions) of the quadratic function. The roots of the TR function are

$$TR = 50Q - 2Q^2$$
$$0 = 50Q - 2Q^2$$
$$0 = Q(50 - 2Q)$$
$$\downarrow \qquad \downarrow$$
$$Q = 0 \qquad \downarrow$$
$$2Q = 50$$
$$Q = 25$$

These roots are illustrated graphically as the points where the TR function intersects the x-axis. The turning point occurs halfway between these points, that is, at $Q = 12.5$.

Substitute $Q = 12.5$ into the TR function and calculate the maximum total revenue as:

$$TR = 50Q - 2Q^2$$
$$= 50(12.5) - 2(12.5)^2$$
$$= 625 - 312.5 = 312.5$$

WORKED EXAMPLE 4.9
CALCULATING BREAK-EVEN POINTS

 Find animated worked examples at **www.wiley.com/college/bradley**

The demand function for a good is given as $Q = 65 - 5P$. Fixed costs are £30 and each unit produced costs an additional £2.

(a) Write down the equations for total revenue and total costs in terms of Q.
(b) Find the break-even point(s) algebraically.
(c) Graph total revenue and total costs on the same diagram; hence, estimate the break-even point(s).

SOLUTION

(a) $TR = P \times Q$. Therefore, if P is written in terms of Q, then TR will also be expressed in terms of Q. The expression for P in terms of Q is obtained from the equation of the demand function

$$Q = 65 - 5P$$
$$5P = 65 - Q$$
$$P = \frac{65 - Q}{5} = 13 - 0.2Q$$

Substitute the expression for P (price per unit) into the equation, $TR = P \times Q$; therefore, $TR = (13 - 0.2Q)Q = 13Q - 0.2Q^2$. Total cost is given as $TC = FC + VC = 30 + 2Q$.

(b) The break-even points occur when $TR = TC$, therefore,

$$13Q - 0.2Q^2 = 30 + 2Q$$
$$0 = 0.2Q^2 - 11Q + 30$$

The reader is expected to solve the quadratic equation for Q correct to one decimal place. The solutions are $Q = 2.91$ and $Q = 52.1$.

(c) A table of values for TR and TC from $Q = 0$ to $Q = 70$ is given in Table 4.6. These points are plotted in Figure 4.10. The break-even points occur at the intersection of the two functions. The break-even points on the graph agree with those calculated in (b).

Table 4.6   Total revenue and total cost

Q	TR	TC
0	0	30
10	110	50
20	180	70
30	210	90
40	200	110
50	150	130
60	60	150
70	−70	170

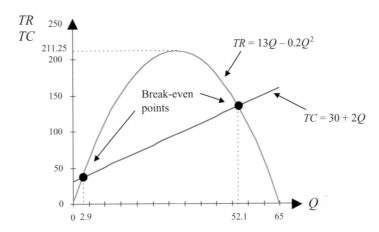

Figure 4.10   Total revenue and total cost: break-even points

PROGRESS EXERCISES 4.3

Quadratic Functions: Applications

1. Find the economically meaningful equilibrium price and quantity (a) graphically and (b) algebraically when the supply and demand functions are given by the equations

$$P = Q^2 - 0.5 \quad \text{and} \quad P = -Q^2 + 4$$

2. The demand function for ringside seats at a particular boxing match is given as $P = 12 - Q$, where P is in £000s.
   (a) Write an expression for TR in terms of Q.
   (b) Graph the TR function, indicating the points of intersection with the axes and turning point. Describe, in words, the meaning of these points.
   (c) Confirm the points of intersection with the axes algebraically.

3. Sketch the quadratic total revenue function which has a maximum value of 1000 at $Q = 20$, and which is zero at $Q = 0$. Hence, locate the second point where $TR = 0$ and find the equation of the total revenue function.

4. Given the demand function of a monopolist as $P = 100 - 2Q$,
   (a) Write down the equation for TR in the form $TR = f(Q)$. Calculate TR when $Q = 10$.
   (b) Write down the equation for TR in the form $TR = f(P)$. Calculate TR when $P = 10$. Comment on your answers in (a) and (b).

5. Sketch the supply function $P_s = 0.5Q^2 + 5$ and the demand function $P_d = 155 - Q^2$ on the same diagram.
   (a) Estimate the equilibrium price and quantity graphically.
   (b) Calculate the equilibrium price and quantity algebraically.

6. A firm's total cost function is given by the equation $TC = 200 + 3Q$, while the demand function is given by the equation $P = 107 - 2Q$.
   (a) Write down the equation of the total revenue function.
   (b) Graph the total revenue function for $0 < Q < 60$. Hence, estimate the output, Q, and total revenue when total revenue is a maximum.
   (c) Plot the total cost function on the diagram in (b). Estimate the break-even point from the graph. Confirm your answer algebraically.
   (d) State the range of values for Q for which the company makes a profit.

7. (See question 6)
   Write down the equation for the profit function (Profit $= TR - TC$). Graph the profit function for $0 < Q < 100$. From the graph estimate:
   (a) The values of Q at which profit is zero.
   (b) Maximum profit and the output, Q, at which profit is a maximum.

8. The demand and supply functions for a good are given by the equations

$$P_d = -(Q+4)^2 + 100, \quad P_s = (Q+2)^2$$

   (a) Sketch each function on the same diagram; hence, estimate the equilibrium price and quantity.
   (b) Confirm the equilibrium using algebra.

9. The demand function for a good is given by the equation $P = 2400 - 8Q$ where P is the price per unit and Q is the number of units sold.
   (a) Total revenue is calculated by multiplying the number of units sold by the price per unit. Write down the equation for the total revenue.
   (b) Find the values of Q at which the total revenue is zero.
   (c) Sketch the total revenue function over the range $Q = -2$ to $Q = 35$. Mark the answer to (b) on the graph.
   (d) From the graph write down the maximum value of the total revenue and the value of Q at which this maximum occurs.

10. An average cost function is given by the equation $AC = 1200 - 128Q + 4Q^2$.
   (a) Find the values of Q at which average cost is zero. Hence explain why this AC function is always positive.
   (b) Sketch the average cost function over the range $Q = 0$ to $Q = 35$.
   (c) From the graph write down the minimum value of the average cost and the value of Q at which this minimum occurs.

11. A total cost function is given by the equation $TC = 800 - 120Q + 5Q^2$.
    (a) Find the values of Q at which the total cost is zero.
    (b) Sketch the average total function over the range $Q = 0$ to $Q = 20$.
    (c) From the graph write down the minimum value of the total cost and the value of Q at which this minimum occurs.
12. (See questions 9 and 11).
    Write down the equation for the profit for the total revenue function in question 9 and the total cost function in question 11.
    (a) Find the values of Q at which the profit is zero.
    (b) Sketch the profit function over the range $Q = 0$ to $Q = 50$.
    (c) From the graph write down the minimum value of the profit and the value of Q at which this minimum occurs.

4.1.4   Cubic functions

A cubic function is expressed by a cubic equation which has the general form

$$ax^3 + bx^2 + cx + d = 0 \quad \text{where } a, b, c \text{ and } d \text{ are constants}$$

There is, unfortunately, no easy general formula for solving a cubic equation, but approximate solutions may be found graphically by plotting the graph and finding the points of intersection with the axes, as in Worked Example 4.10a. In this section, in particular, calculating points by graph plotting is much easier and less time consuming if Excel or other spreadsheets are used.

WORKED EXAMPLE 4.10a
PLOTTING CUBIC FUNCTIONS

Plot the graphs of (a) $y = x^3$ and (b) $y = -x^3$ on the same diagram. From the graph estimate the turning points and the roots of the cubic equation. Compare the graphs of (a) and (b).

SOLUTION
The graphs are plotted in Figure 4.11 from the points calculated in Table 4.7. Figure 4.11 illustrates that the graphs of (a) and (b):

- Have only one root at $x = 0$.
- Have no turning points.
  Graph (a) may be viewed as a reflection of graph (b) in the x-axis or y-axis.

Table 4.7 Calculation of points for graphs of (a) $y = x^3$ and (b) $y = -x^3$

x	$y = x^3$	$y = -x^3$
-3	-27	27
-2	-8	8
-1	-1	1
0	0	0
1	1	-1
2	8	-8
3	27	-27

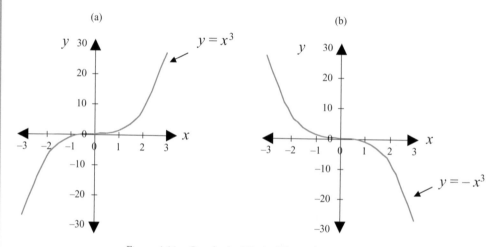

Figure 4.11   Graphs for Worked Example 4.10a

In Worked Example 4.10a, each graph had no turning point and only one root. However, the graphs of more general cubic equations (as in Worked Example 4.10b) will have:

(a) No turning points or two turning points.
(b) Either one root or three roots.

WORKED EXAMPLE 4.10b
GRAPHS OF MORE GENERAL CUBIC FUNCTIONS

Plot the graphs of the following:

(a) $y = 0.5x^3 - 5x^2 + 8.5x + 27$ (b) $y = -0.5x^3 - 5x^2 + 8.5x + 27$.

From the graphs estimate:

(i) The roots.
(ii) The turning points. Compare the graphs of (a) and (b).

SOLUTION

First set up a table of values from which the graphs may be plotted. (See Table 4.8 and Figures 4.12a and 4.12b.) The roots and the turning points are summarised as follows:

(a) has only one root, $x = -1.56$ approximately

(a) has two turning points at $x = 1.0$, $y = 31$ and at $x = 5.5$, $y = 5.6$

(b) has three roots, at $x = -11$, $x = -1.7$ and $x = 2.8$ approximately

(a) has two turning points at $x = -7$, $y = -107$ and at $x = 0.8$, $y = 30$

Table 4.8   Points for plotting graphs in Worked Example 4.10b

x	Graph (a)	Graph (b)
−12	−1659	69
−8	−617	−105
−4	−119	−55
0	27	27
4	13	−51
8	31	−481
2	273	−1455

(a)

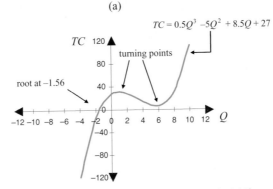

Figure 4.12a   Graph (a) for Worked Example 4.10b

(b)

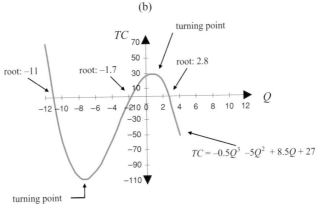

Figure 4.12b   Graph (b) for Worked Example 4.10b

Note: From an economic perspective, only the first quadrant ($x > 0, y > 0$) in Figures 4.12a and 4.12b is of interest. However, for a better understanding of mathematics, it is necessary to analyse the characteristics of the graphs in the other quadrants.

General properties of cubic equations

Cubic equations are continuous curves, which may have:

- One root to three roots.
- No turning point or two turning points.

These properties are illustrated in Figures 4.11, 4.12a and 4.12b.

Polynomials

Quadratic and cubic functions belong to a group of functions called polynomials:

$$f(x) = a_n x^n + a_{n-1} x^{n-1} + a_{n-2} x^{n-2} + \cdots + a_1 x + a_0 \quad a_0, a_1, \ldots, a_n \text{ are constants}$$

The 'degree' of the polynomial is given by the highest power of x in the expression. Therefore, a quadratic is a polynomial of degree 2 and a cubic is a polynomial of degree 3.

We are not going to study polynomials in depth, but you do need to be aware that they are continuous graphs, with no breaks or jumps. Therefore, it is quite safe to plot these graphs by joining the points as calculated and hence use the graphs to estimate roots and turning points. Now for one final worked example in which costs are given by a cubic equation.

WORKED EXAMPLE 4.11
TR, TC AND PROFIT FUNCTIONS

A firm's total cost function is given by the equation, $TC = Q^3$. The demand function for the good is $P = 90 - Q$.

(a) Write down the equations for total revenue and profit. Calculate the break-even points.
(b) Graph the total revenue and total cost functions on the same diagram for $0 \le Q \le 12$, showing the break-even points.
(c) Estimate the total revenue and total costs at break-even.
(d) From the graph estimate the values of Q within which the firm makes (i) a profit, (ii) a loss.
(e) From the graph estimate the maximum profit and the level of output for which profit is maximised.

SOLUTION

(a) $TR = PQ = (90 - Q)Q = 90Q - Q^2$.

Profit $(\pi) = TR - TC = 90Q - Q^2 - Q^3$

The break-even points occur when $TR = TC$, therefore, solve

$$90Q - Q^2 = Q^3$$
$$0 = Q^3 + Q^2 - 90Q$$
$$0 = Q(Q^2 + Q - 90)$$

Therefore, $Q = 0$ and solving the quadratic $Q^2 + Q - 90 = 0$ gives the solutions, $Q = 9$ and $Q = -10$.

(b) To graph TR and TC, calculate a table of values for the specified range using Excel or otherwise. (See Table 4.9 and Figure 4.13.)

(c) From the graph, break-even is at $Q = 0$ and $Q = 9$. (Break-even at $Q = -10$ does not lie within the range of the graph. In fact, $Q = -10$ is not economically meaningful.) At $Q = 0$, $TR = TC = 0$. At $Q = 9$, $TR = TC = 729$.

(d) The firm makes a profit when the TR curve is greater than the TC curve, that is, between $Q = 0$ and $Q = 9$. When Q is greater than 9 the firm makes a loss.

(e) Maximum profit occurs at the point where the vertical distance between TR and TC is greatest. From the graph, this estimate is at $Q = 5.15$. Substitute $Q = 5.15$ into the profit function to calculate profit $= 90(5.15) - (5.15)^2 - (5.15)^3 = 300.4$.

Table 4.9 TR and TC for Worked Example 4.11

Q	TR	TC
0	0	0
3	261	27
6	504	216
9	729	729
12	936	1728

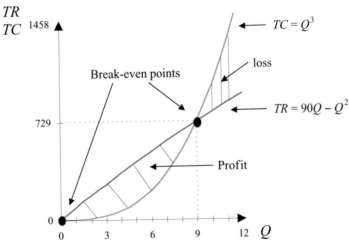

Figure 4.13   Quadratic TR and cubic TC functions

PROGRESS EXERCISES 4.4

Graphical Solution of Cubic Equations

Graph the cubic functions in questions 1–4. From the graph, estimate the roots of each function and estimate the coordinates of the turning point(s).

1. $TC = 0.5Q^3 - 8Q^2 + 200$ over the interval $-5 < Q < 15$.
2. $Q = 3P^3 + 9P^2$ over the interval $-4 < Q < 2$.
3. $P = -4Q^3 + 2Q^2$ over the interval $-0.5 < Q < 0.8$.
4. $TC = 0.5Q^3 - 15Q^2 + 175Q + 1000$ over the interval $-10 < Q < 30$.

4.2 Exponential Functions

At the end of this section you should be able to:

- Define and graph exponential functions
- Simplify and solve exponential equations
- Use exponential functions in various applications.

4.2.1   Definition and graphs of exponential functions

The exponential function has the general form $y = a^x$ or $f(x) = a^x$ where:

- a is a constant and is referred to as the base of the exponential function.
- x is called the index or power of the exponential function; this is the variable part of the function.

For example, assuming $a = 4$, evaluate 4^x for values of $x = 0, 1, 2, 3$. Here 4 is the constant base of the exponential while x is the variable index or power.
On your calculator, exponentials are evaluated using the $[x^y]$ or $[x^{1/y}]$ keys. Using your calculator, your answers should be

$$4^0 = 1 \quad 4^1 = 4 \quad 4^2 = 16 \quad 4^3 = 64$$

The number e

The letter e represents a number which has an unending decimal part, just like the number π. The number $\pi = 22/7 = 3.141\,592\,7\ldots$ arose naturally in circular measurements; the length of the circumference of a circle was shown to be 2π multiplied by the radius. Since the number π is an unending decimal, it is represented by the letter π. (Don't confuse the number π as used here with the symbol for profit, $\pi = TR - TC$.)

The number e arises when growth and decay in all types of systems are described mathematically, for example, growth in populations, growth in investments, growth of current in electrical circuits, decay of radioactive materials. Like π, it is represented by a letter, therefore the user can evaluate it to any required number of decimal places; e is also a natural number in mathematics. As shown later, e^x is the only function which is not changed by differentiation or integration. To evaluate e, find e^1 on the calculator. You should get $e = 2.718\,281\,8$. Correct to two decimal places $e = 2.72$, correct to three decimal places $e = 2.718$, and so on. It takes a bit of practice to get used to treating e as a number, so working through Worked Example 4.13 and Progress Exercises 4.5 should help.

Note: $f(x) = e^x$ is often referred to as the natural exponential function to distinguish it from $f(x) = a^x$, the general exponential function.

Graphs of exponential functions

To find out more about the properties of the exponential function, the analysis begins by looking at its graph.

WORKED EXAMPLE 4.12
GRAPHING EXPONENTIAL FUNCTIONS

(a) Two exponential functions are given by the equations

$$y = (2)^x \tag{4.7}$$

$$y = (2)^{-x} \tag{4.8}$$

Plot these graphs on the same diagram.
(b) Two further functions are given by the equations

$$y = (3.5)^x \tag{4.9}$$

$$y = (e)^x \tag{4.10}$$

Plot the graphs of these functions on the same diagram as the graph $y = (2)^x$ from part (a).

SOLUTION

(a) The function $y = (2)^x$ has base 2 and variable power x. The function $y = (2)^{-x}$ has base 2 and variable power $-x$. To plot these functions, calculate a table of points such as Table 4.10. Using your calculator, confirm that the values in Table 4.10 are correct. These points are then plotted in Figure 4.14.

Table 4.10   Points for the functions $y = 2^x$ and $y = 2^{-x}$

x	$y = 2^x$	$y = 2^{-x}$
-2	0.25	4
-1	0.50	2
0	1	1
1	2	0.50
2	4	0.25

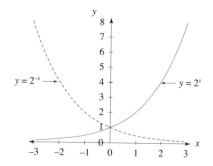

Figure 4.14   Graph for Table 4.10

(b) The function $y = (3.5)^x$ has base 3.5 and variable power x, while the function $y = e^x$ has base e and variable power x. To plot these functions, calculate a table of points as in Table 4.11. These points are plotted in Figure 4.15. Also plotted in Figure 4.15 is the graph for part (a). It is now possible to compare the graphs of the three functions.

Table 4.11   Points for the functions $y = (3.5)^x$ and $y = e^x$

x	$y = (3.5)^x$	$y = e^x$
−2	0.08	0.14
−1	0.29	0.37
0	1	1
1	3.5	2.72
2	12.25	7.29

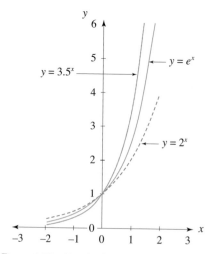

Figure 4.15   Graphs for Tables 4.10 and 4.11

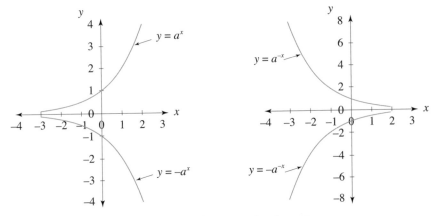

Figure 4.16   Various graphs of $y = a^x$

Properties of exponential functions

The properties of exponential functions $y = a^x$ or $f(x) = a^x$ are now summarised. Here are some features to note from Figures 4.14 to 4.16:

- All curves are continuous (no breaks), and pass through the point $x = 0, y = 1$.
- When the index is positive, the curves are increasing as x increases, provided $a > 1$; these are called **growth curves**.
- When the index is negative, the curves are decreasing as x increases, provided $a > 1$; these are called **decay curves**.
- Exponentials with larger bases increase more rapidly for $x > 0$ and decrease more rapidly for $x < 0$. Compare equations (4.7), (4.9) and (4.10) as graphed in Figure 4.15.
- The graphs of $y = a^x$ and $y = a^{-x}$ are reflections of $y = a^x$ and $y = -a^x$ through the x-axis, hence are always above the x-axis, therefore y is always positive (Figure 4.14). You can check this by evaluating several exponentials on the calculator.
- The graphs of $y = -a^x$ and $y = -a^x$ are reflections of $y = a^x$ and $y = -a^x$ through the x-axis, hence are always below the x-axis, as shown in Figure 4.16.

Remember

That multiplying the RHS of $y = f(x)$ by -1 simply inverts (flips) the graph of the function through the x-axis.

- The x-axis is a horizontal asymptote. An asymptote is a straight line which the graph approaches as x increases. The graphs of the exponentials level off towards the x-axis, but in theory never touch it.

Rules for using exponential functions

Table 4.12 gives the three basic rules for indices: these are the rules for multiplying, dividing exponential functions and raising an exponential function to a power.

Table 4.12   The rules for indices

Statement	Equation	Example illustrating rule
Rule 1 To multiply numbers with the same base, add the indices	$a^m \times a^n \Leftrightarrow a^{m+n}$	By rule 1: $5^2 \times 5^3 = 5^{2+3} = 5^5$ Directly $(5 \times 5) \times (5 \times 5 \times 5) = 5^5$
Rule 2 To divide numbers with the same base, subtract the indices	$\dfrac{a^m}{a^n} \Leftrightarrow a^{m-n}$	By rule 2: $\dfrac{3^4}{3^2} = 3^{4-2} = 3^2$ Directly $\dfrac{3 \times 3 \times 3 \times 3}{3 \times 3} = 3 \times 3 = 3^2$
Rule 3 To raise an exponential to a power, multiply the indices	$(a^m)^k \Leftrightarrow a^{m \times k}$	By rule 3: $(2^3)^2 = 2^{3 \times 2} = 2^6$ Directly $(2 \times 2 \times 2)(2 \times 2 \times 2) = 2^6$

The $\Leftrightarrow$ sign means that the equations can be operated both ways

Notes

1. Anything to the power of zero is one: (any real number)$^0 = 1$.
2. If an exponential in a fraction is moved from below the line to above the line or vice versa, the sign on the index changes:

$$\frac{1}{a^n} = \frac{a^0}{a^n} = a^{0-n} = a^{-n}$$

For example,

$$\frac{1}{3^2} = \frac{3^0}{3^2} = 3^{0-2} = 3^{-2}$$

3. To write roots as indices, everything under the root sign is raised to the power of 1/root: $\sqrt[n]{a} = a^{1/n}$. For example, $\sqrt[3]{9} = 9^{1/3}$.

Also

1. If any product or fraction is raised to a power, the index on each term is multiplied by the power: $(a^x b^y / c^z)^n = a^{nx} b^{ny} / c^{nz}$. For example, $(2a^2/c^5)^2 = 2^2 a^4 / c^{10}$.
2. The above rules apply to products and quotients only. There are no rules for simplifying sums or differences of exponential functions. These rules cannot be used to simplify or reduce $2^x + 2^y$. For example,

$$2^3 + 2^2 \neq 2^5 \qquad 8 + 4 \neq 32$$

WORKED EXAMPLE 4.13
SIMPLIFYING EXPONENTIAL EXPRESSIONS

Simplify each of the following expressions, giving your answers with positive indices only:

(a) $\dfrac{2^5}{2^3 2^{-4}}$ (b) $QQ^{1.5}P$ (c) $\dfrac{\sqrt{3^5}}{3^{-4}3^4}$ (d) $\dfrac{15e^\alpha e^\beta}{e^2}$ (e) $\left(\dfrac{3L^{0.5}}{L^{-2}}\right)^2$ (f) $\left(\dfrac{3e^{0.5}}{e^{-2}}\right)^2$

(g) $\dfrac{p^5 q - p^4 q^2}{p^5 - p^4 q}$ (h) $\dfrac{4e^{\alpha+2\beta} - 2e^\alpha}{2e^\alpha}$

SOLUTION

(a) $\dfrac{2^5}{2^3 2^{-4}} = \dfrac{2^5}{2^{3-4}}$ use rule 1 to add the indices of the base 2 terms below the line

$= \dfrac{2^5}{2^{-1}}$ simplify

$= 2^{5-(-1)}$ use rule 2 and bracket the negative terms

$= 2^{5+1} = 2^6$ use the calculator to evaluate exponentials

$= 64$

(b) $QQ^{1.5}P = Q^1 Q^{1.5} P$ since no index is given on the first Q, it is understood to be 1

$= Q^{1+1.5}P$

$= Q^{2.5}P$

(c) $\dfrac{\sqrt{3^5}}{3^{-4}3^4} = \dfrac{3^{5/2}}{3^{4-4}}$ start by writing $\sqrt{3^5} = (3^5)^{1/2} = (3^{5/1})^{1/2}$

$= \dfrac{3^{5/2}}{3^0}$ $= 3^{5/1 \times 1/2} = 3^{5/2}$

$= 3^{5/2}$ since $3^0 = 1$

$= 15.588$

(d) $\dfrac{15e^\alpha e^\beta}{e^2} = \dfrac{15e^{\alpha+\beta}}{e^2}$ multiplying numbers with same base (rule 1)

$= \dfrac{15e^{\alpha+\beta}e^{-2}}{1}$ bringing up the number from below the line (note 2)

$= 15e^{\alpha+\beta-2}$ multiplying numbers with same base (rule 1)

(e) $\left(\dfrac{3L^{0.5}}{L^{-2}}\right)^2 = \left(\dfrac{3L^{0.5-(-2)}}{1}\right)^2$ simplify the terms inside the bracket first

$= \left(\dfrac{3L^{0.5+2}}{1}\right)^2$

$= \left(\dfrac{3L^{2.5}}{1}\right)^2$

$= \left(\dfrac{3L^{2.5}}{1}\right)\left(\dfrac{3L^{2.5}}{1}\right)$ square out

$= \left(\dfrac{3^2 L^{2.5+2.5}}{1}\right)$

$= 9L^5$

There are usually several different approaches to simplifying; for example, in part (e) we could square out the brackets first then simplify like this:

$$\left(\frac{3^1 L^{0.5}}{L^{-2}}\right)^2 = \frac{3^{1\times2} L^{0.5\times2}}{9L\,L^{-2\times2}}$$

$$= \frac{9L^{-4}}{L^{-4}}$$

square first by multiplying the index on every term (including the 3) by the power 2 outside the bracket

$$= \frac{9L^{1-(-4)}}{1}$$

$$= 9L^5$$

(f) $\left(\dfrac{3e^{0.5}}{e^{-2}}\right)^2 = \left(\dfrac{3e^{0.5}}{e^{-2}}\right)\left(\dfrac{3e^{0.5}}{e^{-2}}\right)$ squaring

$$= \frac{3^2(e^{0.5})^2}{(e^{-2})^2}$$

$$= \frac{3^2 e^1}{e^{-4}}$$ multiplying the indices (rule 3)

$$= 9e^{1-(-4)}$$ subtracting the index of the divisor (rule 2)

$$= 9e^5 = 1335.7184$$ simplifying, then evaluating correct to 4 decimal places

(g) The rules for indices do not apply to sums or differences of the exponential functions. But this expression may be simplified by factorisation, as follows:

$$\frac{p^5 q - p^4 q^2}{p^5 - p^4 q} = \frac{p^4 pq - p^4 qq}{p^4 p - p^4 q}$$ common factors $p^4 q$ in the numerator and p^4 in the denominator

$$= \frac{p^4(p-q)q}{p^4(p-q)}$$ factor out the common factors

$$= \frac{\cancel{p^4}\cancel{(p-q)}q}{\cancel{p^4}\cancel{(p-q)}} = q$$ cancel the factors that are identical above and below the line

(h) $\dfrac{4e^{\alpha+2\beta} - 2e^\alpha}{2e^\alpha} = \dfrac{4e^\alpha e^{2\beta} - 2e^\alpha}{2e^\alpha}$ rule 1: $e^{\alpha+2\beta} = e^\alpha e^{2\beta}$

$$= \frac{2e^\alpha(2e^{2\beta} - 1)}{2e^\alpha}$$ common factor $2e^\alpha$ in the numerator

$$= 2e^{2\beta} - 1$$ cancel the factor that is common in the numerator and denominator

Alternatively

$$\frac{4e^{\alpha+2\beta} - 2e^\alpha}{2e^\alpha} = \frac{4e^{\alpha+2\beta}}{2e^\alpha} - \frac{2e^\alpha}{2e^\alpha} = \frac{4e^{\alpha+2\beta-\alpha}}{2} - \frac{2e^\alpha}{2e^\alpha} = 2e^{2\beta} - 1$$

NON-LINEAR FUNCTIONS AND APPLICATIONS

PROGRESS EXERCISES 4.5

Rules for Indices

1. Use your calculator to evaluate the following:

   (a) 6^2 (b) 3^3 (c) 5^1 (d) 5^3 (e) $(-3)^2$ (f) $(-4)^2$

   (g) 25^0 (h) 5^{-1} (i) 6^{-2} (j) 5^{-3} (k) $(2.5)^{0.5}$ (l) $(1.5)^{-5}$

2. Simplify the following expressions, giving your answer with positive indices only:

   (a) $\dfrac{2^4 2^{-3}}{2^2}$ (b) $\dfrac{(4)3^5 3^{0.8}}{3^4 3^{-2.3}}$

3. (a) $\dfrac{1}{2^3}\dfrac{3^2 5^{-5}}{5^2 3^{-3}}$ (b) $\dfrac{2^x 2^y}{2^x}$

4. (a) $\dfrac{a^x a^y}{a^x}$ (b) $\dfrac{5^x 2^y}{2^x}$

5. (a) $\dfrac{L^\alpha K^\beta}{L}$ (b) $\left(\dfrac{4L^2}{L^{-2}}\right)^2$

6. (a) $\dfrac{4L^2 K^3}{L^{-2}}$ (b) $\dfrac{2L}{(1/L)}$ (c) $5\dfrac{L^2}{(5LK)^{0.5}}$

7. $\dfrac{\left(\dfrac{4(0.6)K^{0.4}L^{-0.4}}{L}\right)}{\left(\dfrac{4(0.4)K^{-0.6}L^{0.6}}{K}\right)}$ 8. $\dfrac{\left(\dfrac{4L^\alpha K^{\beta-1}}{K}\right)}{\left(\dfrac{4L^{\alpha-1}K^\beta}{L}\right)}$

9. (a) $2Q\sqrt{Q}$ (b) $\dfrac{2Q}{3P\sqrt{P}}$

10. (a) $\dfrac{2L^{0.25}K^{0.75}}{4L}$ (b) $\left(\dfrac{2L^{-1}}{4L}\right)^2$

11. $e^2 e^x$ 12. $5e^{5x}e^x$ 13. $\dfrac{e^{2x+3}}{e^{5x-3}}$ 14. $120e^{-0.5t}(1/e^t)$ 15. $e^{0.75t}(1-e^t)$

16. $\left(\dfrac{e^{2t}}{5}\right)^2$ 17. $\left(\dfrac{1}{e^{-5t}}\right)^{0.25}$ 18. $(1+e^t)^2$ 19. $5(e^{2t}+e^{3t})$ 20. $e^{-5t}(e^{2t}+e^{3t})$

21. $\dfrac{P}{Q}-\dfrac{P^2}{Q^{0.5}}$ 22. $\dfrac{x}{x^2\sqrt{2x}}$ 23. $\sqrt{\dfrac{4x^2}{y^{-4}}}$ 24. $\left(\dfrac{xy}{3y^2}\right)-\left(\dfrac{xy^2}{2y}\right)^2$

25. $(x^2-y)^2$ 26. $\dfrac{e^x(1+e^{1-x})}{e}$ 27. $\dfrac{e^{x+y}}{e^{x-y}}$ 28. $\dfrac{p^2q^2-2pq}{pq-2}$ 29. $\dfrac{4L^2K^4-L^2}{2L^2K^2}$

30. $\dfrac{K^6-K^2}{K^4-1}$

4.2.2 Solving equations that contain exponentials

Worked Example 4.14 demonstrates the solution of various exponential equations.

WORKED EXAMPLE 4.14
SOLVING EXPONENTIAL EQUATIONS

 Find animated worked examples at **www.wiley.com/college/bradley**

Solve the equations

(a) $2^x = \dfrac{1}{16}$ (b) $5^{x+3}5^x = 5^4$ (c) $2^P 3^Q = \dfrac{2}{3}$ (d) $(1+K)^{0.4} = 2$ (e) $\dfrac{e^{2x}}{e^{4+x}} = 1$

SOLUTION

The method used for solving exponential equations is straightforward. As always:

- Simplify the exponential equation by writing each side of the equation as (base)^{power}.
- Since the bases on each side of the equation are identical, then if the LHS is to be equal to the RHS, the indices must also be identical, so equate the indices. For example, $5^x = 5^{2.5} \rightarrow x = 2.5$.

 Remember

The solution of an equation is the value of the variable for which the LHS = RHS.

Using these ideas, the exponential equations are solved as follows:

(a)	(b)	(c)	(d)
$2^x = \dfrac{1}{16}$	$5^{x+3}5^x = 5^4$	$2^P 3^Q = \dfrac{2}{3}$	$(1+K)^{0.4} = 2$
$2^x = \dfrac{1}{2^4}$	$5^{x+3+x} = 5^4$	$2^P 3^Q = 2^1 3^{-1}$	$((1+K)^{0.4})^{1/0.4} = 2^{1/0.4}$
$2^x = 2^{-4}$	$5^{2x+3} = 5^4$	equate indices for base 2	raise each side to power 1/0.4
equate indices	equate indices	$P = 1$	
$x = -4$	$2x + 3 = 4$	equate indices for base 3	$(1+K)^{0.4 \times 1/0.4} = 2^{2.5}$
	$2x = 4 - 3 = 1$	$Q = -1$	$1 + K = 5.6569$
	$x = \dfrac{1}{2} = 0.5$	check solutions	$K = 4.6569$
		$2^1 3^{-1} = 2^1 3^{-1}$	

(e)

$$\frac{e^{2x}}{e^{4+x}} = 1$$

$$e^{2x-(4+x)} = 1 \qquad \text{subtract the index of the divisor (rule 2) – careful with } -(4+x)$$

$$e^{x-4} = c^0 \qquad \text{simplify LHS, on RHS } 1 = e^0 \text{ (note 1)}$$

$$x - 4 = 0 \qquad \text{bases the same, so equate the indices}$$

$$x = 4$$

OR

$$\frac{e^{2x}}{e^{4+x}} = 1$$

$$e^{2x} = e^{4+x} \qquad \text{multiply both sides by } e^{4+x}$$

$$2x = 4 + x \qquad \text{equate the indices}$$

$$2x - x = 4 \qquad \text{therefore } x = 4$$

Note: In part (c) other solutions exist, but finding them is beyond the scope of this book. In part (d) the power on the bracket containing K must be 1, in order to gain access to K inside the bracket. The index on the bracket is multiplied by 1/index to obtain the required 1. But as for all equations, both sides must be operated on in the same way, so each side is raised to the power 1/index.

PROGRESS EXERCISES 4.6

Use the Rules for Indices to Solve Certain Equations

Solve the equations in questions 1 to 24 for the variable indicated in square brackets.

1. $2^x = \dfrac{1}{\sqrt{16}}[x]$

2. $a^{0.5Q} = a^4[Q]$

3. $\left(\dfrac{1}{P^3}\right)^4 = (P)^x[x]$

4. $\dfrac{2^x}{4} = 2[x]$

5. $3^{Q+2} = 9[Q]$

6. $125 = 5^x[x]$

7. $\left(\dfrac{1}{K}\right) = 8[K]$

8. $\dfrac{4}{K^{0.5}} = 8[K]$

9. $200 = \dfrac{10^x}{0.5}[x]$

10. $\sqrt[5]{t} = 2[t]$

11. $\dfrac{2}{5^t} = \dfrac{2}{1}[t]$

12. $(1+2t)^{0.5} = 12[t]$

13. $\sqrt{1+2t+t^2} = 4[t]$

14. $16 - \left(\dfrac{1}{8}\right)^t = 0[t]$

15. $16 - \left(\dfrac{1}{8}\right)^{-2t} = 0[t]$

16. $3^x 3^{x+1} = \sqrt{9}[x]$

17. $\sqrt{t-3} = 2[t]$

18. $\dfrac{2}{4^t} = \dfrac{1}{2}[t]$

19. $\dfrac{2}{K} + \dfrac{5}{K} = \dfrac{1}{2}[K]$

20. $\dfrac{4}{L} + L = -4[L]$

21. $e^x = e^5 e^{2.5}[x]$

22. $e^{5x} = \dfrac{1}{e^{x-5}}[x]$

23. $3e^{5t+3} = \dfrac{3}{e^t}[t]$

24. $5 + 2e^t = \dfrac{2}{e^5 e^{t+1}} + 5[t]$

4.2.3 Applications of exponential functions

The laws of growth

Exponential functions to base e describe growth and decay in a wide range of systems, as mentioned above. There are three main laws of growth: unlimited, limited and logistic growth. Each of these is now described.

Unlimited growth

Unlimited growth is modelled by the equation $y(t) = ae^{rt}$, where a and r are constants.
Examples: Investment and some models of population growth.

WORKED EXAMPLE 4.15
UNLIMITED GROWTH: POPULATION GROWTH

The population of a village was 753 in 2010. If the population grows according to the equation

$$P = 753e^{0.03t}$$

where P is the number of persons in the population at time t,

(a) Graph the population equation for $t = 0$ (in 2010) to $t = 30$ (in 2040). From the graph, estimate the population (i) in 2020, and (ii) in 2030.
(b) Confirm your answers algebraically.
(c) In what year will the population reach 1750 persons?

SOLUTION

(a) The general shape of the exponential is known; however, since the graph is being used for estimations, an accurate graph over the required interval, $t = 0$ to $t = 30$, is required. Using the e^x key on your calculator, calculate a table of values for different time periods. A sample of values is given in Table 4.13.

Table 4.13 Population values for different time periods

t	$P = 753e^{0.03t}$
0	753
5	874.9
10	1016.4
15	1180.9
20	1372.1
25	1594.1
30	1852.1

The points in Table 4.13 are used to plot the exponential function in Figure 4.17. From Figure 4.17 you can obtain the required population figures:
  (i) In 2020 ($t = 10$) the population is 1016 persons.
  (ii) In 2030 ($t = 20$) the population is 1372 persons.
(b) The values in (i) and (ii) can be confirmed algebraically as follows. In 2020 $t = 10$, therefore substitute $t = 10$ into equation (4.10) and evaluate P, that is, $P = 753e^{0.03t}$. $P = 753e^{0.03(10)} = 1016.44 = 1016$ since we cannot have 0.44 persons. In 2030, $t = 20$, therefore substitute $t = 20$ into equation (4.10) and evaluate P, that is, $P = 753e^{0.03t}$. Hence $P = 753e^{0.03(20)} = 1372.05 = 1372$ persons.
(c) In Figure 4.17 draw a horizontal line across from population = 1750. When this line cuts the graph, draw a vertical line down to the horizontal axis. The vertical line cuts the horizontal axis at $t = 28.1$. The year is $2010 + 28.1 = 2038.1$, or during 2038.

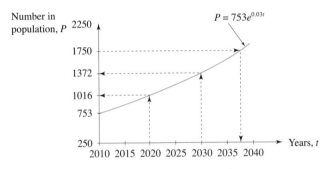

Figure 4.17   Population growth

We now attempt to solve for t algebraically. As P is known, substitute $P = 1750$ into equation (4.10) and solve for t:

$$P = 753e^{0.03t}$$

$$1750 = 753e^{0.03t}$$

$$\frac{1750}{753} = e^{0.03t}$$

$$t = ?$$

We cannot solve for t as t is part of the index. To solve for a variable when the variable is part of the index, logs needs to be used. Logarithmic functions are generally called logs. Logs are covered in Section 4.3.

Limited growth

Limited growth is modelled by the equation $y(t) = M(1 - e^{-rt})$, where M and r are constants.

Examples: consumption functions, amount of random information which can be memorised, sales with advertising (Progress Exercises 4.8, question 2), electrical and mechanical systems.

WORKED EXAMPLE 4.16
LIMITED GROWTH: CONSUMPTION AND CHANGES IN INCOME

A consumption function is modelled by the equation

$$C = 500(1 - e^{-0.3Y})$$

Graph the consumption function over the interval $0 < Y < 20$. Use the graph to describe how consumption changes as income increases.

SOLUTION

Using your calculator, calculate C for several values of Y between 0 and 20. This is illustrated in Table 4.14. These points are used to plot the graph in Figure 4.18, which relates consumption as a function of income. The shape of the graph indicates that as income increases, consumption increases at a decreasing rate towards an upper limit of $C = 500$.

Table 4.14   Consumption values for different income levels

Y	$C = 500(1 - e^{-0.3Y})$
0	0
4	349
8	455
12	486
16	496
20	499

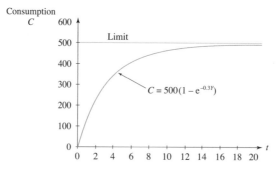

Figure 4.18   Consumption with limited growth

NON-LINEAR FUNCTIONS AND APPLICATIONS

Logistic growth

Logistic growth is modelled by the equation

$$y(t) = \frac{M}{1 + ae^{-rMt}}$$

where M, a and r are constants. Logistic growth is modelled in Figure 4.19.

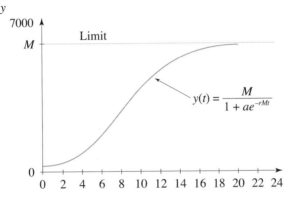

Figure 4.19 Logistic growth

Examples: consumption functions, constrained populations, growth of epidemics, sales. (Progress Exercises 4.8, questions 3.)

PROGRESS EXERCISES 4.7

Graphs of Exponential Functions: Limited and Unlimited Growth

Graph the functions in questions 1 to 6. For each graph from questions 1 to 4, describe how it differs from the graph for $y = e^t$. For questions 1 to 6 state whether the graph displays growth or decay and whether this is limited or unlimited.
   Use Excel to plot the graphs if available.

1. $2^t, e^t, 1^t, 0.5^t$ plot on the same diagram, use intervals of 0.5 for $-1 < t < 2$
2. $e^t, -2e^t, 3e^t$ plot on the same diagram, use intervals of 0.5 for $-1 < t < 2$
3. $e^{0.2t}, e^t, e^{2t}$ plot on the same diagram, use intervals of 0.5 for $-1 < t < 1$
4. $e^{-0.2t}, e^t, -e^{-0.2t}$ plot on the same diagram, use intervals of 0.5 for $-8 < t < 8$
5. $e^{-t}, 1 + e^{-t}, 1 - e^{-t}$ plot on the same diagram, use intervals of 0.5 for $0 < t < 5$
6. $10(1 - e^{-0.2t})$ use intervals of 0.5 for $0 < t < 5$.

PROGRESS EXERCISES 4.8

Solve Limited, Unlimited and Logistic Growth Problems

1. In a given region, the number (000s) in the population aged 60 and over is expected to grow according to the equation

$$P = 125.5e^{0.012t} \quad \text{where } t \text{ is given in years}$$

   (a) If $t = 0$ at the beginning of 2012, calculate the numbers at the beginning of 2012.
   (b) Calculate the numbers in the population at the end of 10, 30, 60, 90, 100 years.
   (c) Graph the population for the years 2012 to 2112. Comment on the general trend in the population.

2. Sales for a new magazine are expected to grow according to the equation

$$S = 200\,000(1 - e^{-0.05t}) \quad \text{where } t \text{ is given in weeks}$$

   (a) Calculate the number of magazines sold after one week.
   (b) Calculate the number of magazines sold after 5, 20, 35, 45, 50, 52 weeks.
   (c) Plot sales over the first 52 weeks. Comment on the general trend in sales.

3. A virus is thought to spread through a chicken farm according to the equation

$$N = 800 \frac{1}{1 + 790e^{-0.1t}}$$

   where N is the number of infected chickens and t is in days.
   (a) How many chickens are infected at $t = 0$?
   (b) Calculate the number of chickens infected after 0, 20, 40, 60, 80, 100 days.
   (c) Will all the chickens become infected eventually? Give reasons for your answer.

4.3 Logarithmic Functions

At the end of this section you should be able to:

- Define the log of a number and convert from log form to exponential form
- Recognise graphs of logarithmic functions and understand their main properties
- Use the rules for logs to simplify expressions containing logs
- Solve equations containing logs.

4.3.1 How to find the log of a number

In Section 4.2 exponential equations, such as $2^x = 2^9$, were solved by equating the indices. This made sense as the bases are already identical, so if the indices are identical, then LHS = RHS. However, in

Worked Example 4.15 we attempted to solve the equation

$$\frac{1750}{753} = e^{0.03t}$$

$$2.3240 = e^{0.03t}$$

At that point, we could not solve for t. First of all, the LHS is not in the form e^{power}, so without identical bases on each side there is no point in equating indices. The problem here is trying to solve for t, when t is part of the power of the exponential. A method is required that will bring the power down onto the line, thus converting the exponential equation to an ordinary equation which can then be easily solved. Logarithms or logs provide such a method.

What is the log of a number?

Logs are powers. In the following example, each number is written as 10^{power}, so the power is the log of the number:

$N = base^{power}$	$10 = 10^1$	$100 = 10^2$	$200 = 10^{2.301\,03}$	$1000 = 10^3$
$\log(N) = power$	$\log(10) = 1$	$\log(100) = 2$	$\log(200) = 2.301\,03$	$\log(1000) = 3$

But there are infinitely many numbers that may be raised to powers of 1, 2, 2.301 03, 3, etc. For example,

$N = base^{power}$	$2 = 2^1$	$4 = 2^2$	$4.928\,09 = 2^{2.301\,03}$	$8 = 2^3$
$\log(N) = power$	$\log 2 = 1$	$\log 4 = 2$	$\log 4.928\,09 = 2.301\,03$	$\log 8 = 3$

So obviously the base of the power plays a role and must be stated when declaring the log of the number, as follows:

$10 = 10^1$	$100 = 10^2$	$200 = 10^{2.301\,03}$	$1000 = 10^3$
$\log_{10} 10 = 1$	$\log_{10} 100 = 2$	$\log_{10} 200 = 2.301\,03$	$\log_{10} 1000 = 3$

$2 = 2^1$	$4 = 2^2$	$4.928\,09 = 2^{2.301\,03}$	$8 = 2^3$
$\log_2 2 = 1$	$\log_2 4 = 2$	$\log_2 4.928\,09 = 2.301\,03$	$\log_2 8 = 3$

Therefore, in general, the log of a number may be described as follows:

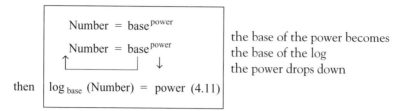

| then | $\log_{\text{base}}(\text{Number}) = \text{power}$ (4.11) |

the base of the power becomes
the base of the log
the power drops down

where the number and base must be real positive numbers (otherwise we may end up dealing with complex numbers).

Logs to base 10 and logs to base e

If logs of all numbers were tabulated for every possible base, there would be endless sets of tables, so for convenience, two sets of tables are available, one for logs to base 10 and one for logs to base e. On your calculator, these may be found on the **log key** (for log to base 10) and on the **ln key** (for log to base e).

Notation

The notation $\log(x)$ is understood to be $\log_{10}(x)$; for example, $\log(7)$ means $\log_{10}(7)$. The notation $\ln(x)$ is understood to be $\log_e(x)$; for example, $\ln(7)$ means $\log_e(7)$. Check on your calculator that $\log 12 = 1.079\,181\,2$ and $\ln 12 = 2.484\,906\,6$.

Note: Rule 4 (below) gives a simple method for taking logs to base n and converting them into logs to base 10 or logs to base e.

PROGRESS EXERCISES 4.9

Logs and Transforming Equations from Index to Log Form

1. Use your calculator to evaluate the following:

   (a) $\log(23)$ (b) $\log(57.98)$ (c) $\log(0.22)$

   (d) $\ln(57.98)$ (e) $\dfrac{\log(57.98)}{\log(21.1)}$ (f) $\ln(0.78)\,\ln(9.15)$

   (g) $\dfrac{\ln(57.98)}{\ln(21.1)}$ (h) $\dfrac{\ln(57.98)}{\log(21.1)}$ (i) $3\log(27) - 2\log(35)$

2. Write each of the following exponential equations in log form using equation (4.11):

   (a) $N = a^{\text{power}}$ (b) $10 = 3^{2.0959}$ (c) $3 = 10^{0.4771}$

   (d) $7.389 = e^2$ (e) $2^x = 10$ (f) $2 + 3^x = 10$

   (g) $5(2^x) = 20$ (h) $2^x 2^{2x} = 15$ (i) $3(5)^Q = 12$

   (j) $(5)^{Q+1} = 10$ (k) $(10)^{4x+3} = 25$ (l) $P^{0.3} = 12$

   (m) $5K^{-0.5} = 28$ (n) $2^{0.3} = 25L^{0.9}$ (o) $(K + 10)^{0.8} = 2$

Solving equations containing exponentials a^x, i.e., where a is any real number

Since logs are tabulated for base 10 and base e, it is now possible to solve certain equations which contain exponentials 10^x or e^x.

WORKED EXAMPLE 4.17
USE LOGS TO SOLVE CERTAIN EQUATIONS

Solve the following equations:

(a) $31 = 10^x$ (b) $25(10)^{2t} = 208$ (c) $38 + 12e^{-0.5t} = 208$

SOLUTION

Step 1: Arrange the exponential equation into the form 'number = base^{power}', where the unknown variable is in the power.
Step 2: Convert from index form to log form, as indicated in equation (4.11), reducing the exponential equation to a linear equation.
Step 3: Solve the linear equation for the unknown.

(a)

Step 1: $31 = 10^x$

Step 2: $\log_{10}(31) = x$

Step 3: $1.4914 = x$

Solution $x = 1.4914$

(b)

$25(10)^{2t} = 208$

$(10)^{2t} = \dfrac{208}{25} = 8.32$

Step 1: $(10)^{2t} = 8.32$

Step 2: $2t = \log_{10}(8.32)$

Step 3: $t = \dfrac{0.9201}{2}$

$= 0.4601$

Solution: $t = 0.4601$

(c)

$38 + 12e^{-0.5t} = 208$

$12e^{-0.5t} = 208 - 38 = 170$

Step 1: $e^{-0.5t} = \dfrac{170}{12} = 14.1667$

Step 2: $-0.5t = \log_e(14.1667)$

$-0.5t = 2.6509$

Step 3: $t = \dfrac{2.6509}{-0.5} = -5.3018$

Solution $t = -5.3018$

Next return to Worked Example 4.15. This was not completed since the mathematical methods covered in the text up to that point could not solve the equation $P = 753e^{0.03t}$ for t when $P = 1750$. Now logs may be used to reduce this equation to an ordinary linear equation in t.

WORKED EXAMPLE 4.18

FINDING THE TIME FOR THE GIVEN POPULATION TO GROW TO 1750

With the definition of a log function, part (c) of Worked Example 4.15 can now be finished. Given $P = 753e^{0.03t}$ find t when the population P is 1750.

SOLUTION

Substitute $P = 1750$ into the equation for population and solve for t.

Step 1	Step 2	Step 3
$1750 = 753e^{0.03t}$	$\log_e(2.324) = 0.03t$	$\dfrac{0.8433}{0.03} = t$
$\dfrac{1750}{753} = e^{0.03t}$	$\ln(2.324) = 0.03t$	$28.11 = t$
$2.234 = e^{0.03t}$	$0.8433 = 0.03t$	

So the year in which the population reaches 1750 persons is $2010 + 28.11 = 2038.11$ or during the year 2038.

PROGRESS EXERCISES 4.10

Solve Equations and Applications with Logs

Solve each of the equations in questions 1–9 for the unknown variable.

1. $28 = 10^x$
2. $5 = 3(10)^x$
3. $5 = 10^{x-4}$
4. $125 = e^t$
5. $125 = 230e^t$
6. $125 = 230e^{-t}$
7. $28 = 5 + 2e^{-0.5t}$
8. $40 = 66(1 - e^{-t})$
9. $5 = 10^x$

10. If the total value of sales, S (in £000s), is given by the equation $S = 500(1 - e^{-0.16t})$, where time t is given in weeks, starting with $t = 0$ when the product is first marketed
    (a) Find the total sales (i) at the beginning of week 1, (ii) at the end of week 1, (iii) at the end of week 5.
    (b) Graph the sales for $0 < t < 10$.
    From the graph, estimate the time, in weeks, taken for sales to reach £400 000. Confirm your answer algebraically. What is the maximum number of sales attainable?

11. Scientists estimate that the maximum carrying capacity of a lake is 6000 and that the number of fish increases according to the equation

$$P(t) = \frac{6000}{1 + 29e^{-0.4t}}$$

where t is given in years and $P(t)$ is the fish population at time t.
(a) Find the fish population in the lake at $t = 0$, $t = 4$, $t = 10$.
(b) Graph the equation for $0 < t < 12$.
(c) From the graph estimate the time taken for the fish population to reach 1000, 3000, 4000. Confirm your answers algebraically.

12. The population of a country is changing according to the equation $P = 5.2e^{-0.001t}$, where t is in years and, P is in millions.
(a) Calculate the number in the population when $t = 0$, 10, 50 years. If $t = 0$ at the beginning of 2012, state the years represented by $t = 0$, 10, 50. Is the population increasing or decreasing in time?
(b) Calculate the number of years from $t = 0$ until the population declines to 4 million. State the year in which this happens.

4.3.2 Graphs and properties of logarithmic functions

The above examples, and the questions in Progress Exercises 4.10, solve exponential equations where the base is 10 or e only. To solve equations for bases other than 10 or e requires a slightly more detailed study of the log functions and the rules for manipulating logs.

The graph of any function reveals many of its characteristics and properties. It is necessary to have an overview of a function and its properties if the function is to be used in equations, and to understand that a function may not exist for all real values of the independent variable, etc.

WORKED EXAMPLE 4.19
GRAPHS OF LOGARITHMIC FUNCTIONS

Graph $\log(x)$ and $\ln(x)$ on the same diagram for $-1 < x < 2$. From the graphs summarise the properties of the log functions. How does the size of the base affect the shape of the graph?

SOLUTION

Calculate a table of values for each function, using your calculator to evaluate the logs or, better still, use Excel. No real value exists for the logs of zero (doesn't exist) or logs of negative numbers (complex). This is indicated on the calculator by $-$ E $-$ and in Excel by #NUM!. Table 4.15 outlines the points used to plot the graphs in Figure 4.20. From Figure 4.20, some

Table 4.15   Values of log(x) and ln(x)

x	log(x)	ln(x)
−0.2	#NUM!	#NUM!
0	#NUM!	#NUM!
0.2	−0.7	−1.61
0.4	−0.4	−0.92
0.6	−0.22	−0.51
0.8	−0.1	−0.22
1	0	0
1.2	0.08	0.18
1.4	0.15	0.34
1.6	0.20	0.47
1.8	0.26	0.59

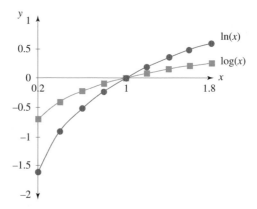

Figure 4.20   Graphs of log(x) and ln(x)

properties of log functions can be deduced:

- There are no real logs of negative numbers.
- Logs of numbers less than one are negative.
- log(1) = 0 for any base.
- logs of numbers greater than one are always positive.

In addition, the graph of $\log_e(x)$ is steeper, that is, it increases and decreases more rapidly than $\log_{10}(x)$.

4.3.3   Rules for logs

Finally, we need to state the rules for combining and manipulating log functions (Table 4.16).

Table 4.16   The rules for logs

Operation	Notation	Example
Rule 1 Add	$\log_b(M) + \log_b(N) \Leftrightarrow \log_b(MN)$	$\ln(4) + \ln(29) = \ln(4 \times 29)$ $1.3863 + 3.3673 = \ln(116)$ $4.7536 = 4.7536$
Rule 2 Subtract	$\log_b(M) - \log_b(N) \Leftrightarrow \log_b\left(\dfrac{M}{N}\right)$	$\log(90) - \log(26) = \log\left(\dfrac{90}{26}\right)$ $1.9542 - 1.4150 = \log(3.4615)$ $0.5392 = 0.539\,26$
Rule 3 Log of an exponential	$\log_b(M^z) \Leftrightarrow z\,\log_b(M)$	$\log(5^3) = 3\,\log(5)$ $\log(125) = 3(0.698\,97)$ $2.096\,91 = 2.096\,91$
Rule 4 Change of base	$\log_b(N) \Leftrightarrow \dfrac{\log_x(N)}{\log_x(b)}$ the new base, x, is usually 10 or e since both are available on the calculator	$\log_2(16) = \dfrac{\log(16)}{\log(2)} = \dfrac{1.2041}{0.3010} = 4.0$ or change to base e $\log_2(16) = \dfrac{\ln(16)}{\ln(2)} = \dfrac{2.7726}{0.6931} = 4.0$

The $\Leftrightarrow$ sign indicates that equations can be operated both ways, as in the following worked examples.

WORKED EXAMPLE 4.20
USING LOG RULES

Use the log rules to simplify the following expressions to a single term, if possible:

(a) $\log_b(25) + \log_b(70) - \log_b(55)$
(b) $4\log_x(7) - 3\log_x(0.85) + \log_x(10)$
(c) $12\log_b(12) + 3\log_x(8.25) - 2\log_b(5)$

Then evaluate each expression if $b = 3$, $x = e$.

SOLUTION

(a) $\log_b(25) + \log_b(70) - \log_b(55)$ all logs have the same base

$= \log_b(25 \times 70) - \log_b(55)$ so use rule 1 to add the first two terms

$= \log_b\left(\dfrac{25 \times 70}{55}\right)$ use rule 2 to divide by the negative term

$= \log_b(31.818)$ this cannot be evaluated unless b is given a value

If $b = 3$ then using rule 4 to change base 3 to base 10, we have

$$\log_3(31.818) = \frac{\log(31.818)}{\log(3)} = \frac{1.502\,67}{0.4771} = 3.149\,46$$

(b) In $4\log_x(7) - 3\log_x(0.85) + \log_x(10)$ start by using rule 3 in reverse, bringing the numbers in that multiply the log term as powers before adding or subtracting using rules 1 and 2. If you look at rules 1 and 2, there are no numbers outside the log terms before adding or subtracting:

$$4\log_x(7) - 3\log_x(0.85) + \log_x(10)$$

$= \log_x(7)^4 - \log_x(0.85)^3 + \log_x(10)$ using rule 3 in reverse

$= \log_x\left(\dfrac{7^4}{0.85^3}\right) + \log_x(10)$ use rule 2

$= \log_x(39\,096.275) + \log_x(10)$

$= \log_x(39\,096.275 \times 10)$ use rule 1

$= \log_x(39\,096.275)$

If $x = e$ we obtain $\ln(39\,096.275) = 10.573\,78$.

(c) In this problem there are two different bases, therefore combine terms with the same base only:

$$12\log_b(12) + 3\log_x(8.25) - 2\log_b(5)$$

$= \log_b(12)^{12} + \log_x(8.25)^3 - \log_b(5)^2$ rule 3 in reverse, bring in the constants as powers

$= \log_b\left(\dfrac{12^{12}}{5^2}\right) + \log_x(561.516)$ rule 2 for base b terms

$= \log_b(3.566\,44 \times 10^{11}) + \log_x(561.516)$ simplify the numbers

This cannot be simplified to a single term as the bases are different. Given $b = 3$ and $x = e$, the expression may be evaluated as

$$\log_3(3.566\,44 \times 10^{11}) + \log_x(561.516) = \frac{\ln(3.566\,44 \times 10^{11})}{\ln(3)} + \ln(561.516) = 30.543$$

4.3.4 Solving equations using the log rules

There are several ways of using logs to solve equations. Worked Example 4.17 solved equations of the form: number $=$ base^{power}, but only for base $= 10$ or base $= e$. The following worked example uses rule 3 for logs (above) to solve such equations when the base is other than 10 or e. The method involves writing the equation in the format: number $=$ (base)^{power} (as in Worked Example 4.17), and then taking logs of both sides of the equation. This is called 'operating' on both sides of the equation. Given an equation

$$\text{number} = (\text{base})^{\text{power}} \quad \text{then} \quad \log(\text{number}) = \log((\text{base})^{\text{power}})$$

Take logs to base 10 or e since they are available on the calculator: if the base on the RHS happens to be e then take logs to base e; otherwise take logs to base 10.

WORKED EXAMPLE 4.21
SOLVING CERTAIN EQUATIONS WITH RULE 3 FOR LOGS

Solve the following equations:
(a) $20 = 3(1.08)^x$ (b) $20 + (2.4)^{2x} = 32.5$

SOLUTION

Method	(a)	(b)
Step 1: Write the equation as number $=$ base^{power}	$20 = 3(1.08)^x$ $20/3 = 1.08^x$	$20 + (2.4)^{2x} = 32.5$ $(2.4)^{2x} = 32.5 - 20 = 12.5$
Step 2: Take logs of both sides	$\log(20/3) = \log(1.08)^x$	$\log(2.4^{2x}) = \log(12.5)$
Step 3: Simplify using rule 3, which says $\log(a^x) = x \log(a)$	$\log(20/3) = x \log(1.08)$	$2x \log(2.4) = \log(12.5)$
Step 4: Solve, using calculator as required	$\dfrac{\log(20/3)}{\log(1.08)} = x$ $\dfrac{0.8239}{0.0334} = x = 24.65$	$2x = \dfrac{\log(12.5)}{\log(2.4)} = \dfrac{1.0969}{0.3802} = 2.8850$ $x = \dfrac{2.8850}{2} = 1.4425$

This method is useful in solving certain equations involving compound interest (Chapter 5). For immediate practice with this method attempt Worked Example 4.17 and try solving questions 1 to 9 of Progress Exercises 4.10.

Solving equations that contain logs

If we attempt to solve the equation

$$\log(x + 2) = 2.5$$

we first solve for $(x + 2)$. But $(x + 2)$ is contained within the expression $\log(x + 2)$. To get at $(x + 2)$ we must undo or reverse the operation by which logs were taken in the first place. This is described as taking **antilogs**.

Taking logs is the process:

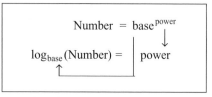

the base of the index becomes the base of the log, then the power drops down

Reversing the log process gives us this:

the base of the log becomes the base of the power, then the power goes up

For example,

$$\log_2(8) = 3$$
$$8 = 2^3$$

This is described as going from log form to index form.
You can see this on your calculator:

The inverse function for the [**log**] key is [10^x]

The inverse function for the [**ln**] key is [e^x]

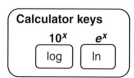

The following worked examples show how to solve equations which contain logs and exponentials.

WORKED EXAMPLE 4.22
SOLVE EQUATIONS CONTAINING LOGS AND EXPONENTIALS

 Find animated worked examples at **www.wiley.com/college/bradley**

Solve the following equations:
(a) $\log(x + 2) = 2.5$ (b) $2\ln(x) - \ln(x + 1) = 0$
(c) $e^{x+5} = 1.56$ (d) $2^x 2^{x+1} = 7$

SOLUTION

(a) $\log(x + 2) = 2.5$ to isolate x, it is necessary to go from log form to index form

$(x + 2) = 10^{2.5}$ $10^{2.5}$ is called the antilog of 2.5; use the second function on the log key

$x = 316.227\,77 - 2$
$\quad = 314.227\,77$

(b) $2\ln(x) - \ln(x + 1) = 0$ rule 3, bring the 2 in as a power
$\ln x^2 - \ln(x + 1) = 0$

$\ln\left(\dfrac{x^2}{x + 1}\right) = 0$ rule 2

$\left(\dfrac{x^2}{x + 1}\right) = e^0 = 1$ going from log form to index form

$x^2 = 1(x + 1)$ multiplying both sides by $(x + 1)$ to avoid fractions
$x^2 - x - 1 = 0$ a quadratic, so solve the quadratic using the formula

The solutions are $x = -0.618$ and $x = 1.618$.

(c) We have $e^{x+5} = 1.56$; x is in the index, so take logs to get x down. There are two possible methods, the second being the more usual. Check the solutions.

Method 1: Go from log to index

$x + 5 = \ln(1.56)$
$x \quad = 0.4447 - 5$
$x \quad = -4.5553$

Method 2: Take logs of both sides

$\ln(e^{x+5}) \quad = \ln(1.56)$
$(x + 5)\ln(e) = 0.4447 \quad \ln(e) = 1$
$x + 5 \quad = 0.4447$
$x \quad = 0.4447 - 5 = -4.5553$

(d) $2^x x^{x+1} \qquad = 7$
$2^{2x+1} \qquad = 7$
$\log(2^{2x+1}) \qquad = \log(7)$ use rule 1 for indices
$\qquad\qquad\qquad\qquad$ take logs of both sides

$(2x + 1)\log(2) \quad = \log(7)$ use rule 3 for logs

$(2x + 1) \qquad = \dfrac{\log(7)}{\log(2)} = \dfrac{0.8451}{0.3010} = 2.8074$ evaluate and simplify

$2x \qquad\qquad = 2.8074 - 1$

$x = 0.9037$ check the solution

PROGRESS EXERCISES 4.11

Transforming Equations from Log to Index Form and Use of Log Rules

Write each of the following log equations in exponential form, if possible:

1. $\log(x + 5) = 1.2$

2. $3 \log(x + 2) = 3.5$

3. $2 + \ln(t - 5) = 2.5$

4. $\dfrac{\log(x + 1)}{2.5} = -0.5$

5. $3 \log(2 - 2x) = 0$

6. $\dfrac{2}{3} \log(x^2 + 1) = \dfrac{1}{5}$

7. $-0.65 = \log(0.5x)$

8. $\log(x(x + 1)) = 1$

9. $2 + \ln(x - 4.5) = 4$

10. $\ln\left(\dfrac{x + 1}{x}\right) = 3.5$

11. Solve each of the equations given in questions 1 to 10, giving your answer correct to four decimal places.

12. Solve each of the following equations:
    (a) $3 = 7(10)^x$ (b) $5 = 12e^x$ (c) $5 = 12(1.05)^x$ (d) $5 = 12(1.05)^{4x}$
    (e) $e^t = 7 - 2e^t$ (f) $80 = 25 + (1.5)^x$ (g) $20 = 40\left(1 - e^{-t}\right)$
    (h) $2.3\,(1.05)^{-2x} = 4$

    Simplify each of the expressions in questions 13 to 17, writing your answers as the log of a single number. Confirm that your answer is correct by evaluating the original expression and the simplified expression.

13. $2 \log(12) + \log(20)$

14. $\ln(125) - \ln(80) + \ln(28)$

15. $3 \ln(138) - 2 \ln(95)$

16. $8 \ln(2) + 3 \ln(8) - 2 \ln(3)$

17. $\ln(4) - 2 \ln(2) + \ln(80)$

    In questions 18 to 23, show that the expression on the LHS simplifies to that on the RHS.

18. $\log(x) + \log(x - 5) = \log(x^2 - 5x)$

19. $2 \log(x - 2) - \log(2x) = \log\left[\dfrac{(x - 2)^2}{2x}\right]$

20. $\ln\left[\dfrac{x^3}{\sqrt{x + 1}}\right] = 3 \ln(x) - 0.5 \ln(x + 1)$

21. $\ln(x^5 e^x) = 5 \ln(x) + x$

22. $0.2 \ln(x) - 0.3 \ln(2x) = \ln\left(\dfrac{1}{2^{0.3} x^{0.1}}\right)$

23. $\log\left(\dfrac{x + 1}{1 - x}\right) + \log\left(\dfrac{1}{x + 1}\right) = -\log(1 - x)$

24. Evaluate the following, giving your answer correct to four decimal places:
    (a) $\log_3(27)$ (b) $\log_5(51)$ (c) $2 \log_{3.5}(30)$

25. Solve the following equations:
    (a) $2 \log(5x) - \log(2x) = 2.5$ (b) $\log(Q) - \left(\frac{Q}{Q+1}\right) = 0.8$

4.4 Hyperbolic (Rational) Functions of the Form $a/(bx + c)$

At the end of this section you should be able to:

- Define and graph hyperbolic functions of the form $a/(bx + c)$
- Define and solve applications which are modelled by hyperbolic functions.

Note: The hyperbolic functions in this section are called rectangular hyperbolic functions and are based on the rectangular hyperbola $y = 1/x$. This class of functions is also referred to as rational functions. There is another class of hyperbolic functions, such as $\sinh x = \frac{1}{2}(e^x - e^{-x})$ and $\cosh x; = \frac{1}{2}(e^x + e^{-x})$, which are not considered in this text.

4.4.1 Define and sketch rectangular hyperbolic functions

A function which is given by the equation $y = a/(bx + c)$ is called a hyperbolic function. The simplest rectangular hyperbolic function is $y = 1/x$. The graph of $y = 1/x$ consists of two separate parts, as illustrated in Figure 4.21.

To plot the graph, calculate a table of points such as Table 4.17. Only one point presents a problem, 1/0. There is no number defined which gives a value for division by zero; no y-value can be defined for

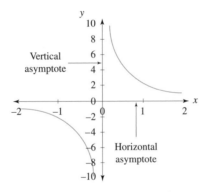

Figure 4.21 Graph of $y = 1/x$

Table 4.17 Calculation of points for Figure 4.21

x	$y = 1/x$
-2	-0.5
-1.5	-0.67
-1	-1
-0.5	-2
0	?
0.5	2
1	1
1.5	0.67
2	0.5

the point $x = 0$, so we cannot plot a point ($x = 0$, $y = ?$). Not only is there a point missing from the graph, but the shape of the graph increases or decreases dramatically on either side of the undefined point. You can see this if you evaluate y for values of x close to the undefined point, as in Table 4.17. In this example the y-axis is a vertical asymptote and the x-axis is a horizontal asymptote. An asymptote is a line which the curve approaches at a distance from the origin.

Functions of the form y = a/(bx + c)

Functions of the form $y = a/(bx + c)$ have a similar shape to $y = 1/x$, except that division by zero no longer occurs at $x = 0$, but at $bx + c = 0$ or $x = -c/b$.

WORKED EXAMPLE 4.23
SKETCHES OF HYPERBOLIC FUNCTIONS

Sketch the functions

(a) $y = \dfrac{1}{x - 0.23}$ (b) $y = \dfrac{1}{x} + 3$

SOLUTION

(a) The function given in (a) is $y = 1/x$ translated horizontally forward by 0.23 units and (b) is $y = 1/x$ translated vertically up by 3 units.

Step 1: Calculate the value of x that gives rise to division by zero. This will determine the vertical asymptote. Division by zero occurs when the denominator is zero.

$$x - 0.23 = 0 \rightarrow x = 0.23$$

Step 2: Calculate the points of intersection with the y-axis, using the fact that $x = 0$ on the y-axis:

$$y = \frac{1}{x - 0.23} \rightarrow y = \frac{1}{0 - 0.23} = -4.35$$

Step 3: To get some idea of the curvature as the graph approaches the vertical asymptote, calculate the coordinates of some points to its left and right:

x	0	0.1	0.2	0.3	0.4
y	-4.35	-0.13	-33.33	14.29	5.88

See Figure 4.22(a).

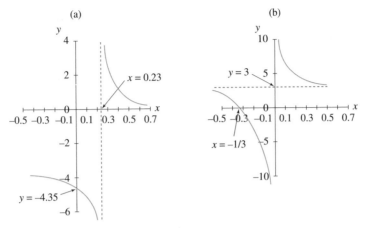

Figure 4.22   Hyperbolic functions

(b) The equation of the vertically translated function is

$$y = \frac{1}{x} + 3$$

The graph is sketched by translating $y = 1/x$ vertically up the y-axis by 3 units as shown in Figure 4.22(b). The point of intersection with the x-axis is easily found by x solving for when y is zero.

The main features of $y = a/(bx + c)$

- The graph has a vertical asymptote at the value of x which results in division by zero.
- The x-axis is a horizontal asymptote.
- The graph has no maximum or minimum values.
- Otherwise the curve increases smoothly at an increasing rate or decreases at a decreasing rate.

PROGRESS EXERCISES 4.12

Graphing Hyperbolic Functions

Graph the following functions, indicating asymptotes and points of intersection with the axis:

1. $y = \frac{20}{Q}$

2. $y = 5 + \frac{20}{Q}$

3. $y = \frac{4}{Q-1}$

4. $y = \frac{4}{Q+1}$

5. $y = 2 + \frac{1}{Q+1}$

6. $y = \frac{4}{1-Q}$

7. $y = 20 - \frac{4}{4+Q}$

8. $y = 20 + \frac{1}{1-2Q}$

9. $y = \frac{1}{3Q-1}$

4.4.2 Equations and applications

Functions of the form $a/(bx + c)$ model average cost, supply, demand and other functions which grow or decay at increasing or decreasing rates. The solutions of equations will involve manipulating fractions (Chapter 1).

WORKED EXAMPLE 4.24
HYPERBOLIC DEMAND FUNCTION

The demand function for a good is given by the equation $Q + 1 = 200/P$ and the supply function is a linear function $P = 5 + 0.5Q$.

(a) Write the demand function in the form $P = f(Q)$.
(b) Sketch the supply and demand functions on the same diagram for $-1 \leq Q \leq 20$. From the graph, estimate the equilibrium point. Confirm your answer algebraically.

SOLUTION

(a)

$$Q + 1 = \frac{200}{P}$$
$$P(Q + 1) = 200$$
$$P = \frac{200}{Q + 1}$$

(b) To sketch the graph, follow these steps:
- **General shape:** the demand function is a hyperbolic function; it has a vertical asymptote at the value of Q which results in division by zero, that is, at $Q = -1$.
- **Intersection with the axis:** the graph cuts the P-axis at $Q = 0$, therefore

$$P = \frac{200}{0 + 1} = \frac{200}{1} = 200$$

The horizontal axis is an asymptote. You can show that the graph never cuts the horizontal axis as follows. Since $P = 0$ at every point on the horizontal axis (Q-axis), substitute $P = 0$ into the equation of the demand curve and find the value of Q at which the graph crosses the Q-axis, therefore

$$0 = \frac{200}{Q + 1} \rightarrow 0(Q + 1) = 200$$

However, there is no value of $Q + 1$ which can be multiplied by 0 and give an answer of 200. Zero multiplied by anything is zero. So there is no solution, or no point where the graph crosses the horizontal axis.

The supply function is a linear function with intercept 5 and slope 0.5. Since two points are sufficient to plot a straight line, we need one other point in addition to the intercept. Choose any value of Q then substitute this value into the equation of the supply function to calculate the corresponding P coordinate; for example, when $Q = 4, P = 7$. The demand and supply functions are sketched in Figure 4.23.

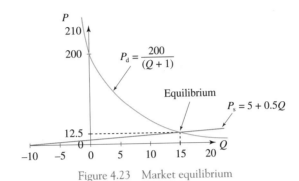

Figure 4.23   Market equilibrium

Algebraically, equilibrium exists when $P_d = P_s$ and $Q_d = Q_s$. Since each equation is in the form $P = f(Q)$, it is easiest to equate prices:

$$\frac{200}{Q+1} = 5 + 0.5Q$$

$$200 = (5 + 0.5Q)(Q + 1)$$

$$200 = 0.5Q^2 + 5.5Q + 5$$

$$0 = 0.5Q^2 + 5.5Q - 195$$

Solve this quadratic for Q. The solutions are $Q = -26$ and $Q = 15$. As the negative quantity is not economically meaningful, substitute $Q = 15$ into either original equation and solve for the corresponding equilibrium price. Therefore market equilibrium is reached when $Q = 15$ and $P = 12.5$.

PROGRESS EXERCISES 4.13

Equations and Applications Based on Hyperbolic Functions

Solve the following equations in questions 1 to 6.

1. $\dfrac{1}{x+1} = 2$

2. $\dfrac{x}{x+4} = 3$

3. $\dfrac{10x - 6}{x^2 + 1} = 2$

4. $\dfrac{5}{Q - 3.5} = 10$

5. $\dfrac{Q + 5}{Q - 5} = Q + 1$

6. $\dfrac{1}{Q} + \dfrac{2}{Q} = 8$

7. The following equation relates the value of a new car (in £000s) to age, in years:

$$V = 1 + \frac{84}{1 + 2t}$$

(a) Graph the value of the car for ages 0 to 10. Describe how the value depreciates over time.
(b) Determine how long (to the nearest year) it will take for the value to drop to £20 000.

8. The demand and supply functions for a brand of tennis shoes are:

$$P_d = \frac{500}{Q + 1}, \quad P_s = 16 + 2Q$$

where P is the price per pair and Q is the quantity in thousands of pairs.
(a) Calculate the equilibrium price and quantity.
(b) Graph the supply and demand functions; hence confirm your answer graphically.

9. A new model of super-light slim laptop computer is gradually replacing an earlier model. The projected sales (in thousands) for the old and new models are given by the equations

$$S_{old} = \frac{9}{t + 3}, \quad S_{new} = \frac{36}{21 - t}$$

where time t is in months; S is the number of laptops sold (old and new models).
(a) Calculate the number of laptops sold when $t = 0, 5, 10$. Graph the sales functions for $t = 0$ to 10. From the graph, estimate the time when the sales of both models are equal.
(b) Calculate the time when sales are equal. Did the introduction of the new model improve overall sales? Give reasons for your answer.

10. Satisfaction derived from solving maths problems is given by the utility function

$$U = 100 - \frac{100}{x + 1}$$

where x is the number of problems solved.
(a) Graph the utility function for $x = 0$ to 100. Hence describe how satisfaction grows with the number of problems attempted. Does satisfaction ever decline?
(b) Calculate the number of problems that should be attempted to achieve a utility of 10, 50, 80, 100. Comment.

4.5 Excel for Non-linear Functions

In this chapter several new functions have been introduced: quadratics; cubics; exponential functions; logarithmic functions; and hyperbolic functions. You should find Excel very useful for graphing these non-linear functions, even for questions set in the progress exercises. This is particularly true for the cubic functions, some of which represent typical total cost functions. Cubic equations and applications involving cubic equations were solved graphically, as the algebraic methods are beyond the scope of this text.

Excel evaluates mathematical functions such as logs and exponentials as follows. Select the function wizard key, $\boxed{f_x}$, from the Excel menu bar. A two-column menu should now appear. From the 'function category' column, select the 'math & trig' option. The second column, 'function name', is a list of names giving the format of various mathematical functions. The functions used most frequently are the following:

Remember

Always type '=' first, to indicate that a formula is being used.

EXP:	for example, type in:	$= \exp(10) \rightarrow$ means e^{10}, etc.
LN:	for example, type in:	$= \ln(4) \rightarrow$ means $\log_e (4)$
LOG:	for example, type in:	$= \log(12,3) \rightarrow$ means $\log_3 (12)$
LOG(10):	for example, type in:	$= \log 43 \rightarrow$ means $\log_{10}(43)$

To enter any other exponential, for example, type in $= 3 \hat{} 2.5 \rightarrow$ means $3^{2.5}$.

The following worked examples use graphs to confirm and illustrate the solutions found by algebraic methods.

WORKED EXAMPLE 4.25
TOTAL COST FUNCTIONS WITH EXCEL

The total cost functions for two soap manufacturers (Plane Soap and Round Soap) are given by equations (i) and (ii) below,

(i) $TC = 12 + 2Q$ (ii) $TC = 0.3Q^3 - 15Q^2 + 250Q$

(a) The demand function for both firms is $P = 115 - 2Q$. Determine the break-even points algebraically. Plot total revenue and TC on the same diagram for each firm.
(b) Comment on the general characteristics of the TC functions.
(c) Determine the break-even points graphically and state the range of values of Q (output) for which the firm makes a profit.

Comment on the profitability of both firms.

SOLUTION
(a)

$$TR = P \times Q = 115Q - 2Q^2$$

For firm (i), the break-even points are calculated algebraically by solving $TC = TR$:

$$12 + 2Q = 115Q - 2Q^2$$
$$2Q^2 - 113Q + 12 = 0$$

The solutions to this quadratic are $Q = 0.11$ and $Q = 56.4$. Therefore, when plotting the graph, choose $0 \le Q \le 60$.

For firm (ii), the break-even points are calculated algebraically by solving $TC = TR$:

$$0.3Q^3 - 15Q^2 + 250Q = 115Q - 2Q^2$$
$$0.3Q^3 - 13Q^2 + 135Q = 0$$

In this text, the methods for solving cubic functions are not covered, unless it is possible to factor the cubic function, thereby reducing the problem to the product of linear and quadratic functions. The above cubic is solved as follows:

$$0.3Q^3 - 13Q^2 + 135Q = 0$$
$$Q(0.3Q^2 - 132Q + 135) = 0 \quad \text{factoring out } Q$$

Therefore, $Q = 0$ and solving the quadratic $0.3Q^2 - 13Q + 135 = 0$, gives $Q = 17.3$ and $Q = 26.1$. Therefore, when plotting the graph, choose $0 \leq Q \leq 30$.

Set up tables for total revenue with each of the TC functions. Make sure that the tables contain the points of intersection and span intervals where $TC < TR$ and $TC > TR$ (unless, of course, the firm never makes a profit, in which case $TC > TR$ always). (See Figures 4.24 and 4.25.)

(b) (i) Total cost is a linear function, therefore, it increases indefinitely. See Figure 4.24.

	A	B	C	D	E	F	G	H
1	Q	0	10	20	30	40	50	60
2	TC	12	32	52	72	92	112	132
3	TR	0	950	1500	1650	1400	750	−300

$TC > TR$ followed by an interval where $TC < TR$ and by a further interval where $TC > TR$

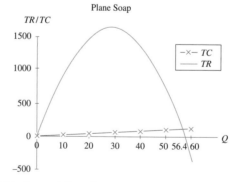

Figure 4.24 Linear TC and quadratic TR functions for Plane Soap Co.

(ii) Initially, the rate of increase of TC is decreasing up to a point, after which the rate of increase of TC is increasing. See Figure 4.25.

	A	B	C	D	E	F	G	H
20	Q	0	5	10	15	20	25	30
21	TC	0	912.5	1300	1387.5	1400	1562.5	2100
22	TR	0	525	950	1275	1500	1625	1650

$TC > TR$ followed by an interval where $TC < TR$ and by a further interval where $TC > TR$

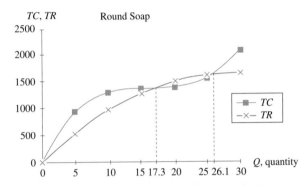

Figure 4.25   Cubic TC and quadratic TR functions for Round Soap Co.

(c) (i) Break-even at $Q = 0.11$ and $Q = 56.4$. Between these levels of output the firm makes a profit.

(ii) Break-even at $Q = 17.3$ and $Q = 26.1$. Between these levels of output the firm makes a profit.

Comment: The Plane Soap Co. makes a much larger profit than the Round Soap Co. since the area between the TR and TC curve is substantially greater.

4.6 Summary

Mathematics

- **Quadratic equations:** solution given by

$$x = \frac{-(b) \pm \sqrt{b^2 - 4ac}}{2a}$$

Also revise graphs.

- **Exponential indices:**
Revise graphs: rules for indices.
1. $a^m \times a^n = a^{m+n}$ **2.** $\frac{a^m}{a^n} = a^{m-n}$ **3.** $(a^m)^k = a^{m \times k}$
Revise graphs.

- **Logarithms**
  (a) Convert from index to log form:

  $\text{Number} = \text{base}^{\text{power}}$

  $$\text{Number} = \text{base}^{\text{power}} \qquad \text{the base of the index becomes the base of the log, the power drops down}$$

  then $\qquad \log_{\text{base}}(\text{Number}) = \text{power}$

  (b) Also, convert from log form to index form.
  (c) Rules for logs:
  1. $\log_b(M) + \log_b(N) \Leftrightarrow \log_b(MN)$ 2. $\log_b(M) - \log_b(N) \Leftrightarrow \log_b\left(\frac{M}{N}\right)$
  3. $\log_b(M^z) \Leftrightarrow z\log_b(M)$ 4. $\log_b(N) \Leftrightarrow \frac{\log_x(N)}{\log_x(b)}$

- **Hyperbolic functions:** $y = \dfrac{a}{(bx + c)}$.

  Revise the graphs, note the vertical asymptote.

www.wiley.com/college/bradley

Go to the website for Problems in Context

Applications

Similar to those in Chapters 2 and 3, but applied to a wider range of functions. Other applications include the laws of growth: limited and unlimited.

- **Excel:** Particularly useful for plotting, to visualise the shape and properties of the new functions introduced in this chapter. Certain cubic equations cannot be solved by any of the methods covered in this text, but graphical solutions are facilitated with Excel.

TEST EXERCISES 4

1. Solve the following equations:
   (a) $x^2 - 25 = 0$ (b) $x^2 + 20x = 0$ (c) $x^2 - 40x + 14 = 0$
2. A firm charges a fixed price of £80 for each shirt sold. The firm has a total cost function: $TC = Q^3 - 136Q$.
   (a) Write down the equation of the total revenue function.
   (b) Determine the break-even point.
3. (a) Simplify to a single term
   (i) $2\log(x) - \log(2x)$ (ii) $e^{5x}e^{-x}e^{5-x}$

(b) A rural population (given in thousands) is projected to decline according to the equation $P = 15e^{-0.1t}$. If $t = 0$ at the beginning of 2008:

  (i) Calculate the numbers in the population at the beginning of 2000; 2010; 2026.

  (ii) Graph the population up to 2060.

  (iii) Calculate the number of years it will take for the population to decline to 10 000.

4. (a) Solve the equations (i) $\dfrac{25}{x+5} = 4$ (ii) $\dfrac{2}{Q+0.5} = 20Q$

  (b) Find the equilibrium price and quantity for the demand and supply functions

$$P_d = \frac{50}{Q+2} \qquad P_s = 10 + 2Q$$

  Confirm your answer graphically

5. (a) Simplify the following expressions to a single term:

  (i) $3\log(x) + \log\left(\dfrac{10}{x}\right) - 2\log(10x)$ (ii) $\dfrac{a^2\sqrt{a}}{2a^{-2}}$

  (b) A new model of a popular television set is gradually replacing an earlier model. The projected sales (in thousands) for the old and new models are given by the equations

$$S_{old} = \frac{4}{t+2} \qquad S_{new} = \frac{21}{12-t}$$

  where time t is in months; S is the number of TVs sold (old and new models).

  (i) Calculate the number of TVs sold when $t = 0, 2, 4, 8$. Graph the sales functions for $t = 0$ to 8. From the graph, estimate the time when the sales of both models are equal.

  (ii) Calculate the time when sales are equal.

6. (a) Simplify the following:

  (i) $\dfrac{a^5\sqrt{a^5}}{2a^{0.3}}$ (ii) $\dfrac{a^x\sqrt{a^x}}{3a^{1-x}}$

  (b) The growth of demand for computer hardware in a college is given by the equation $Q = 200(1.025)^t$, where t is time in months.

  (i) Calculate the demand when $t = 0, 3, 6, 10$ months.

  (ii) Calculate the time taken for the demand to increase to 300.

  Graph the demand for $t = 0$ to 30, hence confirm the answers to (i) and (ii).

7. (a) Solve the equations:

  (i) $\ln(t + 3) = 4.2$ (ii) $2\ln(t + 1) - \ln(t - 1) = 1.504$

  (b) A firm has a total cost function $TC = \sqrt{4 + 6Q}$. If the firm charges a fixed price of 2 per unit sold:

  (i) Write down the equation for total revenue.

  (ii) Determine the break-even point.

FINANCIAL MATHEMATICS 5

5.1 Arithmetic and Geometric Sequences and Series

5.2 Simple Interest, Compound Interest and Annual Percentage Rates

5.3 Depreciation

5.4 Net Present Value and Internal Rate of Return

5.5 Annuities, Debt Repayments, Sinking Funds

5.6 The Relationship between Interest Rates and the Price of Bonds

5.7 Excel for Financial Mathematics

5.8 Summary

Appendix

Chapter Objectives

Arithmetic and geometric series are introduced in this chapter. The geometric series forms the basis for practically all the formulae used in the financial mathematics in the remainder of the chapter. At the end of this chapter you should be able to:

- Solve problems based on arithmetic and geometric series and applications
- Calculate present and future values based on simple interest
- Calculate present and future values based on compound interest using various conversion periods
- Solve for any of the four variables in the compound interest formula when the values of the other three variables are given
- Calculate annual percentage rates
- Calculate the future value of an asset using straight line and depreciating balance depreciation
- Calculate NPV and IRR and use these as investment appraisal techniques
- Calculate the future value for an initial investment P_0 and annual deposits A_0
- Calculate present and future values of ordinary annuities
- Calculate mortgage repayments and related problems
- Calculate the value of sinking funds and related problems
- Define a bond. Calculate the annual repayments and the NPV of investments in bonds
- Use Excel for all of the above financial calculations.

'Households face another hike in mortgage interest rates'
'Banks make record losses'

Such statements are part of our everyday life. Without doubt, the level of interaction between the individual and financial institutions is ever-increasing. The main interaction tends to be through borrowings and savings. Individuals are concerned with the total repayments on their borrowings and the total return received on their savings.

The amount received or repaid in the future is dependent on a number of factors:

- Principal (P_0) the amount borrowed (or invested) initially: **present value**.
- Interest rate (i) interest rate per year, computed as a percentage of the principal.
- Time (t) period of time over which money is borrowed (lent or invested).
- P_t value of the principal after a period of time t: **future value**.

In this chapter, various methods are introduced for calculating the amount received or repaid in the future. The methods are based on arithmetic and geometric sequences and series.

5.1 Arithmetic and Geometric Sequences and Series

At the end of this section you should be familiar with:

- Arithmetic sequences and series
- Geometric sequences and series
- Applications of arithmetic and geometric series.

Definitions

- **A sequence** is a list of numbers which follow a definite pattern or rule.

If the rule is that each term, after the first, is obtained by adding a constant, d, to the previous term, then the sequence is called an **arithmetic sequence**, such as 2, 6, 10, 14, 18, 22, ..., where $d = 4$: d is known as the **common difference**.

If the rule is that each term, after the first, is obtained by multiplying the previous term by a constant, r, then the sequence is called a **geometric sequence**, such as, 4, 16, 64, 256, 1024, ..., where $r = 4$: r is known as the **common ratio**.

- **A series** is the sum of the terms of a sequence.
  A series is **finite** if it is the sum of a finite number of terms of a sequence.
  A series is **infinite** if it is the sum of an infinite number of terms of a sequence.

Arithmetic series (or arithmetic progression denoted by AP)

An arithmetic series is the sum of the terms of an arithmetic sequence, such as 2, 6, 10, 14, 18, 22, ..., with $d = 4$. Denoting the first term of an arithmetic sequence with the value a and progressing by adding the value d to each previous term, the arithmetic sequence can be outlined as in Table 5.1.

Table 5.1   Arithmetic sequence

Sequence	a	$a + d$	$a + 2d$	...	$a + (n-1)d$
Term number	T_1	T_2	T_3	...	T_n

Each element of a sequence can be identified by reference to its term number; for example, the first term of the sequence $T_1 = a = 2$, the second term of the sequence $T_2 = a + d = 6$. The value of any term can be calculated knowing that the nth term of the arithmetic sequence is

$$T_n = a + (n-1)d \tag{5.1}$$

In the above example, the value of the 20th term is $a + (n-1)d = 2 + (20-1)4 = 78$. The sum of the first n terms, S_n, of an arithmetic series is given by the formula

$$S_n = \frac{n}{2}[2a + (n-1)d] \tag{5.2}$$

Formula (5.2) is derived in the Appendix to this chapter.

WORKED EXAMPLE 5.1
SUM OF AN ARITHMETIC SERIES

Find the sum of the first 15 terms of the series: $20 + 18 + 16 + 14 + \ldots$

SOLUTION
This is an arithmetic series since the difference $d = -2$. Therefore, $a = 20, d = -2, n = 15$, and hence, using formula (5.2),

$$S_{15} = \frac{15}{2}[2(20) + (15-1)(-2)]$$
$$= 90$$

Geometric series (or geometric progression denoted by GP)

A geometric series is the sum of the terms of a geometric sequence, such as $4, 16, 64, 256, 1024, \ldots$, with $r = 4$. Denoting the first term of a geometric sequence with the value a and progressing by multiplying the previous term by a common ratio r, the geometric sequence can be outlined as in Table 5.2. In this pattern it can be seen that when any term is divided by the previous term, the result is a common ratio, r.

The nth term of a geometric series is

$$T_n = ar^{n-1} \tag{5.3}$$

Table 5.2   Geometric sequence

Sequence	a	ar	ar^2	$\cdots$	ar^{n-1}
Term number	T_1	T_2	T_3	$\cdots$	T_n

The sum of the first n terms of a geometric series is given by the formula

$$S_n = a + ar + ar^2 + \cdots + ar^{n-1}$$

$$S_n = \frac{a(1 - r^n)}{1 - r} \tag{5.4}$$

$$S_n = \frac{a(r^n - 1)}{r - 1} \tag{5.5}$$

It is more convenient to use formula (5.4) when $r < 1$ and formula (5.5) when $r > 1$. Formula (5.5) is obtained by multiplying formula (5.4) by $-1/-1$. Formula (5.4) is derived in the Appendix to this chapter.

WORKED EXAMPLE 5.2
SUM OF A GEOMETRIC SERIES

Find the sum of the first 12 terms of the series: $4 + 2 + 1 + \dfrac{1}{2} + \dfrac{1}{4} \cdots$

SOLUTION
This is a geometric series since the ratio, $r = 1/2$. Therefore, $r = 1/2$, $a = 4$, $n = 12$, and hence, using formula (5.4),

$$S_{12} = \frac{4\left(1 - \left(\frac{1}{2}\right)^{12}\right)}{\left(1 - \frac{1}{2}\right)} = \frac{4(1 - 2.44 \times 10^{-4})}{\frac{1}{2}} = 8(0.999\,755\,8) = 7.9980$$

The sum of an infinite number of terms of a GP

In Worked Example 5.2 the calculation of r^n gave a very small value,

$$\left(\frac{1}{2}\right)^{12} = 0.000\,244\,1$$

If $r < 1$, then as n becomes larger $r^n \to 0$. For example,

$$\left(\frac{1}{2}\right)^5 = 0.031\,25, \quad \left(\frac{1}{2}\right)^{10} = 0.000\,98, \quad \left(\frac{1}{2}\right)^{15} = 0.000\,031$$

If the series has an infinite number of terms, r^n will become so small that in practice it will be zero. Formula (5.4) reduces to

$$S_\infty = \frac{a(1-0)}{1-r} = \frac{a}{1-r} \qquad\qquad\qquad \textbf{(5.6)}$$

PROGRESS EXERCISES 5.1

APs and GPs

1. (a) State whether the following sequences are APs or GPs.
   State the value of the first term and the common difference/common ratio.
     (i) $4, 6, 8, 10 \ldots$
     (ii) $4, 2, 1 \ldots$
     (iii) $5, 5.5, 6.0, 6.5 \ldots$
     (iv) $5, 4.5, 4, 3.5 \ldots$
     (v) $200, 220, 242 \ldots$
   (b) For each of these series in (a):
       Calculate the value of the 10th term of the series.
       Calculate the sum of the first 20 terms.
       Calculate the term that first exceeds 100.
2. Find the sum of the first eight terms of the series: $26 + 22 + 18 + 14 + \ldots$

3. Find the sum of the first 10 terms of the series: $\dfrac{1}{4} + \dfrac{3}{4} + \dfrac{5}{4} + \dfrac{7}{4} + \cdots$

4. Find the sum of the first eight terms of the series: $\dfrac{1}{3} + \dfrac{1}{9} + \dfrac{1}{27} + \dfrac{1}{81} + \cdots$

5. Find the value of the 50th term of the series: $180 + 162 + 145.8 + 131.22 + \ldots$
   In questions 6 to 8: (i) write down the values of a and d, (ii) use the formula for S_n to determine how many terms of the series must be added to give the sum indicated.
6. $1 + 2 + 3 + \ldots = 210$
7. $2 + 4 + 6 + \ldots = 420$
8. $90 + 87 + 84 + \ldots = 420$
9. The sum of the first 12 terms of an AP is 222, the sum of the first 5 terms is 40. Write out the first 4 terms of the series.
10. Show that the sum of the first n terms of the series: $1 + 3 + 5 + \ldots$ is given by the formula, $S_n = n^2$.
11. The fourth term of a GP is 56 while the sixth term is $\dfrac{7}{8}$. Find the GP and hence S_∞.

12. The first term of a GP represents the number of tins of beans sold in a supermarket the week before an advertising campaign is launched. Sales increase by 10% each week for eight weeks during the campaign.

(a) Write out the weekly sales as a geometric sequence. What is the value of r?

(b) Calculate the increased number of tins sold over the eight weeks of the campaign in terms of a.

(c) Is it worthwhile running this campaign if $a = 800$, the profit on beans is 9p per tin, but the cost of the campaign is £200 per week?

WORKED EXAMPLE 5.3

APPLICATION OF ARITHMETIC AND GEOMETRIC SERIES

A manufacturer produces 1200 computers each week. After week 1, it increases production by:

Scheme I: 80 computers each week.

Scheme II: 5% each week.

(a) (i) Find the output in week 20 under each scheme.

(ii) Find the total output over the first 20 weeks under each scheme.

(b) Calculate for schemes I and II the week in which production first exceeds 8000.

SOLUTION

Scheme I forms an arithmetic series: $a = 1200$ and $d = 80$.

Scheme II is a geometric series: $a = 1200$, $r = 1.05$. How is $r = 1.05$ calculated? If output increases by 5%, then output in any one week is the output in the previous week plus 5% of the output in the previous week, that is,

$$Q_{t+1} = Q_t + \frac{5}{100}Q_t$$

$$= Q_t \left(1 + \frac{5}{100}\right) = Q_t(1.05) = Q_t(r)$$

where Q_t is the output in week t and Q_{t+1} is the output in week $t + 1$.

Note: r is *not* 5, $r = 5\% = \dfrac{5}{100} = 0.05$

(a) (i) The output in week 20 is the value of term number 20.

(ii) The total output over 20 weeks is the sum of the first 20 weekly outputs. Under scheme I, use the AP formulae (5.1) and (5.2) to calculate term number 20 and the sum of the first 20 terms, respectively; while under scheme II, use the GP formulae (5.3) and (5.5) to calculate term 20 and the sum of the first 20 terms, respectively. The calculations are as follows:

Scheme I	Scheme II
(i) Output in week n is $[a + (n-1)d]$, output in week 20 $\quad = [1200 + (20-1)80]$ $\quad = 2720$	(i) Output in week n is ar^{n-1}, output in week 20 $\quad = 1200(1.05)^{19}$ $\quad = 3032.34 = 3032$
(ii) Total output in 20 weeks is $S_n = \dfrac{n}{2}[2a + (n-1)d]$ $S_n = \dfrac{20}{2}[2(1200) + (20-1)80]$ $\quad = 10[2400 + (19)(80)]$ $\quad = 10[2400 + 1520]$ $\quad = 39\,200$	(ii) Total output in 20 weeks is $S_n = \dfrac{a(r^n - 1)}{r - 1}$ $S_{20} = \dfrac{1200(1.05^{20} - 1)}{1.05 - 1}$ $\quad = \dfrac{1200(1.6533)}{(0.05)}$ $\quad = \dfrac{1983.96}{0.05} = 39\,679.2$

(b) Scheme I forms an arithmetic series with $a = 1200$ and $d = 80$.
Each term number in the series represents the corresponding week. The value of each term represents the output for that week, that is,

Week	T_1 = week 1	T_2 = week 2	T_3 = week 3	...	T_n = week n
Output	1200	$1200 + 80$ $= 1280$	$1200 + 2(80)$ $= 1360$	...	$1200 + (n-1)(80)$ $= 8000$

So, solve for the value of n (week n) for which $T_n = 8000$:

$$T_n = a + (n-1)d$$
$$8000 = 1200 + (n-1)(80)$$
$$8000 - 1200 = (n-1)(80)$$
$$\frac{6800}{80} = n - 1$$
$$86 = n$$

The output for week 86 is 8000.

Under scheme II, the series is geometric with $a = 1200$ and $r = 1.05$; therefore, each weekly output corresponds to the terms of the GP:

Week	$T_1 = $ week 1	$T_2 = $ week 2	$T_3 = $ week 3	...	$T_n = $ week n
Output	1200	$1200(1.05)$	$1200(1.05)^2$	...	$1200(1.05)^{n-1}$

So, solve for the value of n (week n) for which $T_n = 8000$:

$$T_n = ar^{n-1}$$
$$8000 = 1200(1.05)^{n-1}$$
$$\frac{8000}{1200} = (1.05)^{n-1} \qquad \text{dividing both sides by 1200}$$
$$\log(2/3) = \log(1.05)^{n-1} \qquad \text{take logs of both sides}$$
$$\log(2/3) = (n-1)\log(1.05) \qquad \text{use rule 3 for logs}$$
$$\frac{\log(2/3)}{\log(1.05)} = n - 1 \qquad \text{divide both sides by } \log(1.05)$$
$$38.8833 = n - 1 \qquad \text{evaluate the logs on the calculator, and divide}$$
$$39.8833 = n$$

During week 39, output reaches 8000 units.

Summary

Characteristics of arithmetic sequences
a is the first term of the sequence
d is the common difference
nth term, T_n, is given by $[a + (n-1)d]$
$S_n = \dfrac{n}{2}[2a + (n-1)d]$

Characteristics of geometric sequences
a is the first term of the sequence
r is the common ratio
nth term, T_n, is given by ar^{n-1}
$S_n = \dfrac{a(1-r^n)}{1-r}$

PROGRESS EXERCISES 5.2

Applications of APs and GPs

1. A TV manufacturer plans to increase its output by 5% each month. If it is now producing 300 TVs per month, calculate, using series,
   (a) Its monthly output in 15 months from now.

    (b) Its total output in 15 months, starting with the present month.

    (c) The month in which its output reaches 500.

2. At a certain time each year, the scaling down of weekly production of sun hats starts.

    (a) If weekly production is 190 000 hats, and is reduced by 15 200 hats each week, use series to calculate:

      (i) The weekly output after 10 weeks of scaling down.

      (ii) The total production during that 10 weeks.

    (b) Calculate the weekly reduction if the production should be scaled down from 190 000 to zero in 12 weeks.

3. A company manufactures deckchairs for a 10-week period each year. Production starts with 600 chairs in week 1 and increases by 50 for each subsequent week. Use series to calculate:

    (a) The number of chairs manufactured in week 7.

    (b) The total number of chairs manufactured during the 10 weeks.

4. Two competing companies produce mobile phones. Company A starts production at 1000 phones per week and plans to increase output by 200 each week. Company B starts production with 500 phones per week and plans to increase output by 20% each week.

    (a) Calculate the weekly production in weeks 1; 5; 10; 15. Plot the graph of weekly output for each producer on the same diagram (use Excel, if available). Estimate the week in which the output is the same for both firms.

    (b) Calculate the total production during the first 15 weeks for each firm.

5. Heating a large building requires 50 000 gallons of oil in 2013. As the building is extended, two forecasts for growth in demand for oil are made:

Forecast 1: increases of 5000 gallons per year.

Forecast 2: increases of 6% each year.

Use series to calculate for each forecast:

    (a) The number of gallons required in 2015, 2020.

    (b) The total consumption of oil from 2015 to 2020 inclusive.

6. The manufacturer of a new cholesterol-reducing yogurt plans to increase output over the coming years. The present output is 5500 units per week.

    (a) If output is increased by 150 per week calculate:

      (i) The weekly output for the 30th week from now.

      (ii) The total output over the first 30 weeks.

      (iii) The number of weeks before the weekly output exceeds 10 000.

    (b) If output is increased by 5% each week, calculate:

      (i) The weekly output in 30 weeks from now.

      (ii) The total output over the first 30 weeks.

      (iii) The week in which output first reaches 10 000.

7. An e-market company is planning to expand into two new markets. Based on extensive market research, the projections for each market are as follows:

Market 1: the number of customers will increase by 800 per year.

Market 2: the number of customers will increase by 7% each year.

If the number of customers at the start of 2007 is 3200 in Market 1 and 7600 in Market 2, use series to calculate for each market:

    (a) The number of new customers in 2020.

    (b) The number of customers in 2020.

5.2 Simple Interest, Compound Interest and Annual Percentage Rates

At the end of this section, you should be able to:

- Calculate simple interest
- Calculate compound interest and derive the compound interest formula
- Solve for unknown variables using the compound interest formula
- Calculate compound interest when compounded several times per year
- Calculate compound interest when compounded continuously
- Calculate the annual percentage rate.

Arithmetic and geometric progressions have many applications in financial mathematics.

Simple interest

Simple interest is a fixed percentage of the principal, P_0, that is paid to an investor each year, irrespective of the number of years the principal has been left on deposit; that is, money invested at simple interest will increase in value by the same amount each year. So, if the investor is paid a fixed annual amount, $i\%$ of P_0, then the amount of simple interest, I, received over t years is given by the formula

$$I = P_0 \times i \times t \tag{5.7}$$

Therefore, the total value after t years, P_t, is the principal plus interest and is given by

$$P_t = P_0(1 + it) \tag{5.8}$$

When the total value (future value), the interest rate and time are known, the principal (present value) may be calculated by rewriting formula (5.8) as

$$P_0(1 + it) = P_t$$

$$P_0 = \frac{P_t}{1 + it} \tag{5.9}$$

Formula (5.9) is often referred to as the 'present value' formula.

Note: The interest rate is always quoted as a percentage, $i\%$. In calculations, you write $i\%$ as the fraction $i/100$ (Section 1.6).

www.wiley.com/college/bradley

Go to the website for the following additional material that accompanies Chapter 5:

Worked Example 5.4 Simple interest calculations.

Compound interest

In the modern business environment, the interest on money borrowed (lent or invested) is usually compounded. Compound interest pays interest on the principal plus on any interest accumulated in previous years. The total value, P_t, of a principal, P_0, when interest is compounded at $i\%$ per annum is

$$P_t = P_0(1+i)^t \qquad \qquad (5.10)$$

Formula (5.10) is derived as follows.

Deriving the compound interest formula

Before proceeding with worked examples on compound interest, the derivation of the compound interest formula is explained briefly:

Principal at start of year	Interest paid each year	Total at end of year = principal + interest	Total at end of year = principal at start of next year
P_0	$i \times P_0$	$P_0 + iP_0 = P_0(1+i)$	$P_1 = P_0(1+i)$
P_1	$i \times P_1$	$P_1 + iP_1 = [P_1] \cdot (1+i) =$ $[P_0(1+i)] \cdot (1+i) = P_0(1+i)^2$	$P_2 = P_0(1+i)^2$
P_2	$i \times P_2$	$P_2 + iP_2 = [P_2] \cdot (1+i) =$ $[P_0(1+i)^2] \cdot (1+i) = P_0(1+i)^3$	$P_3 = P_0(1+i)^3$
$\vdots$	$\vdots$	$\vdots$	$\vdots$
P_{t-1}	$i \times P_{t-1}$	$P_{t-1} + iP_{t-1} = [P_{t-1}] \cdot (1+i) =$ $[P_0(1+i)^{t-1}](1+i)$ $= P_0(1+i)^t$	$P_t = P_0(1+i)^t$

Each year, the interest earned is added to the total amount on deposit (principal, P_0, plus any accumulated interest), at the beginning of that year.

Notice that the amounts due at the end of each year form a geometric progression where $a = P_0$, and r, the common ratio, is $(1+i)$, giving the sequence

$$P_0, \ P_0(1+i)^1, \ P_0(1+i)^2, \ldots, \ P_0(1+i)^t$$

WORKED EXAMPLE 5.5
COMPOUND INTEREST CALCULATIONS

Calculate the amount owed on a loan of £1000 over three years at an interest rate of 8% compounded annually.

SOLUTION

Substitute $P_0 = 1000$, $i = \dfrac{8}{100} = 0.08$, $t = 3$ years into formula (5.10),

$$P_3 = P_0(1+i)^t = 1000(1+0.08)^3$$
$$= 1000(1.08)^3 = 1000(1.259\,712\,0) = 1259.712$$

Present value at compound interest

The present value of a future sum, P_t, is the amount which, when put on deposit at $t = 0$, at i% rate of interest, will grow to the value of P_t after t years. The present value, P_0, is calculated by rearranging the compound interest formula (5.10):

$$P_t = P_0(1+i)^t \rightarrow \frac{P_t}{(1+t)^t} = P_0$$

$$P_0 = \frac{P}{(1+i)^t} \tag{5.11}$$

WORKED EXAMPLE 5.6
FUTURE AND PRESENT VALUES WITH COMPOUND INTEREST

£5000 is invested at 8% compound interest per annum for three years.

(a) Calculate the value of the investment at the end of three years.
(b) Compute the present value of receiving (i) £6298.50, (ii) £15 000 in three years' time when the discount rate is 8%.

SOLUTION

(a) $P_0 = £5000$, $i = 8/100$, $t = 3$. Using formula (5.10), the total amount received after three years is

$$P_3 = 5000(1+0.08)^3$$
$$= 5000(1.2597)$$
$$= 6298.5$$

(b) The present value of receiving £6298.50 and £15 000 in three years at a discount rate of 8% is calculated by using formula (5.11):

Present value of £6298.5

$$P_0 = \frac{P_3}{(1+i)^3}$$
$$= \frac{6298.50}{(1.08)^3} = 5000$$

Present value of £15 000

$$P_0 = \frac{P_3}{(1+i)^3}$$
$$= \frac{15\,000}{(1.08)^3} = 11\,907.48$$

Other applications of the compound interest formula

In the compound interest formula, there are four variables, P_0, P_t, i, and t. If any three of these variables are given, the fourth may be determined. In some cases, you will require the rules for indices and logs. For example, a general expression for i may be derived as

$$P_t = P_0(1+i)^t \quad \rightarrow \quad (1+i)^t = \frac{P_t}{P_0} \quad \rightarrow \quad 1+i = \left(\frac{P_t}{P_0}\right)^{1/t}$$

$$\text{or} \quad i = \sqrt[t]{\frac{P_t}{P_0}} - 1 \qquad\qquad\qquad\qquad\qquad \textbf{(5.12)}$$

However, in the long term, you will find it difficult to remember all these formulae. Therefore, to solve for any one unknown variable, substitute the given values of the other variables into the compound interest formula and solve for the unknown.

WORKED EXAMPLE 5.7
CALCULATING THE COMPOUND INTEREST RATE AND TIME PERIOD

(a) Find the compound interest rate required for £10 000 to grow to £20 000 in six years.
(b) A bank pays 7.5% interest, compounded annually. How long will it take for £10 000 to grow to £20 000?

SOLUTION

(a) Calculate i when given $P_0 =$ £10 000, $P_t =$ £20 000 and $t = 6$.

Using formula (5.12)	Directly from the compound interest formula, (5.10)
$1 + i = \left(\dfrac{P_t}{P_0}\right)^{1/t}$	$20\,000 = 10\,000(1 + i)^6$
$1 + i = \left(\dfrac{20\,000}{10\,000}\right)^{1/6}$	$\dfrac{20\,000}{10\,000} = (1 + i)^6 \rightarrow 2 = (1 + i)^6$
$1 + i = 1.12$	$2^{1/6} = (1 + i)$
$i = 0.12$	$1.12 = 1 + i$
	$0.12 = i$
	$i = 0.12$

An interest rate of 12% is required if this investment is to double in value over a period of six years.

(b) Calculate t when given $P_0 = £10\,000$, $P_t = £20\,000$ and $i = 7.5/100 = 0.075$. Substitute the given values into formula (5.10),

$$20\,000 = 10\,000(1 + 0.075)^t$$
$$\frac{20\,000}{10\,000} = (1 + 0.075)^t \qquad \text{divide both sides by } 10\,000$$
$$2 = (1.075)^t$$
$$\log(2) = \log(1.075)^t = t\,\log(1.075) \qquad \text{take logs of each side of the equation}$$
$$\frac{\log 2}{\log(1.075)} = t \qquad \text{divide both sides by } \log(1.075)$$
$$9.5844 = t \qquad \text{evaluate the logs on the calculator and divide}$$

So, at 7.5% interest it will take over nine years for the investment to double in value.

PROGRESS EXERCISES 5.3

Simple and Compound Interest

1. Suppose £5000 is invested for five years. Calculate the amount accumulated at the end of five years if interest is compounded annually at a nominal rate of (a) 5%, (b) 7%, (c) 10%.
2. A savings account of £10 000 earns simple interest at 5% per annum. Calculate the value of the account (future value) after six years.
3. £2500 is invested at a nominal rate of interest of 5% compounded annually. Calculate the amount accumulated at the end of (a) 1 year, (b) 4.5 years, (c) 10 years, (d) 20 years.
4. Calculate the present value of £6000 that is expected to be received in three years' time with simple interest of 7.5% per annum.

5. How much is a sum of £3500 worth at the end of five years if deposited at (a) 11% simple interest and (b) 11% compound interest, each calculated annually?

6. Calculate the present value of £6500 to be received in two years' time when the interest rate is 8% per annum, compounded annually.

7. Calculate the present value of £10 000 due in five years if interest is compounded annually at 4.5%.

8. Calculate the annual rate of interest required for an investment to double in value in (a) 12 years, (b) 8 years, (c) 5 years. Assume annual compounding.

9. How long should it take for an investment of £3700 to grow to £5000 when interest is compounded annually at 6.5%?

10. How long should it take for an investment of £56 000 to grow to £74 000 when interest is compounded annually at (a) 6% and (b) 3%?

11. How long should it take for an investment to double in value when interest is compounded annually at (a) 6% and (b) 3%?

12. Calculate the compound interest rate required for £5000 to grow to £9000 in four years.

13. Calculate the number of years it will take for a sum of £5000 to grow to £20 000 when invested at 5.5% interest compounded annually.

14. Set up a table of values comparing the growth of an investment of £1 at (a) an annual interest rate of 10% simple interest, (b) an annual interest rate of 10% compounded annually for 0, 5, 10, 15 and 20 years.

15. Suppose £2500 is invested for eight years and grows to £4525 when the annual rate of interest is i%. Calculate i.

When interest is compounded several times per year

So far, it has been assumed that compound interest is compounded once a year. In reality, interest may be compounded several times per year; for example, it may be compounded daily, weekly, monthly, quarterly, semi-annually or continuously. Each time period is known as a conversion period or interest period. The number of conversion periods per year is denoted by the symbol m; the interest rate applied at each conversion is i/m. For example, an investment compounded 12 times per year will have 12 conversion periods; therefore if a five-year investment was compounded 12 times annually, then the investment would have 60 conversion periods; that is,

$$n = m \times t$$

where

n = total number of conversion periods
m = conversion periods per year
t = number of years

The value of the investment at the end of n conversion periods is

$$P_t = P_0 \left(1 + \frac{i}{m}\right)^n = P_0 \left(1 + \frac{i}{m}\right)^{m \times t}$$ (5.13)

WORKED EXAMPLE 5.8
COMPOUNDING DAILY, MONTHLY AND SEMI-ANNUALLY

 Find animated worked examples at **www.wiley.com/college/bradley**

£5000 is invested for three years at 8% per annum compounded semi-annually.
At the end of three years:

(a) Calculate the total value of the investment.
(b) Compare the return on the investment when interest is compounded annually with that when compounded semi-annually.
(c) Calculate the total value of the investment when compounded, (i) monthly, (ii) daily. Assume all months consist of 365/12 days.

SOLUTION

(a) $P_0 = £5000$, $i = 8/100 = 0.08$, $t = 3$ and $m = 2$. Using formula (5.13), the total value after three years with $n = m \times t = 6$ conversion periods is calculated as

$$P_3 = 5000 \left(1 + \frac{0.08}{2}\right)^{(2)(3)}$$

$$= 5000(1 + 0.04)^6$$

$$= 6326.59$$

(b) In Worked Example 5.6, the total value of £5000 after three years compounded annually was £6298.50. When the same investment is compounded semi-annually, the total value is £6326.59, a gain of £28.09 over the value when compounded annually.
(c) The value of the investment at the end of three years for monthly and daily compounding is calculated as:

(i) Monthly compounding	(ii) Daily compounding
$m = 12 \rightarrow n = m \times t = 12(3) = 36$	$m = 365 \rightarrow n = m \times t = 365(3) = 1095$
$P_3 = P_0 \left(1 + \dfrac{i}{m}\right)^{(m)(t)}$	$P_3 = P_0 \left(1 + \dfrac{i}{m}\right)^{(m)(t)}$
$= 5000 \left(1 + \dfrac{0.08}{12}\right)^{(12)(3)}$	$= 5000 \left(1 + \dfrac{0.08}{365}\right)^{(365)(3)}$
$= 5000(1.2702)$	$= 5000(1.2712)$
$= 6351$	$= 6356$
This earns $6351 - 6298.5 = £52.5$ more than when interest is compounded annually.	This earns $6356 - 6298.5 = £57.5$ more than when interest is compounded annually.

Continuous compounding

In the previous example, compounding over several time intervals was illustrated: semi-annually, monthly and daily. In each case, the return on the investment increases as m, the number of compoundings per year, increases. One could also compound continuously, for example, every minute, second, or microsecond. With continuous compounding, the future value is given by the formula

$$P_t = P_0 e^{it} \qquad\qquad (5.14)$$

where $e = 2.718\,281\,828$ (see Chapter 4).

This formula is derived from formula (5.13), by letting the number of compoundings per year, m, become very large so that

$$P_t = P_0 \left(1 + \frac{i}{m}\right)^{(m)(t)} = P_0 \left[\left(1 + \frac{i}{m}\right)^{m}\right]^t = P_0 [e^i]^t = P_0 e^{it}$$

Note: $\left(1 + \frac{i}{m}\right)^m$ approaches e^i as m gets larger. You should check these by evaluating $\left(1 + \frac{i}{m}\right)^m$ for one value of i and various values of m, such as $i = 5/100$, $m = 10, 20, 30, 40, 50, 365$. You should get: 1.051 14, 1.051 205 5, 1.051 227, 1.051 238 3, 1.051 244 8, 1.051 267 5; while $e^i = e^{0.05} = 1.051\,271\,1$.

WORKED EXAMPLE 5.9
CONTINUOUS COMPOUNDING

A financial consultant advises you to invest £5000 at 8% compounded continuously for three years. Find the total value of your investment.

SOLUTION
$P_0 = £5000$, $i = 8/100 = 0.08$ and $t = 3$. Therefore, the total value is given by

$$P_3 = 5000 e^{(0.08)(3)}$$

$$= 5000(1.27)$$

$$= 6356.24$$

When interest was compounded annually in Worked Example 5.6, the value of the investment after three years was £6298.5, which is £57.74 less compared with continuous compounding.

Annual percentage rate (APR)*

Interest rates are usually cited as nominal rates of interest expressed as per annum figures. However, as compounding may occur several times during the year with the nominal rate, the amount owed or accumulated will be different from that calculated by compounding once a year. So, a standard measure is needed to compare the amount earned (or owed) at quoted nominal rates of interest when compounding is carried out several times per year. This standard measure is called the annual percentage rate (APR) or effective annual rate. If interest is compounded once a year at the APR rate, the investment would yield exactly the same return, P_t, at the end of t years, as it would if interest were compounded m times per year at the nominal rate.

For example, consider an investment of £100 at a nominal interest rate of 10% compounded semi-annually, that is, at 5% for six months. The investment at the end of the first six months is worth £[100 + 0.05(100)] = £105. Then invest this £105 for a further six months and it grows to £[105 + 0.05(105)] = £110.25. At the end of the year, the initial investment of £100 has grown to £110.25, giving an annual percentage rate, APR, of 10.25% which is greater than the nominal rate of 10%.

The formula for the annual percentage rate (APR) is derived as follows:

$$P_t = P_0 \left(1 + \frac{i}{m}\right)^{mt} \qquad \text{nominal rate of } i\% \text{ compounded } m \text{ times per year}$$

$$P_t = P_0(1 + APR)^t \qquad \text{APR rate compounded annually}$$

$$P_0(1 + APR)^t = P_0\left(1 + \frac{i}{m}\right)^{mt} \qquad \text{since the yield is the same}$$

$$(1 + APR)^t = \left(1 + \frac{i}{m}\right)^{mt} \qquad \text{dividing both sides by } P_0$$

$$1 + APR = \left(1 + \frac{i}{m}\right)^{m} \qquad \text{taking the } t\text{th root of each side}$$

$$APR = \left(1 + \frac{i}{m}\right)^{m} - 1$$

The reader is asked to explain each step in the derivation of the formula for the APR when interest is compounded continuously as outlined below:

$$P_t = P_0 e^{it} \quad \text{and} \quad P_t = P_0(1 + APR)^t$$

$$P_0(1 + APR)^t = P_0 e^{it}$$

$$(1 + APR)^t = e^{it}$$

$$1 + APR = e^{i}$$

$$APR = e^{i} - 1$$

The APR, therefore, may be calculated by the following formulae:

• In summary, when the nominal rate is compounded m times per year,

$$APR = \left(1 + \frac{i}{m}\right)^{m} - 1 \qquad \textbf{(5.15)}$$

- When the nominal rate is compounded continuously,

$$APR = e^i - 1 \tag{5.16}$$

Again, when calculating the APR, it will be easier, and it will make more sense, to work through the method rather than memorising formulae such as (5.15) and (5.16).

* There are many variations of APR. The definitions and calculations given above are for the basic APR. When you understand the basic APR you should be easily able to deal with any of the numerous variations.

WORKED EXAMPLE 5.10
ANNUAL PERCENTAGE RATES

(a) Find the annual percentage rate on a loan corresponding to 6.0% compounded monthly.
(b) Two banks in a local town quote the following nominal interest rates: bank A pays interest on a savings account at 6.60% compounded monthly and bank B pays 6.65% on a savings account compounded semi-annually. Which bank pays its savers the most interest?

SOLUTION

(a)
$$APR = \left(1 + \frac{i}{m}\right)^m - 1 = \left(1 + \frac{0.06}{12}\right)^{12} - 1 = 1.061\,67 - 1 = 0.061\,67 = 6.17\%$$

The annual percentage rate payable on the loan is 6.17% which is greater than the quoted nominal rate of 6.0%.

(b) Bank A pays 6.60% compounded monthly. The annual percentage rate is given by

$$APR = \left(1 + \frac{i}{m}\right)^m - 1 = \left(1 + \frac{0.066}{12}\right)^{12} - 1 = 1.0680 - 1 = 0.0680 = 6.8\%$$

Bank B pays 6.65% compounded semi-annually. The annual percentage rate is given by

$$APR = \left(1 + \frac{i}{m}\right)^m - 1 = \left(1 + \frac{0.0665}{2}\right)^2 - 1 = 1.0676 - 1 = 0.0676 = 6.76\%$$

Bank A offers the greater annual percentage rate and thus pays its savers more interest despite its lower nominal interest rate.

PROGRESS EXERCISES 5.4

Interest Compounded at Various Intervals: *APR*

1. Suppose £50 000 is invested at a nominal interest rate of 5.5% per annum. Interest is calculated (a) annually as simple interest, (b) annually as compound interest, (c) four times annually as compound interest, (d) continuously.
   (i) Calculate the value of the investment at the end of each year for the first five years.
   (ii) Calculate the number of years it will take for the investment to double in value.

2. Suppose £5000 is invested for five years. Calculate the amount accumulated at the end of five years if interest is compounded continuously at a nominal annual rate of (a) 5%, (b) 7%, (c) 10%.

3. Calculate the *APR* for a 6% per annum nominal rate of interest that is compounded (a) 4 times per year, (b) 12 times per year, (c) continuously.

4. Suppose £2500 is invested at a nominal rate of interest of 5% per annum. Calculate the amount accumulated at the end of 10 years if interest is calculated (a) annually, (b) 4 times annually, (c) 52 times per year, (d) continuously.

5. Calculate the *APR* for a 6% per annum nominal rate of interest compounded 20 times annually.

6. Calculate the APR for an 8% per annum nominal rate of interest compounded four times annually.

7. Calculate the APR for an 8% per annum nominal rate of interest compounded continuously.

8. Calculate the APR for a 3% per annum nominal rate of interest compounded continuously.

9. Calculate *i* if £2500 grows to £5162 when invested at an annual rate of interest of *i*% for 7.5 years.

10. How long should any sum of money be left on deposit at 4.5% APR if the value of the deposit is to increase by 70%?

11. Suppose £5500 is deposited at a nominal rate of interest of 6% per annum. Calculate the value of the investment in three years' time if interest is compounded (a) semi-annually, (b) monthly, (c) daily, (d) continuously.

12. Two banks in a local town quote the following nominal interest rates: bank A charges interest at 8.80% per annum compounded semi-annually and bank B charges 8.75% per annum compounded quarterly. Which bank charges the most interest?

5.3 Depreciation

At the end of this section, you should be able to:

- Calculate straight-line depreciation
- Calculate reducing-balance depreciation.

Depreciation is an allowance made for the wear and tear of equipment during the production process. It involves the deduction of money from the original asset value, A, each year. Two depreciation techniques are analysed: straight-line and reducing-balance depreciation.

Straight-line depreciation

Straight-line depreciation is the converse of simple interest with equal amounts being subtracted from the original asset value each year. For example, if the original value of a machine was £20 000 and after four years its value is estimated to be £8000, then the amount of straight-line depreciation subtracted each year is (£20 000 − £8000) ÷ 4 = £3000.

Reducing-balance depreciation

Reducing-balance depreciation is the converse of compound interest with larger amounts being subtracted from the original asset value each year. The formula for reducing-balance depreciation is given as

$$A_t = A_0(1 - i)^t \tag{5.17}$$

where

A_t = value of the asset after t years taking account of depreciation
A_0 = original value of the asset
i = depreciation rate expressed as a decimal (e.g., 5% is 0.05)
t = number of time periods, usually years.

WORKED EXAMPLE 5.11
FUTURE VALUE OF ASSET AND REDUCING-BALANCE DEPRECIATION

A machine costing £30 000 depreciates by 15% each year. Calculate the value of the machine after five years. What is the total amount of depreciation?

SOLUTION
Write down any information given: A_0 = £30 000, i = 15/100 = 0.15, t = 5. Substitute these values into formula (5.17) to calculate the book value of the machine after five years:

$$A_t = A_0(1 - i)^t = 30\,000(1 - 0.15)^5 = 13\,311.6$$

After five years the book value of the machine is £13 311.16. Therefore, the total amount of depreciation is (30 000 − 13 311.16) = £16 688.84.

WORKED EXAMPLE 5.12
PRESENT VALUE OF ASSET AND REDUCING-BALANCE DEPRECIATION

 Find animated worked examples at **www.wiley.com/college/bradley**

Five years after purchase, a computer has a scrap value of £300. The depreciation rate is 26%. Calculate the value of the computer when it was bought five years ago.

SOLUTION

In this example, you are required to calculate the original value of the computer when first bought. Write down any information given: $A_t = 300$, $i = 26/100 = 0.26$, $t = 5$. Substitute these values into formula (5.17):

$$A_t = A_0(1 - i)^t$$

$$A_0 = \frac{A_t}{(1 - i)^t} \qquad \text{rearranging}$$

$$= \frac{300}{(1 - 0.26)^5} = \frac{300}{0.2219} = 1351.96$$

The original value of the computer five years ago was £1351.96.

5.4 Net Present Value and Internal Rate of Return

At the end of this section you should be able to:

- Use the net present value as an investment appraisal technique
- Use the internal rate of return as an investment appraisal technique
- Compare the two techniques.

Net present value and internal rate of return are two techniques used to appraise investment projects. Each technique is analysed in turn.

Net present value (NPV)

In Section 5.2 the present value of a sum due to be paid in t years' time is calculated by equation (5.11):

$$P_0 = \frac{P_t}{(1 + i)^t}$$

The net present value is the present value of several future sums discounted back to the present.

The net present value technique uses present values to appraise the profitability of investments undertaken by firms. For example, the costs (indicated with a minus sign), and the returns of a project, as estimated by the project team, are given in Table 5.3.

Table 5.3   Cash flows of an investment project

Year	Cash flows (£000s)
0	−400
1	120
2	130
3	140
4	150

The problem confronting the project manager is whether the investment is worthwhile. The decision is made even more difficult since the returns associated with the project will be received in the future, while the costs usually occur at present. Therefore, to make an informed decision, the project manager needs to compare costs and returns, all brought back to present values. These present returns are then compared with present costs to estimate a net present value calculation on the project where

$$NPV = \text{present value of cash inflows} - \text{the present value of cash outflows} \qquad \textbf{(5.18)}$$

A decision rule is

$$NPV > 0 \quad \text{Invest in the project}$$
$$NPV < 0 \quad \text{Don't invest in the project}$$

WORKED EXAMPLE 5.13
CALCULATING NET PRESENT VALUE

Calculate the net present value of the investment project as outlined in Table 5.3 assuming 8% discount rate compounded annually.

Note: When future values are brought back to present values at a given rate of interest, the interest rate is referred to as the discount rate.

SOLUTION

The present value of the cash outflow is the initial outlay of £400 000. The present value of cash inflows is calculated by summing the present value of each cash inflow as follows:

Present value of return in year 1: $P_0 = \dfrac{120\,000}{(1+0.08)^1} = 111\,111$

Present value of return in year 2: $P_0 = \dfrac{130\,000}{(1+0.08)^2} = 111\,454$

Present value of return in year 3: $P_0 = \dfrac{140\,000}{(1+0.08)^3} = 111\,137$

Present value of return in year 4: $P_0 = \dfrac{150\,000}{(1+0.08)^4} = 110\,254$

Total present value over the four years: $443\,956$

The net present value is then

$$NPV = 443\,956 - 400\,000 = 43\,956$$

As the *NPV* is positive, the firm will invest in this project. Thus, the investment is cost effective at 8% discount rate compounded annually.

The size of the *NPV* will become smaller when:

(i) The initial cost outlay increases.
(ii) The discount rate at which *NPV* is computed increases.

Note: The present value of each future cash flow above was calculated by multiplying the cash due in t years by a factor $1/(1 + i)^t$, where i is the discount rate. In fact, tables of discount factors are available for various rates. However, in the following problems, the discounting factors are readily calculated with the use of the calculator, so we shall not avail ourselves of the tables.

Internal rate of return (IRR)

In the previous section, the profitability of a project was assessed on the basis of the *NPV* calculated at a given rate of interest of 8%. The profitability changes as the interest rate changes. If the interest rate was 12.6555%, the *NPV* would be zero, while further increases in the interest rate, such as $r = 15\%$, would result in negative *NPV*s. Table 5.4 shows the calculation of the *NPV* for these three interest rates.

The *IRR* is the interest rate for which the *NPV* is zero. A project is viable if the prevailing interest rate is less than the *IRR*, but not profitable if the prevailing interest rate is greater than the *IRR*. The decision rule is

Accept the project if the *IRR* is greater than the market rate of interest.

Table 5.4   Calculation of *NPV*s for various interest rates

t years	Net cash flow	Dis. factor $i = 8\%$ $\dfrac{1}{(1+0.08)^t}$	Present value (8%)	Dis. factor 12.6555% $\dfrac{1}{(1+0.126\,555)^t}$	Present value (12.6555%)	Dis. factor $i = 15\%$ $\dfrac{1}{(1+0.15)^t}$	Present value (15%)
0	−400	1	−400	1	−400	1	−400
1	120	0.9259	111	0.8877	107	0.8696	104
2	130	0.8573	111	0.7879	102	0.7561	98
3	140	0.7938	111	0.6994	98	0.6575	92
4	150	0.7350	110	0.6209	93	0.5718	86
			43		0		−20

To calculate the IRR for a given project: (a) graphically, (b) by calculation

(a) **The graphical method**: The NPV for several rates of interest is plotted against the interest rates, such that the NPVs range from positive to negative. Then, according to the definition given above, the IRR is the interest rate at which the NPV $= 0$. This is the point at which the curve crosses the horizontal axis.

(b) **By calculation**: As a follow-on from the graphical method described above, the point may be calculated at which the curve of NPV plotted against i crosses the horizontal axis. If two points on the curve are known, one above the axis (i_1, NPV_1) where $NPV_1 > 0$ and the other below the axis (i_2, NPV_2) where $NPV_2 < 0$, the IRR, which is the point of intersection with the horizontal axis, is given by the formula

$$IRR = \frac{(i_1 \times NPV_2) - (i_2 \times NPV_1)}{NPV_2 - NPV_1} \tag{5.19}$$

Formula (5.19) is proved in the Appendix to this chapter.

Note: The IRR is determined here by linear interpolation, that is, the straight line joining the points (i_1, NPV_1) to (i_2, NPV_2). This line is an approximation to a curve. Therefore, the IRR may have slightly different values when calculated using different pairs of points. The best estimate is obtained by taking points on opposite sides, but as close as possible to the horizontal axis.

WORKED EXAMPLE 5.14
IRR DETERMINED GRAPHICALLY (EXCEL) AND BY CALCULATION

A project involves an initial outlay of £550 000. The expected cash flow at the end of the next five years is given as

Year	1	2	3	4	5
Cash flow	108 300	130 500	170 900	200 000	160 000

Determine the IRR

(a) Graphically by plotting the NPV against i for $i = 0.05; 0.07; 0.09; 0.11; 0.13; 0.15; 0.17$.
(b) By calculation. Show that the value of the IRR is slightly different when calculated from different pairs of points.

SOLUTION

(a) **Graphically:** Excel is ideal at this point, since the calculation of NPVs requires the repeated use of formulae. For part (a), several NPVs are required to plot the curve of NPV against i. The formula which calculates the present value of the cash flow at discount rate i, expressed as a decimal, is entered into the appropriate cell as follows:

$$= (\text{cash flow}) * (1/(1+i)\,\hat{}\,t)$$

Since the values of t are always in column A and cash flow in column B (Table 5.5), the addresses of these columns may be entered in the following formula at cell C2:

$$= (\$B2) * (1/(1+i)\,\hat{}\,\$A2)$$

- Enter this formula in cell C2, with $(1+i) = 1.05$. Then copy the formula down column C to calculate the present values for years 1 to 5. Point to cell C8. Then use the Σ button to sum the contents of cells C2 to C7 to find the NPV.
- To calculate the NPV for each of the other discount rates, copy the formula in cell C2 across row 2. The $ sign in front of the A and B means that these references do not change. However, since the discount rate is different for each column, this part of the formula must be changed. For example, in the 11% column, replace $(1+i)$ by 1.11.
- Repeat the 'copy down and sum' procedure described above for columns D, E, F, G, H and I in Table 5.5.

Table 5.5 Excel sheet for calculating NPVs at different interest rates

	A	B	C	D	E	F	G	H	I
			$i = 0.05$	$i = 0.07$	$i = 0.09$	$i = 0.11$	$i = 0.13$	$i = 0.15$	$i = 0.17$
1	t	cash flow	$i = 0.05$	$i = 0.07$	$i = 0.09$	$i = 0.11$	$i = 0.13$	$i = 0.15$	$i = 0.17$
2	0	−550 000	−550 000	−550 000	−550 000	−550 000	−550 000	−550 000	−550 000
3	1	108 300	103 143	101 215	99 358	97 568	95 841	94 174	92 564
4	2	130 500	118 367	113 984	109 839	105 917	102 201	98 677	95 332
5	3	170 900	147 630	139 505	131 966	124 961	118 442	112 370	106 705
6	4	200 000	164 540	152 579	141 685	131 746	122 664	114 351	106 730
7	5	160 000	125 364	114 078	103 989	94 952	86 842	79 548	72 978
8		NPV =	109 045	71 361	36 837	5143	−24 011	−50 881	−75 691

The NPV for each discount rate is plotted in Figure 5.1. The IRR is the value of i at which this graph crosses the horizontal axis. In Figure 5.1, this point is between $i = 11\%$ and 13%, but considerably closer to 11%. $IRR \cong 11.45\%$.

(b) **By calculation:** In Table 5.5, there are several positive and negative NPVs. Therefore, the IRR given by formula (5.19) is calculated from any such pair.
Summarising the points already calculated,

Points	C	D	E	F	G	H	I
i	0.05	0.07	0.09	0.11	0.13	0.15	0.17
NPV	109 045	71 361	36 837	5143	−24 011	−50 881	−75 691

FINANCIAL MATHEMATICS

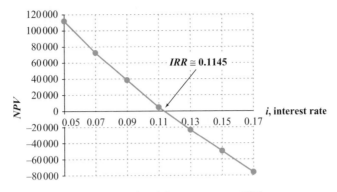

Figure 5.1 Graphical determination of *IRR*

Note: For convenience, the points are labelled according to the column headings in Table 5.5.

(i) From points C and G,

$$IRR = \frac{(i_1 \times NPV_2) - (i_2 \times NPV_1)}{NPV_2 - NPV_1} = \frac{(0.05)(-24\,011) - (0.13)(109\,045)}{-24\,011 - 109\,045}$$

$$= \frac{-15\,376.4}{-133\,056} = 0.115\,56$$

Therefore, *IRR* = 11.556%.

(ii) From points F and I,

$$IRR = \frac{(i_1 \times NPV_2) - (i_2 \times NPV_1)}{NPV_2 - NPV_1} = \frac{(0.11)(-75\,691) - (0.17)(5143)}{-75\,691 - 5143}$$

$$= \frac{-9\,200.32}{-80\,834} = 0.113\,817\,5$$

Therefore, *IRR* = 11.382%.

The two results (i) and (ii) demonstrate that slightly different estimates are calculated from different pairs of points.

Comparison of appraisal techniques: NPV, IRR

When comparing the profitability of two or more projects, the most profitable project would be (a) the project with the largest *NPV*, (b) the project with the largest *IRR*.

The advantage of the *NPV* method is that it gives results in cash terms; it is practical as it discounts net cash flows. A disadvantage is that it relies on the choice of one discount rate; it is possible that a change in the discount rate could lead to a change in the choice of project. The *IRR* method of appraisal does not depend on any external rates of interest, but a major disadvantage of this method is that it does not differentiate between the scale of projects; for example, one project might involve a cash flow in units of £100 000 while another might involve units of £1.

PROGRESS EXERCISES 5.5

Depreciation, *NPV*, *IRR*

1. £150 000 is invested in machinery which depreciates at 8% per annum. How much will the machinery be worth in 10 years by the reducing-balance method?
2. (a) Derive the formula for the net present value factor.
   (b) The net cash flow for two projects, A and B, is as follows:

Year	0	1	2	3	4
Project A	−10 000	−3000	4000	6000	8000
Project B	−5 000	−2000	1000	3000	5000

   (i) Use the net present value criterion to decide which project is the most profitable if a discount rate of 6% is used.
   (ii) Calculate the *IRR*. State the conditions for each project to be profitable.
3. The net cash flow for two projects, A (fast food) and B (amusements), is as follows:

Year	0	1	2	3	4	5
Project A	−420 000	−5 000	122 000	130 000	148 000	150 000
Project B	−95 000	−10 000	−120 000	200 000	110 000	−50 000

   (a) Use the net present value criterion to decide which project is the most profitable if a discount rate of (i) 6% and (ii) 8% is used.
   (b) Estimate the *IRR* of each project. Which project would now be considered more profitable?

5.5 Annuities, Debt Repayments, Sinking Funds

In this section the compound interest problem is extended to annuities, debt repayment and sinking funds. The relevant formulae are derived on the assumption of annual compounding at a rate of $i\%$ per annum, but these formulae are easily adapted to compounding over any interval using equations (5.13) and (5.14), respectively.

5.5.1 Compound interest for fixed deposits at regular intervals of time

In Section 5.2 the standard compound interest formula $P_t = P_0(1 + i)^t$ was derived to calculate the value of an investment of P_0 after t years when interest is compounded annually at $i\%$ per annum. We now consider the situation where, in addition to the initial investment of P_0, a fixed amount A_0 is

FINANCIAL MATHEMATICS

deposited at the end of each year for a period of t years. The following table sets out the calculation of the value of such an investment, year by year.

At the end of	Value of deposit
Year 1	$P_0(1+i) + A_0$
Year 2	$P_0(1+i)^2 + A_0(1+i) + A_0$
Year 3	$P_0(1+i)^3 + A_0(1+i)^2 + A_0(1+i) + A_0$
$\vdots$	$\vdots$
Year t	$P_0(1+i)^t + A_0(1+i)^{t-1} + A_0(1+i)^{t-2} + \cdots + A_0(1+i) + A_0$

$$\text{Value at the} \atop \text{end of } t \text{ years} \quad \underbrace{P_0(1+i)^t}_{P_0 \text{ at } i\% \text{ per annum for } t \text{ years}} + \underbrace{A_0(1+i)^{t-1} + A_0(1+i)^{t-2} + \cdots + A_0(1+i) + A_0}_{A_0 \text{ at } i\% \text{ per annum for years } (t-1),(t-2),\ldots, 2,1,0} \qquad (5.20)$$

The first term, $P_0(1+i)^t$, is simply the value of P_0 after t years with an interest rate of $i\%$ per annum compounded annually. The remainder of the expression is the sum of the sequence of deposits, A_0, made at the end of years $(t-1)$, $(t-2), \ldots, 2,1,0$ with interest rate $i\%$ per annum compounded annually. This series is actually a geometric series, the sum of which may be calculated according to equation (5.4) or (5.5). To show how this is accomplished, rewrite the series

$$A_0(1+i)^{t-1} + A_0(1+i)^{t-2} + \cdots + A_0(1+i) + A_0$$

in reverse order and compare this series with the standard geometric series:

$$\underbrace{A_0}_{a} + \underbrace{A_0(1+i)}_{ar} + \cdots + \underbrace{A_0(1+i)^{t-2}}_{ar^{t-2}} + \underbrace{A_0(1+i)^{t-1}}_{ar^{t-1}} \qquad (5.21)$$

The series (5.21) is a geometric series where $a \equiv A_0$ and $r \equiv (1+i)$. The sum of the first n terms of the standard geometric series is given by (5.5):

$$S_n = \frac{a(r^n - 1)}{r - 1}$$

hence the sum of the first t terms of the series (5.21) is

$$S_t = \frac{A_0[(1+i)^t - 1]}{(1+i) - 1} = A_0 \frac{(1+i)^t - 1}{i} \qquad (5.22)$$

Finally the value (V_t) of the investment at the end of t years is equal to initial investment (P_0) plus the annual investments (A_0) all compounded annually:

$$V_t = P_0(1+i)^t + A_0 \frac{(1+i)^t - 1}{i} \qquad (5.23)$$

This equation forms the basis for a range of applications, such as annuities, debt repayments and sinking funds. In the remainder of this section, formulae are given for annual compounding. It is a simple matter to adjust these formulae for compounding at other intervals, as demonstrated in worked examples.

WORKED EXAMPLE 5.15
COMPOUND INTEREST FOR FIXED PERIODIC DEPOSITS

New members of a club are admitted at the start of each year and pay a joining fee of £2000. Henceforth members pay the annual subscription of £400, which falls due at the end of each year. How much does the club earn from a new member over the first 10 years, assuming an annual interest rate of 5.5%?

SOLUTION

$$P_0 = 2000, \ A_0 = 400, \ i = 5.5/100 = 0.055, \ t = 10$$

Hence

$$V = P_0(1+i)^t + A_0 \frac{(1+i)^t - 1}{i}$$

$$= 2000(1+0.055)^{10} + 400 \frac{(1+0.055)^{10} - 1}{0.055}$$

$$= 3416.2889 + 400 \frac{0.7081}{0.055}$$

$$= 3416.2889 + 400(12.8745)$$

$$= 8566.0889$$

5.5.2 Annuities

An annuity is a series of equal deposits (or withdrawals) made at equal intervals of time; for example, a deposit of £2500 made each year for 20 years towards a pension fund. If the deposit is made at the time of compounding, then the annuity is called an **ordinary annuity**. This is the only type of annuity considered here. To derive the formula for calculating the value of an annuity at the end of t years, consider its value at the end of each year, as follows.

At the end of	Value of annuity
Year 1	A_0
Year 2	$A_0(1+i) + A_0$
Year 3	$A_0(1+i)^2 + A_0(1+i) + A_0$
Year 4	$A_0(1+i)^3 + A_0(1+i)^2 + A_0(1+i) + A_0$
$\vdots$	$\vdots$
Year t	$A_0(1+i)^{t-1} + A_0(1+i)^{t-2} + A_0(1+i)^{t-3} + \cdots + A_0(1+i) + A_0$
Value of annuity after t years	$A_0(1+i)^{t-1} + A_0(1+i)^{t-2} + A_0(1+i)^{t-3} + \cdots + A_0(1+i) + A_0$

This series is identical to the geometric series in equation (5.21). Hence the total amount of the annuity at the end of t years is

$$V_{ANU,t} = A_0 \frac{(1+i)^t - 1}{i} \tag{5.24}$$

Note: This result may be deduced as a special case of (5.23) where $P_0 = 0$.

WORKED EXAMPLE 5.16
ANNUITIES

To provide for future education costs, a family considers various methods of saving. Assume saving will continue for a period of 10 years at an interest rate of 7.5% per annum.

(a) Calculate the value of the fund at the end of 10 years when a single deposit of £2000 is made annually.
(b) How much should be deposited each year if the final value of the fund is £40 000?
(c) How much should be deposited each month if the final value of the fund is £40 000?

SOLUTION

(a) In this case $A_0 = 2000$, $i = 0.075$, $t = 10$ payments. Substitute these values into equation (5.24):

$$V_{ANU,t} = A_0 \frac{(1+i)^t - 1}{i}$$

$$V_{ANU,10} = 2000 \frac{(1+0.075)^{10} - 1}{0.075}$$

$$V_{ANU,10} = 2000 \frac{2.0610 - 1}{0.075} = 2000(14.1467)$$

$$V_{ANU,10} = 28\,293.4$$

The fund will be worth £28 293.4 at the end of 10 years.

(b) $V_{ANU,10} = 40\,000$, $i = 0.075$, $t = 10$. Substitute these values into equation (5.24):

$$40\,000 = A_0 \frac{(1+0.075)^{10} - 1}{0.075}$$

The fraction has been evaluated in (a), so we have

$$40\,000 = 14.1467\,A_0$$

$$A_0 = \frac{40\,000}{14.1467} = 2827.5145$$

To end up with £40 000, the annual deposit should be £2827.5145.

(c) $V_{ANU} = 40\,000$, $\dfrac{i}{12} = \dfrac{0.075}{12}$, $t \times 12 = 10 \times 12 = 120$. Substitute these values into equation (5.24):

$$40\,000 = A_0 \frac{(1 + 0.075/12)^{12 \times 10} - 1}{0.075/12}$$

$$40\,000 = A_0 \frac{1.112\,064\,6}{0.006\,25} = 177.9303\,A_0$$

$$A_0 = \frac{40\,000}{177.9303} = 224.8071$$

Hence saving £224.8071 per month will also provide a fund of £40 000 at the end of 10 years.

The present value of an annuity

Annuities may also be viewed as a series of equal payments (or withdrawals) to be made at equal intervals of time in the future. In such situations we may be interested in the present value of the annuity. In other words, how much should be invested now (V_0) for t years at a given rate of interest (i% per annum) to cover a series of t equal annual payments? If V_0 is invested for t years at i% per annum, its value (V_t) at the end of t years is calculated by the equation $V_t = V_0(1 + i)^t$.

The value of a series of payments, each of A_0, at the end of t years is

$$V_{ANU,t} = A_0 \frac{(1 + i)^t - 1}{i}$$

And if the amount invested is sufficient to provide for the series of payments then

$$V_t = V_{ANU,t}$$

$$V_0(1 + i)^t = A_0 \frac{(1 + i)^t - 1}{i}$$

Solving for V_0 will give the required present value:

$$V_0(1 + i)^t = A_0 \frac{(1 + i)^t - 1}{i}$$

$$V_0 = \frac{1}{(1 + i)^t} A_0 \frac{(1 + i)^t - 1}{i} \qquad \text{dividing both sides by } (1 + i)^t$$

This equation has been used in parts (b) and (c) of Worked Example 5.16. However, for convenience, it is further simplified as follows:

$$V_0 = \frac{A_0}{i} \frac{(1+i)^t - 1}{(1+i)^t} = \frac{A_0}{i} \left(\frac{(1+i)^t}{(1+i)^t} - \frac{1}{(1+i)^t} \right) = \frac{A_0}{i} \left(1 - \frac{1}{(1+i)^t} \right)$$

$$= \frac{A_0}{i} \frac{1 - (1+i)^{-t}}{1}$$

Hence

$$V_0 = A_0 \underbrace{\frac{1 - (1+i)^{-t}}{i}}_{\text{annuity factor}} \tag{5.25}$$

where $\dfrac{1 - (1+i)^{-t}}{i}$ is called the **annuity factor** or **cumulative present value factor**.

WORKED EXAMPLE 5.17
PRESENT VALUE OF ANNUITIES

(a) How much should you pay for an annuity of £1000 a year payable in arrears for 20 years, assuming an interest rate of 6%, if you are to break even?
(b) How much should you pay for an annuity of £1000 a year payable in arrears quarterly for 20 years, assuming an interest rate of 6% per annum.

SOLUTION
Questions (a) and (b) are asking for the present value of the amount of the annuity: How much should be invested now in order to cover the required regular payments? The answers may be calculated by direct application of equation (5.25).

(a) $A_0 = 1000$, $i = 0.06$, $t = 20$, hence

$$V_0 = A_0 \underbrace{\frac{1 - (1+i)^{-t}}{i}}_{\text{annuity factor}}$$

$$V_0 = 1000 \times \frac{1 - (1 + 0.06)^{-20}}{0.06}$$

$$= 1000 \times 11.4699 = 11\,469.92$$

Alternatively, calculate the amount of the annuity first, then its present value:

$$V_{ANU,t} = A_0 \frac{(1+i)^t - 1}{i}$$

$$V_{ANU,20} = 1000 \frac{(1+0.06)^{20} - 1}{0.06} = 36\,785.5912$$

The present value of $V_{ANU,20}$ is given by

$$V_0 = \frac{V_{ANU,20}}{(1+i)^{20}} = \frac{36\,785.5912}{(1.06)^{20}} = 11\,469.9212$$

(b) For quarterly payments,

$$A_0 = \frac{1000}{4} = 250, \; \frac{i}{4} = \frac{0.06}{4} = 0.015, \, t \times 4 = 20 \times 4 = 80$$

And with these adjustments to equation (5.25) for quarterly payments,

$$V_0 = A_0 \underbrace{\frac{1 - (1+i/4)^{-t \times 4}}{i/4}}_{\text{annuity factor}}$$

$$V_0 = 250 \times \frac{1 - (1+0.015)^{-80}}{0.015}$$

$$= 250 \times 46.4073 = 11\,601.8309$$

5.5.3 Debt repayments

A loan is said to be **amortised** if both principal and interest are to be repaid by a series of equal payments made at equal intervals of time, assuming a fixed rate of interest throughout. For such a repayment scheme, the value of the loan (L) and interest rate are usually known but the amount to be repaid at each interval must be calculated. To deduce the formula for such calculations, return to equation (5.23) and make the following substitutions:

$$\underbrace{V_t}_{\substack{\text{At the end of } t \text{ years,} \\ \text{loan is repaid, } V_t = 0}} = \underbrace{P_0(1+i)^t}_{\substack{\text{Debt at end of } t \text{ years assuming} \\ \text{no repayments where } -L \equiv P_0}} + \underbrace{A_0 \frac{(1+i)^t - 1}{i}}_{\substack{\text{Value of a series of regular} \\ \text{payments at the end of } t \text{ years}}} \qquad \textbf{(5.23)}$$

Since the loan L is a debt, it is negative, hence

$$0 = -L(1+i)^t + A_0 \frac{(1+i)^t - 1}{i} \qquad \textbf{(5.26)}$$

(i) Equation (5.26) may be solved for the size of the loan, L, given A_0, i and t:

$$L = \frac{A_0}{i}\left(1 - \frac{1}{(1+i)^t}\right) \tag{5.27}$$

This equation is identical to (5.25).

(ii) Equation (5.26) may be solved for the size of the repayments, A_0, given L, i and t:

$$A_0 = \frac{Li}{1 - \dfrac{1}{(1+i)^t}} = L\,\underbrace{\frac{i}{1-(1+i)^{-t}}}_{\text{capital recovery factor}} \tag{5.28}$$

Here $\dfrac{i}{1-(1+i)^{-t}}$ is called the capital recovery factor.

Compare the capital recovery factor with the annuity factor (5.25) and you will observe that

$$\text{capital recovery factor} = \frac{1}{\text{annuity factor}}$$

Note: Tables of net present values, capital recovery factors, annuity or cumulative present value factors are published. However, these factors are easily evaluated with basic calculators. Tables will not be used in this text.

WORKED EXAMPLE 5.18
MORTGAGE REPAYMENTS

A mortgage of £200 000 is to be repaid over a 25-year period at a fixed interest rate of 4.5%. Calculate the monthly repayments.

SOLUTION

Here we are given that $L = 200\,000$, $\dfrac{i}{12} = \dfrac{0.045}{12} = 0.003\,75$ per month, $t = 25 \times 12 = 300$ payments. Substitute these values into equation (5.28), adjusted for monthly repayments, then solve for the monthly repayment, A_0:

$$A_0 = L\,\underbrace{\frac{i/12}{1-(1+i/12)^{-t\times12}}}_{\text{capital recovery factor}}$$

$$= 200\,000 \times \frac{0.003\,75}{1-(1.003\,75)^{-300}}$$

$$= 200\,000 \times 0.005\,558\,3$$

$$= 1111.6650$$

Repayments per month $= £\,1111.6650$.

How much of the repayment is interest?

When a loan is repaid by a series of equal instalments over equal intervals of time, the proportion of interest being repaid decreases steadily. The following worked example illustrates this behaviour.

WORKED EXAMPLE 5.19
HOW MUCH OF THE REPAYMENT IS INTEREST?

A loan of £5000 is to be paid off in four equal payments at the end of each quarter. Assuming an interest rate of 20%, calculate (a) the amount of each payment, (b) the present value of the loan, (c) the total amount of interest paid, (d) the amount of interest paid each quarter.

SOLUTION

$$L = 5000, \frac{i}{4} = \frac{0.20}{4} = 0.05 \text{ per quarter}, t = 1 \times 4 \text{ payment intervals}$$

(a)
$$A_0 = L \underbrace{\frac{i/4}{1 - (1 + i/4)^{-t\times4}}}_{\text{capital recovery factor}}$$

$$A_0 = 5000 \times \frac{0.05}{1 - (1.05)^{-4}}$$

$$= 5000 \times 0.2820 = 1410.06$$

(b) The present value of the loan is $V_0 = 5000$. Check by calculation:

$$V_0 = A_0 \underbrace{\frac{1 - (1 + i/4)^{-t\times4}}{i/4}}_{\text{annuity factor}} = 1410.06 \frac{1 - (1.05)^{-4}}{0.05} = 5000$$

(c) Total payments $= 1410.06 \times 4 = 5640.24$

Total interest $=$ total payments $-$ value of loan $= 5640.24 - 5000 = 640.24$

(d) At the start of quarter 1, the borrower has the loan of 5000 in his account. Calculate the value of the account at the end of each quarter, where

$$\begin{pmatrix} \text{value at start} \\ \text{of quarter} \end{pmatrix} + \begin{pmatrix} \text{interest for} \\ \text{quarter} \end{pmatrix} - \begin{pmatrix} \text{quarterly} \\ \text{payment} \end{pmatrix} = \begin{pmatrix} \text{value at end} \\ \text{of quarter} \end{pmatrix}$$

Quarterly calculations are set out as follows.

FINANCIAL MATHEMATICS

Value at start (VSQ)	+Interest (VSQ × 0.05)	−Payment (−1410.06)	Value at end (VSQ + VSQ × 0.05 − 1410.06)
5000.000	250.000	−1410.06	3839.940
3839.940	191.997	−1410.06	2621.877
2621.877	131.103	−1410.06	1342.920
1342.920	67.146	−1410.06	0.006
	640.246	−5640.24	

Sinking funds

A sinking fund is created by putting aside a fixed sum each year for the purpose of paying debts, replacing equipment, etc. In other words, an annuity is set up to repay the debt. If a fixed sum, A_0, is set aside at the start of each year and interest is compounded annually at $i\%$, the fund will grow year by year as follows.

At the end of	Value of sinking fund
Year 1	$A_0(1+i) + A_0$
Year 2	$A_0(1+i)^2 + A_0(1+i) + A_0$
Year 3	$A_0(1+i)^3 + A_0(1+i)^2 + A_0(1+i) + A_0$
⋮	⋮
Year t	$A_0(1+i)^t + A_0(1+i)^{t-1} + A_0(1+i)^{t-2} + \cdots + A_0(1+i) + 0$
Value of sinking fund after t years	$A_0(1+i)^t + A_0(1+i)^{t-1} + A_0(1+i)^{t-2} + \cdots + A_0(1+i)$

Note: No deposit is made at the end of year t. The fund has matured and is sufficient to repay the debt. The value of the fund is

$$V_{SK,t} = A_0(1+i)^t + A_0(1+i)^{t-1} + A_0(1+i)^{t-2} + \cdots + A_0(1+i)$$

This series is a simple geometric series whose sum is readily calculated by equation (5.5), after some slight simplifications. Start by writing the series in reverse order:

$$V_{SK,t} = A_0(1+i) + \cdots + A_0(1+i)^{t-2} + A_0(1+i)^{t-1} + A_0(1+i)^t$$

Factor out $A_0(1+i)$, which is common to all terms:

$$V_{SK,t} = A_0(1+i)\underbrace{\left[1 + (1+i) + (1+i)^2 + \cdots + (1+i)^{t-1}\right]}_{\text{geometric series with } a=1,\, r=(1+i)} \tag{5.29}$$

The series in the square bracket is geometric, $a = 1, r = (1 + i)$. The sum of the series is calculated by (5.5) with these values for a and r:

$$1 + (1 + i) + (1 + i)^2 + \cdots + (1 + i)^{t-1} = \frac{(1 + i)^t - 1}{(1 + i) - 1}$$

Hence equation (5.29) simplifies to

$$V_{SK,t} = A_0(1 + i)\frac{(1 + i)^t - 1}{(1 + i) - 1} = A_0(1 + i)\frac{(1 + i)^t - 1}{i} \tag{5.30a}$$

It is left as an exercise to show that equation (5.30a) may also be deduced from equation (5.23) by substituting $P_0 = A_0$ and subtracting A_0 from the series, since there is no payment at the end of year t. Note that a sinking fund is a form of annuity where deposits are made at the beginning of each interval. Alternatively,

$$V_{SK,t} = A_0(1/i + 1)\left[(1 + i)^t - 1\right] \tag{5.30b}$$

WORKED EXAMPLE 5.20
SINKING FUNDS

A taxi service must replace cars every five years at a cost of £450 000. At an 8% rate of interest, calculate:

(a) the size of the fund if £4000 is deposited at the beginning of each month;
(b) the size of each quarterly payment necessary to meet this target.

SOLUTION

(a) Given $A_0 = 4000, \dfrac{i}{12} = \dfrac{0.08}{12} = 0.0067, t = 5 \times 12 = 60$ instalments.
   Hence by

$$V_{SK,t} = A_0(1 + i/12)\frac{(1 + i/12)^t - 1}{i/12} \tag{5.29}$$

$$V_{SK,60} = 4000(1 + 0.0067)\frac{(1.0067)^{60} - 1}{0.0067}$$

$$= 4000(1.0067)\frac{0.4928}{0.0067}$$

$$= 4000(1.0067)(73.5522)$$

$$= 296\,179.9989$$

Not sufficient to meet target of £450 000.

(b) Substitute the information given into equation (5.29):

$$V_{SK} = 450\,000, \frac{i}{4} = \frac{0.08}{4} = 0.02, t \times 4 = 5 \times 4 = 20 \text{ payment intervals}$$

$$V_{SK,t} = A_0(1 + i/4)\frac{(1 + i/4)^{t \times 4} - 1}{i/4}$$

$$450\,000 = A_0(1 + 0.02)\frac{(1 + 0.02)^{20} - 1}{0.02}$$

$$450\,000 = A_0(1.02)(24.2974)$$

$$450\,000 = A_0(24.7833)$$

$$A_0 = \frac{450\,000}{24.7833} = 18\,157.3882$$

Quarterly payments of £18 157.3882 are required to meet the target.

PROGRESS EXERCISES 5.6

Annuities, Debt Repayments, Sinking Funds

1. A company operates a savings scheme (interest rate 4% per annum) for its employees. (a) Calculate the amount saved at the end of one year when £200 is deposited at the end of each month. (b) How much should be saved from a monthly salary in order to have £5000 at the end of the year?
2. A child is given a savings account with £2000. A further £500 is deposited in the account at the end of each year. Assuming an annual interest rate of 6%, calculate the value of the account (a) at the end of 10 years, (b) at the end of 20 years.
3. How much should you pay for an annuity of quarterly payments of £1500 for five years, assuming interest rates of 5% per annum?
4. A loan of £100 000 is to be repaid in annual payments over 10 years. Assuming a fixed interest rate of 10% per annum, calculate (a) the amount of each annual repayment, (b) the total interest paid, (c) the amount of interest paid in the first and second repayment.
5. A loan of £50 000 is to be repaid in equal quarterly instalments over a period of five years. If interest is 7.5% per annum, calculate (a) the amount of each quarterly payment, (b) the total amount of interest paid, (c) the interest paid each quarter in the first year of repayments.
6. A loan is repaid at the rate of £200 per week for 15 years. If interest is 6.65% per annum, calculate (a) the total amount repaid, (b) the size of the loan, (c) the total amount of interest paid.
7. A company must replace machinery every two years at a cost of £100 000. It is decided to set aside equal amounts at the beginning of each quarter. If interest is 5.75% per annum, calculate the size of the quarterly deposits.

8. Starting on her 35th birthday, a woman saves £200 at the end of each month with a view to retiring on her 50th birthday.
   (a) What is the value of her savings if the interest rate is 8% per annum?
   (b) From the age of 50, how much should she withdraw each month if the fund is to last for the next 15 years, assuming the interest rate is 9% per annum?

9. An education fund of £40 000 is set up for a child on his 4th birthday. The fund is to mature on his 18th birthday. Assuming an interest rate of 10% per annum, calculate (a) the present value of the fund assuming annual compounding, (b) the size of monthly deposits (in arrears) required to achieve the fund of £40 000, (c) the size of quarterly deposits required to achieve the fund of £40 000.

10. A student registers for a four-year course. She has a fund of £40 000 to cover expenses over the next four years. Calculate the amount of periodic withdrawals (in arrears) if she withdraws money (a) weekly, and (b) monthly. Assume the interest rate is 7.5% per annum.

11. A company has the option of leasing equipment for an annual fee of £3500 over four years or buying it for £12 500. If the equipment is worthless at the end of four years, which option is preferable if the annual rate of interest is (a) 5%, and (b) 4%?

5.6 The Relationship between Interest Rates and the Price of Bonds

This section analyses the relationship between interest rates, the speculative demand for money and government bonds. It involves everything from simple interest, present values to *NPVs*.

A **bond** is a cash investment made to the government, usually in units of £1000 for an agreed number of years. In return, the government pays the investor a fixed sum at the end of each year; in addition, the government repays the original value (face value) of the bond to the investor with the final payment.

To make bonds attractive to investors, the size of the fixed annual payments (sometimes referred to as the coupon) must be based on the prevailing rate of interest (i) at the time of purchase. The fixed annual payment is calculated as follows:

$$\text{Annual payment} = i \times (\text{price of bond}) \tag{5.31}$$

For example, a £1000 bond is bought when the prevailing interest rate is 8%, then

$$\text{Annual payment} = i \times (\text{price of bond}) = 0.08 \times 1000 = 80$$

The annual payment, £80, is fixed for the duration of the life of the bond. On maturity, the face value of £1000 is repaid to the investor. (In effect, this is an application of simple interest.)

So why are bonds attractive as an investment? If interest rates do not change, it certainly does not make sense to invest when the growth of the investment is based on simple interest. Compound interest is much more attractive! However, interest rates *do* change. To assess the attractiveness of the future fixed payments (cash flow) as interest rates increase or decrease, we return to *NPVs*. The future cash

FINANCIAL MATHEMATICS

flows are discounted to the present, hence, comparisons may be easily made; in addition, the value of the bond itself, due with the final payment, is also discounted to the present, to determine whether a capital gain or loss is made on the cash investment. These calculations are carried out in Worked Example 5.21.

WORKED EXAMPLE 5.21
THE INTEREST RATE AND THE PRICE OF BONDS

A five-year government bond valued at £1000 is purchased when the market rate of interest is 8%.

(a) Calculate the annual repayment (value of the coupon) made to the investor at the end of each year.
(b) Calculate the NPV of the agreed cash flow when interest rates change immediately to each of the following (the original 8% is included for comparison): 6.5%, 7%, 7.5%, 8%, 8.5%, 9%, 9.5%, 10%.
(c) Calculate the present value of the bond which is due to be repaid at the end of the five-year lifetime.
Comment on:
  (i) The relationship between interest rates and the attractiveness of bonds as an investment.
  (ii) The return from bonds when the interest does not change compared with investing the same amount at interest rates compounded annually.

SOLUTION

(a) When the interest rate is 8%, the annual payment (or coupon) is $(8/100) \times 1000 = 80$, that is, £80 is received by the investor at the end of each year.
(b) Table 5.6 outlines the calculation of the NPVs for the future cash flows at the given interest rates. (Excel is very useful for these calculations.)

Table 5.6   The rates and the NPV for the cash flow for a £1000 bond

End of year	Cash flow	0.065	0.07	0.075	0.08	0.085	0.09	0.095	0.1
1	80	75.12	74.77	74.42	74.07	73.73	73.39	73.06	72.73
2	80	70.53	69.88	69.23	68.59	67.96	67.33	66.72	66.12
3	80	66.23	65.30	64.40	63.51	62.63	61.77	60.93	60.11
4	80	62.19	61.03	59.90	58.80	57.73	56.67	55.65	54.64
5	80	58.39	57.04	55.72	54.45	53.20	51.99	50.82	49.67
	NPV	332.45	328.02	323.67	319.42	315.25	311.17	307.18	303.26
P_0 of bond		729.88	712.99	696.56	680.58	665.05	649.93	635.23	620.92
(P_0 of bond + NPV)		1062.34	1041.00	1020.23	1000.00	980.30	961.10	942.40	924.18

(c) The present value of a redeemed bond due at the end of five years is calculated by equation (5.11), $P_0 = P_t/(1 + i)^t$, where $P_t = £1000$

$$i = 6.5\% = 0.065 \rightarrow P_0 \frac{1000}{(1 + 0.065)^5} = £729.88$$

$$i = 7.0\% = 0.07 \rightarrow P_0 \frac{1000}{(1 + 0.07)^5} = £712.99$$

and similarly for each of the other interest rates. The results are summarised in Table 5.6.

Comment

(i) From Table 5.6 it is evident from a comparison of the present value of the annual payments together with the present value of the returned bond that when interest rates:
  • decrease, the present value of the cash flow and the present value of the bond increase;
  • increase, the present value of the cash flow and the present value of the bond decrease.
  Hence the inverse relationship between the attractiveness of bonds as an investment and the interest rate.

(ii) If the interest rate does not change, the investment holds its value (see Table 5.6). The outcome would be exactly the same if £1000 were put on deposit at 8% interest compounded annually when £80 is withdrawn at the end of each year. Comparisons for the returns from bonds and investments in which interest is compounded annually for rates above and below the purchase rate are left as an exercise for the reader in Progress Exercises 5.7, question 4.

With the above information, we are now able to explain in more detail the relationship between the interest rate and the speculative demand for money which was introduced in Chapter 3.

If the prevailing interest rate is higher than the accepted normal rate for an economy, one expects the rate to fall towards normal in the future. Therefore, at present, investors would buy bonds hoping to make a gain (from the fixed cash flow and a capital price gain on the present value of the bond) in the future. Therefore, speculative balances are low as investors put their money into bonds.

If the prevailing interest rate is lower than the accepted normal rate for an economy, one expects the rate to rise towards normal in the future. Therefore, at present, investors would not buy bonds since they are likely to incur a loss (from the fixed cash flow and a capital price loss on the present value of the bond). Therefore, speculative balances are high as investors do not put their money into bonds.

FINANCIAL MATHEMATICS

PROGRESS EXERCISES 5.7

Bonds and Interest Rates

1. A five-year government bond valued at £5000 is purchased when the market rate of interest is 20%.
   (a) Calculate the annual repayment (value of the coupon) made to the investor at the end of each year.
   (b) Calculate the NPV of the agreed cash flow when interest rates change immediately to each of the following (the original 20% is included for comparison): 5%, 10%, 15%, 20%, 25%, 30%.
   (c) Calculate the present value of the bond which is due to be repaid at the end of the five-year lifetime when interest rates are 5%, 10%, 15%, 20%, 25% or 30%.
   Comment on the relationship between interest rates and the attractiveness of bonds as an investment.
2. A 10-year government bond valued at £10 000 was purchased when the market rate of interest was 9%.
   (a) Calculate the annual repayment (value of the coupon) made to the investor at the end of each year.
   (b) Calculate the NPV of the agreed cash flow when interest rates change immediately to each of the following (include the original 9% for comparison): 6.5%, 9%, 12%.
   (c) After a period of five years, the bond is sold to another investor for £10 000. Determine whether the purchase is profitable for the remaining five years, given interest rates of (i) 6.5%, (ii) 12%.
3. A bond is purchased for £5000 when the interest rate is 8%.
   (a) Calculate the fixed annual payments.
   (b) Calculate the capital gain if the interest rate immediately (i) falls to 6.25%, (ii) increases to 10%.
4. A £5000 bond is purchased for a five-year period when interest rates are 20%. At the same time £5000 is put on deposit at a bank and £1000 withdrawn from the account at the end of each year. Compare the return from the investment in the bond with that from the bank if the bank offers:
   (a) A 20% rate of interest compounded annually.
   (b) A 5% rate of interest compounded annually.
   (c) A 30% rate of interest compounded annually.

5.7 Excel for Financial Mathematics

In this chapter, Excel is particularly useful for calculating tables of values, such as in Worked Example 5.14, where several NPVs were calculated at different discount rates. Worked Example 5.22 compares the growth of an investment for the same nominal rate of interest, but using different methods of compounding.

WORKED EXAMPLE 5.22
GROWTH OF AN INVESTMENT USING DIFFERENT METHODS OF COMPOUNDING (EXCEL)

£1 is invested at a nominal rate of interest of 50%.

(a) Set up a table comparing the growth of the investment over a 20-year period if interest is compounded (i) annually, (ii) quarterly, (iii) monthly, (iv) continuously.
(b) Graph the growth of the investment by all four methods given in (a) on the same diagram for the last five years.

SOLUTION
(a) Set up a table in Excel similar to Table 5.7. The formula for each method of compounding is entered into cells B4, C4, D4 and E4 respectively as follows:

Table 5.7   Different methods of compounding

	A	B	C	D	E
			Method of compounding		
1	A	B	C	D	E
2	Year	Annually	Quarterly	Monthly	Continuous
3	0.00	1.00	1.00	1.00	1.00
4	1.00	1.50	1.60	1.63	1.65
5	2.00	2.25	2.57	2.66	2.72
6	3.00	3.38	4.11	4.35	4.48
7	4.00	5.06	6.58	7.10	7.39
8	5.00	7.59	10.55	11.58	12.18
9	6.00	11.39	16.89	18.90	20.09
10	7.00	17.09	27.06	30.85	33.12
11	8.00	25.63	43.34	50.35	54.60
12	9.00	38.44	69.42	82.17	90.02
13	10.00	57.67	111.20	134.11	148.41
14	11.00	86.50	178.12	218.88	244.69
15	12.00	129.75	285.31	357.23	403.43
16	13.00	194.62	457.02	583.03	665.14
17	14.00	291.93	732.05	951.56	1 096.63
18	15.00	437.89	1 172.60	1 553.03	1 808.04
19	16.00	656.84	1 878.28	2 534.70	2 980.96
20	17.00	985.26	3 008.65	4 136.86	4 914.77
21	18.00	1 477.89	4 819.27	6 751.75	8 103.08
22	19.00	2 216.84	7 719.55	11 019.50	13 359.73
23	20.00	3 325.26	12 365.22	17 984.87	22 026.47

(i) Annual compounding: $P_t = P_0(1 + i)^t$ for $t = 1$ to 20.

With the cursor (or pointer) in cell B4, type $= \text{B\$ } 3*1.5\char94A4$.

This formula states: principal $\times$ 1.5 raised to the power of time, when rate $= 0.5$.

The \$ sign in front of the number 3 is required since we want to reference cell B3 (the same principal), when this formula is copied down column B. If the \$ sign is not used, then as the formula is copied down, B3 would become B4, B4 becomes B5, etc.

(Of course, with a £1 deposit, it is not necessary to reference any cell, but once the table is set up, the table will recalculate itself if the initial value of the investment is changed. Try it!)

(ii) Quarterly compounding:

$$P_t = P_0 \left(1 + \frac{i}{4}\right)^{4t} \quad \text{for } t = 1 \text{ to } 20$$

With the cursor in cell C4, type $= \text{C\$ } 3*1.125\char94(A4*4)$, then copy this formula down column C. This formula states: principal $\times$ (1 + rate/4) raised to the power of (time $\times$ 4), when rate $= 0.5$.

(iii) Monthly compounding:

$$P_t = P_0 \left(1 + \frac{i}{12}\right)^{12t} \quad \text{for } t = 1 \text{ to } 20$$

With the cursor in cell D4, type $= \text{D\$ } 3*1.041667\char94(A4*12)$, then copy this formula down column D.

This formula states: principal $\times$ (1 + rate/12) raised to the power of (time $\times$ 12), when rate $= 0.5$.

(iv) Continuous compounding: $P_t = P_0 e^{it}$.

With the cursor in cell E4, type $= \text{E\$ } 3*\exp(0.5*A4)$, then copy this formula down column E. This formula states: principal $\times$ e raised to the power of (time $\times$ rate), when rate $= 0.5$.

The results of the calculations are given in Table 5.7.

(b) Since we require the graph for the last five years only, copy the row of column titles to a new area on the sheet. Immediately below the row of titles copy the last five rows of Table 5.7. The new table is given as Table 5.8. Now go through the usual steps of plotting the graph. The final graph should look like Figure 5.2.

From the graphs, the continuous method of compounding leads to the greatest cumulative total after 20 years, while the annual method gives the smallest sum. Furthermore, the longer the investment is left on deposit, the wider the differences in its value when compounded by methods (i), (ii), (iii) and (iv).

Table 5.8   Data for plotting Figure 5.2

Year	Annually	Quarterly	Monthly	Continuous
16.00	656.84	1 878.28	2 534.70	2 980.96
17.00	985.26	3 008.65	4 136.86	4 914.77
18.00	1 477.89	4 819.27	6 751.75	8 103.08
19.00	2 216.84	7 719.55	11 019.50	13 359.73
20.00	3 325.26	12 365.22	17 984.87	22 026.47

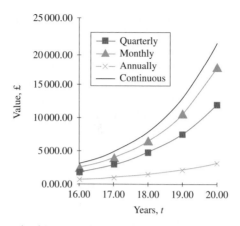

Figure 5.2   Growth of £1 at $i = 50\%$ using different compounding methods

5.8   Summary

Series

Characteristics of arithmetic sequences
a is the first term of the sequence
d is the common difference
nth term, T_n, is given by $[a + (n - 1)d]$

$$S_n = \frac{n}{2}[2a + (n - 1)d]$$

Characteristics of geometric sequences
a is the first term of the sequence
r is the common ratio
nth term, T_n, is given by ar^{n-1}

$$S_n = \frac{a(1 - r^n)}{1 - r}$$

Financial mathematics

- **Amount due after t years (future value)** – bringing forward a single payment.
  Simple interest: $P_t = P_0(1 + it)$
  Compound interest (annual): $P_t = P_0(1 + i)^t$
  Compound m times annually: $P_t = P_0\left(1 + \frac{i}{m}\right)^{mt}$
  Continuous compounding: $P_t = P_0 e^{it}$
- **Present value** – of a single payment due in t years from now.

  Simple discounting: $P_0 = \dfrac{P}{(1 + it)} = P_t(1 + it)^{-1}$

  Compound discounting: $P_0 = \dfrac{P_t}{(1 + i)^t} = P_t(1 + i)^{-t}$

  Continuous discounting: $P_0 = P_t e^{-it}$
- **Annual percentage rate**
  (a) When the nominal rate is compounded m times per year,

  $$\text{APR} = \left(1 + \frac{i}{m}\right)^m - 1$$

  (b) When the nominal rate is compounded continuously,

  $$\text{APR} = e^i - 1$$

FINANCIAL MATHEMATICS

- **Depreciation**
  Reducing-balance depreciation, $A_t = A_0 (1 - i)^t$
  Straight-line depreciation

  Present value: $$A_0 = \frac{A_t}{(1 - i)^t}$$

- **Net present value and *IRR***
  NPV: present value of a future cash flow, discounted at a given discount rate r.
  IRR: the discount rate for which *NPV* = 0. The *IRR* may be estimated graphically or by the formula

  $$IRR = \frac{(i_1 \times NPV_2) - (i_2 \times NPV_1)}{NPV_2 - NPV_1}$$

- **Annuities, debt repayments, sinking funds** In the text, the basic time interval was assumed to be one year. Calculations for other time intervals are adjusted accordingly. The following assume t time periods, interest compounded at $i\%$ per time interval.
  The value V_t for an initial investment of P_0 and t periodic investments of A_0 is

  $$V_t = P_0(1 + i)^t + A_0 \frac{(1 + i)^t - 1}{i}$$

  The value of an annuity is

  $$V_{ANU,t} = A_0 \frac{(1 + i)^t - 1}{i}$$

  The present value of an annuity is

  $$V_0 = A_0 \underbrace{\frac{1 - (1 + i)^{-t}}{i}}_{\text{annuity factor}}$$

  The amount of periodic repayments on a loan L is

  $$A_0 = \frac{Li}{1 - \frac{1}{(1+i)^t}} = L \times \underbrace{\frac{i}{1 - (1 + i)^{-t}}}_{\text{capital recovery factor}}$$

- The value of a sinking fund, payments A_0 made at the start of each year, is

  $$V_{SK,t} = A_0(1 + i) \frac{(1 + i)^t - 1}{i}$$

Excel

Excel is particularly useful for calculating tables of values where the calculations are based on formulae for graphing cash flows. Besides that, Excel has a variety of financial functions.

www.wiley.com/college/bradley

Go to the website for Problems in Context

Appendix

- **Formula (5.2): Sum of terms of an AP**

$$S_n = a + (a+d) + (a+2d) + \cdots + [a+(n-2)d] + [a+(n-1)d]$$

$$S_n = a + (a+d) + (a+2d) + \cdots + (a+nd-2d) + (a+nd-d)$$

$$\underline{S_n = (a+nd-d) + (a+nd-2d) + \cdots + (a+2d) + (a+d) + a}$$

$$2S_n = a + (2a+nd) + (2a+nd) + \cdots + (2a+nd) + (2a+nd) + a$$

There are (n) $2a$-terms $\rightarrow 2a \times n = 2an$.
There are $(n-1)$ nd-terms $\rightarrow nd \times (n-1) = nd(n-1)$; therefore,

$$2S_n = 2an + nd(n-1)$$

$$2S_n = n[2a + (n-1)d]$$

$$S_n = \frac{n}{2}[2a + (n-1)d]$$

- **Formula (5.4): Sum of terms of a GP**

$$S_n = a + ar + ar^2 + \cdots + ar^{n-2} + ar^{n-1}$$

$$r\,S_n = \underline{ar + ar^2 + \cdots + ar^{n-2} + ar^{n-1} + ar^n}$$

$$S_n - r\,S_n = a + 0 + 0 + \cdots + 0 + 0 - ar^n$$

$$S_n(1-r) = a(1-r^n)$$

$$S_n = \frac{a(1-r^n)}{1-r}$$

- **Formula (5.19): Estimating the IRR**
  Equation of line joining (x_1, y_1) to (x_2, y_2) is

$$y - y_1 = m(x - x_1)$$

$$y - y_1 = \frac{y_2 - y_1}{x_2 - x_1}(x - x_1)$$

This line cuts the horizontal axis at $y = 0$; hence, solve for x. Therefore

$$-y_1 = \frac{y_2 - y_1}{x_2 - x_1}(x - x_1)$$

$$-y_1(x_2 - x_1) = (y_2 - y_1)(x - x_1)$$

$$-x_2 y_1 + x_1 y_1 = x(y_2 - y_1) - x_1 y_2 + x_1 y_1$$

$$\frac{x_1 y_2 - x_2 y_1}{y_2 - y_1} = x \text{ See Figure 5.3}$$

$$\frac{i_1 NPV_2 - i_2 NPV_1}{NPV_2 - NPV_1} = IRR$$

FINANCIAL MATHEMATICS

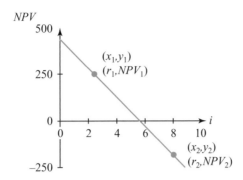

NPV

Figure 5.3   Estimating the internal rate of return

TEXT EXERCISES 5

1. (a) Find the sum of the first seven terms of the series:

(i) $3 + 9 + 27 + 81 + \ldots$ (ii) $2 + 1 + \dfrac{1}{2} + \ldots$

(b) A food-processing company gradually closes down the production of potato chips when the strawberry season begins. If the quantity of chips produced is 10 500 bags per week and reduction is at the rate of 20% each week and 200 kg of strawberries is processed in the first week of the season, increasing by 100 kg each for eight weeks, calculate:

   (i) The number of bags of chips and kg of strawberries produced in week 5 of the strawberry season.
   (ii) The number of weeks taken for the production of chips to drop to 500 bags per week.
   (iii) The total weight of strawberries processed in eight weeks.

2. Deduce the compound interest formula. If £20 000 is invested at 6.5% compound interest compounded annually calculate:

   (a) The value of the investment at the end of 10 years.
   (b) The number of years that the investment should be left on deposit if it is to double in value.

3. An investor is quoted an interest rate of 7.5%. Determine the APR if interest is compounded
   (a) six times annually, (b) 50 times annually, (c) continuously.

4. Projects A and B involve the following net cash flows:

		Net cash flows			
	Initial cost	Year 1	Year 2	Year 3	Year 4
Project A	600 000	−300	650 000	885 000	−2000
Project B	4045	2000	2500	3000	3500

Decide which project is the most profitable by:
(a) Determining the NPV at a discount rate of 6.5%.
(b) The IRR.

 Go to the website **www.wiley.com/college/bradley** for questions 5, 6 and 7.

DIFFERENTIATION AND APPLICATIONS

6

6.1   Slope of a Curve and Differentiation

6.2   Applications of Differentiation, Marginal Functions, Average Functions

6.3   Optimisation for Functions of One Variable

6.4   Economic Applications of Maximum and Minimum Points

6.5   Curvature and Other Applications

6.6   Further Differentiation and Applications

6.7   Elasticity and the Derivative

6.8   Summary

Chapter Objectives

In this chapter you are introduced to differentiation. Since the derivative is the instantaneous rate of change, finding derivatives and applications of differentiation are extremely important in all areas of business and economics. At the end of this chapter you should be able to:

- Calculate the derivatives of a range of functions
- Calculate marginal and average functions
- Determine the maximum and minimum values
- Determine maximum and minimum values of revenue, profit, cost and other economic functions
- Determine points of inflection and use these to describe curvature
- Use points of inflection in applications such as the point of diminishing returns
- Determine the point elasticity of demand and relationships between price elasticity of demand and marginal revenue, total revenue and price changes.

6.1 Slope of a Curve and Differentiation

At the end of this section you should be able to:

- Show how to measure the average slope of a curve between two points
- Show how to calculate the slope of a curve at a point approximately
- Use differentiation to determine the equation for slope from the equation of the curve
- Differentiate functions containing x^n using the power rule
- Use rules for differentiating sums and differences of several functions
- Calculate higher derivatives
- Use differentiation to determine whether a function is increasing or decreasing.

6.1.1 The slope of a curve is variable

In Section 2.1 the slope of a line was defined as the change in height (Δy) per unit increase in the horizontal distance (Δx). It was also noted that the slope of a straight line is the same when measured between any two points on the line. The situation is, however, different for a curve (smooth continuous curve). To see how the slope varies along a curve, the curve in Figure 6.1 is approximated by a series of straight lines or chords. A chord is a line joining any two points on a curve, such as A to B to give chord AB; B to C to give chord BC. The slope of the curve along any interval is given approximately by the slope of the chord along that same interval. It is obvious that the slope of the curve is changing as the slope of each chord is different.

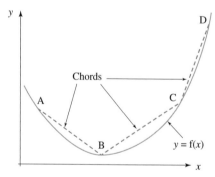

Figure 6.1   The slope of a curve is variable

The slope of a curve varies at every point on the curve.

6.1.2 Slope of a curve and turning points

Consider the chord BC in Figure 6.2(a) as a magnification of a very small chord, where the point B has coordinates (x, y), very close to the point C $(x + \Delta x, y + \Delta y)$, whose coordinates are only slightly different from those of B by small amounts Δx and Δy, respectively. The slope of the chord BC is a reasonable approximation to the slope of the curve over this interval:

$$\text{Slope of chord } BC = \frac{\text{change in vertical height}}{\text{change in horizontal distance}} = \frac{\Delta y}{\Delta x}$$

However, if the chord is made progressively shorter (Figure 6.2(b)) by moving the point C closer to the point B, the slopes of the chord and curve become closer as Δx and Δy become smaller. Ultimately, when the point C reaches the point B, the slopes of chord and curve are the same and the straight line that touches the curve B is called the tangent to the curve at B. At this stage the horizontal distance Δx is almost zero but not exactly zero (otherwise we have division by zero, which is not defined). We describe this by saying 'Δx tends to zero' or mathematically $\Delta x \to 0$. The slope at a point is then described as 'the limit of Δy over Δx as Δx tends to zero', written $\lim_{\Delta x \to 0} \Delta y/\Delta x$. A more concise way of writing this fraction is dx/dy:

$$\lim_{\Delta x \to 0} \frac{\Delta y}{\Delta x} = \frac{dy}{dx}$$

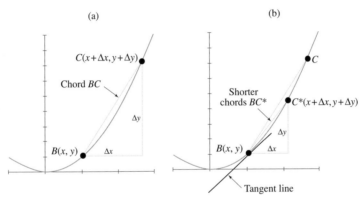

Figure 6.2 (a) Slope of chord approximates slope of curve. (b) Point $C \to C^*$ as C moves towards B. Slope of tangent at $B =$ slope of curve at B

The area of mathematics which deals with infinitely small changes is called **calculus**. The process of finding dx/dy for a given curve, $y = f(x)$, by first determining $\Delta y/\Delta x$ and then allowing $\Delta x \to 0$, i.e., determining $\lim_{\Delta x \to 0} \Delta y/\Delta x$, is called differentiation from first principles. This process yields a general equation for the slope of the curve $y = f(x)$. When the equation for the slope is available, the value of the slope at any point is obtained by substituting the x-coordinate of that point. The following worked example should clarify these concepts.

WORKED EXAMPLE 6.1
EQUATION FOR THE SLOPE OF $y = x^2$
FROM FIRST PRINCIPLES

(a) Derive the equation for the slope of the curve $y = x^2$ from first principles, i.e., use $dy/dx = \lim_{\Delta x \to 0} \Delta y/\Delta x$.

(b) For the curve $y = x^2$ at $x = 1.5$ and $x = -1$, (i) calculate the corresponding y-coordinates and (ii) the value of the slopes of the curve, using the formula for slope derived in (a). Illustrate the answers to (i) and (ii) by calculating and drawing tangent lines to the curve at $x = -1$ and $x = 1.5$.

SOLUTION

(a) To determine $\lim_{\Delta x \to 0} \Delta y / \Delta x$ we need to find an expression for $\Delta y / \Delta x$. Consider any point B with coordinates (x, y) on the curve $y = x^2$ and a point C with coordinates $(x + \Delta x, y + \Delta y)$, very close to B. Δx and Δy are very small distances (Figure 6.2).

$y = x^2$	equation of curve satisfied by coordinates of B
$(y + \Delta y) = (x + \Delta x)^2$	equation of curve satisfied by coordinates of C
$(y + \Delta y) = x^2 + 2x\,\Delta x + (\Delta x)^2$	squaring the RHS
$\Delta y = (x^2 - y) + 2x\,\Delta x + (\Delta x)^2$	bring y over to RHS
$\Delta y = (x^2 - x^2) + 2x\,\Delta x + (\Delta x)^2$	since $y = x^2$
$\Delta y = 2x\,\Delta x + (\Delta x)^2$	since $(x^2 - x^2) = 0$
$\dfrac{\Delta y}{\Delta x} = \dfrac{2x\,\Delta x}{\Delta x} + \dfrac{(\Delta x)^2}{\Delta x}$	dividing everything on both sides by Δx
$\dfrac{\Delta y}{\Delta x} = 2x + \Delta x$	simplifying

Now let $\Delta x \to 0$, i.e., we are getting the limit as $\Delta x \to 0$:

$$\lim_{\Delta x \to 0} \frac{\Delta y}{\Delta x} = \lim_{\Delta x \to 0} (2x + \Delta x)$$

$$\frac{dy}{dx} = 2x$$

So the slope of the curve $y = x^2$ is given by the equation $dy/dx = 2x$.

(b) (i) To find y for a given value of x, substitute the value of the x-coordinate of the point into the equation of the curve, $y = f(x)$ (see table below).

(ii) To find the value of the slope dy/dx at a given point, substitute the value of the x-coordinate of the point into the equation for the slope, $dy/dx = f'(x)$ (see table below).

The values of y and dy/dx at $x = 1.5$ and $x = -1$ are calculated below:

x-coordinate	Equation of curve, $y = x^2$	Equation of slope, $dy/dx = 2x$
$x = 1.5$	$y = x^2 = (1.5)^2 = 2.25$	$dy/dx = 2x = 2(1.5) = 3$
$x = -1$	$y = x^2 = (-1)^2 = 1$	$dy/dx = 2x = 2(-1) = -2$

The slope of the curve at $x = 1.5$ and $x = -1$ is the same as the slope of the tangents at these points (Figure 6.3).

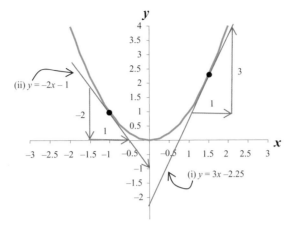

Figure 6.3   Slope of curve $y = x^2$ at $x = -1$ and $x = 1.5$

Hence, using equation (2.8), $y - y_1 = m(x - x_1)$, the equation of:

(i) the tangent through $(1.5, 2.25)$ is $y - 2.25 = 3(x - 1.5) \rightarrow y = 3x - 2.25$
(ii) the tangent through $(-1, 1)$ is $y - (1) = -2\{x - (-1)\} \rightarrow y = -2x - 2$.

6.1.3   The derivative

Here is a brief summary of the ideas and terminology introduced in the previous section:

- dy/dx is called the derivative of y with respect to x.
- The process of finding dy/dx is called differentiation.
- dy/dx is the equation for the slope of the curve at any point (x, y) on the curve.
- $\Delta y/\Delta x$ is the slope of the chord over a small interval Δx.
- For very small intervals Δx, the slope of the curve is approximately equal to the slope of the chord, i.e., $dy/dx \approx \Delta y/\Delta x$ (the symbol $\approx$ means approximately equal to).

The power rule for differentiation

- Worked Example 6.1 demonstrated how calculus was used to derive the equation for the slope of the curve $y = x^2$ from first principles.
- First principles may be used to derive the derivatives for a wide variety of other functions, such as x^n and e^x.
- From first principles, the derivative of $y = x^n$ is $dy/dx = nx^{n-1}$, where n is any real number.
- From first principles, the derivative of $y = e^x$ is $dy/dx = e^x$.

Once the general formula for the derivative of a function is deduced, the formula may be applied to this and related functions, for example:

$$y = x^n \rightarrow \frac{dy}{dx} = nx^{n-1} \text{ is called the power rule} \qquad (6.1)$$

Worked Example 6.1 was a particular case of the power rule with $n = 2$:

$$y = x^2 \rightarrow \frac{dy}{dx} = 2x^{2-1} = 2x$$

6.1.4 How to use the power rule for differentiation

The power rule is stated again:

$$\text{If } y = x^n \quad \text{then} \quad \frac{dy}{dx} = nx^{n-1}.$$

The use of the power rule could be described as

$$\begin{array}{c} n \text{ drops down} \\ \downarrow \\ \frac{dy}{dx} = nx^{(n-1)} \leftarrow \leftarrow \text{power loses 1} \end{array}$$

The power, n, may be any real number: positive, negative, decimal or a fraction. For example, if $y = x^5$, then the power is $n = 5$. According to the power rule, the slope of the curve is given by the equation

$$\text{Slope} = \frac{dy}{dx} = 5x^{5-1} = 5x^4$$

Worked Example 6.2 illustrates how this rule is used for various powers, and also for variables other than x and y.

WORKED EXAMPLE 6.2
USING THE POWER RULE

(a) $y = x^3, n = 3 \quad \rightarrow \quad$ then $\dfrac{dy}{dx} = 3x^{3-1} = 3x^2$

(b) $y = x^4, n = 4 \quad \rightarrow \quad$ then $\dfrac{dy}{dx} = 4x^{4-1} = 4x^3$

(c) $P = Q^{2.6}, n = 2.6 \quad \rightarrow \quad$ then $\dfrac{dP}{dQ} = 2.6\, Q^{2.6-1} = 2.6\, Q^{1.6}$

(d) $C = Q^2, n = 2 \quad \rightarrow \quad$ then $\dfrac{dC}{dQ} = 2Q^{2-1} = 2Q$

(e) $C = Q^{-1}, n = -1 \quad \rightarrow \quad$ then $\dfrac{dC}{dQ} = -1Q^{-1-1} = -\dfrac{1}{Q^2}$

Some important points to note before using the power rule

- Differentiation is with respect to the variable on the RHS of each equation. In Worked Example 6.2, the variable is x in the first two cases and Q in the other three.
- Always simplify the expression to be differentiated before starting **and** whenever possible, at any stage, while working through a problem. The problem may simplify to a much easier one, which should save you a good deal of time, and mistakes are less likely. For example, differentiate C, where $C = Q^6/Q^{-4.5}$. When simplified, the expression for C is much easier to differentiate:

$$C = \frac{Q^6}{Q^{-4.5}} = \frac{Q^{6-(-4.5)}}{1} = \frac{Q^{10.5}}{1} \Rightarrow \frac{dC}{dQ} = 10.5Q^{9.5}$$

- Before using the power rule, write all base^n terms above the line, then state the value of n. For example,

$$P = \frac{1}{Q^5} = Q^{-5} \quad n \text{ has value} - 5$$

$$\frac{dP}{dQ} = -5Q^{-5-1} = -5Q^{-6}$$

- Write root as indices; for example,

$$y = \frac{1}{\sqrt[5]{x}} = \frac{1}{(x)^{1/5}} = \frac{1}{x^{0.2}} = x^{-0.2}$$

$$\frac{dy}{dx} = (-0.2x^{-0.2-1}) = -0.2x^{-1.2}$$

The slope of a curve at a point

To find the slope of a curve $y = f(x)$ at a given point, simply substitute the x-coordinate of the point into the equation for dy/dx. This is also the slope of the tangent at the given point.

The following are other notations that may be used for the derivative of $y = f(x)$:

$$\frac{dy}{dx}, \ \frac{df(x)}{dx}, \ \frac{df}{dx}, \ f'(x), \ y', \ Dy, \ D(f(x))$$

Practical problems

In most practical problems, functions such as total cost are the sum or difference of several terms. For example, a total cost function is given by

$$TC = 0.2Q^3 - 15Q^2 + 175Q + 1000$$

where the Q-terms are multiplied by constants and there is a constant term, 1000. So how do we differentiate terms such as these? The working rules listed in Section 6.1.5 deal with these problems.

6.1.5 Working rules for differentiating sums and differences of several functions

Working rule	Notation	Examples/comment
Rule 1: The derivative of a constant term is zero	$y = 10 \rightarrow y = 10x^0$ $\dfrac{dy}{dx} = 10(0x^{0-1}) = 0$	**Rule 1:** The derivative is the rate of change and the rate of change of a constant term is zero; 10 is a constant term but, for example, $10x$ is not a constant term – it varies with x
Rule 2: The derivative of $Kf(x)$ or $\dfrac{f(x)}{K}$ is the derivative of $f(x)$ multiplied or divided by K. (Note: $Kf(x)$ and $f(x)/K$ are *not* constant terms)	$y = Kx^n$ $\dfrac{dy}{dx} = K\dfrac{d(x^n)}{dx}$ $= K(nx^{n-1})$ $y = x^n/K : \dfrac{dy}{dx} = \dfrac{nx^{n-1}}{K}$	**Rules 2 and 3:** (a) $y = 25x^{0.85} + \dfrac{x^{2.4}}{12}$ $\dfrac{dy}{dx} = 25(0.85x^{0.85-1}) + \dfrac{2.4x^{1.4}}{12}$ $= -21.25x^{-0.15} + 0.2x^{1.4}$
Rule 3: To differentiate the sum or difference of several terms, simply differentiate each term separately	$y = K_1x^m + x^n/K_2$ $\dfrac{dy}{dx} = K_1(mx^{m-1}) + (nx^{n-1})/K_2$ $= K_1mx^{m-1}$ $+ nx^{n-1}/K_2$	(b) $TC = 0.5Q^3 - 12Q^2 + 300$ $\dfrac{d(TC)}{dQ} = 0.5(3Q^2) - 12(2Q) + 0$ $= 1.5Q^2 - 24Q$

Note: Sometimes functions may be written in such a way that it may be more convenient to initially find dx/dy rather than dy/dx. In order to find dy/dx, simply invert dx/dy, that is,

$$\frac{dy}{dx} = \frac{1}{\left(\dfrac{dx}{dy}\right)}$$

For example,
$$x = 0.75y^{-1.8}$$

$$\frac{dx}{dy} = 0.75(-1.8y^{-1.8-1}) = -1.35y^{-2.8} = -\frac{1.35}{y^{2.8}}$$

$$\frac{dy}{dx} = -\frac{y^{2.8}}{1.35} \quad \text{inverting both sides}$$

WORKED EXAMPLE 6.3
MORE DIFFERENTIATION USING THE POWER RULE

Find the derivatives of each of the following using the power rule.

(a) $y = 5x^7$

(b) $Q = \dfrac{12}{L^{2.5}}$

(c) $y = \dfrac{2}{3\sqrt{x}}$

(d) $AC = 5 + \dfrac{2}{Q}$

SOLUTION

(a) $y = 5x^7$

$\dfrac{dy}{dx} = 5(7x^{7-1}) = 5(7x^6) = 35x^6$ the derivative of x^7 is multiplied by the constant 5

(b) $Q = \dfrac{12}{L^{2.5}}$

$Q = 12(L^{-2.5})$ bringing the L-term above the line

$\dfrac{dQ}{dL} = 12(-2.5L^{-2.5-1}) = 12(-2.5L^{-3.5}) = -30L^{-3.5}$

$\dfrac{dQ}{dL} = \dfrac{-30}{L^{3.5}}$

(c) $y = \dfrac{2}{3\sqrt{x}}$

$y = \dfrac{2}{3}(x)^{-0.5}$ arrange the x-term above the line, use rules for indices

$\dfrac{dy}{dx} = \dfrac{2}{3}(-0.5x^{-0.5-1})$ differentiating the x-term by the power rule

$= -\dfrac{2}{3}\dfrac{(0.5x^{-1.5})}{1}$ simplifying by writing signs first, numbers, then variables

$= \dfrac{-1}{3x^{1.5}}$

(d) $AC = 5 + \dfrac{2}{Q}$

$AC = 5 + 2Q^{-1}$ rules for indices

$\dfrac{d(AC)}{dQ} = 0 + 2(-1Q^{-1-1})$ the derivative of a constant term is zero

$= \dfrac{-2}{Q^2} = -2Q^{-2}$

6.1.6 Higher derivatives

Differentiating the first derivative gives the second derivative, differentiating the second derivative gives the third, and so on. The same rules for differentiation apply always, whether the required derivative is first, second or higher.

Notation: The second derivative is written as

$$\frac{d}{dx}\left(\frac{dy}{dx}\right) = \frac{d^2 y}{dx^2} = y'' \tag{6.2}$$

Second and third derivatives will be used later to find and confirm maximum and minimum points, and points of inflection.

WORKED EXAMPLE 6.4
CALCULATING HIGHER DERIVATIVES

Determine the first, second and third derivatives of the following demand function:

$$P = 25Q^4 - 10Q^2 + 200$$

SOLUTION

$$\frac{dP}{dQ} = 25(4Q^3) - 10(2Q) + 0 = 100Q^3 - 20Q$$

$$\frac{d^2 P}{dQ^2} = 100(3Q^2) - 20(Q^0) = 300Q^2 - 20 \quad \text{since} \quad Q^0 = 1$$

$$\frac{d^3 P}{dQ^3} = 300(2Q) - 0 = 600Q$$

PROGRESS EXERCISES 6.1

Introduction to Differentiation: The Power Rule

1. (a) Plot the curve, $y = x^2$, for $x = -2.5; -2; -1.5; -1; -0.5; 0; 0.5; 1; 1.5; 2$ and 2.5.
   (b) Draw tangents to the curve at $x = -1.5$, $x = 0$ and $x = 2$; then, measure the slope of the tangents at each point.
   (c) Find the slope of the curve at $x = -1.5$, $x = 0$ and $x = 2$ using differentiation. Compare your answers with those obtained in (b).

2. Differentiate each of the following, giving your answer with positive indices only:

(a) $y = x^5$ (b) $y = Q^{5.5}$ (c) $y = 10Q^8$ (d) $y = 20$

(e) $y = 20 + 3Q^2$ (f) $P = 15.2Q^{0.2}$ (g) $C = 10 - 2Y^{0.7}$

3. Differentiate each of the following, giving your answer with positive indices only:

(a) $y = \dfrac{1}{x^2}$ (b) $y = \dfrac{5}{x^2}$ (c) $y = 10 + 5x + \dfrac{1}{x^2}$

(d) $y = 10 + \dfrac{5}{x^2} + \dfrac{1}{x^2}$ (e) $P = \dfrac{Q^3}{3} + 70Q - 15Q^2$

4. Differentiate the following with respect to x giving your answers with positive indices.

(a) $x^2 + 7x + 5$ (b) $2 - 3x - 3x^{0.5}$ (c) $4 + x + 2\sqrt{x}$ (d) $x + \dfrac{1}{2x} + \dfrac{1}{x^2}$

(e) $x - \dfrac{2}{x} - \dfrac{3}{4x^2}$ (f) $x - 4\dfrac{1}{x} + 8\dfrac{2}{x^2}$

5. Differentiate each of the following, giving your answer with positive indices only:

(a) $y = 5\sqrt{x}$ (b) $y = \dfrac{1}{\sqrt{x}}$ (c) $y = \dfrac{1}{\sqrt{Q}}$ (d) $y = 7 + \dfrac{7}{\sqrt{Q}}$

(e) $y = 3\sqrt{Q^5}$ (f) $y = \dfrac{1}{3}\sqrt[3]{Q}$ (g) $P = \dfrac{5Q + 2}{5}$ (h) $P = \dfrac{5Q + 2}{Q}$

6. Simplify each of the following so that differentiation may be carried out using the power rule:

(a) $\dfrac{Q^{10}}{Q^2}$ (b) $4\dfrac{\sqrt{Q}}{Q^2}$ (c) $\dfrac{x^2 + x}{x^3}$ (d) $\sqrt{5Q^3}$ (e) $3\left(\dfrac{Q^2 + 2Q}{Q}\right) + 2$

7. Differentiate each of the functions given in question 6.

8. Find the first and second derivatives of the following functions:

(a) x^5 (b) $3Q^2$ (c) $10Q + Q^{0.5}$ (d) $Q^{1/3}$ (e) $\dfrac{1}{x}$ (f) $Y^4 - \dfrac{1}{Y^4}$

9. Find the first, second and third derivatives of the following total cost function:

$$TC = \frac{Q^3}{5} - 8Q^2 + \frac{5Q}{2} + 180$$

10. Find dP/dQ and dQ/dP when $30P + 3Q - 81 = 0$.

11. Find dy/dx and dx/dy in terms of x, when $x^2 = 25y$.

12. Find dS/di when $480i^2 + 40i^5 - 48i = S$.

13. Find dx/dt and dt/dx in terms of x when $x^3 6t = 18$.

14. Find the first and second derivatives of the total revenue function

$$TR = (150 - 2Q)Q$$

www.wiley.com/college/bradley

Go to the website for the following additional material that accompanies Chapter 6:

6.1.7 Use differentiation to determine whether a function or its rate of change (slope) is increasing or decreasing
- Use the first derivative to determine whether a function is increasing or decreasing
- Use the second derivative to determine whether the rate of change (slope) of a function is increasing or decreasing
- What is an increasing or decreasing slope (or rate of change?)
  Table 6.1 The coordinates and slope of $y = x^2$ at selected x-values

Figure 6.4 Slope increasing $\dfrac{d}{dx}$ (slope) $= \dfrac{d}{dx}\left(\dfrac{dy}{dx}\right) = \dfrac{d^2y}{dx^2} = y'' > 0$ throughout

Table 6.2 The coordinates and slope of $y = 6x - x^2$ at selected x-values

Figure 6.5 Slope increasing $\dfrac{d}{dx}$ (slope) $= \dfrac{d}{dx}\left(\dfrac{dy}{dx}\right) = \dfrac{d^2y}{dx^2} = y'' < 0$ throughout

Figure 6.6
Worked Example 6.5 Determining the rate of change of the slope of a curve: $y = 3x^2 - 0.1x^3$
Progress Exercise 6.2 Use differentiation to determine whether a function or its rate of change (slope) is increasing or decreasing
Solutions to Progress Exercises 6.2

6.2 Applications of Differentiation, Marginal Functions, Average Functions

At the end of this section you should be familiar with the following concepts:

- Marginal functions: an introduction, cost and revenue
- Average functions: an introduction, cost and revenue
- Marginal and average product of labour
- Marginal and average labour cost
- Marginal propensity to consume and marginal propensity to save
- Marginal utility and the law of diminishing marginal utility.

6.2.1 Marginal functions: an introduction

In Section 6.1 the derivative of a function was described as the equation for the rate of change of that function. For example, given the function, $y = x^3$, the equation $dy/dx = 3x^2$ describes the rate of change in y per unit increase in x. The derivative of certain economic variables such as TR, TC, profit, etc, is called the **marginal function**.

The marginal revenue is the rate of change in total revenue per unit increase in output, Q.

$$\text{Marginal revenue: } MR = \frac{d(TR)}{dQ} \tag{6.5}$$

The marginal cost is the rate of change in total cost per unit increase in output, Q.

$$\text{Marginal cost: } MC = \frac{d(TC)}{dQ} \tag{6.6}$$

To determine an expression for a marginal function the steps are as follows:

Step 1: Determine an expression for the total function.
Step 2: Differentiate the total function.

Marginal revenue

To determine the equation for marginal revenue one must first obtain the equation for total revenue: $TR = P \times Q$, where P, the price, is expressed in terms of Q through the equation of the demand function.

WORKED EXAMPLE 6.6
CALCULATING MARGINAL REVENUE GIVEN THE DEMAND FUNCTION

Given the demand function $P = 6 - 0.5Q$ find the value of MR for $Q = 1, 2, 3, 4, 5, 6, 7$.

SOLUTION

Step 1: Determine an expression for total revenue:

$$TR = P \times Q = (6 - 0.5Q)Q = 6Q - 0.5Q^2$$

Step 2: Differentiate TR to deduce the equation for MR:

$$TR = 6Q - 0.5Q^2$$

$$MR = \frac{d(TR)}{dQ} = 6(1) - 0.5(2Q) \rightarrow MR = 6 - Q$$

Table 6.3a shows the values of TR and MR calculated at points $Q = 1$ to $Q = 7$.

Table 6.3a   Total revenue and marginal revenue calculated by differentiation

Q	1	2	3	4	5	6	7
$TR = 6Q - 0.5Q^2$	5.5	10.0	13.5	16.0	17.5	18.0	17.5
$MR = 6 - Q$	5	4	3	2	1	0	−1

Note: In introductory economics textbooks, MR is often defined as the change in TR per unit change in output:

$$MR = \frac{\Delta TR}{\Delta Q} \tag{6.7}$$

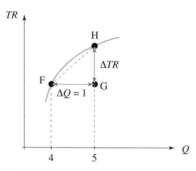

Figure 6.7   Marginal revenue measured along a chord

Equation (6.7) gives an expression for MR over a small interval, ΔQ, not MR at a point. This is illustrated in Figure 6.7 where MR is the change in TR over the interval of length ΔQ between points F and H, and not the rate of change in TR at either point F or at point H.

When the TR function is non-linear, equation (6.7) provides a reasonable approximation to equation (6.5), provided the interval is small, hence,

$$\frac{\Delta TR}{\Delta Q} \cong \frac{d(TR)}{dQ} \rightarrow \frac{\Delta TR}{\Delta Q} \cong MR$$

Therefore

$$\Delta TR \cong MR \times \Delta Q \tag{6.8}$$

That is, the change in total revenue is approximately equal to marginal revenue multiplied by the change in output. Similarly, for marginal cost:

$$\frac{\Delta TC}{\Delta Q} \cong MC \rightarrow \Delta TC \cong MC \times \Delta Q \tag{6.9}$$

WORKED EXAMPLE 6.7
CALCULATING MARGINAL REVENUE OVER AN INTERVAL

Given the demand function, $P = 6 - 0.5Q$, find the value of MR for $1 \leq Q \leq 7$ using the definition in (6.7): $MR = \dfrac{\Delta TR}{\Delta Q}$

SOLUTION

In this case, MR is calculated as the change in total revenue per unit change in output,

$$MR = \frac{\Delta TR}{\Delta Q}$$

The total revenue function is, $TR = P \times Q = (6 - 0.5Q)Q = 6Q - 0.5Q^2$.

Note: The price, as always, is defined by the demand function.
    The marginal revenue is calculated for consecutive values of Q in Table 6.3b. For example, over the interval $Q = 4$ to $Q = 5$, $MR = \frac{\Delta TR}{\Delta Q} = \frac{17.5 - 16.0}{5 - 4} = \frac{1.5}{1} = 1.5$

Table 6.3b   Total revenue and marginal revenue calculated over an interval, $\Delta Q = 1$

Q	1	2	3	4	5	6	7
$TR = 6Q - 0.5Q^2$	5.5	10.0	13.5	16.0	17.5	18.0	17.5
$MR = 6 - Q$	5	4	3	2	1	0	−1
$\Delta TR = TR_2 - TR_1$	10 − 5.5	13.5 − 10	16 − 13.5	17.5 − 16	18 − 17.5	17.5 − 18	
$MR = \Delta TR/\Delta Q$	4.5	3.5	2.5	1.5	0.5	−0.5	

Comment: Compare the values of MR in Tables 6.3a and 6.3b. MR over an interval is the average of the point MR at the start and end of the interval.
    For example, MR over the interval $Q = 4$ to $Q = 5$ is 1.5. This is the average of MR at points $Q = 4$ and at $Q = 5$: $\frac{(MR \text{ at } Q = 4) + (MR \text{ at } Q = 5)}{2} = \frac{2 + 1}{2} = 1.5$

Note: This is a feature of quadratic total revenue functions. It applies intervals of any width, not just to $\Delta Q = 1$ as above.

Marginal cost

Marginal cost, MC, is defined as the derivative of total cost with respect to output. In turn, total cost, TC, is the sum of fixed cost, FC, and variable cost, VC. Since FC is constant, MC may be shown to equal marginal variable costs, MVC, as follows:

$$MC = \frac{d(TC)}{dQ} = \frac{d(FC + VC)}{dQ} = \frac{d(VC)}{dQ} = MVC \qquad \textbf{(6.10)}$$

since the derivative of fixed costs (a constant) is zero.

WORKED EXAMPLE 6.8

DERIVE MARGINAL COST EQUATION FROM TOTAL COST FUNCTION

(a) Given the total cost function, $TC = 10 + 4Q$:
 (i) Derive an equation for MC. Does MC vary with output?
 (ii) Show that the derivative of total costs and variable costs with respect to output provide the same answer.
(b) Given the total cost function, $TC = \frac{1}{3}Q^3 - 8Q^2 + 120Q$:
 (i) Deduce the equation for MC. Does MC vary with output?
 (ii) Estimate the approximate change in TC as output, Q, increases from 15 to 16 units using equation (6.9).

SOLUTION

(a) (i) Since the TC function is given, differentiate TC with respect to output to obtain the equation for MC:

$$TC = 10 + 4Q$$

$$MC = \frac{d(TC)}{dQ} = 0 + 4(1) = 4$$

Comment: MC is a constant. It does not vary with output; that is, if output increases, MC remains at £4.

(ii) From the total cost function, variable cost $VC = 4Q$. The derivative of VC with respect to output is

$$VC = 4Q$$

$$MVC = \frac{d(VC)}{dQ} = 4$$

Comment: MC and MVC are the same.

(b) (i) MC for this cubic TC function is derived as

Step 1: $TC = \dfrac{1}{3}Q^3 - 8Q^2 + 120Q$

Step 2: $MC = \dfrac{d(TC)}{dQ} = Q^2 - 16Q + 120$

Comment: MC does vary with output, for example,
At $Q = 10$, $MC = Q^2 - 16Q + 120 = (10)^2 - 16(10) + 120 = 60$.
At $Q = 20$, $MC = Q^2 - 16Q + 120 = (20)^2 - 16(20) + 120 = 200$.

(ii) As Q increases from 15 to 16 units ($\Delta Q = 1$), the approximate change in TC is

$$\Delta TC \approx MC \cdot \Delta Q \text{ (see equation (6.9))}$$
$$= [(15)^2 - 16(15) + 120](1) = 105$$

where MC is evaluated at $Q = 15$. So, when output increases from 15 to 16 units, TC increases by approximately £105. That is, when the level of production is at 15 units, the total cost of producing another unit is approximately £105.

Note: The exact change in $TC = £112.33$. This is calculated by evaluating TC at $Q = 15$ and $Q = 16$ and subtracting.

6.2.2 Average functions: an introduction

From the above discussion, the equations for marginal functions, such as MR, may be used to determine the rate at which TR changes per unit increase in Q at any point. The **average function** gives an expression for the average value of an economic variable throughout an interval. The average function is defined as

$$\frac{\text{Total function}}{Q}$$

Average revenue (AR)

AR is defined as average revenue per unit for the first Q successive units sold. AR is determined by dividing total revenue by the quantity sold, Q:

$$\text{Average revenue: } AR = \frac{TR}{Q} \qquad (6.11)$$

Multiplying each side of equation (6.11) by Q gives another equation for total revenue:

$$TR = AR \times Q \qquad (6.12)$$

The relationship between AR and price

Total revenue has already been defined as

$$TR = P \times Q$$

but $TR = AR \times Q$, therefore

$$AR \times Q = P \times Q$$
$$P = AR \qquad (6.13)$$

The AR function is equal to price, P, where P is given by the demand function. $P = AR$ for any demand function, whether it is horizontal or downward sloping.

Average cost

AC is total cost divided by the level of output produced, that is,

$$\text{Average cost: } AC = \frac{TC}{Q} \tag{6.14}$$

Average cost is also the sum of average fixed cost, AFC, and average variable cost, AVC:

$$\text{Average cost: } AC = \frac{TC}{Q} = \frac{FC + VC}{Q} = \frac{FC}{Q} + \frac{VC}{Q} = AFC + AVC \tag{6.15}$$

where AFC and AVC are defined as

$$AFC = \frac{FC}{Q} \text{ and } AVC = \frac{VC}{Q}$$

In Worked Example 6.9, marginal and average functions are derived. A comparison is then made between the marginal and average functions of a perfectly competitive firm and a monopolist.

WORKED EXAMPLE 6.9
MR, AR FOR A PERFECTLY COMPETITIVE FIRM AND A MONOPOLIST

(a) Given a perfectly competitive firm's demand function, $P = 20$, find expressions for the marginal and average revenue functions.
(b) A monopolist is faced with a linear demand function $P = 50 - 2Q$. Find expressions for the marginal and average revenue functions. Hence:
(c) Calculate MR and AR for $Q = 0, 5, 10, 12.5, 15, 20, 25$. Comment on the relationship between MR and AR in (a) and (b).

SOLUTION

(a) To find expressions for average and marginal revenue, first write down the equation for total revenue:

Step 1: $TR = P \cdot Q = 20Q$

Step 2: Marginal revenue is $\qquad\qquad$ Average revenue is

$$MR = \frac{d(TR)}{dQ} = \frac{d(20Q)}{dQ} = 20 \quad \bigg| \quad AR = \frac{TR}{Q} = \frac{20Q}{Q} = 20$$

Comment: The AR function is always equal to the demand function; however, for a perfectly competitive firm, the MR function is also equal to the demand function. That is, $P = AR = MR = 20$. This condition is graphically demonstrated in Figure 6.8.

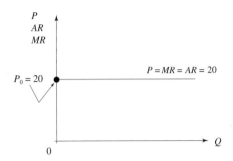

Figure 6.8   A perfectly competitive firm's AR and MR functions

(b) For the monopolist whose demand function is $P = 50 - 2Q$
   Step 1: $TR = P \cdot Q = (50 - 2Q)Q = 50Q - 2Q^2$
   Step 2:

Marginal revenue is	Average revenue is

$$MR = \frac{d(TR)}{dQ} = \frac{d(50Q - 2Q^2)}{dQ} \qquad AR = \frac{TR}{Q} = \frac{50Q - 2Q^2}{Q}$$

$$= 50 - 4Q \qquad\qquad\qquad\qquad = 50 - 2Q$$

(c) As expected, the AR function is the same as the demand function; however, for the monopolist, the slope of the MR function (slope = 4) is double that of the demand function (slope = 2). Table 6.4 illustrates numerical values for the monopolist's average and marginal revenue functions. The functions are graphically demonstrated in Figure 6.9.

Table 6.4   Marginal revenue and average revenue for a monopolist

(1) Q	(2) $P = 50 - 2Q$	(3) $TR = 50Q - 2Q^2$	(4) $MR = 50 - 4Q$	(5) $AR = 50 - 2Q$
0	50	0	50 (at $Q = 0$)	50
5	40	200	30 (at $Q = 5$)	40 (for each unit $Q = 1$ to $Q = 5$)
10	30	300	10 (at $Q = 10$)	30 (for each unit $Q = 1$ to $Q = 10$)
12.5	25	312.5	0 (at $Q = 12.5$)	25 (for each unit $Q = 1$ to $Q = 12.5$)
15	20	300	−10 (at $Q = 15$)	20 (for each unit $Q = 1$ to $Q = 15$)
20	10	200	−30 (at $Q = 20$)	10 (for each unit $Q = 1$ to $Q = 20$)
25	0	0	−50 (at $Q = 25$)	0 (for each unit $Q = 1$ to $Q = 25$)

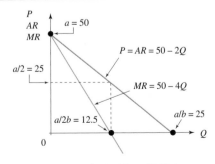

Figure 6.9   A monopolist's AR and MR functions

Marginal and average revenue functions for a perfectly competitive firm and a monopolist: a summary

The relationship between the marginal and average revenue functions for a perfectly competitive firm and a monopolist are summarised in Table 6.5. Columns (2) and (4) illustrate the results as calculated in Worked Example 6.9 while columns (3) and (5) give general expressions for each type of market.

The following points are of interest:

- For monopoly, $MR < AR$ by the amount bQ. To prove this statement, write down the definition of MR in column (5) of Table 6.5:

$$MR = a - 2bQ = (a - bQ) - bQ = AR - bQ, \text{ since } AR = (a - bQ)$$

Hence, $MR = AR - bQ$ or $MR + bQ = AR$
- The monopolist would not be interested in producing output where MR is negative. Thus the monopolist should only produce output up to the point where $MR = 0$. To find this point proceed as follows:

Table 6.5   Average revenue and marginal revenue functions: a summary

Functions	Perfect competition		Monopoly	
(1)	(2)	(3)	(4)	(5)
Demand function	$P = 20$	$P = P_0$	$P = 50 - 2Q$	$P = a - bQ$
Total revenue	$TR = 20Q$	$TR = P_0Q$	$TR = 50Q - 2Q^2$	$TR = aQ - bQ^2$
Average revenue	$AR = 20$	$AR = P_0$	$AR = 50 - 2Q$	$AR = a - bQ$
Marginal revenue	$MR = 20$	$MR = P_0$	$MR = 50 - 4Q$	$MR = a - 2bQ$

Example
$MR = 50 - 4Q$
$0 = 50 - 4Q$ $MR = 0$
$4Q = 50$
$Q = 12.5$

General expression
$MR = a - 2bQ$
$0 = a - 2bQ$ $MR = 0$
$2bQ = a$
$Q = \dfrac{a}{2b}$

Therefore, the monopolist should not produce any more than 12.5 units of output. (See Figure 6.9.)

Total cost from average cost

Average cost is total cost divided by the level of output. Therefore, it can be shown that total cost is average cost multiplied by output, Q:

$$AC = \frac{TC}{Q} \rightarrow TC = AC \times Q \tag{6.16}$$

WORKED EXAMPLE 6.10
DERIVE MARGINAL COST FROM AVERAGE COST

Find an expression for the MC function given the following average cost functions:

(a) $AC = 2Q + 5 + \dfrac{30}{Q}$

(b) $AC = 3Q^2 - 4Q + 6 + \dfrac{100}{Q}$

In each case (i) state the value of fixed cost and variable cost, and (ii) calculate the value of marginal cost when $Q = 50$.

SOLUTION

(a) Since the AC function is given, derive the equation for TC from the definition of AC:

$$AC = \frac{TC}{Q} \rightarrow AC(Q) = TC$$

$$TC = AC \times Q = \left(2Q + 5 + \frac{30}{Q}\right)Q = 2Q^2 + 5Q + 30 \qquad \textbf{(6.17)}$$

The MC function is then deduced by differentiating the total cost function,

$$MC = \frac{d(TC)}{dQ} = \frac{d(2Q^2 + 5Q + 30)}{dQ} = 4Q + 5 \qquad \textbf{(6.18)}$$

(i) From the total cost function (equation 6.17), $FC = 30$ and $VC = 2Q^2 + 5Q$.
(ii) Substitute $Q = 50$ into equation (6.18). Therefore, $MC = 4(50)+5 = 205$.

(b) $TC = AC(Q)$

$$TC = \left(3Q^2 - 4Q + 6 + \frac{100}{Q}\right)Q = 3Q^3 - 4Q^2 + 6Q + 100 \qquad \textbf{(6.19)}$$

$$MC = \frac{d(TC)}{dQ} = \frac{d(3Q^3 - 4Q^2 + 6Q + 100)}{dQ} = 9Q^2 - 8Q + 6 \qquad \textbf{(6.20)}$$

(i) From the total cost function (equation 6.19) $FC = 100$ and $VC = (3Q^3 - 4Q^2 + 6Q)$.
(ii) Substitute $Q = 50$ into equation (6.20), therefore, $MC = 9(50)^2 - 8(50) + 6 = 22\,106$.

PROGRESS EXERCISES 6.3

Applications of Differentiation: MR, AR, MC, AC

1. The demand function for a monopolist is $Q = 120 - 3P$.
   (a) Find expressions for TR, MR and AR. Evaluate TR, MR and AR at $Q = 15$. Hence explain in words the meaning of each function at $Q = 15$.
   (b) Calculate the value of Q for which AR = 0. Calculate the value of MR when AR = 0. Is it sensible for a monopolist to sell this value of Q?
   (c) Graph TR, MR and AR on the same diagram.
2. The demand function for a good is $P = 125 - Q^{1.5}$.
   (a) Find expressions for TR, MR and AR. Is the slope of the MR curve twice the slope of the AR curve?
   (b) Evaluate TR, MR and AR at $Q = 10$ and at $Q = 25$. Hence, explain in words the meaning of each function at $Q = 10, Q = 25$.
   (c) Calculate the value of Q for which MR = 0.
   Calculate the value of Q for which AR = 0.
   At what point (value of Q) does the sale of further units first start to reduce total revenue? Explain.
   (d) Graph TR, MR and AR on the same diagram.
3. A firm's fixed costs are 1000 and variable costs are given by $3Q$.
   (a) Write down the equation for TC. Calculate the value of TC when $Q = 20$.
   (b) Write down the equation for MC. Calculate the value of MC when $Q = 20$. Describe, in words, the meaning of MC for this function.
4. A firm has the following average cost function:

$$AC = 50 + \frac{10}{Q}$$

   (a) Show by differentiation that AC decreases indefinitely as Q increases. Give an economic interpretation of this phenomenon.
   (b) Write down the equation for total cost. Hence, write down the equation for total variable cost and average variable costs. State the value of fixed costs.
   (c) Write down the equation for marginal costs. Comment on the relationship between TC and MC in this example.
5. A firm has an average cost function,

$$AC = Q^2 - 9Q + \frac{150}{Q} + 75$$

   (a) Find an expression for the TC function and calculate TC when $Q = 15$.
   (b) Write down the equations for FC and TVC.
   (c) Find an expression for the MC function.
   Show that MC > 0 for all Q. Plot the MC and AC curves on the same diagram.
6. Given the AR functions,
   (i) $AR = 180 - 12Q$ (ii) $AR = 12$
   (a) Write down the equations for the total revenue functions.
   (b) Write down the equations for the marginal revenue functions.
   (c) Comment on the relationship between MR and AR.
   (d) Graph TR, MR and AR on the same diagram.

7. (a) Determine the first and second derivatives of the following:

    (i) $y = 3x^2 - 4x + \dfrac{2}{x}$

    (ii) $TR = 200Q - 4Q^2$

    (iii) $TC = \dfrac{Q^3}{3} - 12Q^2 + 164Q + 100$

    (b) From the equations for total revenue (TR) and total cost (TC) given in (a) above derive the formulae for:

    (i) marginal revenue, (ii) average revenue, (iii) marginal cost, (iv) average cost, (v) profit (where profit $= TR - TC$).

8. The demand function for a monopolist is given by the equation
$P = a - bQ$ where a and b are positive constants.

    (a) Write down the equations for (i) TR, (ii) MR, (iii) AR.

    (b) Write down expressions for (i) TR, (ii) MR, (iii) AR at $Q = 0$.

    (c) Show that $AR = a/2$ when $MR = 0$.

9. A total cost function is given by the equation

$$TC = aQ^3 - bQ^2 + cQ + d$$

where a, b, c and d are positive constants.

    (a) Write down expressions for:

      (i) total variable cost,   (ii) fixed cost,   (iii) marginal cost,   (iv) average cost.

    (b) Write down the equation for marginal cost when $a = 1/3, b = 2, c = 3$ and $d = 84$.

    Hence determine the values of Q for which marginal cost is (i) zero, (ii) a minimum. Calculate the minimum value of marginal cost. Plot the marginal cost function to confirm the results above graphically.

6.2.3   Production functions and the marginal and average product of labour

Firms, through the production process, transform inputs (or factors of production) into units of output. There are a variety of inputs: labour, L; physical capital (buildings, machinery), K; raw materials, R; technology (including information technology), T_e; land, S; and enterprise, E.

    A production function illustrates the relationship between input and output. Therefore, a production function may take the general form

$$Q = f(L, \ K, \ R, \ T_e, \ S, \ E \ldots)$$

This expression states that the level of output is dependent on the amounts of inputs used in the production process.

    In the short run, the inputs K, R, T_e, S and E can be assumed to be fixed so the level of output, Q, is then only a function of labour, L. That is, $Q = f(L)$, such as $Q = -0.5L^3 + 20L^2$. The derivation of the marginal and average products of labour follow the methods outlined for the goods market in Section 6.2.1.

    The **marginal product of labour** (MP_L) is the rate of change in total output, Q, with respect to labour input, L:

$$\text{Marginal product of labour: } MP_L = \frac{d(Q)}{dL} \qquad\qquad \textbf{(6.21)}$$

The average product of labour (APL) is total output divided by the number of units of labour employed:

$$\text{Average product of labour: } APL = \frac{Q}{L} \tag{6.22}$$

Note: APL is a measure of the average output per unit of labour, that is, it measures labour productivity.

WORKED EXAMPLE 6.11
DEDUCE THE EQUATION FOR THE MARGINAL AND AVERAGE PRODUCT OF LABOUR FROM A GIVEN PRODUCTION FUNCTION

Given the short-run production function $Q = 15L^2 - 0.5L^3$:

(a) Deduce the equation for the marginal product of labour. Calculate and comment on the marginal product of labour when 10 units of labour are employed.
(b) Derive the equation for the average product of labour. Calculate and comment on the average product of labour for the first 10 units of labour employed.

SOLUTION

(a) Since the total output, Q, is given, differentiate Q with respect to L to obtain the equation of the marginal product of labour:

$$Q = 15L^2 - 0.5L^3 \quad \text{total production}$$

$$MP_L = \frac{dQ}{dL} = 15(2L) - 0.5(3L^2)$$

$$= 30L - 1.5L^2$$

When $L = 10$, $MP_L = 30(10) - 1.5(10)^2 = 150$.

Comment: At the point where 10 units of labour are employed, production is increasing at the rate of 150 units of output per additional unit of labour employed.

(b) The average product of labour is derived as

$$APL = \frac{Q}{L} = \frac{15L^2 - 0.5L^3}{L} = 15L - 0.5L^2$$

When $L = 10$, $APL = 15(10) - 0.5(10)^2 = 100$.

Comment: The average productivity per unit of labour is 100 units of output for each of the first 10 units of labour employed. Therefore, the total output of the first 10 units of labour is $APL \times L$, since

$$\frac{Q}{L} = APL$$

$$Q = APL \cdot L = (15L - 0.5L^2) \cdot L = [15(10) - 0.5(10)^2] \cdot 10 = 100$$

www.wiley.com/college/bradley

Go to the website for the following additional material that accompanies Chapter 6:

6.2.4 Perfectly competitive and monopsony labour markets

- Marginal and average labour costs

  Worked Example 6.12 Find *TLC, MLC, ALC* given the labour supply functions

  Worked Example 6.13 Find *TLC, MLC, ALC* for a perfectly competitive firm and a monopsonist

  Figure 6.10 A perfectly competitive firm's *ALC* and *MLC* functions

  Figure 6.11 A monopsonist's *ALC* and *MLC* functions

- Marginal and average labour costs for a perfectly competitive firm and monopsonist: a summary

  Table 6.6 Marginal and average labour cost functions: a summary

6.2.5   Marginal and average propensity to consume and save

Marginal propensity to consume and save

Recall that when income is defined as consumption plus savings

$$Y = C + S$$

The marginal propensity to consume, *MPC*, is defined as the change in consumption per unit change in income, and, likewise, the marginal propensity to save, *MPS*, is defined as the change in savings per unit change in income. These were expressed in Chapter 3 as

$$MPC = \frac{\Delta C}{\Delta Y} \quad \text{and} \quad MPS = \frac{\Delta S}{\Delta Y}$$

Hence, $1 = MPC + MPS$.

These formulae are only exact when the consumption and savings functions are linear. When the consumption and savings functions are non-linear, the derivative should be used to measure *MPC* and *MPS*.

Summary of definitions for MPC, APC, MPS, APS for the consumption function $C = C_0 + bY$

Function	Consumption: $C = C_0 + bY$		Savings: $S = Y - C = Y - C_0 - bY$	
Marginal function	$MPC = \dfrac{d(C)}{dY}$	(6.25)	$MPS = \dfrac{d(S)}{dY}$	(6.27)
$0 < MPC < 1$	$MPC = \dfrac{dC}{dY} = \dfrac{d(C_0 + bY)}{dY} = b$		$MPS = \dfrac{dS}{dY} = \dfrac{d(Y - C_0 - bY)}{dY}$ $= 1 - b$	
$0 < MPS < 1$				
Average function	$APC = \dfrac{C}{Y}$		$APS = \dfrac{S}{Y}$	
	$APC = \dfrac{C}{Y} = \dfrac{C_0 + bY}{Y} = \dfrac{C_0}{Y} + b$		$APS = \dfrac{S}{Y} = \dfrac{Y - C_0 - bY}{Y}$ $= 1 - \dfrac{C_0}{Y} - b$	

Note

- Since $Y = C + S$, then

$$\frac{dY}{dY} = \frac{dC}{dY} + \frac{dS}{dY} \rightarrow 1 = MPC + MPS$$

Similarly, $APC + APS = 1$.
- Any reference to consumption and savings assumes planned consumption and planned savings, respectively.
- $APC > MPC$ since

$$\frac{C_0}{Y} + b > b$$

- $MPS > APS$ since

$$1 - b > 1 - b - \frac{C_0}{Y}$$

WORKED EXAMPLE 6.14
MPC, MPS, APC, APS

Given the consumption function $C = 20 + 3Y^{0.4}$

(a) Write down the equations for MPC and MPS.
(b) Write down the equations for APC and APS.
(c) Verify the inequalities $APC > MPC$ and $MPS > APS$ by comparing the values of MPC, APC, MPS and APS at $Y = 10$.

SOLUTION

(a) $C = 20 + 3Y^{0.4}$

$$MPC = \frac{dC}{dY} = 0 + 3(0.4Y^{0.4-1})$$

$$MPC = 1.2Y^{-0.6} = \frac{1.2}{Y^{0.6}}$$

$$MPS = 1 - MPC = 1 - \frac{1.2}{Y^{0.6}}$$

(b) $C = 20 + 3Y^{0.4}$

$$APC = \frac{C}{Y} = \frac{20}{Y} + \frac{3Y^{0.4}}{Y}$$

$$= \frac{20}{Y} + \frac{3}{Y^{0.6}}$$

$$APS = 1 - APC = 1 - \frac{20}{Y} - \frac{3}{Y^{0.6}}$$

(c) Now evaluate MPC, APC, MPS, APS at y = 10.

Marginal function at $Y = 10$

$$MPC = \frac{1.2}{(10)^{0.6}} = 0.30$$

$$MPS = 1 - MPC = 1 - 0.30 = 0.70$$

Average function at $Y = 10$

$$APC = \frac{20}{10} + \frac{3}{(10)^{0.6}} = 2.75$$

$$APS = 1 - APC = 1 - 2.75 = -1.75$$

These results verify the given inequalities APC > MPC and MPS > APS.

www.wiley.com/college/bradley

Go to the website for the following additional material that accompanies Chapter 6:

6.2.6 Marginal utility
Worked Example 6.15 Calculating the marginal utility

PROGRESS EXERCISES 6.4

Production, Labour Supply, Consumption, Savings and Utility

(Graphs may be plotted in Excel)

1. Given the labour supply functions,
    (i) $w = 180$, perfectly competitive,
    (ii) $w = 200 + 5L$, a monopsony:
        (a) Find expressions for TLC, ALC and MLC.
        (b) Draw the ALC and MLC functions. Comment on the relationship between them.
2. Given the production functions, (i) $Q = 20L - L^2$, (ii) $Q = 225L - \frac{1}{3}L^3$:
    (a) Find an expression for MP_L. Show that MP_L decreases indefinitely as labour input, L, increases (i.e., the law of diminishing returns applies).
    (b) Find an expression for APL.
    (c) Plot MP_L and APL. State the relationship between them.
3. Given the APL functions, (i) $APL = 150$, (ii) $APL = 30 - 2L$:
    (a) Deduce an expression for the production function, Q.

(b) Derive the equations for MP_L.

(c) What is the relationship between MP_L and APL?

4. Given the consumption functions, (i) $C = 40 + 0.8Y$, (ii) $C = 50 + 0.5Y^{0.8}$:

   (a) Write down expressions for the marginal propensity to consume and the average propensity to consume. Show, in general, that $APC > MPC$. Confirm this statement by evaluating APC and MPC at $Y = 10, 20$.

   (b) Deduce expressions for the marginal propensity to save and the average propensity to save. Show that $MPS > APS$. Confirm this statement by evaluating APS and MPS at $Y = 10, 20$.

   (c) Plot the graphs of APC, APS, MPC and MPS on the same diagram. Comment.

5. The benefit derived from study (at one sitting) is given by the utility function

$$U = 5x^{0.8}$$

where x is the number of hours spent studying.

   (a) Find an expression for the marginal utility, MU.

   (b) Show, by differentiation, that the level of utility decreases as the units of x increase (the law of diminishing marginal utility applies here).

   (c) Does MU ever become negative? Explain your answer.

6.3 Optimisation for Functions of One Variable

At the end of this section you should be able to:

- Use the slope of a curve to find turning points
- Use the change in slope and second derivatives to determine whether a turning point is a maximum or minimum
- Find the intervals along which a function is increasing or decreasing
- Outline the key features of curve sketching.

6.3.1 Slope of a curve and turning points

Optimisation is mainly concerned with finding maximum and minimum points, also known as stationary points, on a curve. Applications include finding maximum and minimum values for functions such as profit, cost, utility and production. In Section 6.1 the slope of a curve was shown to change at every point on the curve. Graphically, to 'see' the slope of a curve at a point, draw the tangent line to the curve at that point.

Remember

The slope of the curve at a point is the same as the slope of the tangent at that point. Figure 6.12 shows four turning points, two minima and two maxima, with the tangents drawn at these points. Note these are called 'local' minimum and maximum points as opposed to 'global' or absolute minimum and maximum points.

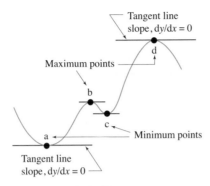

Figure 6.12   Turning points

In Figure 6.12 the tangents drawn at the turning points are horizontal, so they have no slope; that is, slope $= 0$. One can prove, with calculus, that **at turning points, the slope of a curve is always zero**. This fact provides the key to finding the exact coordinates of turning points. The argument is as follows:

- We know that dy/dx gives the equation for slope at any point (x, y).
- We know slope $= 0$ at turning points.
- So, $dy/dx = 0$ at turning points.

Therefore, to find the x-coordinates of the turning points on a curve $y = f(x)$, the following method is used:

Step 1: Find dy/dx for the given curve $y = f(x)$.
Step 2: Solve the equation $dy/dx = 0$.

The solution of this equation gives the x-coordinates of the turning points.

Notes
(i) If the equation $dy/dx = 0$ has no solution, then there is no turning point.
(ii) While it is true that at turning points we can say $dy/dx = 0$, the converse is not always true. If $dy/dx = 0$ at a point, it may indeed be a turning point but it could be a point of inflection. Points of inflection are discussed in Section 6.5.2, but until then all examples and questions will deal with turning points only.

Before proceeding to worked examples, Table 6.7 briefly outlines some other terminology that is commonly used in optimisation.

Table 6.7   Some terminology used in optimisation

Terminology	Reference point, Figure 6.12	Explanation
Local minimum	c	Lowest point in a given interval
Local maximum	b	Highest point in a given interval
(Absolute) minimum	a	Lowest point along the entire curve
(Absolute) maximum	d	Highest point along the entire curve
Stationary points	a, b, c, d	Point at which slope of curve is zero

WORKED EXAMPLE 6.16
FINDING TURNING POINTS

Find the turning points for the following functions:

(a) (i) $y = 3x^2 - 18x + 34$ (b) $y = 1/x$
     (ii) $AC = 3Q^2 - 18x + 34$

SOLUTION

(a) (i) and (ii) are identical equations in different variables.

Step 1: $y = 3x^2 - 18x + 34$

slope $= \dfrac{dy}{dx} = 6x - 18$

Step 2: Solve $dy/dx = 0$

$0 = 6x - 18$ $dy/dx = 0$ at turning point
$-6x = -18$

$x = \dfrac{-18}{-6} = 3$ at turning point

When $x = 3$, $y = 3(3^2) - 18(3) + 34 = 7$

The turning point exists at $x = 3$, $y = 7$

This is identical to the AC curve
   (Figure 6.13) with $y \equiv AC$ and $x \equiv Q$

Step 1: $AC = 3Q^2 - 18Q + 34$

slope $= \dfrac{d(AC)}{dQ} = 6Q - 18$

Step 2: Solve $d(AC)/dQ = 0$

$0 = 6Q - 18$ $dy/dx = 0$ at turning point
$-6Q = -18$

$Q = \dfrac{-18}{-6} = 3$ at turning point

When $Q = 3$, $AC = 3(3)^2 - 18(3) + 34 = 7$

The turning point exists at $Q = 3$, $AC = 7$

The AC function is shown in Figure 6.13

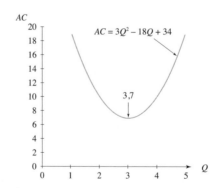

Figure 6.13   Locating turning points

(b) **Step 1:** $y = \dfrac{1}{x} = x^{-1}$ hence $\dfrac{dy}{dx} = -1\dfrac{x^{-1-1}}{1} = -\dfrac{x^{-2}}{1} = -\dfrac{1}{x^2}$

Step 2: Solve $dy/dx = 0 \rightarrow 0 = -\dfrac{1}{x^2} \rightarrow 0(x^2) = -1$

There is no value of x^2 (or x) which when multiplied by zero will have a value of -1.

 Remember

$0 \times$ (any real number) $= 0$. Therefore, no value of x exists for which slope dy/dx is zero, so there is no turning point. The graph of $y = 1/x$ is illustrated in Figure 6.14.

 Remember

the rectangular hyperbola of Chapter 4.

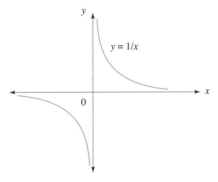

Figure 6.14   Graph of $y = 1/x$

Note: In general, points at which $dy/dx = 0$ are called stationary points. However, $dy/dx = 0$ at points of inflection, which are not turning points: see Section 6.5.2. Therefore, strictly speaking, the value(s) of x for which $dy/dx = 0$ are 'possible' turning points.

PROGRESS EXERCISES 6.5

Locate Turning Points

Locate the possible turning point(s) for the following functions:

1. $y = x^2 - 6x + 6$
2. $y = 2x^3 - 3x^2$
3. $P = -2Q^2 + 8Q$
4. $C = Y + 100$
5. $Q = 120 - 2.5P$
6. $y = x^3 - 3x^2 - 9x$
7. $TC = 144Q + \dfrac{1}{Q}$
8. $y = 8 - x^2 + x$
9. $U = 8x^{0.6}$
10. $y = x^4 - 2x^3$
11. $TC = \dfrac{Q^3}{3} - 2Q^2 + 3Q + 84$
12. $TR = (200 - 2Q)Q$

6.3.2   Determining maximum and minimum turning points

The next step is to determine whether each turning point is a maximum or a minimum. We could of course plot the graphs, since we know the precise turning points, as in Figure 6.13, but this is time consuming and not practical in the long term. Before proceeding to methods that will enable us to determine whether a point is a maximum or a minimum, here are some points to note.

When y is increasing, the derivative or rate of change of y, i.e., dy/dx, is positive

When dy/dx is increasing, the derivative or rate of change of dy/dx is positive, i.e.,

$$\frac{d}{dx}\left(\frac{dy}{dx}\right) = \frac{d^2y}{dx^2} \text{ is positive}$$

When y is decreasing, the derivative or rate of change of y, i.e., dy/dx, is negative

When dy/dx is decreasing, the derivative or rate of change of dy/dx is negative, i.e.,

$$\frac{d}{dx}\left(\frac{dy}{dx}\right) = \frac{d^2y}{dx^2} \text{ is negative}$$

To summarise

When y is increasing then y' is positive

When y' is increasing then y'' is positive

When y is decreasing then y' is negative

When y' is decreasing then y'' is negative

For further detail on increasing and decreasing slopes, see Section 6.1.7 on the website.

Testing for minimum and maximum points

Figure 6.15 illustrates how slope changes as we move through maximum and minimum points.

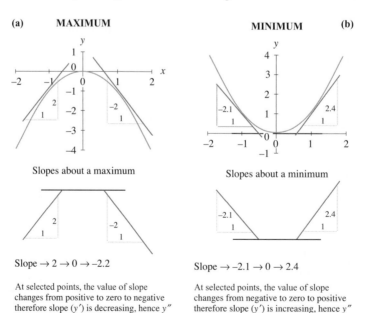

Slope $\to 2 \to 0 \to -2.2$

At selected points, the value of slope changes from positive to zero to negative therefore slope (y') is decreasing, hence y'' is negative in the region of a maximum

Slope $\to -2.1 \to 0 \to 2.4$

At selected points, the value of slope changes from negative to zero to positive therefore slope (y') is increasing, hence y'' is positive in the region of a minimum

Figure 6.15   (a) Maximum point. (b) Minimum point

- As we move through a maximum point, illustrated in Figure 6.15(a), the slope y' changes from being positive before the maximum (as y increases towards the maximum) to zero at the maximum and negative after the maximum (as y decreases down from the maximum).
- As we move through a minimum point, illustrated in Figure 6.15(b), the slope y' changes from being negative before the minimum (as y decreases down to the minimum) to zero at the minimum and positive after (as y increases away from the minimum).

This behaviour provides two methods of determining whether a turning point is a maximum or a minimum. Start by calculating the value of x at which $dy/dx = 0$. Call this value x_0. Also determine d^2y/dx^2 and evaluate it at x_0. This is written symbolically as

$$\left.\frac{d^2y}{dx^2}\right|_{x_0}$$

Method A: slope test	Method B: second-derivative test
TEST FOR A MAXIMUM	
Evaluate slope, dy/dx, at values of x immediately before and after x_0. If the slope changes from positive (before x_0) to negative (after x_0), the point is a maximum. See Figure 6.15(a)	Since the value of slope changes from positive through zero to negative as we move through a maximum, the value of slope (y') is decreasing, hence the derivative of slope (y'') is negative
	If $\left.\dfrac{d^2y}{dx^2}\right|_{x_0}$ is negative, the point is a maximum
TEST FOR A MINIMUM	
Evaluate slope dy/dx at values of x immediately before and after x_0	Since the value of slope changes from negative through zero to positive as we move through a minimum, the value of the slope (y') is increasing, hence derivative of slope (y'') is positive
If the slope changes from negative (before x_0) to positive (after x_0), the point is a minimum. See Figure 6.15(b)	If $\left.\dfrac{d^2y}{dx^2}\right|_{x_0}$ is positive, the point is a minimum

The second-derivative sign test is the simplest to use and works for most common functions. However, if the second derivative is zero, then the test provides no information

$$\left.\frac{d^2y}{dx^2}\right|_{x_0} = 0$$

It is then necessary to use the slope test to decide whether the point is a turning point. If the slope does not change direction (or sign) immediately before and after the point, it may be a point of inflection (Section 6.5.2). The **max/min** method may be summarised as follows:

Step 1: Find dy/dx and d²y/dx²; both derivatives will be required.
Step 2: Solve the equation dy/dx = 0; the solution of this equation gives the x-coordinates of the possible turning points.
Step 2a: Calculate the y-coordinate of each turning point by substituting the x-coordinate of the turning point (from step 2) into the equation of the curve.
Step 3: Determine whether the turning point is a maximum or minimum by substituting the x-coordinate of the turning point (from step 2) into the equation of the second derivative (method B):

- The point is a maximum if the value of d²y/dx² is negative at the point.
- The point is a minimum if the value of d²y/dx² is positive at the point.

Note: If the second derivative is zero, use the slope test to determine the nature of the stationary point (method A).

WORKED EXAMPLE 6.17
MAXIMUM AND MINIMUM TURNING POINTS

Find the turning points for the curve, $y = -x^3 + 9x^2 - 24x + 26$.
Determine which point is a maximum and which is a minimum by using the second derivatives.

SOLUTION

The graph of $y = -x^3 + 9x^2 - 24x + 26$ is drawn in Figure 6.16 to confirm and show the answers found using the following method.

Step 1: Find first and second derivatives,

$$y = -x^3 + 9x^2 - 24x + 26$$

$$\text{Slope:} \quad \frac{dy}{dx} = -3x^2 + 18x - 24$$

$$\text{Second derivative:} \quad \frac{d^2y}{dx^2} = -6x + 18$$

Step 2: At turning points, slope is zero, therefore, solve the equation dy/dx = 0,

$$\frac{dy}{dx} = 0 \quad \text{at turning points}$$

$$-3x^2 + 18x - 24 = 0$$

$$x^2 - 6x + 8 = 0 \quad \text{dividing both sides by } -3$$

$$(x - 4)(x - 2) = 0$$

$$x = 4 \text{ or } x = 2$$

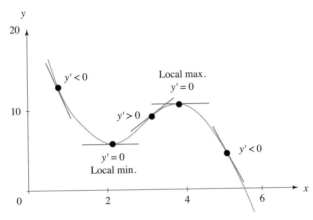

Figure 6.16   Graph of $y = -x^3 + 9x^2 - 24x + 26$

(If it is difficult to find the factors for a quadratic, use the 'minus b' method to find the roots.) Therefore, there are two turning points, one at the point whose x-coordinate is $x = 4$ and the other at the point whose x-coordinate is $x = 2$.

Step 2a: Find y.

Since we know the x-coordinate, find y from the equation of the curve:

$$y = -x^3 + 9x^2 - 24x + 26$$

$$y = -(2)^3 + 9(2)^2 - 24(2) + 26 \quad \text{substituting } x = 2$$

$$= -8 + 36 - 48 + 26 = 6 \therefore y = 6 \text{ when } x = 2$$

$$y = -(4)^3 + 9(4)^2 - 24(4) + 26 \quad \text{substituting } x = 4$$

$$= -64 + 144 - 96 + 26 = 10 \therefore y = 10 \text{ when } x = 4$$

Step 3: Determine whether each turning point is a maximum or a minimum from either the 'slope test' or the 'second-derivative test'. Both methods are illustrated below.

Turning point at $x = 2$
Method A: slope test

Evaluate slope: $\frac{dy}{dx} = -3x^2 + 18x - 24$ at any convenient value of x before $x = 2$, at $x = 2$ and after $x = 2$, the turning point

For example, take $x = 0$ as the point before $x = 2$ and $x = 3$ as the point after $x = 2$

The slope at each x-value is given below followed by a rough sketch:

	at $x = 1$	at $x = 2$	at $x = 3$
dy/dx	−9	0	3
Direction	Negative	Zero	Positive

Hence there is a minimum point at $x = 2$

Method B

Evaluate the second derivative at $x = 2$

$$\frac{d^2y}{dx^2} = -6x + 18$$
$$= -6(2) + 18 = 6$$

POSITIVE

Therefore, a minimum at $x = 2$
(2, 6) is a minimum point

Turning point at $x = 4$
Method A: slope test

Evaluate slope: $\frac{dy}{dx} = -3x^2 + 18x - 24$ at any convenient value of x before $x = 4$, at $x = 4$ and after $x = 4$, the turning point

For example, take $x = 3$ as the point before $x = 4$ and $x = 5$ as the point after it

The slope at each x-value is given below followed by a rough sketch:

	at $x = 3$	at $x = 2$	at $x = 5$
dy/dx	3	0	−9
Direction	Positive	Zero	Negative

Hence there is a maximum point at $x = 2$

Method B

Evaluate the second derivative at $x = 4$

$$\frac{d^2y}{dx^2} = -6x + 18$$
$$= -6(4) + 18 = -6$$

NEGATIVE

Therefore,
(4, 10) is a maximum point

PROGRESS EXERCISES 6.6

Determine Whether a Turning Point is Maximum or Minimum

Find the maximum and/or minimum values (if any) for each of the following functions (these are the same functions as given in Progress Exercises 6.5).

1. $y = x^2 - 6x + 6$

2. $y = 2x^3 - 3x^2$

3. $P = -2Q^2 + 8Q$

4. $C = Y + 100$

5. $Q = 120 - 2.5P$

6. $y = x^3 - 3x^2 - 9x$

7. $TC = 144Q + \dfrac{1}{Q}$

8. $y = 8 - x^2 + x$

9. $U = 8x^{0.6}$

10. $y = 12x^3 - x^4$

11. $TC = \dfrac{Q^3}{3} - 2Q^2 + 3Q + 84$

12. $TR = (200 - 2Q)Q$

For questions 13 to 22, find the maximum and/or minimum values (if any) for each of the following functions.

13. $y = x^2 - 20x + 20$

14. $y = \dfrac{1}{3}t^3 - 2t^2 - 5t + 8$

15. $y = x^2 - 8x + 6$

16. $y = \dfrac{1}{3}x^3 - x^2 - 3$

17. $y = 4x^3 - 6x^2$

18. $Q = 64P + \dfrac{1}{P}$

19. $P = 200 - 2Q^2$

20. $P = 200 - 2(Q - 4)^2$

21. $C = -2Q^2 + 8Q$

22. $Q = 2P + 3$

6.3.3 Intervals along which a function is increasing or decreasing

Now that maximum and minimum points can be located exactly, stating the intervals along which a curve is increasing or decreasing is a very simple matter.

- If a point is a maximum point, then y must increase in the interval immediately before the maximum and decrease in the interval immediately after the maximum.
- If a point is a minimum point, then y must decrease in the interval immediately before the minimum and increase in the interval immediately after the minimum.

So, when the turning points are found, the intervals along which y is increasing or decreasing may be stated precisely. In Figure 6.16, y is decreasing towards a minimum at $x = 2$. Between the minimum and the maximum point at $x = 4$, y is increasing.

WORKED EXAMPLE 6.18
INTERVALS ALONG WHICH A CURVE IS INCREASING OR DECREASING

State the range of values of Q for which the average cost function

$$AC = -9Q + 0.5Q^2 + 43$$

is (a) increasing, (b) decreasing.

SOLUTION

Before any statements are made about whether the function is increasing or decreasing, we must find the turning point(s), and also decide if they are a maximum or minimum. Work through the max/min method as follows:

Step 1: Find first and second derivatives:

$$AC = -9Q + 0.5Q^2 + 43$$

$$\frac{d(AC)}{dQ} = -9 + Q$$

$$\frac{d^2(AC)}{dQ^2} = 1$$

Note: When the second derivative is a constant, there is only one turning point. The turning point is a minimum if the constant is positive, a maximum if the constant is negative.

Step 2: Solve slope $= 0$ as slope is zero at turning point(s):

$$\frac{d(AC)}{dQ} = 0$$

$$-9 + Q = 0 \rightarrow Q = 9$$

The turning point is at $Q = 9$.

Step 3: Determine whether the turning point is a maximum or minimum. Since the second derivative is a positive constant, then the only turning point is a minimum.

 Remember

Conclusion: AC is decreasing until $Q = 9$ (the minimum), increasing after $Q = 9$ as shown in Figure 6.17.

DIFFERENTIATION AND APPLICATIONS

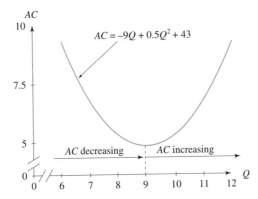

Figure 6.17 Interval along which AC is decreasing or increasing

6.3.4 Graphs of y, y', y'': derived curves

Earlier in this section it was stated that:

- In the region of a maximum, the first derivative (slope) changes from positive to negative (y' decreases), so the second derivative is negative.
- In the region of a minimum, the first derivative (slope) changes from negative to positive (y' increases), so the second derivative is positive.

These statements are illustrated graphically in Figure 6.18, Worked Example 6.19.

WORKED EXAMPLE 6.19
DERIVED CURVES

(a) Find the turning points for the curve, $y = 3x^2 - 0.1x^3$. Hence, determine the intervals along which y is increasing and decreasing.
(b) Plot the graphs of y against x, dy/dx against x and d^2y/dx^2 against x showing all turning points. Do the graphs of the first and second derivative confirm the rules stated above for finding optimum points?

SOLUTION

(a) **Step 1:** Derive expressions for dy/dx and d^2y/dx as follows:

$$y = 3x^2 - 0.1x^3$$

$$\frac{dy}{dx} = 6x - 0.3x^2$$

$$\frac{d^2y}{dx^2} = 6 - 0.6x$$

Step 2: Find the turning points by solving the equation $dy/dx = 0$:

$$\text{slope} = \frac{dy}{dx} = 6x - 0.3x^2 = 0$$

$$x(6 - 0.3x) = 0 \text{ factoring the LHS}$$

$$x = 0$$

$$\text{and } 6 - 0.3x = 0 \rightarrow x = \frac{6}{0.3} = 20$$

There are two turning points at $x = 0$ and at $x = 20$.

Step 2a: For each turning point, find the y-coordinate.
At $x = 0$, $y = 0$.
At $x = 20$, $y = 400$.

Step 3: Find the nature of the turning point by evaluating the second derivative at each point.
At $x = 0$, $y'' = 6 - 0.6(0) = 6$, positive, so there is a minimum at $x = 0$, $y = 0$.
At $x = 20$, $y'' = 6 - 0.6(20) = -6$, negative, so there is a maximum at $x = 20$, $y = 400$.
These points are illustrated in Figure 6.18. Therefore, the curve is decreasing for all values of x in the interval $x < 0$ and in the interval $x > 20$. The curve is increasing in the interval $x = 0$ to $x = 20$ (between the minimum and maximum).

(b) The graphs to be plotted are

$$y = 3x^2 - 0.1x^3$$

$$\frac{dy}{dx} = 6x - 0.3x^2$$

$$\frac{d^2y}{dx^2} = 6 - 0.6x$$

To show all the turning points, the graphs are plotted over the interval $x = -10$ to $x = 30$. You should set up a table to calculate several points for each graph, and plot the graphs. Use Excel, if it is available (see Table 6.8).

Comments: From Figure 6.18(b) slope is negative before and positive after the minimum point at $x = 0$, while slope is positive before and negative after the maximum point at $x = 20$. From Figure 6.18(c) the second derivative is positive in the region of the minimum. At $x = 10$, the second derivative is zero and thereafter becomes negative and remains negative throughout the region of the maximum at $x = 20$.

One further comment about the point $x = 10$: From Figure 6.18(b) slope is positive and decreasing up to the point at $x = 10$, and negative, but decreasing afterwards. So, at $x = 10$, the slope changes from being a positive slope to being a negative slope: the second derivative $y'' = 0$. This point is called a point of inflection. Points of inflection are studied later in the chapter.

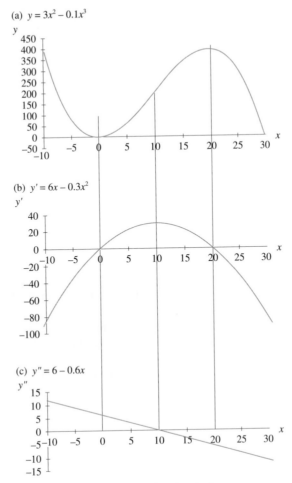

Figure 6.18   Graphs of y, y', y''

Table 6.8   Selected points for graphs in Figure 6.18

x	−10	−5	0	5	10	15	20	25	30
(a) $y = 3x^2 - 0.1x^3$	400	87.5	0	62.5	200	337.5	400	312.5	0
(b) $\dfrac{dy}{dx} = 6x - 0.3x^2$	−90	−37.5	0	22.5	30	22.5	0	−37.5	−90
(c) $\dfrac{d^2y}{dx^2} = 6 - 0.6x$	12	9	6	3	0	−3	−6	−9	−12

PROGRESS EXERCISES 6.7

Determine the Intervals Along which a Curve Increases/Decreases

Determine the intervals along which each of the following curves is increasing or decreasing (consider the positive half of the plane, $x \geq 0$).

1. $y = x^2$

2. $AC = Q^2 - 20Q + 120$

3. $TR = 50Q - Q^2$

4. $AR = 50 - 2Q$

5. $MC = Q^2 - 18Q + 11$

6. $P = 10 - 2Q^{0.4}$

7. $TC = \dfrac{Q^3}{3} + 20$

8. $C = 90 + \sqrt{Y}$

6.3.5   Curve sketching and applications

Sketching curves outlines the general shape of the curve, but it also shows specific points such as turning points, points of intersection with the axes, etc. Thus, an economist and business person should be able to take most equations, such as a total cost function or a function outlining projected company sales, and determine when costs and sales increase or decrease; when the rate is increasing or decreasing; whether cost and sales ever reach a maximum or minimum, etc.

However, if a curve $y = f(x)$ is plotted by joining several points, key features of the curve may be missed out completely. For example, if the graph of $y = 1/(x - 0.23)$ is plotted as in Figure 6.19 from the points calculated in Table 6.9, a totally incorrect representation of the function in the interval $x = 0$ to $x = 1$ is obtained. The correct representation is shown in Figure 6.20.

Table 6.9   Points for Figure 6.19

x	-2	-1	0	1	2	3
$y = \dfrac{1}{(x - 0.23)}$	-0.448	$-0.813\,01$	$-4.347\,83$	$1.298\,701$	$0.564\,972$	$0.361\,01$

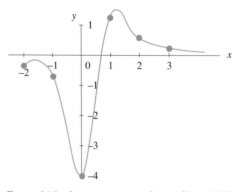

Figure 6.19   Incorrect curve of $y = 1/(x - 0.23)$

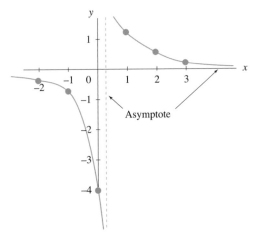

Figure 6.20   Correct curve of $y = 1/(x - 0.23)$

 Remember

Figure 6.19 is incorrect as:

- It loses the key feature of the correct graph, namely the asymptote at $x = 0.23$.
- It shows turning points when there are no turning points in the correct graph. You can prove this by differentiating the function.

It is therefore important to get an overview or sketch of a function at first, then the curve can be plotted accurately, point by point, over any interval if required.

Some key features to look for when sketching a curve

The following points act as a useful guide to graph sketching:

1. Is the curve similar to any standard curve, for example, a quadratic or polynomial, exponential, log, etc?
2. Is there any value that causes division by zero? If there is, then there will be a vertical asymptote at this point (see Chapter 4).
3. Check that the curve exists in every region of the plane; for example, $y = \sqrt{x}$ is not a real number when x is negative, so there is no graph in the negative half of the plane.
4. Find the points of intersection with the axes.
5. Find turning points and points of inflection (Section 6.5.2).
6. Define the curvature over any interval if required. This is particularly important for graphs that do not have turning points, such as exponential graphs. See Worked Example 6.16(b) and Worked Example 6.20(b).

The examples which follow should demonstrate how points 1 to 6 are used in the general approach to graph sketching. Points 1, 2 and 3 are normally checked first, that is, state whether the function in question is similar to any of the well-known curves, whether there are any vertical asymptotes as a result of division by zero, and whether there are regions of the plane where there are no graphs. Then points 4 and 5 are checked: the usual points of intersection with axes and turning points are calculated, and, finally, check for points of inflection and curvature, particularly if there are no turning points (see Section 6.5). A number of points may be plotted, particularly if there is no turning point or point of intersection with the axes, or if you need convincing.

WORKED EXAMPLE 6.20
SKETCHING FUNCTIONS

Sketch the following functions over the stated interval.

(a) $Q = 100 - P^2$ $-1 < P < 10$

(b) $AC = \dfrac{5}{Q}$ $0 < Q < 4$

SOLUTION

(a) Plot Q on the vertical, P on the horizontal as in Figure 6.21. This is a polynomial, therefore there are no problems with discontinuities and we proceed as follows:

Where does the graph cross the axes?	**Find the turning point(s)**
The graph crosses the Q-axis at	$Q = 100 - P^2$
$P = 0$, that is, $Q = 100$.	$\dfrac{dQ}{dP} = -2P$
The graph crosses the P-axis at	$= 0$ when $P = 0$, $Q = 100$
$Q = 0$, that is, $P^2 = 100$, that is,	$\dfrac{d^2Q}{dP^2} = -2$, negative, $\therefore$ a maximum
at $P = 10$ and $P = -10$.	

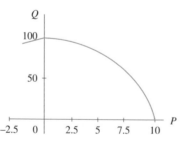

Figure 6.21   Graph of $Q = 100 - P^2$

(b) Plot AC on the vertical, Q on the horizontal as in Figure 6.22. First, this is a familiar graph (for example, see Worked Example 6.16(b), also Chapter 4, Hyperbolic functions). There is division by zero when $Q = 0$, so the AC-axis is a vertical asymptote. Also there are no turning points, as shown below. Proceed as follows:

Where does the graph cross the axes?

Graph crosses the Q-axis at $AC = 0$, i.e.,

$$AC = \frac{5}{Q}$$

$$0 = \frac{5}{Q}$$

$$0(Q) = 5$$

There is no value, Q, that can be multiplied by zero to give 5; therefore, there is no solution, so the graph does not cut the Q-axis.

Graph crosses the AC-axis at $Q = 0$, i.e.,

$$AC = \frac{5}{Q}$$

$$AC = \frac{5}{0}$$

$$AC(0) = 5$$

Again, there is no solution, so the graph does not cross the AC-axis. Division by zero gives the vertical asymptote.

Show that there is no turning point(s):

Step 1: Find derivatives

$$AC = \frac{5}{Q}$$

$$\frac{d(AC)}{dQ} = -\frac{5}{Q^2}$$

$$\frac{d^2(AC)}{dQ^2} = \frac{10}{Q^3}$$

Step 2: Find turning points: Solve slope $= 0$

$$\frac{d(AC)}{dQ} = 0$$

$$\frac{-5}{Q^2} = 0$$

$$-5 = 0(Q^2) = 0$$

No solution, therefore no turning point and, hence, no need to do step 3 to determine whether point is a maximum or minimum.

Table 6.10 gives some points for $AC = 5/Q$, which are then plotted as Figure 6.22.

Table 6.10   Points for Figure 6.22

Q	0	0.5	1	2	3	4
$AC = \dfrac{5}{Q}$	?	10	5	2.5	1.67	1.25

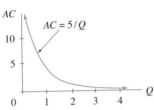

Figure 6.22   Graph of $AC = 5/Q$

PROGRESS EXERCISES 6.8

Sketch Curves, Indicating Turning Points and Intersection with Axes

The following curves were given in Progress Exercises 6.5 and 6.6. Sketch each by finding the points of intersection with the axes and using the maximum and minimum points already calculated in Progress Exercises 6.6.

1. $y = x^2 - 6x + 6$

2. $y = 2x^3 - 3x^2$

3. $P = -2Q^2 + 8Q$

4. $C = Y + 100$

5. $Q = 120 - 2.5P$

6. $y = x^3 - 3x^2 + 9x$

7. $TC = 144Q + \dfrac{1}{Q}$

8. $y = 8 - x^2 + x$

9. $U = 8x^{0.6}$

10. $y = x^4 - 2x^2$

Confirm each of the above sketches by plotting the graphs in Excel, if available.

6.4 Economic Applications of Maximum and Minimum Points

At the end of this section you should be able to:
- Sketch curves
- Find the value of output at which TR is at a maximum
- Find the value of output at which profit (π) is at a maximum
- Understand the use of the rule: $MC = MR$ in profit optimisation
- Show the profit-maximising point of a price-discriminating firm
- Show the profit-maximising points for a perfectly competitive firm and a monopoly in the goods market.

In Section 6.2 the first derivatives of economic functions were called marginal functions. Therefore, the optimum value of functions, such as revenue, profit, cost, etc., will all occur when the corresponding marginal function (first derivative) is zero as will be shown in the following worked examples.

WORKED EXAMPLE 6.21
MAXIMUM *TR* AND A SKETCH OF THE *TR* FUNCTION

The demand function for a good is given as $P = 50 - 2Q$.

(a) Write down expressions for the TR and MR functions.
(b) Calculate the output at which TR is a maximum and confirm that marginal revenue is zero at this point.
(c) Sketch the TR and MR functions.

▨ SOLUTION

(a)
$$TR = P \times Q = (50 - 2Q)Q$$
$$= 50Q - 2Q^2$$
$$MR = \frac{d(TR)}{dQ} = 50 - 4Q$$

(b) To find the output level at which *TR* is at a maximum, work through the max/min method for finding turning points as follows:

Step 1: Get derivatives	**Step 2:** Find turning point(s)	**Step 3:** Max or min?
$TR = 50Q - 2Q^2$ $$\frac{d(TR)}{dQ} = 50 - 4Q$$ $$\frac{d^2(TR)}{dQ^2} = -4$$	$\dfrac{d(TR)}{dQ} = 0$ at turning point $$50 - 4Q = 0$$ $$Q = \frac{50}{4} = 12.5$$ **Step 2a:** At $Q = 12.5$, $$TR = [50 - 2(12.5)]12.5 = 312.5$$	The second derivative is a negative constant, therefore *TR* has a maximum value. $$TR = 312.5$$ at $$Q = 12.5$$

In general, *TR* is maximised when

$$\frac{d(TR)}{dQ} = MR = 0 \tag{6.28}$$

and

$$\frac{d^2(TR)}{dQ^2} = \frac{d(MR)}{dQ} = (MR)' < 0$$

Note: When $MR = 0$, and $(MR)' > 0$, then *TR* is a minimum. This situation does not arise in this example.
To confirm these rules in this example, calculate *MR* when $Q = 12.5$, the value at which *TR* is maximised:

$$MR = 50 - 4Q = 50 - 4(12.5) = 0$$

(c) The *TR* and *MR* functions are sketched in Figure 6.23.
Sketching the total revenue function: *TR* is a quadratic and was sketched in Worked Example 4.7, Figure 4.8. Therefore, find

(1) Points of intersection with the axes. TR crosses the horizontal axis when $TR = 0$, that is, at $Q = 0$ and $Q = 25$. See Worked Example 4.7. Therefore, the points of intersection with the horizontal axis are $(0, 0)$ and $(25, 0)$.

(2) The turning point for the total revenue function, which was calculated in part (b), at $TR = 312.5$ and $Q = 12.5$.

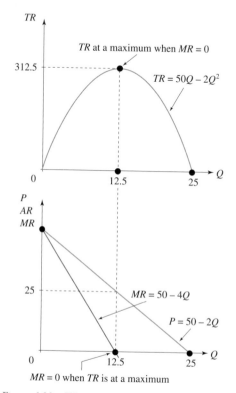

Figure 6.23 TR is at a maximum when $MR = 0$

Remember

In Chapter 4 it was stated that the turning point for a quadratic occurred halfway between the points of intersection with the horizontal axis.

Sketching the marginal revenue function: $MR = 50 - 4Q$ is a linear function, therefore two points are required to sketch its graph. The MR function intersects the vertical axis at $Q = 0$, therefore $MR = 50 - 4(0) = 50$. The MR function intersects the horizontal axis when $MR = 0$, therefore $0 = 50 - 4Q \rightarrow Q = 12.5$. The demand function is also sketched in Figure 6.23.

WORKED EXAMPLE 6.22
BREAK-EVEN, PROFIT, LOSS AND GRAPHS

▶ Find animated worked examples at **www.wiley.com/college/bradley**

The demand function for a good is given by the equation $P = 50 - 2Q$, while total cost is given by $TC = 160 + 2Q$.

(a) Write down the equation for (i) total revenue and (ii) profit.
(b) (i) Sketch the total cost and total revenue functions on the same diagram.
  (ii) From the graph, estimate, in terms of Q, when the firm breaks even, makes a profit and makes a loss.
  Confirm these answers algebraically.
(c) Determine the maximum profit and the value of Q at which profit is a maximum. Sketch the profit function.
(d) Compare the levels of output at which profit and total revenue are maximised.

SOLUTION

(a) (i) $TR = PQ = (50 - 2Q)Q = 50Q - 2Q^2$
   (ii) $\pi = TR - TC = 50Q - 2Q^2 - (160 + 2Q)$
    $= -2Q^2 + 48Q - 160$
    $= -2(Q^2 - 24Q + 80)$

(b) (i) **Sketching the total cost function:** $TC = 160 + 2Q$ is a linear function, therefore two points are required to sketch its graph. For example, at $Q = 0$, $TC = 160 + 2(0) = 160$ and taking any other Q-value, say $Q = 20$, then $TC = 160 + 2(20) = 200$. The total cost function is drawn in Figure 6.24.

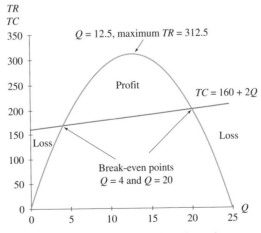

Figure 6.24 Total revenue and total cost functions

Sketching the total revenue function: This total revenue function is the same as that sketched in Worked Example 6.21, Figure 6.23. *TR* and *TC* are graphed in Figure 6.24.
(ii) From the graph, break-even occurs when $TC = TR$, at $Q = 4$ and $Q = 20$.
The break-even points are calculated algebraically as follows:

$$TC = TR$$
$$160 + 2Q = 50Q - 2Q^2$$
$$2Q^2 - 48Q + 160 = 0$$
$$Q^2 - 24Q + 80 = 0 \qquad \text{dividing both sides by 2}$$
$$(Q - 4)(Q - 20) = 0$$

Therefore the break-even quantities are $Q = 4$ and $Q = 20$.
As shown in Figure 6.24, losses are incurred below break-even point $Q = 4$ (where $TC >$ TR) and above break-even point $Q = 20$ (where $TC > TR$). Profits are made between the two break-even points since $TC < TR$ in this interval.

(c) Maximum profits are calculated by finding the turning point(s) for the profit function. This is done using the three-step max/min method:

Step 1: Get derivatives	Step 2: Find turning point(s)	Step 3: Max or min?
$\pi = -2Q^2 + 48Q - 160$ $$\frac{d\pi}{dQ} = -4Q + 48$$ $$\frac{d^2\pi}{dQ^2} = -4$$ at turning point	$$\frac{d\pi}{dQ} = 0$$ at turning point $$-4Q + 48 = 0$$ $$Q = \frac{48}{4} = 12$$ **Step 2a:** At $Q = 12$, $$\pi = -2(12)^2 + 48(12)$$ $$-160 = 128$$	The second derivative is a negative constant (-4) therefore profits are a maximum $\pi = 128$ at $Q = 12$

The profit function has only one turning point and this is a maximum.
Sketching the profit function: The profit function is sketched in Figure 6.25. To sketch the profit function, find

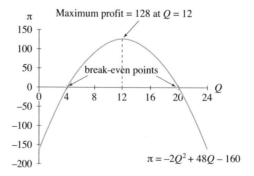

Figure 6.25   Profit function

(1) the points of intersection with the axes and (2) the turning point.

(1) The graph cuts the Q-axis when $\pi = 0$.

$$\pi = 0$$
$$-2(Q^2 - 24Q + 80) = 0$$
$$Q^2 - 24Q + 80 = 0 \quad \text{dividing both sides by } -2$$
$$(Q - 4)(Q - 20) = 0$$
$$\text{therefore} \quad Q = 4 \quad \text{and} \quad Q = 20$$

These values of Q where $\pi = 0$ are also the break-even points.
(2) We have just shown that profit has a maximum value of 128 when $Q = 12$. This provides another point for sketching the profit function.
(d) Maximum profit of 128 occurs when $Q = 12$.

WORKED EXAMPLE 6.23
MAXIMUM AND MINIMUM OUTPUT FOR A FIRM OVER TIME

The output for a firm over time is given by the function

$$Q = \frac{t^3}{30} - \frac{t^2}{5} + \frac{3t}{10} + 120$$

Determine the years (t) in which output is at a maximum and a minimum.

SOLUTION

In order to find the years in which output is at a maximum and minimum the turning points of the function must be calculated. The three steps for finding the turning points are outlined briefly as follows:

Step 1: Find derivatives

$$Q = \frac{t^3}{30} - \frac{t^2}{5} + \frac{3t}{10} + 120$$

$$\frac{dQ}{dt} = \frac{3t^2}{30} - \frac{2t}{5} + \frac{3}{10}$$

$$\frac{d^2Q}{dt^2} = \frac{6t}{30} - \frac{2}{5}$$

$$= 0.2t - 0.4$$

Step 2: Find turning point(s)

$dQ/dt = 0$ at turning point

$$\frac{3t^2}{30} - \frac{2t}{5} + \frac{3}{10} = 0$$

$$\frac{t^2}{10} - \frac{4t}{10} + \frac{3}{10} = 0$$

$$\frac{1}{10}(t^2 - 4t + 3) = 0$$

$$t^2 - 4t + 3 = 0$$

$$(t - 3)(t - 1) = 0$$

Turning points at
$t = 3$ and at $t = 1$

Step 3: Max or min?

$$\frac{d^2Q}{dt^2} = 0.2t - 0.4$$

At $t = 3$,

$$\frac{d^2Q}{dt^2} = 0.2(3) - 0.4$$

$$= 0.2$$

Positive, so a minimum.
At $t = 1$,

$$\frac{d^2Q}{dt^2} = 0.2(1) - 0.4$$

$$= -0.2$$

Negative, so a maximum.

Turning points: minimum at $t = 3, Q = 120.000$, maximum at $t = 1, Q = 120.13$.

The firm produces maximum output in year 1 and minimum output in year 3. It may have been first on the market with a new good. Initially, its sales are high; however, other firms are quick to enter the market supplying alternative goods. The first firm incurs a drop in sales as there is a slump in the demand for its good or maybe some of the rival firms have produced a better alternative good.

Sketching Q as a function of t: Next, find where the graph crosses the axes. The graph crosses the vertical axis when $t = 0$, that is, at $Q = 120$. The graph crosses the horizontal axis at $Q = 0$, that is, at

$$\frac{t^3}{30} - \frac{t^2}{5} + \frac{3t}{10} + 120 = 0$$

This is a cubic equation, which is not easily solved. So we resort to calculating several points on the curve between $t = 0$ and $t = 5$ as in Table 6.11. The graph is then sketched in Figure 6.26.

Table 6.11 Points for sketching the Q function in Figure 6.26

t		0	1	2	3	4	5
$Q = \dfrac{t^3}{30} - \dfrac{t^2}{5} + \dfrac{3t}{10} + 120$		120	120.13	120.067	120	120.13	120.67

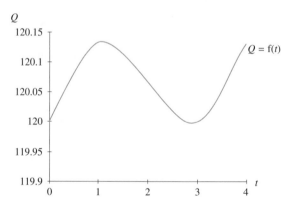

Figure 6.26 A firm's output function over time

To show that MR = MC and (MR)′ < (MC)′ when profit is maximised

In the previous worked examples, profit, π, was maximised when $(\pi)' = 0$ and $(\pi)'' < 0$. However, an alternative and sometimes simpler method based on the marginal revenue and marginal cost function is derived as follows. To derive the rule, work through the usual max/min method, but write the first derivatives of TR and TC as MR and MC as follows:

$$\pi = TR - TC$$

$$\frac{d(\pi)}{dQ} = \frac{d(TR)}{dQ} - \frac{d(TC)}{dQ} \quad \text{or} \quad (\pi)' = MR - MC$$

$$\frac{d^2(\pi)}{dQ^2} = \frac{d(MR)}{dQ} - \frac{d(MC)}{dQ} \quad \text{or} \quad (\pi)'' = (MR)' - (MC)'$$

Apply the usual conditions for a maximum: $(\pi)' = 0$ and $(\pi)'' < 0$. That is,

$$(\pi)' = MR - MC = 0 \rightarrow MR = MC \qquad \textbf{(6.29)}$$

and apply the second-order condition

$$(\pi)'' = (MR)' - (MC)' < 0 \rightarrow (MR)' < (MC)' \qquad \textbf{(6.30)}$$

Equation (6.30) gives the second-order condition for profit maximisation: the derivative of MC is greater than the derivative of MR: $(MC)' > (MR)'$. In Worked Example 6.24, equations (6.29) and (6.30) provide a simpler alternative method for finding maximum profit.

Price discrimination

In the previous worked examples it was assumed, rather unrealistically, that a firm charges all consumers the same price for its good or services. Worked Example 6.24 shows that a firm may make a greater profit when different prices are charged in markets with different demand functions.

WORKED EXAMPLE 6.24
PROFIT MAXIMISATION AND PRICE DISCRIMINATION

A hotel charges different prices for the same meal in two different markets. The demand function in each market is given by the respective equations

$$P_1 = 50 - 4Q_1 \quad \text{and} \quad P_2 = 80 - 3Q_2$$

The company's cost function is given by the equation

$$TC = 120 + 8Q$$

where $Q = Q_1 + Q_2$.

(a) For each market, write down the marginal revenue and marginal cost.
(b) Determine the price and quantity at which profit is maximised. Calculate the overall profit from the two markets.

(c) If price discrimination is declared illegal, what price should the company charge to maximise profits? Calculate the overall profit. How does it compare with the profit made under price discrimination?

SOLUTION

(a)

Market 1
$$TR = (50 - 4Q_1)Q_1$$
$$MR = 50 - 8Q_1$$
$$MC = 8$$

Market 2
$$TR = (80 - 3Q_2)Q_2$$
$$MR = 80 - 6Q_2$$
$$MC = 8$$

(b) Profit is maximised when $MR = MC$, subject to second-order conditions, namely $(MR)' < (MC)'$. Alternatively, profit is maximised when $(\pi)' = 0$ and $(\pi)'' < 0$.
In this example, the first method will be used to find maximum profit.

Market 1
$$MR = MC$$
$$50 - 8Q_1 = 8$$
$$42 = 8Q_1$$
$$Q_1 = 5.25 \quad (P_1 = 29)$$
Second-order conditions confirm a maximum:
$(MR)' = -8$: $(MC)' = 0$
therefore $(MR)' < (MC)'$, as required.

Market 2
$$MR = MC$$
$$80 - 6Q_2 = 8$$
$$72 = 6Q_2$$
$$Q_2 = 12 \quad (P_2 = 44)$$
Second-order conditions confirm a maximum:
$(MR)' = -6$: $(MC)' = 0$
therefore $(MR)' < (MC)'$, as required.

Maximum profit from both markets:

$$\pi = TR - TC = P_1Q_1 + P_2Q_2 - (120 + 8Q) \text{ since } Q = Q_1 + Q_2$$
$$= 29(5.25) + 44(12) - 120 - 8(12 + 5.25) = 422.25$$

(c) If price discrimination is not permitted, then $P_1 = P_2 = P$. The overall demand is the sum of the two separate demands: $Q = Q_1 + Q_2$.
So, let $P_1 = P_2 = P$ in the two separate demand functions. From these demand functions find an expression for $Q_1 + Q_2$, then replace it with Q. This is the equation of the overall demand function:

$$3P = 3(50) - 12Q_1 \qquad \text{demand function (1)} \times 3$$
$$4P = 4(80) - 12Q_2 \qquad \text{demand function (2)} \times 4$$
$$7P = 470 - 12(Q_1 + Q_2) \quad \text{adding}$$
$$7P = 470 - 12Q$$
$$P = 67.14 - 1.714Q$$

This overall demand function may now be used to find maximum profit by the same method as in part (b):

$$MR = MC \rightarrow 67.14 - 3.428Q = 8$$

$$Q = 17.25 \ (P = 37.57)$$

Second-order conditions are satisfied: $(MR)' - (MC)' = -3.428 - 0 < 0$. Overall profit $= TR - TC = (37.57)(17.25) - (120 + 8(17.25)) = 390.08$. Overall profit with price discrimination $= 422.25$ from part (b).

The profits from no price discrimination, £390.08, are less than those from price discrimination, £422.25.

Price discrimination pays.

Profit maximisation in perfect competition and monopoly (goods market)

In the following worked examples, some calculations are left to the reader as similar problems have already been covered. However, the answers are given.

WORKED EXAMPLE 6.25
PROFIT MAXIMISATION FOR A PERFECTLY COMPETITIVE FIRM

A perfectly competitive firm has a demand function $P = 121$ and costs given as

$$TVC = \frac{1}{2}Q^3 - 15Q^2 + 175Q, \quad TFC = 500$$

(a) Write down the equations for TC, TR and π.
(b) Find the output, Q, at which profit is minimised and maximised. Comment on the relationship between MC and MR at these points.
(c) Show that $MC = MR$ is a necessary but not sufficient condition for maximum profit.
(d) Sketch TC, TR, π on one diagram, and MR, MC on a diagram vertically below it.

SOLUTION

(a) $TC = TFC + TVC$

$$TC = \frac{1}{2}Q^3 - 15Q^2 + 175Q + 500$$

$$TR = 121Q$$

$$\pi = TR - TC = 121Q - \frac{1}{2}Q^3 + 15Q^2 - 175Q - 500$$

$$= -\frac{1}{2}Q^3 + 15Q^2 - 54Q - 500$$

(b) Work through the max/min method:

Step 1: Get derivatives	**Step 2:** Solve $d\pi/dQ = 0$	**Step 3:** Max or min?
$\pi = -\frac{1}{2}Q^3 + 15Q^2$ $-54Q - 500$ $\dfrac{d\pi}{dQ} = -\frac{3}{2}Q^2$ $+30Q - 54$ $\dfrac{d^2\pi}{dQ^2} = -\frac{6}{2}Q + 30$ $= -3Q + 30$	$-\frac{3}{2}Q^2 + 30Q - 54 = 0$ Multiplying both sides by -2 gives $3Q^2 - 60Q + 108 = 0$ Solve this quadratic for Q $Q = \dfrac{-(-60) \pm \sqrt{(-60)^2 - 4(3)(108)}}{2(3)}$ $= \dfrac{60 \pm 48}{6}$ $Q = 18$ or $Q = 2$	$\dfrac{d^2\pi}{dQ^2} = -3Q + 30$ At $Q = 18$, $\pi'' = -3(18) + 30$ $= -24$ Negative, so a max at $Q = 18, \pi = 472$ At $Q = 2$ $\pi'' = -3(2) + 30$ $= 24$ Positive, so a min at $Q = 2, \pi = -552$

- The profit function has a minimum value at $Q = 2$, $\pi = -552$. Note, at $Q = 2$, $MR = MC = 121$.
- The profit function has a maximum value at $Q = 18$, $\pi = 472$. Note, at $Q = 18$, $MR = MC = 121$.

(c) Therefore, following from part (b), $MR = MC$ for both maximum and minimum profit. This is a necessary but not a sufficient condition for profit maximisation. For a maximum point, the second derivative, π'', must be negative at that point or $(MR)' - (MC)' < 0$ from equation (6.30). Descriptively, in this example, the maximum and minimum profit points are distinguished by the slope of the MC curve at the point of intersection with the MR curve. Figure 6.27 clearly illustrates that profit is maximised at the point where the slope of the MC curve is greater than the slope of the MR curve.

(d) The TC, TR, MR, MC and π functions are plotted in Figure 6.27 from the data in Table 6.12. (Again, Excel may be used to calculate and plot these functions.)

Table 6.12   Points for profit maximisation of a PC firm

Q	TC	TR	π	MC	MR
0	500	0	−500	175	121
5	1062.5	605	−457.5	62.5	121
10	1250	1210	−40	25	121
15	1437.5	1815	377.5	62.5	121
20	2000	2420	420	175	121
25	3312.5	3025	−287.5	362.5	121
30	5750	3630	−2120	625	121

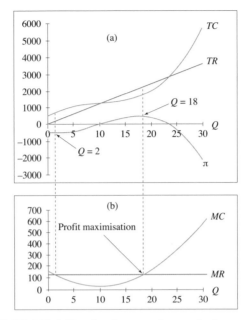

Figure 6.27   *TR*, *TC*, *MR*, *MC* and π functions for a perfectly competitive firm

WORKED EXAMPLE 6.26

PROFIT MAXIMISATION FOR A MONOPOLIST

 Find animated worked examples at **www.wiley.com/college/bradley**

A monopolist faces a demand function $P = 152.5 - 3Q$ and has costs given as

$$TVC = \frac{1}{2}Q^3 - 15Q^2 + 175Q, \quad TFC = 300$$

(a) Write down the equations for *TC*, *TR* and π.
(b) Find the output, *Q*, at which profit is minimised and maximised.
(c) Confirm the condition *MC* = *MR* for maximum profit.
(d) Sketch *TC*, *TR*, π on one diagram, and *MR*, *MC* on a diagram vertically below it.

SOLUTION

(a) $TC = \dfrac{1}{2}Q^3 - 15Q^2 + 175Q + 300$

$TR = (152.5 - 3Q)Q$

$\pi = TR - TC = (152.5Q - 3Q^2) - \left(\dfrac{1}{2}Q^3 - 15Q^2 + 175Q + 300 \right)$

$= -\dfrac{1}{2}Q^3 + 12Q^2 - 22.5Q - 300$

(b) The turning points of the profit function are calculated using the max/min method.

Step 1: Get derivatives	**Step 2: Solve $d\pi/dQ = 0$**	**Step 3: Max or min?**
$\pi = -\dfrac{1}{2}Q^3 + 12Q^2$ $- 22.5Q - 300$	$-\dfrac{3}{2}Q^2 + 24Q - 22.5 = 0$	$\dfrac{d^2\pi}{dQ^2} = -3Q + 24$
$\dfrac{d\pi}{dQ} = -\dfrac{3}{2}Q^2 + 24Q$ -22.5	Multiplying both sides by -2 gives	When $Q = 15$,
$\dfrac{d^2\pi}{dQ^2} = -\dfrac{6}{2}Q + 24$ $= -3Q + 24$	$3Q^2 - 48Q + 45 = 0$ Solve this quadratic for Q	$(\pi)'' = -3(15) + 24$ $= -21$ Negative, so a max at $Q = 15, \pi = 375$
	$Q = \dfrac{-(-48) \pm \sqrt{(-48)^2 - 4(3)(45)}}{2(3)}$ $= \dfrac{48 \pm 42}{6}$ $Q = 15 \text{ or } Q = 1$	When $Q = 1$, $(\pi)'' = -3(1) + 24$ $= 21$ Positive, so a min at $Q = 1, \pi = -311$

(c) In Worked Example 6.25 it was shown that profit is either maximised or minimised when $MR = MC$. We can solve for the value of Q at which profit is a maximum as follows:

$$MR = MC$$
$$152.5 - 6Q = \frac{3}{2}Q^2 - 30Q + 175$$
$$-\frac{3}{2}Q^2 + 24Q - 22.5 = 0$$
$$3Q^2 - 48Q + 45 = 0 \quad \text{multiplying both sides by } -2$$

This is the same quadratic as in part (b), step 2 where profit is maximised or minimised at $Q = 15$ and $Q = 1$. So $MR = MC$ is a necessary but not sufficient condition for profit maximisation. To determine which value of Q gives the maximum profit, use the second-order condition, or be practical! Calculate profit at each value of Q: at $Q = 15$, $\pi = 375$ and at $Q = 1$, $\pi = -311$. It is obvious that profit is maximised at $Q = 15$.

(d) The TC, TR, MR, MC and π functions are sketched in Figure 6.28 from the data in Table 6.13. (Again, Excel can be easily used to plot these functions.)

Table 6.13   Points for profit maximisation of a monopolist

Q	TC	TR	π	MC	MR
0	300	0	−300	175	152.5
5	862.5	687.5	−175	62.5	122.5
10	1050	1225	175	25	92.5
15	1237.5	1612.5	375	62.5	62.5
20	1800	1850	50	175	32.5
25	3112.5	1937.5	−1175	362.5	2.5

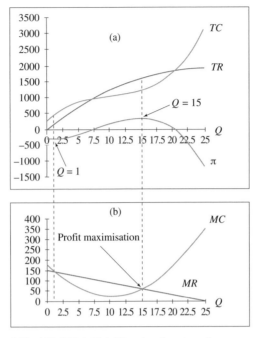

Figure 6.28 TR, TC, MR, MC and π functions for a monopolist

Summary

The use of first and second derivatives to locate and determine the nature of turning points is summarised in Figure 6.29. y' and y'' mean the first and second derivatives, respectively. The three steps are summarised as follows:

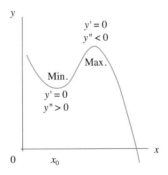

Figure 6.29 Summary of turning points

| **Step 1:** Find y' and y'' | **Step 2:** Solve $y' = 0$ to get the x-coordinate of the turning point(s), (x_0).

Step 2a: For each (x_0) find the y-coordinate. | **Step 3:** Evaluate y'' at each (x_0). If $y'' < 0$, then the point at (x_0) is a maximum. If $y'' > 0$, then the point at (x_0) is a minimum. |

PROGRESS EXERCISES 6.9

Maximum/Minimum Economic Applications

1. A firm's total revenue function is given by the equation $TR = 90Q - 3Q^2$.
   (a) Find the value of Q for which TR is maximised, hence calculate the maximum TR.
   (b) Write down the equations of the average revenue and marginal revenue functions. Describe how AR and MR change before and after maximum TR.
   (c) Sketch TR, MR and AR on the same diagram. From the graph, write down the values of AR and MR when TR is zero and when TR is a maximum.

2. Given the demand functions,
   (i) $Q = 150 - 0.5P$ (ii) $P = 80 - 2Q$ (iii) $P = 45$
   (a) Write down the equations for TR.
   (b) Calculate the number of units which must be sold to maximise TR. Calculate the maximum TR.
   (c) State the values of Q at which MR and AR are zero.

3. A shop which sells T-shirts has a demand function and a total cost function given by the equations

$$P = 240 - 10Q \quad \text{and} \quad TC = 120 + 8Q$$

   (a) Write down the equations for TR and profit.
   (b) Calculate the number of T-shirts which must be sold to maximise (i) profit, (ii) total revenue.
   (c) Write down the equations for MR and MC.
      Show that $MR = MC$ when profit is maximised.
   (d) Plot the graphs of:
      (i) TR and TC on the same diagram. From the graph estimate the break-even points (confirm your answer algebraically).
      (ii) MR and MC on the same diagram. What is the significance of the point of intersection of these two graphs?

4. The average cost and average revenue functions for a particular brand of mobile phone are given by the equations

$$AC = 15 + \frac{8000}{Q} \quad \text{and} \quad AR = 25$$

   (a) Write down the equations for TR, TC, MR, MC.
   (b) How many mobile phones must be made and sold to break even?
   (c) Write down the equation of the profit function. Show, by differentiation, that neither profit nor revenue ever reach a maximum. Use MR and MC to explain why there is no maximum.
   (d) Plot the graphs of TR, TC and the profit function on one diagram. Plot MR and MC on a separate diagram. Comment on the relationship between the two diagrams.

5. A company supplying fitted kitchens is considering two different pricing strategies:
   I: $P = 2374$ II: $P = 5504 - 0.8Q$
   The total cost function is $TC = 608\,580 + 120Q$.

(a) Write down the equations for MR, MC and the profit functions. Calculate the price and quantity for which profit is maximised/minimised by:
   (i) using the condition $MR = MC$;
   (ii) differentiating the profit function.
   If a maximum profit is not possible give an explanation.
(b) Calculate the profit for both pricing strategies at the value of Q for which profit in market II is maximised.
(c) Graph the TC function and both TR functions on the same diagram. Estimate the break-even points. Estimate the price and quantity at which the company makes the same total revenue by both pricing schemes.
(d) Graph both profit functions on the same diagram. Estimate the point at which the company would make the same profit by both pricing schemes. Compare your answer with part (c). Comment. Briefly compare the pricing strategies.

6. (a) Sketch each of the following, indicating turning points and points of intersection with the axis:
   (i) $y = (x - 2)^2$ (ii) $y = 5e^x$ (iii) $y = \dfrac{1}{x + 1}$
   Show, by calculus, that (ii) and (iii) have no turning point.
   (b) The demand function for a good is $P = -2.5Q + 500$.
      (i) Write down the equation for the total revenue function.
      (ii) Sketch the graph of the total revenue function.
      (iii) Use calculus to determine the maximum total revenue.
      (iv) Determine the price elasticity of demand when total revenue is at a maximum
      (the price elasticity of demand is defined as $\varepsilon = \dfrac{dQ}{dP}\dfrac{P}{Q}$).

7. The demand function and total cost function for a product are
$$6P = 660 - 3Q, \quad TC = 80 - 20Q^2 + 600Q$$
   (a) Write down expressions for:
      (i) average cost, (ii) marginal cost, (iii) total revenue, (iv) marginal revenue, (v) profit.
   (b) Determine the values of Q for which total cost is a maximum.
   Calculate the maximum costs.
   Calculate the profit when total costs are a maximum.
   (c) Determine the value of Q for which profit is a minimum. What is the minimum profit?
   State the range of values of Q for which profit is increasing.

8. (a) Explain why profit is maximised when marginal revenue = marginal cost.
   (b) A producer can charge different prices in two markets A and B. The demand function in each market is $Q_a = 45 - 3P_a$ and $Q_b = 80 - 5P_b$, respectively.
   The cost function is: $TC = 50 + 6Q$ where $Q = Q_a + Q_b$.
   What price should be charged in each market to maximise profits, if price discrimination is used?
   Calculate the total cost when profit is maximised.

9. The equation of a total cost function is given as $TC = aQ^3 - bQ^2 + cQ + d$ where a, b, c, d are positive constants.
   Write down the equations for marginal cost and average variable cost.
   Show that AVC is a minimum at $Q = \dfrac{b}{2a}$.

Calculate the values of MC and AVC at $Q = \dfrac{b}{2a}$.

What do you conclude?

10. The demand function for a good is given by the equation $P = a - bQ$ while the total cost function is $TC = dQ^2 + eQ + f$ where a, b, d, e, f are positive constants

Write down the equation for profit.

Derive an expression for the value of Q for which profit is maximised.

6.5   Curvature and Other Applications

At the end of this section you should be familiar with the following:

- The second derivative and curvature
- Finding the point of inflection
- The relationship between Q, MP_L and APL
- The relationship between TC, TVC and MC
- The relationship between MC, AVC, AC.

6.5.1   Second derivative and curvature

The first and second derivatives (y' and y'') have already been used to find maximum and minimum points of various functions. In this section, both y' and y'' will be used to describe curvature and find points of inflection.

In Section 6.3, Worked Example 6.19, the first derivative and second derivative of the graph $y = 3x^2 - 0.1x^3$ were plotted in Figure 6.18 to demonstrate that first derivatives are zero at turning points, while the second derivative is positive in the region of a minimum, negative in the region of a maximum and zero at a point of inflection.

In the region of a minimum, where $y'' > 0$, the graph curves upwards and is described as concave up, while in the region of a maximum, where $y'' < 0$, the graph curves downwards and is described as concave down. Hence, the second derivative may also be used to define the 'curvature' of a curve as follows:

The curvature in the interval about a minimum (when $y'' > 0$) is described as concave up.	$y'' > 0$ concave up
The curvature in the interval about a maximum (when $y'' < 0$) is described as concave down.	concave down $y'' < 0$

To determine the curvature of a curve along any interval, determine whether second derivatives are negative or positive in the required interval.

The rules which use the first and second derivatives to determine maximum points, minimum points, points of inflection and define curvature are extremely important. The results from Worked Example 6.19, Figure 6.18(a), (b) and (c) demonstrate graphically,

- How y, y' and y'' work together to determine maximum and minimum points.
- How to use y'' to define the curvature of a curve within an interval.
- How to use y'' to find the point of inflection.

DIFFERENTIATION AND APPLICATIONS

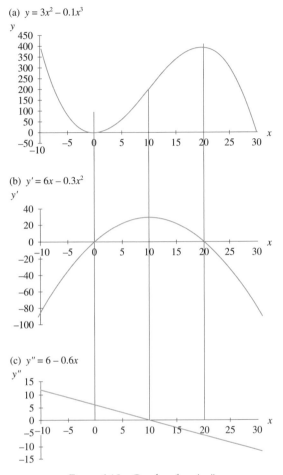

(a) $y = 3x^2 - 0.1x^3$

(b) $y' = 6x - 0.3x^2$

(c) $y'' = 6 - 0.6x$

Figure 6.18 Graphs of y, y', y''

Figure 6.18 is reproduced here for convenience.

- In Figure 6.18(a), y has a minimum at $x = 0$ and a maximum at $x = 20$.
- In Figure 6.18(b), the slope, y', is zero at $x = 0$ and at $x = 20$.
- In Figure 6.18(c), $y'' > 0$ around the minimum point at $x = 0$ and $y'' < 0$ around the maximum point at $x = 20$.
- In Figure 6.18(c), $y'' > 0$ in the region surrounding the minimum point where the graph curves up: curvature is described as concave up. Curvature is described as concave down in the region surrounding the maximum point, where $y'' < 0$.
- The second derivative, y'', is zero at $x = 10$. This is called a **point of inflection**. It is the point at which the curvature changes from concave up to concave down (or vice versa). Points of inflection will be discussed in more detail later in this chapter.

A summary of these points is given in Table 6.14.

Note: It is beyond the scope of this book to deal with cases where the second derivative is zero or non-existent.

Table 6.14   Summary of relationship between y, y' and y''

	At $x = 0$	At $x = 20$
Graph (a) $y = 3x^2 - 0.1x^3$	y has a minimum point	y has a maximum point
Graph (b) $\dfrac{dy}{dx} = 6x - 0.3x^2$	y' is zero	y' is zero
Graph (c) $\dfrac{d^2y}{dx^2} = 6 - 0.6x$	y'' is positive	y'' is negative

Curvature in economics

In Figure 6.18 curvature in the region of a minimum, where $y'' > 0$, is described as concave up, while that in the region of a maximum, where $y'' < 0$, is described as concave down. These rules for determining curvature based on the sign of the second derivative apply whether the interval contains an actual turning point or not. The only requirement is that slope is increasing, hence $y'' > 0$, for the graph to curve upwards, and that slope is decreasing, hence $y'' < 0$, for the graph to curve downwards as shown in Figures 6.30 and 6.31. See also Figures 6.4 and 6.5.

In economic applications, curves are usually economically meaningful in the first quadrant only. In such applications, curves are frequently described as convex towards the origin instead of concave up and concave towards the origin instead of concave down.

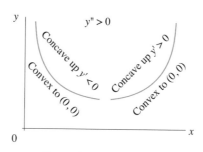

Figure 6.30   Concave up

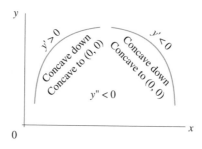

Figure 6.31   Concave down

WORKED EXAMPLE 6.27
CURVATURE OF CURVES: CONVEX OR CONCAVE TOWARDS THE ORIGIN

Use the first and second derivatives to determine the curvature of each of the following curves:

$$\text{(a) } y = 3x^4 + 20 \quad \text{(b) } Q = \frac{25}{L}$$

in terms of whether:
   (i) The curve is increasing or decreasing.
  (ii) Curvature is concave up or concave down.
 (iii) Curvature is convex or concave towards the origin (in first quadrant only).

   Make a rough sketch of each curve.

SOLUTION

In each case, find the first and second derivatives. Then check whether the derivatives are positive or negative.

(a) $y = 3x^4 + 20$

   (i) $y' = 12x^3 \begin{cases} \text{when } x > 0, \ y' > 0 \rightarrow y \text{ is increasing as } x \text{ increases} \\ \text{when } x < 0, \ y' < 0 \rightarrow y \text{ is increasing as } x \text{ increases} \end{cases}$

  (ii) $y'' = 36x^2$ positive $\rightarrow$ concave up for any value of x.

 (iii) The curve is graphed in the first quadrant of Figure 6.32, and it may be described as convex towards the origin.

 (iv) Plot about four or five points as shown in Table 6.15 to find the general shape of the curve.

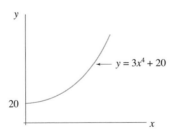

Figure 6.32   Graph of $y = 3x^4 + 20$

Table 6.15   Points for Figure 6.32

x	0	1	2	3	4	5
$y = 3x^4 + 20$	20	23	68	263	788	1895

(b) $Q = \dfrac{25}{L} = 25L^{-1}$

(i) $Q' = 25(-L^{-2}) = -25L^{-2} = -\dfrac{25}{L^2}$

$\rightarrow$ negative for all $L \rightarrow$ curve is falling

(ii) $Q'' = -25(-2L^{-3}) = 50L^{-3} = -\dfrac{50}{L^3}$

$\rightarrow$ positive for all $L > 0 \rightarrow$ concave up

(iii) In the first quadrant this curve is described as convex towards the origin.

(iv) To plot the curve, calculate some points as shown in Table 6.16 which are then plotted in Figure 6.33.

Table 6.16   Points for Figure 6.33

L	0	1	2	3	4	5
$Q = \dfrac{25}{L}$	?	25	12.5	8.33	6.25	5

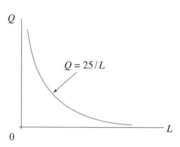

Figure 6.33   Graph of $Q = 25/L$

Again note **division by zero** when $x = 0$, hence a vertical asymptote.

6.5.2   Points of inflection

In Figure 6.18(c) the point at which $y'' = 0$ was called **a point of inflection (Pol)**. This occurs at the point where curvature changes from concave up ($y'' > 0$) to concave down ($y'' < 0$), or vice versa.

Logically, if curvature changes from concave down to concave up, y'' changes from being negative to positive. This change can only occur at $y'' = 0$. Two points of inflection are shown on different curves in Figure 6.34, one at point A1 and the other at point A2.

> At points of inflection, $y'' = 0$ *and* y'' changes sign
> when evaluated before and after the point of inflection.

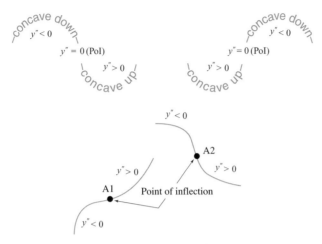

Figure 6.34   Point of inflection

Stationary points of inflection

If both y' and y'' are zero, then the point of inflection is said to be stationary. See Figure 6.35, points B1 and B2.

Points of inflection in economics

A point of inflection may also be described as the point at which slope (rate of change) changes from being an increasing slope, as indicated when $y'' > 0$, to being a decreasing slope when $y'' < 0$, or vice versa. This description is useful in economics, where slope is the marginal function. Therefore, the PoI is the point at which the marginal function (slope) changes from being an increasing marginal function (slope) with $y'' > 0$ to being a decreasing marginal function (slope) with $y'' < 0$, or vice versa.

Before proceeding with further worked examples, the general method for finding points of inflection is outlined:

Step 1: Find the derivatives y', y'' and y'''.
Step 2: Find points of inflection. Solve $y'' = 0$ for the x-coordinate of the point of inflection.
Step 3: Confirm that the point in step 2 is actually a point of inflection by checking for change in the sign of y'' immediately before and after the point in question.

A simple alternative test for confirming a PoI for polynomials of degree 3 is to show that y''' is not zero at the point. Since y'' must change sign before and after a PoI, then y'' is either increasing or decreasing, so the rate of change of y'', which is y''', is either positive or negative, but not zero.

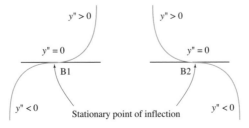

Figure 6.35   Stationary points of inflection

WORKED EXAMPLE 6.28
LOCATE THE POINT OF INFLECTION, POI = POINT AT WHICH MARGINAL RATE CHANGES

Show that the curve $y = 3x^2 - 0.1x^3$ given in Worked Example 6.19 has a point of inflection at $x = 10$.

SOLUTION

The first, second and third derivatives are

$$y = 3x^2 - 0.1x^3 \rightarrow \frac{dy}{dx} = 6x - 0.3x^2 \rightarrow \frac{d^2y}{dx^2}$$

$$= 6 - 0.6x \rightarrow \frac{d^3y}{dx^3} = -0.6$$

To find the point of inflection solve the equation, second derivative equal to 0:

$$6 - 0.6x = 0 \rightarrow x = 10$$

Figure 6.18(c) confirms that the second derivative is zero at $x = 10$ indicating a possible point of inflection. To confirm that this is a point of inflection, from Figure 6.18(c) we see that the second derivative is positive for values of x before $x = 10$ and negative for values of x after $x = 10$. Alternatively, y''' is not zero at $x = 10$, confirming that this is a point of inflection.

PROGRESS EXERCISES 6.10

Curvature, Points of Inflection

For each of the following:

(a) Determine the curvature along which the interval $x > 0$.
(b) Find the points of inflection (if any).
(c) Sketch the function (use Excel, if available).

1. $y = x^2 + 2x$
2. $y = -x^3 + 5x + 6$
3. $y = 3x^4 + x^2$
4. $L = 0.5K^{0.25}$
5. $TC = Q^3 - 6Q^2 + 3Q$
6. $TC = -Q^4 + 200$
7. $AC = Q^2 + \dfrac{2}{Q} + 8$
8. $TR = Q(120 - 0.8Q)$
9. $y = \dfrac{(x + 2)}{x^2} - \dfrac{2x}{x^3}$
10. $AFC = \dfrac{100}{Q}$

Go to the website for the following additional material that accompanies Chapter 6:

- Points of inflection and curvature for production functions
  Worked Example 6.29 Relationship between APL and MP_L functions
  Table 6.17 Points for plotting short-run production function and APL and MP_L functions
  Figure 6.36 Short-run production function MP_L and APL functions
  Worked Example 6.30 Point of inflection on the production function: law of diminishing returns

Points of inflection and curvature for total cost functions

Many total cost functions do not contain maximum or minimum points. The point of inflection defines the level of output at which marginal cost changes from decreasing to increasing.

WORKED EXAMPLE 6.31
RELATIONSHIP BETWEEN *TC* AND *MC*

Given the total cost function

$$TC = \frac{1}{2}Q^3 - 15Q^2 + 175Q + 1000$$

find the point of inflection and give an economic interpretation in terms of total costs and marginal costs.

SOLUTION

The point of inflection for the total cost function occurs when the second derivative is zero, so find the first, second and third derivatives:

$$TC = \frac{1}{2}Q^3 - 15Q^2 + 175Q + 1000$$

$$MC = \frac{d(TC)}{dQ} = \frac{1}{2}(3Q^2) - 15(2Q) + 175(1) = 1.5Q^2 - 30Q + 175$$

$$\frac{d^2(TC)}{dQ^2} = 1.5(2Q) - 30 = 3Q - 30$$

$$\frac{d^3(TC)}{dQ^3} = 3$$

Equate the second derivative to zero, hence solve for Q:

$$3Q - 30 = 0, \quad \text{therefore} \quad Q = 10$$

The third derivative is not zero, so there is a point of inflection at $Q = 10$, $TC = 1750$.
To give an economic interpretation, it is necessary to determine whether marginal cost (slope) is increasing, $TC'' > 0$, or decreasing, $TC'' < 0$, for values of $Q < 10$ and $Q > 10$. $TC'' = 3Q - 30$, therefore:

- For $Q < 10$, TC'' is negative, indicating that MC, or slope, is decreasing. The cost of producing an extra unit of output is decreasing, as in Figure 6.37(b).
- For $Q > 10$, TC'' is positive, indicating that MC, or slope, is increasing. The cost of producing an extra unit of output is increasing, as in Figure 6.37(b).

At $Q = 10$, the point of inflection, the marginal costs change from decreasing to increasing. At $Q = 10$, marginal costs are a minimum.
The point of inflection is shown in Figure 6.37(a). As the TC and TVC functions are simply vertical translations of each other, then the point of inflection of the TVC function is also at $Q = 10$.

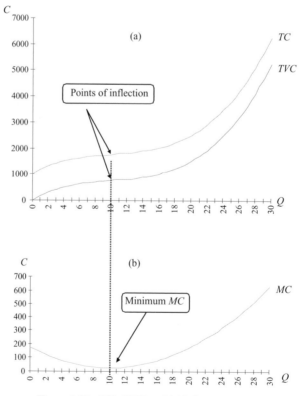

Figure 6.37 TC, TVC and MC functions

WORKED EXAMPLE 6.32
RELATIONSHIP BETWEEN *AC, AVC, AFC* AND *MC* FUNCTIONS

The total variable cost, *TVC*, and total fixed cost, *TFC*, functions for a good are given as

$$TVC = \frac{Q^3}{2} - 15Q^2 + 175Q, \quad TFC = 1000$$

(*TC* = *TVC* + *TFC*. This is the same *TC* function as in Worked Example 6.32.)

(a) Write down the equations for:
   (i) Total costs, TC.
   (ii) Average costs, AC.
   (iii) Average variable cost, AVC.
   (iv) Average fixed cost, AFC.
   (v) Marginal cost, MC.
(b) Find the values of Q for which MC, AVC, AFC and AC are minimised and sketch the graphs of these functions.
(c) Show that the MC curve passes through the minimum points of the AC and AVC curves. (That is, MC = AVC when AVC is at a minimum and MC = AC when AC is at a minimum.)

SOLUTION

(a)

$$TC = TVC + TFC = \frac{Q^3}{2} - 15Q^2 + 175Q + 1000$$

$$AC = \frac{TC}{Q} = \frac{\frac{Q^3}{2} - 15Q^2 + 175Q + 1000}{Q} = \frac{Q^2}{2} - 15Q + 175 + \frac{1000}{Q}$$

$$AVC = \frac{TVC}{Q} = \frac{\frac{Q^3}{2} - 15Q^2 + 175Q}{Q} = \frac{Q^2}{2} - 15Q + 175$$

$$AFC = \frac{TFC}{Q} = \frac{1000}{Q}$$

$$MC = \frac{d(TC)}{dQ} = \frac{3Q^2}{2} - 30Q + 175 \text{ or } MC = \frac{d(TVC)}{dQ} = \frac{3Q^2}{2} - 30Q + 175$$

Notice that AC = AVC + AFC.

(b) The turning points for MC, AVC, AFC and AC are now calculated.
   • To find the minimum values for the MC and AVC functions equate the first derivatives to zero. The second derivative should be positive at each minimum point.

$$MC = \frac{3Q^2}{2} - 30Q + 175$$

Step 1: Find first and second derivatives

$$\frac{d(MC)}{dQ} = \frac{3(2Q)}{2} - 30(1) = 3Q - 30$$

$$\frac{d^2(MC)}{dQ^2} = 3$$

Step 2: Find turning points by solving first derivative = 0.

$3Q - 30 = 0 \rightarrow Q = 10$

Step 2a: When $Q = 10$, $MC = 25$

Step 3: Confirm the minimum.

The second derivative is a positive constant, so the only turning point is a minimum.

$$AVC = \frac{Q^2}{2} - 15Q + 175$$

Step 1: Find first and second derivatives

$$\frac{d(AVC)}{dQ} = \frac{(2Q)}{2} - 15(1) = Q - 15$$

$$\frac{d^2(AVC)}{dQ^2} = 1$$

Step 2: Find turning points by solving first derivative = 0.

$Q - 15 = 0 \rightarrow Q = 15$

Step 2a: When $Q = 15$, $AVC = 62.5$

Step 3: Confirm the minimum.

The second derivative is a positive constant, so the only turning point is a minimum.

The graphs of the AVC and MC functions are plotted in Figure 6.38(b) from the points in columns (1), (6) and (8) of Table 6.18.

• The graph of $AFC = 1000/Q$ is a hyperbolic curve with the vertical axis as asymptote.

Step 1: Get derivatives:

$$AFC = 1000Q^{-1}$$

$$\frac{d(AFC)}{dQ} = 1000(-Q^{-2}) = -1000Q^{-2} = -\frac{1000}{Q^2}$$

$$\frac{d^2(AFC)}{d^2Q} = -1000(-2Q^{-3}) = 2000Q^{-3} = -\frac{3000}{Q^3}$$

Step 2: At a minimum, the first derivative is zero. Solve this equation for Q:

$$-\frac{1000}{Q^2} = 0 \rightarrow -1000 = 0(Q^2) = 0$$

This statement, $-1000 = 0$, is a contradiction, therefore no solution. To put it another way, there is no value of Q^2 which can be multiplied by 0 to give -1000. Therefore, since we cannot find a value of Q for which slope is zero, there is no turning point.

Curvature: Slope

$$\frac{d(AFC)}{dQ} = -\frac{1000}{Q^2}$$

is negative for all values of Q, therefore this is a decreasing curve. The second derivative,

$$\frac{d^2(AFC)}{dQ^2} = \frac{2000}{Q^3}$$

is positive for all values of $Q > 0$, therefore the curvature is concave up or convex towards the origin.

- The point where AC is at a minimum is not so easily calculated. Equating the derivative of AC to zero and solving for Q produces a cubic equation, as shown below:

$$AC = 0.5Q^2 - 15Q + 175 + 1000Q^{-1}$$

$$\frac{d(AC)}{dQ} = 0.5(2Q) - 15(1) + 1000(-Q^{-2}) = Q - 15 - 1000Q^{-2}$$

At the minimum AC, the first derivative is zero,

$$Q - 15 + \frac{1000}{Q^2} = 0$$

$$\frac{Q}{1} - \frac{15}{1} + \frac{1000}{Q^2} = 0$$

$$\frac{Q^3 - 15Q^2 + 1000}{Q^2} = 0$$

$$Q^3 - 15Q^2 + 1000 = 0 \quad \text{multiplying both sides by } Q^2$$

The solution to cubic equations is beyond the scope of this text, unless it is possible to factor the cubic function, thereby reducing the problem to the product of linear and quadratic functions. However, you are expected to be resourceful, so when all else fails, sketch the graph by plotting the points from columns (1) and (7) in Table 6.18, and estimate from the graph that minimum AC is approximately at $Q = 18.1$ and AC = 123. In this case, it is quite safe to plot points, since polynomials, like this one, have no infinite jumps, etc. The AC function is plotted in Figure 6.38(b).

Since the minima for the MC and AVC are at $Q = 10$ and $Q = 15$, respectively, sketch the graphs from $Q = 0$ to $Q = 30$ to get a reasonable picture of the main points of all curves. Use Excel, if available, to calculate the tables of points and plot the graphs.

(c) From Figure 6.38(b), we can deduce the following:
- **The MC curve passes through the minimum point of the AVC curve** (MC = AVC when AVC is at a minimum).
The minimum value of AVC is at $Q = 15$, AVC = 62.5. To show that MC and AVC curves intersect at this point, evaluate MC at $Q = 15$:

$$MC = 1.5Q^2 - 30Q + 175$$

$$= 1.5(15)^2 - 30(15) + 175$$

$$= 337.5 - 450 + 175 = 62.5$$

So both curves pass through the point $Q = 15$, MC = AVC = 62.5. This confirms that MC = AVC at minimum AVC.

Table 6.18   Points for plotting the total, average and marginal cost functions

(1) Q	(2) TFC	(3) TVC	(4) TC	(5) AFC	(6) AVC	(7) AC	(8) MC
0	1000	0	1000	–	–	–	–
3	1000	403.5	1403.5	333.33	134.5	467.83	98.5
6	1000	618	1618	166.66	103	269.66	49
9	1000	724.5	1724.5	111.11	80.5	191.61	26.5
10	1000	750	1750	100	75	175	25
12	1000	804	1804	83.33	67	150.33	31
15	1000	937.5	1937.5	66.66	62.5	129.16	62.5
18	1000	1206	2206	55.55	67	122.55	121
18.1	1000	1218.22	2218.22	55.24	67.30	122.54	123.41
21	1000	1690.5	2690.5	47.61	80.5	128.11	206.5
24	1000	2472	3472	41.66	103	144.66	319
27	1000	3631.5	4631.5	37.03	134.5	171.53	458.5
30	1000	5250	6250	33.33	175	208.33	625

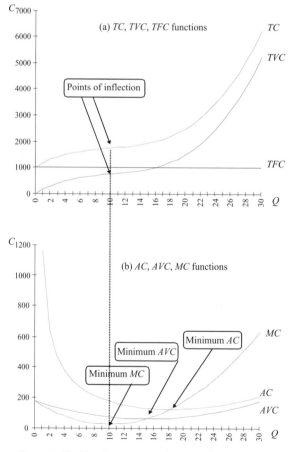

Figure 6.38   Total, average and marginal cost functions

- **The MC curve passes through the minimum point of the AC curve** (MC = AC when AC is a minimum).

  AC is at a minimum at $Q = 18.1$ and $AC = 123$. To show that MC and AC curves intersect at the point $Q = 18.1$, evaluate MC at $Q = 18.1$:

  $$MC = 1.5Q^2 - 30Q + 175$$
  $$= 1.5(18.1)^2 - 30(18.1) + 175$$
  $$= 491 - 543 + 175 = 123$$

This confirms that MC = AC when AC is a minimum.

To summarise:

When AC is at a minimum, MC = AC.
When AC is falling, MC < AC; when AC is rising, MC > AC.
When AVC is at a minimum, MC = AVC.
Minimum MC < Minimum AVC < Minimum AC.

PROGRESS EXERCISES 6.11

Applications of Differentiation, Curvature and Points of Inflection MP_L, APL, TC, MC, etc. Relationships between MC/AVC/AFC

1. The quantity of sandwiches, Q, made in a small coffee shop is given by the equation

$$Q = -2L^3 + 12L^2$$

where L is the number of labour-hours hired.
   (a) Write down the equations for the marginal product of labour (MP_L) and average product of labour (APL). Give a verbal description of MP_L and APL at $L = 1.5$.
   (b) Calculate the units of labour at which MP_L and APL are maximised, and plot both graphs on the same diagram. Confirm algebraically that MP_L and APL are equal when APL is at a maximum.
   (c) Find the turning points and point of inflection for the production function. Plot the production function.
   (d) Use the graphs and any other results calculated in (a) and (b) to describe how productivity (Q = number of sandwiches) and the rate of change in productivity as the number of labour units employed (L) increase.
   Is there any relationship between the turning points and points of inflection in the graphs plotted in (b) and (c)?
2. Consider a firm whose total cost function is

$$TC = \frac{1}{3}Q^3 - 30Q^2 + 2800Q + 900$$

   (a) Write down the equations for MC, AC, AVC and AFC.
   (b) Find the minimum MC.

(c) Find the minimum AVC. Show that the MC and AVC curves intersect at the minimum point on the AVC curve.

(d) Plot the MC, AFC and AVC curves on the same diagram. Comment on the points at which these curves intersect.

3. The total cost and total revenue functions for a company are

$$TC = \frac{1}{3}Q^3 - 15Q^2 + 480Q + 750 \quad TR = Q(536 - 2Q)$$

(a) Plot the graph of TC and TR for $Q = 0$ to 50. From the graph write down the range of values of Q for which the firm makes a profit.

(b) Write down the equation for the profit function. Hence,

  (i) Determine the level of output, Q, for which profit is maximised and minimised.

  (ii) State an alternative method for locating the optimum profit.

(c) Write down the equation of the marginal profit curve. Hence find the maximum value of marginal profit.

4. Given the total cost function $TC = \frac{1}{3}Q^3 - 36Q^2 + 1500Q + 9900$ and a total revenue function $TR = 1360Q$:

(a) Plot the graph of TC and TR for $Q = 0$ to 110. From the graph, write down the range of values of Q for which the firm makes a profit.

(b) Write down the equation for the profit function. Hence determine the level of output, Q, for which profit is maximised and minimised.

6.6 Further Differentiation and Applications

At the end of this section you should be able to:

- Differentiate other standard functions
- Use the chain rule to differentiate functions of functions
- Use the product and quotient rules
- Use differentiation to describe and analyse other economic models.

6.6.1 Derivatives of other standard functions

So far we have only used the general power rule for differentiation which is restated here as

$$y = x^n \quad \text{then} \quad \frac{dy}{dx} = nx^{n-1}$$

However, when more versatile models are developed, differentiation must be extended and used in the analysis of these models. We begin by writing down the rules for finding the derivatives of other standard functions which are frequently used in economics, management and finance. The differentiation of these functions is very straightforward; it simply involves following the rules which are summarised in Table 6.19.

Table 6.19   Rules for finding derivatives

	$f(x)$	$f'(x)$
Power rule	x^n	nx^{n-1}
Exponential, base e, power x	e^x	e^x
Exponential, base e, power ax where a is a constant	e^{ax}	ae^{ax}
Log to base e	$\ln(x)$	$\dfrac{1}{x}$

WORKED EXAMPLE 6.33

DERIVATIVES OF EXPONENTIALS AND LOGS

Find the first, second and third derivatives of the following functions:

(a) $C = 100(1 - e^Y)$ (b) $TC = 10 + \ln Q$ (c) $y = 4e^{2.5x}$

(a) $C = 100(1 - e^Y)$	(b) $TC = 10 + \ln Q$	(c) $y = 4e^{2.5x}$
Solution	**Solution**	**Solution**
$C = 100 - 100e^Y$	$\dfrac{d(TC)}{dQ} = \dfrac{1}{Q} = Q^{-1}$	$y = 4e^{2.5x}$
$\dfrac{dC}{dY} = -100(e^Y)$	$\dfrac{d^2(TC)}{dQ^2} = -Q^{-1-1}$ $= -Q^{-2} = \dfrac{-1}{Q^2}$	$\dfrac{dy}{dx} = 4\left(2.5e^{2.5x}\right) = 10e^{2.5x}$
$\dfrac{d^2C}{dY^2} = -100(e^Y)$	$\dfrac{d^3(TC)}{dQ^3} = -1\left(-2Q^{-2-1}\right)$ $= 2Q^{-3} = \dfrac{2}{Q^3}$	$\dfrac{d^2y}{dx^2} = 10\left(2.5e^{2.5x}\right) = 25e^{2.5x}$
$\dfrac{d^3C}{dY^3} = -100(e^Y)$		$\dfrac{d^3y}{dx^3} = 25\left(2.5e^{2.5x}\right) = 62.5e^{2.5x}$

PROGRESS EXERCISES 6.12

Differentiation of Logs and Exponentials

Determine the first derivative in questions 1 to 12.

1. $y = 10e^x$

2. $y = \dfrac{e^x}{10}$

3. $TVC = 243\dfrac{1}{x} + \ln(x)$

4. $P = 20 + 20\ln Q$

5. $Q = \sqrt[3]{L} + 4$

6. $P = 150(1 - e^t)$

7. $P = 87(1.2)\dfrac{e^{2t}}{e^t}$

8. $AC = \ln Q + \dfrac{2}{Q}$

9. $Q = \dfrac{\ln(P)}{12} - \dfrac{12}{\sqrt{4P}}$

10. $x = 10.85(e^t)$

11. $C = \dfrac{150}{Y} + 0.8e^Y$

12. $C = x^2 - 5e^x + \ln(x)$

Determine the indicated derivative in questions 13 to 16.

13. $T - \dfrac{12}{Y} = e^y$ find $\dfrac{dT}{dY}$, $\dfrac{d^2T}{dY^2}$ 14. $e^{Y+12} = x$ find $\dfrac{dY}{dx}$

15. $\ln(Y) = 15 + t$ find $\dfrac{dY}{dt}$, $\dfrac{d^2Y}{dt^2}$ 16. $\ln(Y+10) = x+5$ find $\dfrac{dY}{dx}$, $\dfrac{d^2Y}{dx^2}$

Determine the first derivative in questions 17 to 22.

17. $4e^t + e^{2t}$ 18. $\dfrac{e^x}{2} + \dfrac{x}{2}$ 19. $12 + e^{4x}$

20. $x\left(2 + \dfrac{1}{x} - 2x\right)$ 21. $e^{2t}(1 - e^t)$ 22. $4(1 - e^{-0.4t})$

6.6.2 Chain rule for differentiating a function of a function

What is a function of a function?

In Chapter 2 a function was loosely defined as a rule for operating on x (resulting in one y-value). For example,

$5x + 7$: is a function of x $\rightarrow$ multiply x by 5, then add 7
e^x: is a function of x $\rightarrow$ raise x to the base e
$\ln(x)$: is a function of x $\rightarrow$ get the natural log of x

If a function of x is operated on again, we then have a function of a function of x. For example, if the first function is $(5x + 7)$

(a) $(5x + 7)^{-4}$ $\rightarrow$ (**first function**) raised to the power of (-4)
(b) $e^{(5x+7)}$ $\rightarrow$ raise the (**first function**) to base e
(c) $\ln(5x + 7)$ $\rightarrow$ get the natural log of the (**first function**)

To differentiate a function of a function, do so in stages as summarised below in Worked Example 6.34.

Stages of the chain rule

Stage 1	Stage 2
Call the first function (or inside function) u	Multiply (or chain) the results together so that the intermediate step, u, cancels out:
($u =$ **first function in terms of x**).	
Then write out y, the main function, in terms of u (y in terms of u). Differentiate each function separately to get	$$\boxed{\dfrac{dy}{dx} = \dfrac{dy}{du}\dfrac{du}{dx}} \qquad \textbf{(6.31)}$$
$$\dfrac{du}{dx} \quad \text{and} \quad \dfrac{dy}{du}$$	

WORKED EXAMPLE 6.34
USING THE CHAIN RULE FOR DIFFERENTIATION

Find the first derivative of each of the following:

(a) $(5x + 7)^{-4}$ (b) $e^{(5x+7)}$ (c) $\ln(5x + 7)$

SOLUTION

(a) $y = (5x + 7)^{-4}$

This is essentially a base to a power, so treat the base as the single variable, u. The stages are outlined as follows:

Stage 1	Stage 2

Stage 1

Treat the base as the variable, u, therefore,

hence
$$u = 5x + 7$$
$$y = u^{-4}$$

We have two equations, so differentiate:
$$\frac{du}{dx} = 5$$
and
$$\frac{dy}{du} = -4u^{-4-1} = -4u^{-5}$$

Result

If $y = (5x + 7)^{-4}$ then
$$\frac{dy}{dx} = -\frac{20}{(5x + 7)^5}$$

Stage 2

Chain the two results together so that u cancels using equation (6.31):

$$\frac{dy}{dx} = \frac{dy}{du}\frac{du}{dx}$$
$$= (-4u^{-5})(5) = -20(u)^{-5}$$
$$= -20(5x + 7)^{-5} \quad \text{substitute}$$
$$\qquad\qquad\qquad 5x + 7 = u$$
$$= -\frac{20}{(5x + 7)^5}$$

(b)

Stage 1

$$y = e^{5x+7}$$
$$y = e^u \quad \text{where} \quad u = 5x + 7$$
$$\frac{dy}{du} = e^u \quad \text{and} \quad \frac{du}{dx} = 5$$

Result

If $y = e^{5x+7}$ then
$$\frac{dy}{dx} = 5e^{5x+7}$$

Stage 2

$$\frac{dy}{dx} = \frac{dy}{du}\frac{du}{dx}$$
$$= (e^u)(5)$$
$$= 5e^{5x+7}$$

(c)

Stage 1

$$y = \ln(5x + 7)$$
$$y = \ln(u) \quad \text{where} \quad u = 5x + 7$$
$$\frac{dy}{du} = \frac{1}{u} \quad \text{and} \quad \frac{du}{dx} = 5$$

Result

If $y = \ln(5x + 7)$ then
$$\frac{dy}{dx} = \frac{5}{5x + 7}$$

Stage 2

$$\frac{dy}{dx} = \frac{dy}{du}\frac{du}{dx}$$
$$= \left(\frac{1}{u}\right)\left(\frac{5}{1}\right)$$
$$= \left(\frac{5}{u}\right)$$
$$= \left(\frac{5}{5x + 7}\right)$$

PROGRESS EXERCISES 6.13

Differentiation with the Chain Rule

Differentiate each of the following:

1. $y = (2x - 5)^7$

2. $y = (4 - 5x)^3$

3. $y = (1 - 0.8x)^{-5}$

4. $y = \dfrac{1}{\sqrt{5x + 12}}$

5. $P = \sqrt{Q^2 + 12}$

6. $TC = \sqrt{Q^5 + 3Q^2}$

7. $Q = 0.85\sqrt[3]{L^2 + 3L}$

8. $Q = 120\left(\dfrac{10}{P + 5}\right)$

9. $y = 12e^{0.81x}$

10. $P = 1223e^{1.09t}$

11. $P = 800e^{-1.4Q}$

12. $S = 185(1 - e^{-1.2t})$

13. $y = (1 - e^{2.5t})$

14. $P = 27 + \dfrac{80}{e^{4t}}$

15. $y = \dfrac{15}{1 + e^t}$

16. $T = 200 + 30e^{0.88Y}$

17. $C = 1240 + \ln(Y^2 + 4)$

18. $Q = 500\ln(K^3 + 8K)$

19. $y = 12e^{-1.57} + \dfrac{2}{x + 2} - \ln(2x + 2)$

20. $P = \ln\left(\dfrac{Q^2 - 2Q}{Q}\right)$

21. $P = 8023(1.25e^{-05t})$

22. $C = 425 + 1.2\ln\left(\dfrac{Y}{Y + 2}\right)$

23. $S = 248(1 - e^{-0.65Y})$

24. $P = \dfrac{4}{\sqrt{Q}} + 5\ln(Q)$

6.6.3 Product rule for differentiation

In differentiation, a 'product' refers to two distinct, independent functions of x multiplied together. For example:

(a) $y = x^2 e^x$ → x^2 is multiplied by e^x or the product of x^2 and e^x

(b) $P = 2\sqrt{Q}(Q + 5)$ → $2\sqrt{Q}$ is multiplied by $(Q + 5)$ or the product of $2Q^{0.5}$ and $(Q + 5)$

(c) $C = (y + 4)\ln(y)$ → $(y + 4)$ is multiplied by $\ln(y)$ or product of $(y + 4)$ and $\ln(y)$

Product rule

The above products are differentiated according to the product rule (which can be derived from first principles, calculus again) given as equation (6.32).

If $y = u(x)v(x)$, usually written as $y = uv$, then

$$\frac{dy}{dx} = v\frac{du}{dx} + u\frac{dv}{dx} \tag{6.32}$$

where u is the first function of x and v is the second function of x.

Stages of the product rule

Stage 1	Stage 2
State the equation for u. Find du/dx. State the equation for v. Find dv/dx.	Fill in the right-hand side of the product rule given in equation (6.32) and simplify your answer: $\dfrac{dy}{dx} = v\dfrac{du}{dx} + u\dfrac{dv}{dx}$

WORKED EXAMPLE 6.35

USING THE PRODUCT RULE FOR DIFFERENTIATION

Differentiate each of the following:

(a) $y = x^2 e^x$ (b) $P = 2\sqrt{Q}(Q + 5)$ (c) $C = (Y + 4)\ln(Y)$

SOLUTION

(a) **Stage 1**

State u and v and find du/dx and dv/dx.

$y = x^2 e^x$

$u = x^2$ and $v = e^x$

$\dfrac{du}{dx} = 2x$ and $\dfrac{dv}{dx} = e^x$

Result

$y = x^2 e^x$, then $dy/dx = xe^x(2 + x)$

Stage 2

Fill in the product rule, equation (6.32)

$$\frac{dy}{dx} = v\frac{du}{dx} + u\frac{dv}{dx}$$
$$= (e^x)(2x) + (x^2)(e^x)$$
$$= e^x(2x + x^2)$$
$$= xe^x(2 + x)$$

(b) **Stage 1**

State u and v and find du/dQ and dv/dQ.

$P = 2\sqrt{Q}(Q + 5)$
$= 2(Q)^{1/2}(Q + 5)$

$u = 2Q^{(0.5)}$ and $v = Q + 5$

$\dfrac{du}{dQ} = 2(0.5Q^{0.5-1})$ and $\dfrac{dv}{dQ} = 1$
$= 2(0.5Q^{-0.5})$
$= Q^{-0.5}$

Stage 2

Fill in the product rule, equation (6.32)

$$\frac{dP}{dQ} = v\frac{du}{dQ} + u\frac{dv}{dQ}$$
$$= (Q + 5)(Q^{-0.5}) + (2Q^{0.5})1$$
$$= Q^{0.5} + 5Q^{-0.5} + 2Q^{0.5}$$
$$= 3Q^{0.5} + 5Q^{-0.5}$$

Result

$P = 2\sqrt{Q}(Q + 5)$ then
$$\frac{dP}{dQ} = 3Q^{0.5} + \frac{5}{Q^{0.5}}$$

(c) **Stage 1**

State u and v and find du/dY and dv/dY.

$C = (Y + 4)\ln(Y)$

$u = (Y + 4)$ and $v = \ln(Y)$

$\dfrac{du}{dY} = 1$ and $\dfrac{dv}{dY} = \dfrac{1}{Y}$

Stage 2

Fill in the product rule, equation (6.32)

$$\frac{dC}{dY} = v\frac{du}{dY} + u\frac{dv}{dY}$$
$$= \ln(Y)(1) + (Y + 4)\frac{1}{Y}$$
$$= \ln(Y) + \frac{Y + 4}{Y}$$
$$= \ln(Y) + 1 + \frac{4}{Y}$$

Result

$C = (Y + 4)\ln(Y)$	$\dfrac{dC}{dY} = \ln(Y) + 1 + \dfrac{4}{Y}$

PROGRESS EXERCISES 6.14

Differentiation with the Product Rule

Find the first derivatives for each of the following:

1. $y = x\ln(x)$

2. $TF = (50 - 2\sqrt{Q})Q$

3. $y = (t^2)e^t + 10$

4. $AC = \dfrac{1}{Q} - Q\ln(Q)$

5. $Q = L\sqrt{L+5}$

6. $y = x^{1.5}\sqrt{9x}$

7. $P = 15 - Qe^Q$

8. $C = Q\sqrt{1+2Q}$

9. $S = 100Ye^{-0.5Y}$

10. $I = \dfrac{1}{Y}\ln(Y+8)$

11. $C = 200 + \sqrt{Y}\ln(8Y)$

12. $P = \dfrac{3}{20}xe^{x-5}$

13. $TC = \dfrac{Q^3}{15} + Q\sqrt{3Q-1} - Q + 250$

14. $P = 20 - \sqrt{5Q}\ln Q$

15. $S = t(3t - 5)^{-4}$

16. $x = y^3\ln(y^3)$

6.6.4 Quotient rule for differentiation

In differentiation, a 'quotient' refers to a fraction in which a function of x is divided by another function in x. For example,

(a) $\dfrac{x^2}{1+e^x}$ $\rightarrow$ x^2 is divided by $(1 + e^x)$

(b) $\dfrac{Y+4}{\ln Y}$ $\rightarrow$ $(y + 4)$ is divided by $\ln(Y)$

(c) $\dfrac{Q}{3Q+5}$ $\rightarrow$ Q is divided by $(3Q + 5)$

Quotient rule

These quotients are differentiated according to the quotient rule given in equation (6.33):

$$\text{If} \quad y = \frac{u(x)}{v(x)}, \text{ usually written as } y = \frac{u}{v}, \text{ then}$$

$$\frac{dy}{dx} = \frac{v\dfrac{du}{dx} - u\dfrac{dv}{dx}}{v^2} \tag{6.33}$$

where u is the function above the line and v is the function below it.

Stages of the quotient rule

Stage 1	Stage 2
State the equation for u. Find du/dx. State the equation for v. Find dv/dx.	Fill in the right-hand side of the quotient rule given in equation (6.33) and simplify your answer. $$\frac{dy}{dx} = \frac{v\dfrac{du}{dx} - u\dfrac{dv}{dx}}{v^2}$$

WORKED EXAMPLE 6.36
USING THE QUOTIENT RULE FOR DIFFERENTIATION

Differentiate each of the following:

(a) $y = \dfrac{x^2}{1 + e^x}$

(b) $C = \dfrac{Y + 4}{\ln(Y)}$

(c) $P = \dfrac{Q}{3Q + 5}$

SOLUTION

(a)

Stage 1	Stage 2
State u and v and find du/dx and dv/dx.	Fill in the quotient rule, equation (6.33)
$$y = \frac{x^2}{1 + e^x}$$	$$\frac{dy}{dx} = \frac{v\dfrac{du}{dx} - u\dfrac{dv}{dx}}{v^2}$$
$u = x^2 \qquad v = 1 + e^x$	$$= \frac{(1 + e^x)(2x) - x^2(e^x)}{(1 + e^x)^2}$$
$$\frac{du}{dx} = 2x \qquad \frac{dv}{dx} = e^x$$	$$= \frac{x(2 + 2e^x - xe^x)}{1 + 2e^x + e^{2x}}$$

> **Result**
>
> If $y = \dfrac{x^2}{(1 + e^x)}$ then $\dfrac{dy}{dx} = \dfrac{x(2 + 2e^x - xe^x)}{1 + 2e^x + e^{2x}}$

(b)

Stage 1	Stage 2
State u and v and find du/dY and dv/dY.	Fill in the quotient rule, equation (6.33)

$$C = \frac{Y + 4}{\ln(Y)}$$

$$u = (Y + 4) \text{ and } v = \ln(Y)$$

$$\frac{du}{dY} = 1 \quad \frac{dv}{dY} = \frac{1}{Y}$$

Result

If $C = \dfrac{Y + 4}{\ln(Y)}$ then

$$\frac{dC}{dY} = \frac{\ln(Y) - 1 - \dfrac{4}{Y}}{(\ln Y)^2}$$

$$\frac{dC}{dY} = \frac{v\dfrac{du}{dY} - u\dfrac{dv}{dY}}{v^2}$$

$$= \frac{(\ln Y)(1) - (Y + 4)\dfrac{1}{Y}}{(\ln Y)^2}$$

$$= \frac{\ln(Y) - 1 - \dfrac{4}{Y}}{(\ln Y)^2}$$

(c)

Stage 1	Stage 2
State u and v and find du/dQ and dv/dQ.	Fill in the quotient rule, equation (6.33)

$$P = \frac{Q}{3Q + 5}$$

$$u = Q \quad v = 3Q + 5$$

$$\frac{du}{dQ} = 1 \quad \frac{dv}{dQ} = 3$$

Result

If $P = \dfrac{Q}{3Q + 5}$ then $\dfrac{dP}{dQ} = \dfrac{5}{(3Q + 5)^2}$

$$\frac{dP}{dQ} = \frac{v\dfrac{du}{dQ} - u\dfrac{dv}{dQ}}{v^2}$$

$$= \frac{(3Q + 5)(1) - (Q)(3)}{(3Q + 5)^2}$$

$$= \frac{3Q + 5 - 3Q}{(3Q + 5)^2}$$

$$= \frac{5}{(3Q + 5)^2}$$

PROGRESS EXERCISES 6.15

Differentiation with the Quotient Rule

Find the first derivative for each of the following and simplify your answer.

1. $y = \dfrac{\ln(x)}{x}$

2. $P = \dfrac{50 - Q}{50 + Q}$

3. $y = \dfrac{t^2 + 1}{t^2 - 1}$

4. $AC = \dfrac{\sqrt{Q + 1}}{Q}$

5. $P = 100 - \dfrac{Q}{Q + 1}$

6. $C = 919\left(1 - \dfrac{e^{-0.8Y}}{Y}\right)$

7. $y = \dfrac{2x \ln(x) + 1}{x}$

8. $T = 205\dfrac{\ln(Y)}{Y}$

WORKED EXAMPLE 6.37

FIND *MC* GIVEN A LOGARITHMIC *TC* FUNCTION

A total cost function is given by the equation $TC = 120 \ln(Q + 10)$.

(a) Write down the value of *TFC*. Sketch the *TC* function.
(b) Write down the equation for marginal cost.
(c) Show that *TC* and *MC* do not have turning points.

SOLUTION

(a) $TC = 120 \ln(Q + 10)$. Therefore, when $Q = 0$, $TFC = 120 \ln 10 = 276.3$. The *TC* function is plotted in Figure 6.39 from the points in Table 6.20.

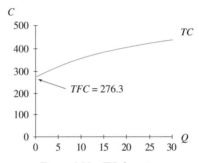

Figure 6.39   *TC* function

Table 6.20   Points for plotting $TC = 120 \ln(Q + 10)$

Q	0	5	10	15	20	25	30
TC	276.3	324.9	359.5	386.3	408.1	426.6	442.7

(b) $MC = \dfrac{d(TC)}{dQ} = 120\dfrac{1}{Q + 10} = \dfrac{120}{Q + 10}$

(c) *TC* has turning points if $d(TC)/dQ = 0$, that is, $120/Q + 10 = 0 \rightarrow 120 = 0(Q + 10) \rightarrow 120 = 0$. This is a contradiction, so there is no turning point. *MC* has turning points if $d(MC)/dQ = 0$, that is,

$$\frac{d}{dQ}\left(\frac{120}{Q + 10}\right) = 0 \rightarrow -\frac{120}{(Q + 10)^2} = 0$$

$$\rightarrow -120 = 0(Q + 10)^2 \rightarrow -120 = 0$$

Again a contradiction, therefore, no turning point.

WORKED EXAMPLE 6.38
DEMAND, *TR, MR* EXPRESSED IN TERMS OF EXPONENTIALS

A firm's demand function is given by the equation $P = 150/e^{0.02Q}$.

(a) Sketch the demand function.
(b) Write down the equations for *TR* and *MR*.
(c) Determine the output Q at which *TR* is a maximum.

SOLUTION

(a) The demand function is sketched in Figure 6.40 from the points in Table 6.21.
(b) $TR = PQ = 150Qe^{-0.02Q}$.
    To find *MR*, differentiate *TR*. As *TR* is a product, use the product rule to differentiate w.r.t. Q.
    Let $u = 150Q$ and $v = e^{-0.02Q}$, hence,

$$\frac{du}{dQ} = 150 \quad \text{and} \quad \frac{dv}{dQ} = -0.02e^{-0.02Q}$$

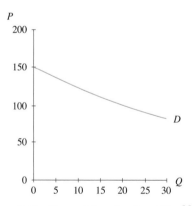

Figure 6.40 Demand function $P = 150e^{-0.02Q}$

Table 6.21 Points for plotting $P = 150e^{-0.02Q}$

Q	0	5	10	15	20	25	30
P	150	135.7	122.8	111.1	100.5	91.0	82.3

Now, substitute u, du/dQ, v and dv/dQ into the equation for the product rule, equation 6.32:

$$\frac{d(TR)}{dQ} = v\frac{du}{dQ} + u\frac{dv}{dQ}$$

$$= e^{-0.02Q}(150) + 150Q(-0.02e^{-0.02Q})$$

$$= 150e^{-0.02Q}(1 - 0.02Q) \quad \text{factoring out the common terms, } 150e^{-0.02Q}$$

Therefore, $MR = 150e^{-0.02Q}(1 - 0.02Q)$.

(c) Work through the usual max/min method to find the turning points on the TR curve.

Step 1: Get derivatives,

$$\frac{d(TR)}{dQ} = 150e^{-0.02Q}(1 - 0.02Q)$$

$d^2(TR)/dQ^2$ is found by differentiating $d(TR)/dQ$ which is a product. The reader is asked to show that

$$\frac{d^2(TR)}{dQ^2} = 150e^{-0.02Q}(0.0004Q - 0.04)$$

Step 2: Solve $\dfrac{d(TR)}{dQ} = 0$:

$$150e^{-0.02Q}(1 - 0.02Q)$$

$$\underbrace{150e^{-0.02Q}}_{\neq 0} \; \underbrace{(1 - 0.02Q)}_{1-0.02Q=0 \to Q=50} = 0$$

Step 3: Max or min?
Evaluate the second derivative at $Q = 50$:

$$TR'' = 150e^{-0.02(50)}[0.0004(50) - 0.04] = 55.182[-0.02] = -1.1036 < 0$$

Since $TR'' < 0$, TR is a maximum at $Q = 50$, $TR = 2759.1$.

PROGRESS EXERCISES 6.16

Further Applications, Using Chain, Product and Quotient Rule

1. A demand function is given by the equation $P = 24 - 6\ln(Q)$.
   (a) Write down the equation for TR. Determine the value of Q at which TR is maximised.
   (b) Write down the equation for MR. Show, by differentiation, that MR decreases but never reaches zero: that MR is concave up.

2. A consumption function is given by the equation $C = 800(1 - e^{-0.2y})$.
   (a) Write down the equation for the MPC. Hence describe how consumption changes as income increases.
   (b) Use differentiation to show that consumption has no maximum value.
   (c) Plot the consumption function. How would you describe the curvature of the consumption function? Use the second derivative to confirm your answer.

3. A utility function is given by the equation $U = 20xe^{-0.1x}$, where x is the number of glasses of wine consumed.
   (a) Show that this utility function has a maximum value and calculate the maximum utility.
   (b) Describe how marginal utility changes for glasses of wine consumed after the maximum utility is reached. Do you consider this reasonable? Give an explanation.

4. A firm's average revenue function is $AR = 50/e^{0.05Q}$.
   (a) Determine the equations for TR and MR.
   (b) Find the value of Q for which TR is a maximum. Calculate the price and TR when TR is maximised.

5. The enjoyment derived from watching a particular film is given by the equation $U = 95 - \ln(t + 5)$, where t is the viewing time in minutes.
   (a) Graph the utility function for the first two hours of viewing. Hence describe how enjoyment changes as the film progresses.
   (b) Determine the equation for marginal utility and plot its graph.

6. Given

$$AC = \frac{100e^{0.05(Q-10)}}{Q+1}$$

   (a) Find the minimum AC and graph the AC function for $Q = 0$ to $Q = 25$.
   (b) Write down the equation of TC.

7. (a) Differentiate the following:
   (i) $y = 4 + x^{4.5} + \dfrac{2x}{\sqrt{x}}$ (ii) $y = 80\,e^{\frac{1}{8}t}$

   (b) The demand function for safari holidays is given by the equation $P = 8000 - 20Q$, where P is the price of the holiday and Q is the number of travellers:
       (i) Write down the equations for total revenue and marginal revenue.
       (ii) Calculate the number of travellers on a safari (Q) for which total revenue is maximised. Calculate the maximum revenue.
       (iii) Hence or otherwise graph the total revenue function to confirm the answers in (ii).

8. (a) Differentiate the following:
   (i) $y = 20(1 - e^{-2t})$ (ii) $Q = 3L^{0.8} + \dfrac{2}{3\sqrt{L}}$

   (b) The demand function for hand-painted T-shirts is given by the equation $P = 1600 - 4Q$ while the average cost function is $AC = \dfrac{500}{Q} + 3$:
       (i) Write down the equations for total revenue and marginal revenue.
       (ii) Calculate the quantity which must be sold to maximise total revenue. Calculate the maximum revenue.
       (iii) Confirm your answer in (ii) by graphing the total revenue function.

(c) (i) Write down the equations for total cost and marginal cost.
(ii) Write down the equation for profit.
(iii) Calculate the maximum profit. How many T-shirts must be produced and sold to maximise profit?

9. (a) Differentiate the following:
(i) $y = \ln(x^2 + 5x)$ (ii) $y = (3 + x)e^x$
Hence determine the maximum and minimum points using the 'slope' test.

(b) A firm has an average cost function

$$AC = \frac{125}{Q} + \frac{Q^2}{16} - 4$$

(i) Calculate the value of Q for which average costs are a minimum. Hence state the range of values for which average costs are decreasing.
(ii) Write down the equation for total cost. Hence calculate the value of Q for which total costs are a minimum.

(c) If a government subsidy of 3 per unit produced is given, write down the equation of the total cost function which includes the subsidy.
Use calculus to determine the minimum total cost, with the subsidy.

10. See question 2, Progress Exercises 4.8.
(a) Differentiate the following:

$$S = 200\,000(1 - e^{-0.05t})$$

(b) The sales of a magazine are expected to grow according to the equation in (a) where t is in weeks. Show by differentiation that sales never reach an exact maximum value. What is the limiting sales value approached?
Give a verbal explanation of the growth in sales

11. See question 3 Progress Exercises 4.8.
(a) Differentiate the following:

$$N = \frac{800}{1 + 790e^{-0.1t}}$$

(b) A virus is thought to spread through a chicken farm according to the equation given in (a) where time t is given in days. Show by differentiation that the virus never reaches a maximum value but tends towards a maximum value. Calculate the time at which the growth curve reaches a point of inflection. Give a verbal explanation of the spread of the virus through the chicken farm.

6.7 Elasticity and the Derivative

At the end of this section you should be familiar with:

- The use of the derivative to calculate the coefficient of point elasticity of demand
- The constant elasticity demand function
- The relationship between price elasticity of demand, marginal revenue, total revenue and price changes.

6.7.1 Point elasticity of demand and the derivative

In Section 2.6 the concept of elasticity was introduced. Price elasticity of demand was defined as

$$\varepsilon_d = \frac{\%\ \text{change in quantity}}{\%\ \text{change in price}} \rightarrow \varepsilon_d \times (\%\ \text{change in } P) = (\%\ \text{change in } Q) \tag{6.34}$$

The formula for price elasticity of demand measured at a point on the demand function was given as

$$\varepsilon_d = \frac{\Delta Q}{\Delta P} \cdot \frac{P}{Q} = -\frac{1}{b} \cdot \frac{P}{Q}$$

This formula is exact when the demand function is linear. When the demand function is non-linear, the derivative dQ/dP must be used instead of ΔQ/ΔP, that is,

$$\varepsilon_d = \frac{dQ}{dP} \cdot \frac{P}{Q} \tag{6.35}$$

Note: If the demand function is linear, then both

$$\varepsilon_d = -\frac{1}{b} \cdot \frac{P}{Q} \quad \text{and} \quad \varepsilon_d = \frac{dQ}{dP} \cdot \frac{P}{Q}$$

can be used interchangeably as either formula provides the same answer (slope = ΔQ/ΔP is a constant value).

In the worked examples which follow, the price elasticity of demand will be determined by the following method:

Step 1: From the equation of the demand function determine an expression for dP/dQ or dQ/dP, whichever is most convenient. Since the definition of elasticity requires dQ/dP, this is easily obtained by inverting dP/dQ, since

$$\frac{dQ}{dP} = \frac{1}{\dfrac{dP}{dQ}}$$

Step 2: The expression for ε_d is obtained by substituting the expression for dQ/dP into equation (6.35). Frequently, the expression for ε_d may be simplified to an expression in Q only or P only. See Worked Example 6.40.

Step 3: To evaluate ε_d at a given price (P) or a given quantity (Q), find the simplest expression for ε_d. Then substitute the given value(s) of Q or P or both into the simplified form of equation (6.35).

WORKED EXAMPLE 6.39
EXPRESSIONS FOR POINT ELASTICITY OF DEMAND IN TERMS OF P, Q OR BOTH FOR LINEAR AND NON-LINEAR DEMAND FUNCTIONS

Given the demand functions

$$P = 100 - 2Q \quad \text{and} \quad Q = 80 - 10 \ln(P)$$

(a) For each function, derive an expression for the price elasticity of demand, ε_d, in terms of
  (i) Q only
  (ii) P only.

  Hence evaluate ε_d at
  (i) $Q = 20$
  (ii) $P = 60$

(b) Calculate the percentage change in demand in response to a 5% price increase by:
  (i) Using the definition of ε_d in equation (6.34).
  (ii) Calculating the exact percentage change in Q.
  Give reasons for the different answers obtained in (i) and (ii).

SOLUTION

(a) (i) Q only

$$P = 100 - 2Q$$

Step 1: $\dfrac{dP}{dQ} = -2 \rightarrow \dfrac{dQ}{dP} = \dfrac{1}{-2} = -\dfrac{1}{2}$

Steps 2 and 3:

$$\varepsilon_d = \frac{dQ}{dP}\frac{P}{Q} = -\frac{1}{2}\frac{P}{Q} = \frac{-0.5}{1}\frac{P}{Q}$$

$$\varepsilon_d = \frac{-0.5}{1} \times \frac{100 - 2Q}{Q} = \frac{-50 + Q}{Q}$$

$$\varepsilon_d = 1 - \frac{50}{Q}$$

This is ε_d expressed in terms of Q.

At $Q = 20$,

(a) (i) Q only

$$Q = 80 - 10\ln(P)$$

Step 1: $\dfrac{dQ}{dP} = -\dfrac{10}{P}$

Steps 2 and 3:

$$\varepsilon_d = \frac{dQ}{dP}\frac{P}{Q} = -\frac{10}{P}\frac{P}{Q} = -\frac{10}{Q}$$

$$\varepsilon_d = -\frac{10}{Q}$$

This is ε_d expressed in terms of Q.

At $Q = 20$,

$$\varepsilon_d = 1 - \frac{50}{Q} = 1 - \frac{50}{20} = -1.5 \qquad \bigg| \qquad \varepsilon_d = -\frac{10}{Q} = -\frac{10}{20} = -0.5$$

(a) (ii) P only $\qquad\qquad\qquad$ (a) (ii) P only

$$\varepsilon_d = \frac{dQ}{dP}\frac{P}{Q} = -\frac{1}{2}\frac{P}{Q} \qquad\qquad \varepsilon_d = \frac{dQ}{dP}\frac{P}{Q} = \frac{-10}{P}\frac{P}{Q} = \frac{-10}{Q}$$

$$= -\frac{1}{2} \times \frac{P}{50 - 0.5P} = \frac{P}{-100 + P} \qquad = \frac{-10}{80 - 10\ln(P)} \times \frac{-0.1}{-0.1}$$

$$= \frac{P}{P - 100} \qquad\qquad\qquad\qquad = \frac{1}{\ln(P) - 8}$$

$$\varepsilon_d = \frac{P}{P - 100} \qquad\qquad\qquad\qquad \varepsilon_d = \frac{1}{\ln(P) - 8}$$

This is ε_d expressed in terms of P. $\qquad$ This is ε_d expressed in terms of P.

At $P = 60$, $\qquad\qquad\qquad\qquad\qquad$ At $P = 60$,

$$\varepsilon_d = \frac{P}{P - 100} = \frac{60}{60 - 100} = -1.5 \qquad \varepsilon_d = \frac{1}{\ln(P) - 8} = \frac{1}{\ln(60) - 8} = -0.256$$

(b) (i) To calculate the percentage change in Q when P increases by 5%, use the definition of elasticity given in equation (6.34)

$$\varepsilon_d = \frac{\% \text{ change in quantity}}{\% \text{ change in price}} \rightarrow \varepsilon_d \times (\% \text{ change in } P) = (\% \text{ change in } Q)$$

$\varepsilon_d \times (5\%) = (\% \text{ change in } Q) \qquad\quad \varepsilon_d \times (5\%) = (\% \text{ change in } Q)$

$(-1.5)(5) = -7.5\% \qquad\qquad\qquad\quad (-0.256)(5) = -1.28\%$

So, demand drops by 7.5% when price $\qquad$ So, demand drops by 1.28% when price
increases by 5%. $\qquad\qquad\qquad\qquad$ increases by 5%.

(ii) When price increases by 5%, the new price is $P = \frac{105}{100} \times 60 = 63$.

The corresponding value of Q for each function is

$P = 100 - 2Q \rightarrow Q = 50 - 0.5P \qquad Q = 80 - 10\ln(P)$

At $P = 60$, $Q = 50 - 0.5(60) = 20.0 \qquad$ At $P = 60$, $Q = 80 - 10\ln(60) = 39.057$

At $P = 63$, $Q = 50 - 0.5(63) = 18.5 \qquad$ At $P = 63$, $Q = 80 - 10\ln(63) = 38.569$

Therefore, the percentage change in Q

$$= \frac{\text{change in } Q}{\text{initial value of } Q} \times 100$$

$$= \frac{18.5 - 20}{20} \times 100 = -7.5\%$$

This is exactly the same result as that calculated using elasticity.

Therefore, the percentage change in Q

$$= \frac{\text{change in } Q}{\text{initial value of } Q} \times 100$$

$$= \frac{38.569 - 39.057}{39.057} \times 100$$

$$= -1.249\%$$

This is *not* exactly the same result as that calculated using elasticity. The reason: for non-linear demand functions, the change over a small interval, $P = 60$ to 63, is approximately equal to the change at a point.

Therefore, for non-linear demand functions, the equation

$$\varepsilon_d = \frac{\% \text{ change in quantity}}{\% \text{ change in price}} \rightarrow \varepsilon_d \times (\% \text{ change in } P) = (\% \text{ change in } Q)$$

may be used to calculate the approximate percentage change in demand, Q, as a result of small percentage changes in P.

WORKED EXAMPLE 6.40
POINT ELASTICITY OF DEMAND FOR NON-LINEAR DEMAND FUNCTIONS

(a) Given the demand function $P = 60 - Q^2$, calculate the coefficient for point elasticity of demand at $P = 44$.

(b) Given the demand function $Q = 45e^{-0.04P}$, calculate the coefficient for point elasticity of demand at $P = 10$.

SOLUTION

(a) $P = 60 - Q^2$

Step 1: Find dQ/dP. Since the demand function is written as $P = 60 - Q^2$, find dP/dQ.

$$\frac{dP}{dQ} = -2Q, \quad \text{then} \quad \frac{dQ}{dP} = \frac{1}{\dfrac{dP}{dQ}} = \frac{1}{-2Q}$$

(b) $Q = 45e^{-0.04P}$

Step 1: Find dQ/dP. Since the demand function is written as $Q = 45e^{-0.04P}$,

$$\frac{dQ}{dP} = 45(-0.04)e^{-0.04P} = -1.8e^{-0.04P}$$

Step 2: Derive an expression for ε_d in terms of P.

$$\varepsilon_d = \frac{dQ}{dP} \cdot \frac{P}{Q} = -\frac{1}{2Q}\frac{P}{Q} = -\frac{P}{2Q^2}$$

$$= -\frac{P}{2(60 - P)}$$

Step 3: Evaluate ε_d at $P = 44$:

$$\varepsilon_d = -\frac{P}{2(60 - P)} = -\frac{44}{2(60 - 44)}$$

$$= -1.375$$

Since $\varepsilon_d < -1$, demand is elastic.

Step 2: Derive an expression for ε_d in terms of P.

$$\varepsilon_d = \frac{dQ}{dP} \cdot \frac{P}{Q} = \frac{-1.8e^{-0.04P}}{1} \cdot \frac{P}{45e^{-0.04P}}$$

$$\varepsilon_d = -0.04P$$

Step 3: Evaluate ε_d at $P = 10$:

This one is easy!

$$\varepsilon_d = -0.04P = -0.04(10) = -0.4$$

Since $\varepsilon_d > -1$, demand is inelastic.

6.7.2 Constant elasticity demand function

In Section 2.6 it was illustrated that the coefficient of price elasticity of demand varies along a linear demand function. The same is true for a non-linear demand function. For example, choose several values for P and prove that the coefficient of point elasticity varies along the demand function, $P = 60 - Q^2$.

There is, however, a case when the price elasticity of demand is constant along a non-linear demand function. This is known as the **constant elasticity demand function**.

WORKED EXAMPLE 6.41
CONSTANT ELASTICITY DEMAND FUNCTION

Show that a demand function of the form $Q = a/P^c$, where a and C are constants, has a constant elasticity of demand $\varepsilon_d = -c$, that is, for every value of (P, Q), $\varepsilon_d = -c$. Hence, show that $Q = 200/P^2$ has a constant elasticity of demand, $\varepsilon_d = -2$.

SOLUTION

General expression

Step 1:

$$Q = \frac{a}{P^c} = aP^{-c}$$

$$\frac{dQ}{dP} = -caP^{-c-1}$$

Step 2:

$$\varepsilon_d = \frac{dQ}{dP} \cdot \frac{P}{Q}$$

Example

Step 1:

$$Q = \frac{200}{P^2} = 200P^{-2}$$

$$\frac{dQ}{dP} = (-2)200P^{-2-1}$$

Step 2:

$$\varepsilon_d = \frac{dQ}{dP} \cdot \frac{P}{Q}$$

$$= -\frac{ca\,P^{-c-1}}{1}\,\frac{P}{Q}$$

$$= -\frac{ca\,P^{-c}}{Q} \quad \text{adding indices on } P$$

$$= -\frac{cQ}{Q} \quad \text{since } Q = a\,P^{-c}$$

$$= -c \quad Qs \text{ cancel}$$

$$= \frac{-2(200)P^{-3}}{1}\,\frac{P}{Q}$$

$$= \frac{-2(200)P^{-2}}{Q} \quad \text{adding indices on } P$$

$$= \frac{(-2)Q}{Q} \quad \text{since } Q = 200P^{-2}$$

$$= -2 \quad Qs \text{ cancel}$$

6.7.3 Price elasticity of demand, marginal revenue, total revenue and price changes

- The relationship between price elasticity of demand and marginal revenue may be expressed by equation (6.36).
- The relationship between price elasticity of demand and the change in total revenue with respect to price may be expressed by equation (6.37).

$$MR = P\left(1 + \frac{1}{\varepsilon_{\mathrm{d}}}\right) \tag{6.36}$$

$$\frac{\mathrm{d}(TR)}{\mathrm{d}P} = Q(1 + \varepsilon_{\mathrm{d}}) \tag{6.37}$$

Equations (6.36) and (6.37) are derived as follows.

Regard TR as a product, $(P \times Q)$, and differentiate using the product rule. $TR = PQ$, let $u = P$ and $v = Q$.

Differentiate w.r.t. Q

$$MR = \frac{\mathrm{d}(TR)}{\mathrm{d}Q} = \frac{\mathrm{d}(PQ)}{\mathrm{d}Q}$$

let $u = P$, $v = Q$, hence

$$\frac{\mathrm{d}u}{\mathrm{d}Q} = \frac{\mathrm{d}P}{\mathrm{d}Q} \quad \text{and} \quad \frac{\mathrm{d}v}{\mathrm{d}Q} = \frac{\mathrm{d}Q}{\mathrm{d}Q} = 1$$

substitute for u, v, $\dfrac{\mathrm{d}u}{\mathrm{d}Q}, \dfrac{\mathrm{d}v}{\mathrm{d}Q}$

$$MR = \frac{\mathrm{d}(TR)}{\mathrm{d}Q} = u\frac{\mathrm{d}v}{\mathrm{d}Q} + v\frac{\mathrm{d}u}{\mathrm{d}Q}$$

$$MR = P\frac{\mathrm{d}Q}{\mathrm{d}Q} + Q\frac{\mathrm{d}P}{\mathrm{d}Q}$$

Differentiate w.r.t. P

$$MR = \frac{\mathrm{d}(TR)}{\mathrm{d}P} = \frac{\mathrm{d}(PQ)}{\mathrm{d}P}$$

let $u = P$, $v = Q$, hence

$$\frac{\mathrm{d}u}{\mathrm{d}P} = \frac{\mathrm{d}P}{\mathrm{d}P} = 1 \quad \text{and} \quad \frac{\mathrm{d}v}{\mathrm{d}P} = \frac{\mathrm{d}Q}{\mathrm{d}P}$$

substitute for u, v, $\dfrac{\mathrm{d}u}{\mathrm{d}Q}, \dfrac{\mathrm{d}v}{\mathrm{d}Q}$

$$\frac{\mathrm{d}(TR)}{\mathrm{d}P} = u\frac{\mathrm{d}v}{\mathrm{d}P} + v\frac{\mathrm{d}u}{\mathrm{d}P}$$

$$\frac{\mathrm{d}(TR)}{\mathrm{d}P} = P\frac{\mathrm{d}Q}{\mathrm{d}P} + Q\frac{\mathrm{d}P}{\mathrm{d}P}$$

$$MR = P + Q \cdot \frac{dP}{dQ} \qquad \text{(6.38)}$$

Now: $\dfrac{\varepsilon_d}{1} = \dfrac{dQ}{dP} \cdot \dfrac{P}{Q}$

$\dfrac{1}{\varepsilon_d} = \dfrac{dP}{dQ} \cdot \dfrac{Q}{P}$ inverting both sides

Multiply $Q\dfrac{dP}{dQ}$ in equation (6.38) by $\dfrac{P}{P}$.

$$MR = P + \frac{P}{1}\left(\frac{Q}{P} \cdot \frac{dP}{dQ}\right) = P + P\left(\frac{1}{\varepsilon_d}\right)$$

$$MR = P\left(1 + \frac{1}{\varepsilon_d}\right), \text{ i.e., equation (6.36)}$$

$$\frac{d(TR)}{dP} = P\frac{dQ}{dP} + Q \qquad \text{(6.39)}$$

Now: $\dfrac{\varepsilon_d}{1} = \dfrac{dQ}{dP} \cdot \dfrac{P}{Q}$

Multiply $P\dfrac{dQ}{dP}$ in equation (6.39) by $\dfrac{Q}{Q}$.

$$\frac{d(TR)}{dP} = Q + \frac{Q}{1}\left(\frac{P}{Q} \cdot \frac{dQ}{dP}\right) = Q + Q(\varepsilon_d)$$

$$\frac{d(TR)}{dP} = Q(1 + \varepsilon_d), \text{ i.e., equation (6.37)}$$

Equation (6.36) may also be written as

$$MR = P + \frac{P}{\varepsilon_d} \rightarrow \frac{MR - P}{1} = \frac{P}{\varepsilon_d} \rightarrow \frac{1}{MR - P} = \frac{\varepsilon_d}{P} \rightarrow \varepsilon_d = \frac{P}{MR - P} \qquad \text{(6.40)}$$

There are various approaches to demonstrating the relationship between ε_d, MR and TR as P varies. An explanation based on equation (6.40) and Figure 6.41 will now be presented.

Remember

Price, $P = AR$.

www.wiley.com/college/bradley

Go to the website for the following additional material that accompanies Chapter 6:

- Price elasticity of demand, TR, MR and P (price)
  Figure 6.41
  Table 6.22 Price elasticity of demand, marginal and total revenue
  Worked Example 6.42 Elasticity, marginal revenue and total revenue
  Table 6.23 Elasticity, total revenue and marginal revenue
  Figure 6.42 Elasticity, total revenue and marginal revenue

PROGRESS EXERCISES 6.17

Differentiation and Point Elasticity

1. (a) Define the price elasticity of demand ε_d.
   (b) The price elasticity of demand for a good is -0.8. What is the percentage change in the quantity demanded when price increases by 5%? Write down the general expression of the function which has a constant elasticity of demand of -0.8.
2. Given the demand functions
   (i) $P = 80 - 2Q$ (ii) $Q = 120 - 4P$ (iii) $P = 432$ (iv) $P = a - bQ$
   (a) Determine expressions for the price elasticity of demand in terms of
      (i) P and (ii) Q.
   (b) Evaluate the price elasticity of demand at
      (i) $P = 50$ and (ii) $Q = 30$.
3. The demand for family membership of a sports club is given by the equation $P = 500e^{-0.01Q}$, where P is the monthly fee.
   (a) Derive an expression for the price elasticity of demand in terms of Q. Evaluate ε_d when $Q = 100$. Describe how demand changes for price increases when $Q < 100$ and when $Q > 100$.
   (b) If the club has 150 members, calculate the fee per membership (price). If the fee increases by 10%, use elasticity to calculate the approximate percentage change in demand. Why is the answer approximate?
4. The demand for seats at a cup final is given by the equation $Q = 192 - P^2$.
   (a) Derive an expression for the price elasticity of demand in terms of P. Calculate the price which should be charged when $\varepsilon_d = -1$. How many seats are available at this price?
   (b) If 10 more seats become available when $\varepsilon_d = -1$, calculate the approximate percentage change in price. Calculate the new price.
5. (a) Show that the function $Q = a/P^c$, where a and c are constants, has a constant elasticity of demand: $\varepsilon_d = -c$.
   (b) (i) Determine the elasticity of demand for train journeys on a given route when the demand function is $Q = 1200/P^{1.2}$, where Q is the number of fares in thousands.
      (ii) If the fare increases by 5%, use elasticity to calculate the percentage change in demand.
      (iii) If the fare decreases by 5%, use elasticity to calculate the percentage change in demand.
6. (See question 5)
   (a) If the company charges £30 per fare, calculate the corresponding number of fares demanded to the nearest fare.
   (b) If the fare increases by 5%, calculate the percentage change in demand exactly. How does this compare with 5(b)(ii) above? Give reasons for the difference.
7. The demand for soft drinks is given by the equation: $Q = 100 - 2P$ where P pence is the price per can and Q is the number of cans demanded.
   (a) Write down the equations for TR, MR, AR.
   (b) Determine the price and quantity at which revenue is maximised.

(c) Derive an expression for the price elasticity of demand in terms of

   (i) P and (ii) Q.

(d) Show that TR is a maximum, MR is zero when $\varepsilon_d = -1$.

8. (See question 7)
   (a) Plot the graphs of TR, MR, AR (suitable for Excel).
   (b) Find the value of Q at which $\varepsilon_d = -1$. Hence indicate on the diagram the intervals along which $\varepsilon_d < -1$ and $\varepsilon_d > -1$.
   Comment on the relationship between P, Q, TR, MR and elasticity.

9. The demand for a certain wine is given by the equation: $P = 1500e^{-0.025Q}$ where P (Euros) is the price per bottle and Q is the number of bottles demanded.
   (a) Write down the equations for TR, MR, AR.
   (b) Determine the price and quantity at which revenue is maximised.
   (c) Derive an expression for the price elasticity of demand in terms of

   (i) P (ii) Q

   (d) Show that $\varepsilon_d = -1$ when TR is maximised.

10. Derive the equation

$$MR = P \left(1 + \frac{1}{\varepsilon_d} \right)$$

   (a) The demand function is of the form $P = a - bQ$. Would you expect ε_d to have the same value at different prices? Justify your answer.
   (b) If the price of a new car is £45 400 and the price elasticity of demand is -0.9:
      (i) Calculate the marginal revenue.
      (ii) Deduce the equation of the demand function $P = a - bQ$, if it is known that $P = 800$ when $Q = 90$. Hence,
      (iii) Use elasticity, or otherwise, to determine the price and quantity at which TR is maximised.

11. (a) Show that when profit is maximised

$$P = \frac{MC}{\left(1 + \dfrac{1}{\varepsilon_d} \right)}$$

   (b) A PC centre has a demand and total cost function for printers given by the respective equations

$$P = 300 - Q, \quad TC = 800 + 2Q$$

   Find the price, quantity and ε_d at which profit is maximised. Verify that the equation given in (a) holds when profit is maximised but not at other values of P and Q on the demand function.

6.8 Summary

Mathematics

Slope of the straight line $y = mx + c$ is m

Slope of a curve $y = f(x)$ over a small horizontal distance Δx is $\Delta y / \Delta x$

Slope of a curve $y = f(x)$ at a point (x_0, y_0) is dy/dx evaluated at $(x = x_0, y = y_0)$

Derivatives of standard functions:

Function:	Derivative:
$y = f(x)$	$\dfrac{dy}{dx} = f'(x)$
x^n	nx^{n-1}
e^x	e^x
e^{ax}	ae^{ax}
$\ln(x)$	$\dfrac{1}{x}$
K	0
Kx^n	$K(nx^{n-1})$
$K_1 x^n + K_2 x^m$	$K_1(nx^{n-1}) + K_2(m\ x^{m-1})$

Chain rule: $y = f(g(x))$. Let $u = g(x)$, hence $y = f(u)$ then

$$\frac{dy}{dx} = \frac{dy}{du}\frac{du}{dx}$$

Useful derivatives to remember:

1. $y = e^{f(x)} \rightarrow \dfrac{dy}{dx} = f'(x) \times e^f(x)$

2. $y = \ln(f(x)) \rightarrow \dfrac{dy}{dx} = \dfrac{f'(x)}{f(x)}$

Product rule: $y = f(x)g(x) = uv$ then

$$\frac{dy}{dx} = v\frac{du}{dx} + u\frac{dv}{dx}$$

Quotient rule:

$$y = \frac{f(x)}{g(x)} = \frac{u}{v}, \quad \text{then} \quad \frac{dy}{dx} = \frac{v\dfrac{du}{dx} - u\dfrac{dv}{dx}}{v^2}$$

Maximum/minimum and points of inflection

	First derivative $\dfrac{dy}{dx}$	$\dfrac{d^2y}{dx^2}$ evaluated at (x_0, y_0)
Maximum point at (x_0, y_0)	$\dfrac{dy}{dx} = 0$	$\left.\begin{array}{l} \dfrac{d^2y}{dx^2} < 0 \\[1em] \dfrac{d^2y}{dx^2} > 0 \end{array}\right\}$ or check for changes in sign of dy/dx
Minimum point at (x_0, y_0)	$\dfrac{dy}{dx} = 0$	
Point of inflection at (x_0, y_0)	$\dfrac{dy}{dx} = 0$ for a horizontal PoI	$\dfrac{d^2y}{dx^2} = 0$ and
	$\dfrac{dy}{dx} \neq 0$ for a non-horizontal PoI	$\dfrac{d^2y}{dx^2}$ changes sign

Curvature

The curvature along an interval is **concave up** if $d^2y/dx^2 > 0$ along that interval.
The curvature along an interval is **concave down** if $d^2y/dx^2 < 0$ along that interval.

Applications

Marginal function is the derivative of the total function:

$$MC = \frac{d(TC)}{d(Q)} : MR = \frac{d(TR)}{d(Q)} : MP_L = \frac{d(Q)}{d(L)}$$

Revise the relationships between marginal and average functions.

Average function: $AC = \dfrac{TC}{Q} : AR = \dfrac{TR}{Q} : APL = \dfrac{Q}{L}$

Optimisation: particular rules to note

1. Maximum profit when $MR = MC$ and

$$\frac{d(MR)}{dQ} < \frac{d(MC)}{dQ}$$

2. MC and AVC intersect at the minimum point on the AVC curve.
3. APL and MP_L intersect at the maximum point on the APL curve.

DIFFERENTIATION AND APPLICATIONS

Elasticity: point elasticity of demand

1. Definition: $P = f(Q)$, then

$$\varepsilon_{\mathrm{d}} = \frac{\%\Delta Q}{\%\Delta P} = \frac{\mathrm{d}Q/Q}{\mathrm{d}P/P} = \frac{\mathrm{d}Q}{\mathrm{d}P}\frac{P}{Q}$$

2. The function $Q = a/P^c$ has a constant elasticity of demand $\varepsilon_{\mathrm{d}} = -c$

3. Relationship between MR and ε_{d}: $MR = P\left(1 + \dfrac{1}{\varepsilon_{\mathrm{d}}}\right)$

Production and labour
A short-run production function $Q = f(L)$
Costs:

$$TLC = wL$$

$$ALC = \frac{TLC}{L}$$

$$MLC = \frac{\mathrm{d}(TLC)}{\mathrm{d}L}$$

www.wiley.com/college/bradley

Go to the website for Problems in Context

TEST EXERCISES 6

1. (a) Determine the first derivative of the following:
   (i) $y = 7x^4$ (ii) $y = 1/x^2$ (iii) $y = \sqrt{x} + 4x - 8$
   (b) A firm has the following total cost and total revenue functions:

$$TC = \frac{1}{3}Q^3 - 9Q^2 + 200Q + 5050, \quad TR = Q(120 - 10Q)$$

   Deduce the equations for the following functions:
   (i) marginal cost (ii) marginal revenue (iii) average cost (iv) average revenue.
2. The output from a pottery is related to the number of labour units employed according to the production function $Q = 9L^2 - 0.1L^3$.
   (a) Determine the number of labour units which should be employed to maximise output, Q.
   (b) Find the point of inflection.
3. (See question 2)
   (a) Derive the equation for the marginal product of labour and the average product of labour.
   (b) Hence determine the number of labour units which maximise

   (i) MP_L and (ii) APL.

(c) Show that the MP_L and APL curves intersect at the maximum point on the APL curve.
4. (See questions 2 and 3)
Graph the production functions MP_L and APL, indicating the relationship between turning point(s) and point of inflection.
5. A firm has an average cost function $AC = 10 - 3Q + Q^2$.
(a) Write down the equations for TC, MC.
(b) Determine the values of Q at which (i) MC and (ii) AC are minimised.
(c) Plot the AC and MC curves on the same diagram. Confirm algebraically that the curves intersect at the minimum point on the AC curve.
6. (a) Show that profit is maximised/minimised when $MR = MC$. How would you determine whether profit is a maximum and not a minimum when $MR = MC$?
(b) The demand and total cost functions for a good are given by the equations

$$P = 80 - 4Q, \quad TC = \frac{1}{3}Q^3 - 18Q^2 + 240Q + 1500$$

(i) Write down the equations for MR, MC and profit.
(ii) Determine the number of goods which must be produced and sold to maximise profit.

7. (a) Write down the definition for the price elasticity of demand, ε_d. If the demand function is of the form $P = a - bQ$, show that the value of ε_d is not influenced by the value of slope.
(b) A firm faces different demand functions for its product in two separate markets

Market I: $P = 120 - 3Q$ Market II: $P = 180e^{-0.6Q}$
For each market:

(i) Determine the price elasticity of demand when $P = 10$.
(ii) Calculate the response to a 5% increase in price when $P = 10$ using:
(i) the definition of elasticity, (ii) direct calculation. Comment.
8. A firm may sell its product in two different markets in which the demand functions are
Market I: $P = 120 - 2Q$ Market II: $P = 492 - 10Q$
The total cost function is given by the equation $TC = 15 + 12Q$.
(a) Calculate the price which should be charged in each market to maximise the firm's profit.
(b) Calculate the price elasticity of demand when profit is maximised in each market.
What is the relationship between price elasticity of demand and the price charged in different markets?
9. (See question 8)
(a) Calculate the price charged for the product if profits are maximised with no price discrimination.
(b) Compare the profit made with and without price discrimination.
10. (a) Determine the first derivatives of the following:
(i) $y = x - x \ln(x)$ (ii) $y = 2xe^{-0.1x}$
(b) The demand function for a good is given by the equation $P = 200e^{-0.1Q}$.

(i) Write down the equation for total revenue. (ii) Hence determine the value of Q at which total revenue is a maximum.
11. Determine the first and second derivatives of the following:

(a) $y = \dfrac{x}{x + 10}$ (b) $y = 10xe^{5x}$ (c) $P = (20Q - 2)^{0.5}$

FUNCTIONS OF SEVERAL VARIABLES

7

7.1   Partial Differentiation

7.2   Applications of Partial Differentiation

7.3   Unconstrained Optimisation

7.4   Constrained Optimisation and Lagrange Multipliers

7.5   Summary

Chapter Objectives

In this chapter you will discover that partial differentiation is used in applications ranging from isoquants, indifference curves, marginal functions, incremental changes, constrained and unconstrained optimisation of functions of two variables. At the end of this chapter you should be able to:

- Calculate first- and second-order partial derivatives
- Calculate differentials and incremental changes
- Calculate marginal functions and the law of diminishing returns
- Show that a Cobb–Douglas function is homogeneous degree r and determine whether the function exhibits constant, decreasing or increasing returns to scale
- Use partial derivatives to analyse the properties of production functions and utility functions
- Calculate partial elasticities
- Calculate multipliers for the linear national income model
- Locate and determine the nature of stationary points for functions of several variables
- Use Lagrange multipliers to determine maximum and minimum values for functions of two variables subject to a constraint.

In Chapter 6, differentiation was confined to functions of one independent variable only. However, more realistic economic models are usually functions of more than one variable. For example, in Chapter 6 the short-run production function $Q = f(L)$ was described as a function of only one input: labour. In the long run, a more realistic model assumes that output, $Q = f(L, K)$, is a function of two inputs: labour and capital.

In this chapter, those basic differentiation techniques introduced in Chapter 6 are extended to the differentiation of functions of two or more variables, with applications such as optimisation, rates of change, etc.

7.1 Partial Differentiation

At the end of this section you should be able to:

- Understand the term functions of two or more variables, and graph such functions
- Find first-order partial derivatives of functions of two or more variables
- Find second-order partial derivatives of functions of two or more variables
- Determine differentials and small changes.

7.1.1 Functions of two or more variables

Functions of two variables are written in general form as

$$z = f(x, y)$$

for example,

$$z = x + 2y + 4 \tag{7.1}$$

where x and y are the independent variables and z is the dependent variable, that is, its value depends on the values of x and y.

Graphically, functions of two variables are represented by a three-dimensional diagram. Equation (7.1) is graphed in Figure 7.1, where both x and y are represented on the two horizontal axes and z on the vertical axis. To plot the function, values are assigned to both x and y in order to calculate the corresponding value of the dependent variable z, such as those given in Table 7.1.

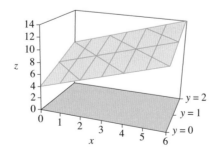

Figure 7.1 $z = x + 2y + 4$: a three-dimensional plane

FUNCTIONS OF SEVERAL VARIABLES

Table 7.1   Calculate points for the function $z = x + 2y + 4$

Point	A	B	C	D	E	F	G
x	0	1	2	3	4	5	6
y	0	1	2	1	2	1	2
$z = x + 2y + 4$	4	7	10	9	12	11	14

Graphically, the function, $z = x + 2y + 4$, is a plane in three dimensions. A plane in three dimensions is represented by the general linear function in two variables:

$$z = ax + by + c$$

Otherwise, if there is any power on either x or y, or products of x- and y-terms, the function is a surface in three dimensions which may contain humps (relative maxima), hollows (relative minima), etc.

Graphical representation of functions of two variables in economics

Diagrams in three dimensions are difficult to draw and to visualise. In economics, functions of two variables are frequently plotted by fixing one of the variables, thereby reducing the problem of graphical representation and mathematical analysis to a series of two-dimensional curves.

To see how this two-dimensional representation comes about, consider a production function whose equation is

$$Q = 5LK \tag{7.2}$$

That is, output is a function of the amounts of inputs of labour, L and capital, K.

If output, Q, is fixed at some constant value, for example, at $Q = 5$, then equation (7.2) may be rearranged to express K in terms of L:

$$Q = 5LK \quad \rightarrow \quad 5 = 5LK \quad \rightarrow \quad 1 = LK \quad \rightarrow \quad K = \frac{1}{L}$$

This is a curve in a plane defined by an L and a K-axis. In effect, it is a single slice of the three-dimensional diagram along the plane $Q = 5$. This downward-sloping curve is a hyperbolic function as described in Chapter 4. The graph is plotted in Figure 7.3 below. In economics, it is called an **isoquant** (isoquant means equal quantity). It shows the various combinations of inputs L and K for which quantity (production) is constant at $Q = 5$.

WORKED EXAMPLE 7.1

PLOT ISOQUANTS FOR A GIVEN PRODUCTION FUNCTION

Given the production function $Q = 5LK$, write down the equations for the isoquants defined by (i) $Q = 5$, (ii) $Q = 10$, (iii) $Q = 15$, (iv) $Q = 18$.

(a) Calculate a table of points for each isoquant for values $L = 0.02, 0.04, \ldots, 0.16$, hence show that any pair of points on a given isoquant give the same output, Q.
(b) Graph the isoquants on the same diagram.
(c) Deduce an expression for the slope of the isoquants. Use the second derivative to show that the isoquants are convex towards the origin.

SOLUTION

Substitute the fixed value of Q into the equation of the production function, $Q = 5LK$, then write the resulting equation in the form $K = f(L)$. The equations of the corresponding isoquants are given in Table 7.2.

Table 7.2   Production functions and isoquants

Output, Q	Equation of production function, given Q	Equation of isoquant
$Q = 5$	$5 = 5LK$	$K = \dfrac{1}{L}$
$Q = 10$	$10 = 5LK$	$K = \dfrac{2}{L}$
$Q = 15$	$15 = 5LK$	$K = \dfrac{3}{L}$
$Q = 18$	$18 = 5LK$	$K = \dfrac{3.6}{L}$

(a) The points for each isoquant are calculated in Table 7.3.

Note: When Q is calculated for any point (L,K) on a given isoquant, as expected, the result is the given fixed value of Q. For example, on the isoquant $Q = 10$, calculate $Q = 5LK$ for selected points,

$(L = 0.02, K = 100) \rightarrow Q = 5(0.02)(100) = 10$

$(L = 0.14, K = 14.29) \rightarrow Q = 5(0.14)(14.29) = 10.003$

(the error of 0.003 is due to rounding $2/K = 2/0.14 = 14.285\,714\,29$ to 14.29)

$(L = 0.16, K = 12.50) \rightarrow Q = 5(0.16)(12.5) = 10$

Table 7.3   Calculation of points of isoquants

L	Q = 5 K	Q = 10 K	Q = 15 K	Q = 18 K
0.02	50.00	100.00	150.00	180.00
0.04	25.00	50.00	75.00	90.00
0.06	16.67	33.33	50.00	60.00
0.08	12.50	25.00	37.50	45.00
0.10	10.00	20.00	30.00	36.00
0.12	8.33	16.67	25.00	30.00
0.14	7.14	14.29	21.43	25.71
0.16	6.25	12.50	18.75	22.50

(b) The isoquants are graphed in three dimensions in Figure 7.2, using Excel. The graphs are all in parallel *LK* planes. In most economics textbooks, isoquants will normally be drawn in the *LK* plane as two-dimensional diagrams such as shown in Figure 7.3. The standard convention is to plot *L* on the horizontal axis and *K* on the vertical one. Furthermore, note that isoquants that lie further in a northeasterly direction represent higher levels of quantity.

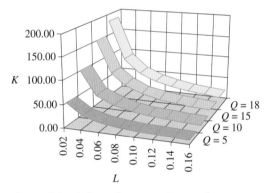

Figure 7.2   A three-dimensional view of isoquants

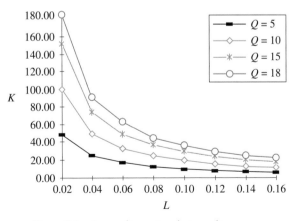

Figure 7.3   A two-dimensional view of isoquants

(c) The expression for the slope of any function, $y = f(x)$, is found by differentiation. The equation of each isoquant is of the same general form:

$$Q_0 = 5LK \rightarrow K = \frac{Q_0}{5L} \quad \text{where } Q_0 \text{ is constant}$$

Therefore, the expression for slope is simply the equation of the first derivative:

$$K = \frac{Q_0}{5L} = \frac{1}{5}Q_0 L^{-1}$$

$$\frac{dK}{dL} = \frac{1}{5}Q_0(-1L^{-2}) = -\frac{1}{5}\frac{Q_0}{L^2} > 0 \quad \text{for } L > 0$$

So the slope of any isoquant is negative, therefore K is decreasing as L increases.

Furthermore, the second derivative is positive, indicating that the curve is convex towards the origin, agreeing with the graphs in part (b):

$$\frac{dK}{dL} = -\frac{1}{5}\frac{Q_0}{L^2} \quad \rightarrow \quad \frac{d^2K}{dL^2} = \frac{2}{5}\frac{Q_0}{L^3} > 0 \quad \text{for } L > 0$$

7.1.2 Partial differentiation: first-order partial derivatives

Partial differentiation is equivalent to differentiating along an isoquant: one of the three variables is fixed, then differentiation proceeds as normal with the two remaining variables.

For example, in Worked Example 7.1, $Q = Q_0$:

$$K = \frac{Q_0}{5L} = \frac{Q_0}{5}L^{-1}$$

$$\frac{dK}{dL} = \frac{Q_0}{5}(-1L^{-2}) = -\frac{Q_0 L^{-2}}{5} = -\frac{Q_0}{5L^2}$$

However, it is necessary to indicate that the derivative is part of a three-dimensional set-up in which the other variable, Q, was held constant. Such a derivative is called a partial derivative, denoted by using the symbol '∂' instead of 'd' for differentiation:

$$\frac{dK}{dL} \rightarrow \frac{\partial K}{\partial L} \quad Q \text{ is fixed}$$

It may take some time to get used to treating variables as constants in one line, then treating them as variables again in the next line. You may find it useful to put the temporarily 'fixed variable' in square brackets in the first line before differentiation, and then drop the square brackets when differentiation is completed. For example,

$$K = \frac{[Q]}{5L} = \frac{[Q]}{5}L^{-1}$$

$$\frac{dK}{dL} = \frac{Q}{5}(-1L^{-2}) = -\frac{QL^{-2}}{5} = -\frac{Q}{5L^2}$$

WORKED EXAMPLE 7.2
PARTIAL DIFFERENTIATION: A FIRST EXAMPLE

Consider the function, $z = x + 2y + 4$. Use the following method to differentiate z partially with respect to (a) x and (b) y.

Method

(a) To differentiate z partially with respect to x, treat y as a constant, and then differentiate z w.r.t. x using the usual rules of differentiation as outlined in Chapter 6. To indicate partial differentiation is being used, the symbol '∂' is used instead of 'd'; that is,

$$\boxed{\frac{\partial z}{\partial x} \text{ denotes partial differentiation of } z \text{ w.r.t. } x}$$

$(\partial z / \partial x)$ means the partial derivative of z with respect to x.

$$z = x + 2[y] + 4 \rightarrow \frac{\partial z}{\partial x} = 1 + 0 + 0 = 1 \quad \text{since } 2y \text{ is held constant and } 4 \text{ is a constant}$$

(b) To differentiate z partially with respect to y, treat x as a constant, and then differentiate z w.r.t. y. In this case,

$$\boxed{\frac{\partial z}{\partial y} \text{ denotes partial differentiation of } z \text{ w.r.t. } y}$$

$$z = [x] + 2y + 4 \rightarrow \frac{\partial z}{\partial y} = 0 + 2 + 0 = 2 \quad \text{since } x \text{ is held constant and } 4 \text{ is a constant}$$

Abbreviations: The first-order partial derivatives may be abbreviated as

$$\frac{\partial z}{\partial x} \equiv z_x, \quad \frac{\partial z}{\partial y} \equiv z_y$$

Some examples of first-order partial derivatives are outlined in Worked Example 7.3 where the normal rules for differentiation apply: power rule, chain rule, product rule, quotient rule, etc.

WORKED EXAMPLE 7.3
DETERMINING FIRST-ORDER PARTIAL DERIVATIVES

Find the first-order partial derivatives for each of the following functions.

(a) $z = 2x^2 + 3xy + 5$
(b) $Q = 10L^{0.7}K^{0.3}$
(c) $U = x^2 y^5$

SOLUTION

(a) To find $\partial z/\partial x$, differentiate partially w.r.t. x, therefore, treat y as if it were a constant (like 10, 12, any number!) and proceed with the following steps:

Step 1: Since we are differentiating w.r.t. x and treating y as if it were a constant, examine each term in turn, look for x in each term, then decide which rules of differentiation apply. In this example,

$2x^2$: Use the power rule to differentiate this term.

$3[x]y$: This term is simply a constant, $3y$ (since y is considered constant) multiplied by x; therefore, differentiation requires the power rule.

5: This is a constant term so its derivative is zero.

Step 2: Differentiate partially w.r.t. x:

$$z = 2x^2 + 3[y]x + 5$$
$$\frac{\partial z}{\partial x} = 2(2x) + 3y(1) + 0 \quad \text{differentiating partially w.r.t. } x$$
$$\frac{\partial z}{\partial x} = 4x + 3y$$

Now find $\partial z/\partial y$, the first-order partial derivative of z w.r.t. y (treat x as constant).

Step 1: Examine each term in turn, look for y in each term, and decide which rule for differentiation applies. In this example,

$2[x^2]$: There is no y in this term, therefore treat it as a constant term, so its derivative is zero.
$3[x]y$: y is multiplied by $[3x]$, which is considered constant, so the power rule applies.
5: This is a constant term so its derivative is zero.

Step 2: Differentiate partially w.r.t. y

$$z = 2[x^2] + 3[x]y + 5$$
$$\frac{\partial z}{\partial y} = 0 + 3x(1) + 0 \quad \text{differentiating partially w.r.t. } y$$
$$\frac{\partial z}{\partial y} = 3x$$

(b) $Q = 10L^{0.7}K^{0.3}$

Find $\partial Q/\partial L$, differentiate partially w.r.t. L	Find $\partial Q/\partial K$, differentiate partially w.r.t. K
Step 1: To differentiate Q partially w.r.t. L, treat K as a constant. Since there is only one term in Q, we are differentiating $L^{0.7}$ multiplied by a constant. Apply power rule.	**Step 1:** This time, L is treated as a constant, so we are differentiating a constant multiplied by $K^{0.3}$. The power rule applies.
Step 2: Differentiate partially w.r.t. L	**Step 2:** Differentiate partially w.r.t. K

$$Q = 10[K^{0.3}]L^{0.7}$$
$$\frac{\partial Q}{\partial L} = 10K^{0.3}(0.7L^{0.7-1})$$
$$= 7L^{-0.3}K^{0.3}$$

$$Q = 10[L^{0.7}]K^{0.3}$$
$$\frac{\partial Q}{\partial K} = 10L^{0.7}(0.3K^{0.3-1})$$
$$= 3L^{0.7}K^{-0.7}$$

(c) $U = x^2y^5$. This problem is very similar to part (b).

Note: This is an equation of the utility function written in the form $U = f(x,y)$. U is the level of utility or satisfaction derived from consuming goods X and Y.

Find $\partial U/\partial x$, differentiate partially w.r.t. x	Find $\partial U/\partial y$, differentiate partially w.r.t. y
Step 1: Treat y as a constant. Since there is only one term in U, we are differentiating x^2 multiplied by a constant. Apply power rule.	**Step 1:** This time x is treated as a constant so we are differentiating a constant multiplied by y^5. Power rule applies.
Step 2: Differentiate partially w.r.t. x	**Step 2:** Differentiate partially w.r.t. y

$$U = [y^5]x^2$$
$$\frac{\partial U}{\partial y} = y^5(2x)$$
$$= 2xy^5$$

$$U = [x^2]y^5$$
$$\frac{\partial U}{\partial y} = x^2(5y^4)$$
$$= 5x^2y^4$$

PROGRESS EXERCISES 7.1

First-order Partial Differentiation and Isoquants

Find the first-order partial derivatives of the functions in questions 1 to 11:

1. $z = x^2 + 2xy$
2. $z = 5x^2y^2$
3. $Q = \ln x + 3\ln y$
4. $Q = 10L^{0.8}K^{0.2}$
5. $P = 120e^{0.8t}$
6. $U = 4x^2y^2$
7. $z = x^2(1 + 2y)$
8. $z = 5xy + \dfrac{y}{x}$
9. $Q = \ln 5x + 3x\ln y$
10. $Q = 10L^{0.8}K^{0.2} + 30 - 6L - 4K$

11. $C = 120(1 + e^{-0.8Q})$.

12. Find Z_L, Z_K, Z_λ, where $Z = L^{0.5}K^{0.5} - \lambda(50 - 2L - 3K)$.

13. If $Q = L^{0.5}K^{0.5}$, show that

$$Q = L\frac{\partial Q}{\partial L} + K\frac{\partial Q}{\partial K}$$

14. A production function is given by the equation $Q = 2LK$.
    (a) When production is fixed at $Q = 400$, write down the equation of the corresponding isoquant in the form $K = f(L)$.
    (b) Calculate the values of K when the number of units of labour, $L = 5$; 10; 15; 20 on the isoquant given in (a).
    Confirm that each of these points lies on the isoquant $Q = 400$.
    (c) Graph the isoquant, $Q = 400$.

15. Repeat question 14 for the production function $Q = 2LK^{0.5}$ and production, $Q = 20$.

7.1.3   Second-order partial derivatives

To find the second-order partial derivatives, simply differentiate the first-order partial derivatives again (the same idea as in Chapter 6). The only problem is that there are several possibilities. If there are two first-order partial derivatives, $\partial z/\partial x$, $\partial z/\partial y$, each first-order partial derivative could be differentiated w.r.t. the same variable again, giving two straight second-order partial derivatives:

$$\frac{\partial}{\partial x}\left(\frac{\partial z}{\partial x}\right) = \frac{\partial^2 z}{\partial x^2}, \quad \frac{\partial}{\partial y}\left(\frac{\partial z}{\partial y}\right) = \frac{\partial^2 z}{\partial y^2}$$

Furthermore, each first-order partial derivative could also be differentiated w.r.t. the opposite variable giving two mixed second-order partial derivatives:

$$\frac{\partial}{\partial x}\left(\frac{\partial z}{\partial y}\right) = \frac{\partial^2 z}{\partial x \partial y}, \quad \frac{\partial}{\partial y}\left(\frac{\partial z}{\partial x}\right) = \frac{\partial^2 z}{\partial y \partial x}$$

However, these two mixed second-order partial derivatives are equal for most functions met in practice; that is,

$$\frac{\partial^2 z}{\partial x \partial y} = \frac{\partial^2 z}{\partial y \partial x}$$

as is demonstrated in the following worked examples.

> The **straight second-order partial derivatives** are calculated by differentiating the first-order partial derivatives again with respect to the same variable.
>
> The **mixed second-order partial derivatives** are calculated by differentiating the first-order partial derivatives again with respect to the other variable.

The actual differentiation follows the same basic rules given in Chapter 6.

Abbreviations: The second-order partial derivatives may be abbreviated as

$$\frac{\partial^2 z}{\partial x^2} \equiv z_{xx}, \quad \frac{\partial^2 z}{\partial y^2} \equiv z_{yy}, \quad \frac{\partial^2 z}{\partial x \partial y} \equiv z_{xy}, \quad \frac{\partial^2 z}{\partial y \partial x} \equiv z_{yx}$$

WORKED EXAMPLE 7.4
DETERMINING SECOND-ORDER PARTIAL DERIVATIVES

(a) Find the second-order partial derivatives for the following functions:
   (i) $z = 2x^2 + 3xy + 5$
   (ii) $Q = 10L^{0.7}K^{0.3}$
   (iii) $U = x^2 y^5$
(b) Show that the mixed second-order partial derivatives are equal.

SOLUTION

To calculate the second-order derivatives (partial or otherwise), find the first-order derivatives, and simply differentiate again. The first-order derivatives for the above functions were calculated in Worked Example 7.3.

(a) (i) Find the straight second-order partial derivatives of $z = 2x^2 + 3xy + 5$

To find $\partial^2 z/\partial x^2$:	To find $\partial^2 z/\partial y^2$:
1. Determine the first-order partial derivative; this has already been done in Worked Example 7.3:	**1.** Determine the first-order partial derivative; this has already been done in Worked Example 7.3:
$$\partial z/\partial x = 4x + 3y$$	$$\partial z/\partial y = 3x$$
2. Differentiate partially w.r.t. x again.	**2.** Differentiate partially w.r.t. y again.
Step 1: The first term is x, the power rule applies. The second term is 3[y], considered a 'constant' term, so its partial derivative w.r.t. x is zero.	**Step 1:** The only term is 3[x], considered a 'constant' term, so its derivative w.r.t. y is zero.

Step 2: Differentiate partially w.r.t. x

$$\frac{\partial z}{\partial x} = 4x + 3[y]$$

$$\frac{\partial^2 z}{\partial x^2} = 4(1) + 0 = 4$$

Step 2: Differentiate partially w.r.t. y

$$\frac{\partial z}{\partial y} = 3[x]$$

$$\frac{\partial^2 z}{\partial y^2} = 0$$

Now find the mixed second-order partial derivatives, $\partial^2 z/\partial y \partial x$, $\partial^2 z/\partial x \partial y$.

To find $\partial^2 z/\partial y \partial x$

1. $\partial z/\partial x = 4x + 3y$, Worked Example 7.3
2. Now differentiate partially w.r.t. y.

Step 1: The second term is y, the power rule applies. The first term is $4[x]$, considered a 'constant' term, so its partial derivative w.r.t. y is zero.

Step 2: Differentiate partially w.r.t. y

$$\frac{\partial z}{\partial x} = 4[x] + 3y$$

$$\frac{\partial^2 z}{\partial y \partial x} = 0 + 3(1) = 3$$

To find $\partial^2 z/\partial x \partial y$

1. $\partial z/\partial y = 3x$, Worked Example 7.3
2. Now differentiate partially w.r.t. x.

Step 1: The only term is $3x$, so differentiating partially w.r.t. x is the same as ordinary differentiation w.r.t. x.

Step 2: Differentiate partially w.r.t. x

$$\frac{\partial z}{\partial y} = 3x$$

$$\frac{\partial^2 z}{\partial x \partial y} = 3(1) = 3$$

(ii) Find the straight second-order partial derivatives of $Q = 10L^{0.7} K^{0.3}$.

To find $\partial^2 Q/\partial L^2$

1. Determine the first-order partial derivative; this has been done in Worked Example 7.3:

$$\partial Q/\partial L = 7L^{-0.3}K^{0.3}$$

2. Differentiate partially w.r.t. L again.

Step 1: The only term is $7L^{-0.3}K^{0.3}$ with $7[K^{0.3}]$ as a 'constant', multiplied by $L^{-0.3}$. Use the power rule.

Step 2: Differentiate partially w.r.t. L.

$$\frac{\partial Q}{\partial L} = 7L^{-0.3}[K^{0.3}]$$

$$\frac{\partial^2 Q}{\partial L^2} = 7(-0.3L^{-0.3-1})K^{0.3}$$
$$= -2.1L^{-1.3}K^{0.3}$$

To find $\partial^2 Q/\partial K^2$

1. Determine the first-order partial derivative; this has been done in Worked Example 7.3:

$$\partial Q/\partial K = 3L^{0.7}K^{-0.7}$$

2. Differentiate partially w.r.t. K again.

Step 1: The only term is $3L^{0.7} K^{-0.7}$ with $3[L^{0.7}]$ as a 'constant', multiplied by $K^{-0.7}$. Use the power rule.

Step 2: Differentiate partially w.r.t. K.

$$\frac{\partial Q}{\partial K} = 3[L^{0.7}]K^{-0.7}$$

$$\frac{\partial^2 Q}{\partial K^2} = 3L^{0.7}(-0.7K^{-0.7-1})$$
$$= -2.1L^{0.7}K^{-1.7}$$

Now find the mixed second-order partial derivatives, $\partial^2 Q/\partial K \partial L$, $\partial^2 Q/\partial L \partial K$.

$$\frac{\partial^2 Q}{\partial K \partial L} \qquad\qquad \frac{\partial^2 Q}{\partial L \partial K}$$

The first-order partial derivative w.r.t. L is | The first-order partial derivative w.r.t. K is

$$\frac{\partial Q}{\partial L} = 7L^{-0.3}K^{0.3} \qquad\qquad \frac{\partial Q}{\partial K} = 3L^{0.7}K^{-0.7}$$

Now differentiate partially w.r.t. K. | Now differentiate partially w.r.t. L.

Step 1: The only term is $7L^{-0.3}K^{0.3}$ with $7[L^{-0.3}]$ as a 'constant', multiplied by $K^{0.3}$. Use the power rule. | **Step 1:** The only term is $3L^{0.7}K^{-0.7}$ with $3[K^{-0.7}]$ as a 'constant', multiplied by $L^{0.7}$. Use the power rule.

Step 2: Differentiate partially w.r.t. K | **Step 2:** Differentiate partially w.r.t. L

$$\frac{\partial Q}{\partial L} = 7[L^{-0.3}]K^{0.3} \qquad\qquad \frac{\partial Q}{\partial K} = 3L^{0.7}K^{-0.7}$$

$$\frac{\partial^2 Q}{\partial K \partial L} = 7L^{-0.3}(0.3K^{0.3-1}) \qquad \frac{\partial^2 Q}{\partial L \partial K} = 3(0.7L^{0.7-1})K^{-0.7}$$

$$= 2.1L^{-0.3}K^{-0.7} \qquad\qquad\qquad = 2.1L^{-0.3}K^{-0.7}$$

(iii) The straight second-order partial derivatives for $U = x^2 y^5$ are outlined briefly, the comments being left to the reader.

$$\frac{\partial^2 U}{\partial x^2} \qquad\qquad\qquad \frac{\partial^2 U}{\partial y^2}$$

The first-order partial derivative w.r.t. x is | The first-order partial derivative w.r.t. y is

$$\frac{\partial U}{\partial x} = 2xy^5 \qquad\qquad \frac{\partial U}{\partial y} = 5x^2 y^4$$

Step 1: Which rule is required for differentiating partially w.r.t. x? | **Step 1:** Which rule is required for differentiating partially w.r.t. y?

Step 2: Differentiate partially w.r.t. x: | **Step 2:** Differentiate partially w.r.t. y:

$$\frac{\partial U}{\partial x} = 2x[y^5] \qquad\qquad \frac{\partial U}{\partial y} = 5[x^2]y^4$$

$$\frac{\partial^2 U}{\partial x^2} = 2(1)y^5 = 2y^5 \qquad \frac{\partial^2 U}{\partial y^2} = 5x^2(4y^3) = 20x^2 y^3$$

Now, find the mixed second-order partial derivatives $\partial^2 U/\partial y \partial x$, $\partial^2 U/\partial x \partial y$.

$$\frac{\partial U}{\partial x} = 2xy^5 \qquad\qquad \frac{\partial U}{\partial y} = 5x^2 y^4$$

Step 1: Which rule is required for differentiating partially w.r.t. y? | **Step 1:** Which rule is required for differentiating partially w.r.t. x?

Step 2: Differentiate partially w.r.t. y | **Step 2:** Differentiate partially w.r.t. x

$$\frac{\partial U}{\partial x} = 2[x]y^5 \qquad\qquad \frac{\partial U}{\partial y} = 5x^2[y^4]$$

$$\frac{\partial^2 U}{\partial y \partial x} = 2x(5y^4) \qquad\qquad \frac{\partial^2 U}{\partial x \partial y} = 5(2x)y^4$$

$$= 10xy^4 \qquad\qquad\qquad = 10xy^4$$

(b) In (i), (ii) and (iii) above it is clear that each pair of mixed second-order partial derivatives are equivalent. This is true for all the functions which are used in this text.

PROGRESS EXERCISES 7.2

Second-order Partial Differentiation

Find the second-order partial derivatives for questions 1 to 12 from Progress Exercise 7.1, which are reproduced below.

1. $z = x^2 + 2xy$ 2. $z = 5x^2y^3$

3. $Q = \ln x + 3\ln y$ 4. $Q = 10L^{0.8}K^{0.2}$

5. $P = 120e^{0.8t}$ 6. $U = 4x^2y^2$

7. $z = x^2(1 + 2y)$ 8. $z = 5xy + \dfrac{y}{x}$

9. $Q = \ln 5x + 3x \ln y$ 10. $Q = 10L^{0.8}K^{0.2} + 30 - 6L - 4K$

11. $C = 120(1 + e^{-0.8Q})$ 12. $Z = L^{0.5}K^{0.5} - \lambda(50 - 2L - 3K)$

7.1.4 Differentials and small changes (incremental changes)

The rate of change of y w.r.t. x (where $y = f(x)$) has already been defined many times as $\Delta y/\Delta x$

$$\frac{\Delta y}{\Delta x} = \frac{\Delta y}{\Delta x} \rightarrow \Delta y = \left(\frac{\Delta y}{\Delta x}\right)\Delta x$$

and, as Δx becomes infinitesimal, these small changes, Δx, Δy, are written as 'dx' and 'dy' respectively, so the differential of y may be written as

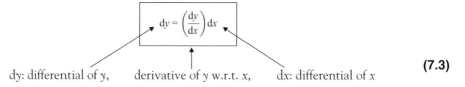

$$dy = \left(\frac{dy}{dx}\right)dx$$

where   dy: differential of y,   derivative of y w.r.t. x,   dx: differential of x   **(7.3)**

Incremental changes

Incremental changes are small changes in the dependent variable (y), which result from changes in the independent variable(s). The formula for incremental changes is derived from equation (7.3), where the differentials dx and dy are replaced by Δx and Δy, for small changes, but not infinitesimally small changes in x and y:

$$\Delta y \cong \left(\frac{dy}{dx}\right)\Delta x, \qquad\qquad \textbf{(7.4)}$$

This formula, often referred to as 'the small changes' formula or 'incremental changes' formula, which gives the approximate change in y as a result of a small change in x, is illustrated graphically in Figure 7.4.

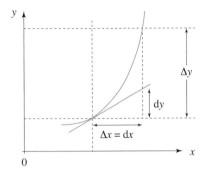

Figure 7.4 Differentials and small changes

WORKED EXAMPLE 7.5
DIFFERENTIALS FOR FUNCTIONS OF ONE VARIABLE

A supply function is given by the equation $P = Q^2$.

(a) Find the derivative of P.
(b) Find the differential of P.
(c) If Q is increased by 3%, use differentials to find the approximate change in P.

SOLUTION

(a) The derivative of P simply involves ordinary differentiation as studied in Chapter 6:

$$P = Q^2$$
$$\frac{dP}{dQ} = 2Q$$

(b) Using equation (7.3), where $y \to P$ and $x \to Q$, the differential of P is defined as

$$dP = \left(\frac{dP}{dQ}\right) dQ$$
$$dP = (2Q)dQ$$

(c) The increase in Q is written as

$$\Delta Q = \frac{3}{100} Q$$

Note: 3% of Q is never just 3/100, it must be 3% of Q. If you are given a 3% increase in pay, would you be satisfied with a rise of 3/100 or would you prefer to get a rise of 3/100 × present pay?

Applying equation (7.4),

$$\Delta P \cong \left(\frac{dP}{dQ} \right) \Delta Q$$

$$\Delta P \cong 2Q \left(\frac{3Q}{100} \right) \cong \frac{2Q}{1} \frac{3Q}{100} \cong \frac{6}{100} Q^2 \cong \frac{6}{100} P$$

So, a 3% increase in Q will result in a 6% increase in P, approximately.

Differentials for functions of two variables

The argument for the differential of a function of one variable may be extended to functions of two variables, so the total differential of $z = f(x,y)$, a function of two variables, is written as

$$dz = \left(\frac{\partial f}{\partial x} \right) dx + \left(\frac{\partial f}{\partial y} \right) dy \qquad \textbf{(7.5)}$$

or for small changes (incremental changes),

$$\Delta z = \left(\frac{\partial f}{\partial x} \right) \Delta x + \left(\frac{\partial f}{\partial y} \right) \Delta y \qquad \textbf{(7.6)}$$

WORKED EXAMPLE 7.6a
DIFFERENTIALS AND INCREMENTAL CHANGES

A company's revenue is given by the equation

$$TR = 5W^2 A^3$$

where W is the payment on wages for sales people and A is the amount of advertising expenditure.

(a) Find the differential of TR.
(b) Use differentials to find the approximate percentage change in TR if spending on:
    (i) Wages is increased by 5%, no change in advertising expenditure.
    (ii) Wages is increased by 5% and advertising expenditure is decreased by 2%.

SOLUTION

(a) Applying equation (7.5), the differential of TR is calculated as

$$dz = \left(\frac{\partial f}{\partial x}\right) dx + \left(\frac{\partial f}{\partial y}\right) dy$$

$$d(TR) = \left(\frac{\partial TR}{\partial W}\right) dW + \left(\frac{\partial TR}{\partial A}\right) dA \quad \text{where} \quad \frac{\partial TR}{\partial W} = 5(2W)A^3; \frac{\partial TR}{\partial A} = 5W^2(3A^2)$$

$$= 5(2W)A^3 \, dW + 5W^2(3A^2) \, dA = 10WA^3 \, dW + 15W^2A^2 \, dA$$

(b) (i) Wages increase by 5%, then $\Delta W = 5W/100$. The change in advertising expenditure is zero, so $\Delta A = 0$. Substitute ΔW, ΔA and the partial derivatives $\partial TR/\partial W$ and $\partial TR/\partial A$ into equation (7.6) for incremental changes as follows:

$$\Delta(TR) \cong \left(\frac{\partial TR}{\partial W}\right)\Delta W + \left(\frac{\partial TR}{\partial A}\right)\Delta A$$

$$\cong 5(2W)A^3 \, \Delta W + 5W^2(3A^2) \, \Delta A$$

$$\cong \frac{5(2WA^3)}{1} \times \frac{5W}{100} + 0 \qquad \text{since } \Delta A = 0$$

$$\cong \frac{5W^2A^3}{1} \frac{10}{100} \qquad \text{note: the numbers are not multiplied out, but } 5W^2A^3 = TR \text{ is}$$
$$\text{factored out since we require the change as a percentage of } TR$$

$$\cong \frac{TR}{1} \frac{10}{100}$$

So TR has increased by 10%.

(b) (ii) The change in W is $\Delta W = \dfrac{5W}{100}$ and the change in A is $\Delta A = -\dfrac{2A}{100}$. Substitute these into equation (7.6) for small changes,

$$\Delta(TR) \cong \left(\frac{\partial TR}{\partial W}\right)\Delta W + \left(\frac{\partial TR}{\partial A}\right)\Delta A$$

$$\cong 5(2W)A^3 \, \Delta W + 5W^2(3A^2)\Delta A$$

$$\cong \frac{5(2WA^3)}{1} \times \frac{5W}{100} + \frac{15W^2A^2}{1} \times \frac{-2A}{100}$$

$$\cong \frac{5W^2A^3}{1} \frac{10}{100} - \frac{5W^2A^3}{1} \frac{6}{100} \qquad \text{factor out } 5W^2A^3 = TR, \text{ since we require}$$
$$\text{the answer as a percentage change in } TR$$

$$\cong \frac{TR}{1} \frac{10}{100} - \frac{TR}{1} \frac{6}{100}$$

$$\cong TR\left(\frac{10}{100} - \frac{6}{100}\right) \cong TR\frac{4}{100}$$

TR has increased by approximately 4%.

WORKED EXAMPLE 7.6b

INCREMENTAL CHANGES

The Cobb–Douglas production function $Q = 10L^{0.7}K^{0.3}$, where Q is the quantity produced, L is the number of units of labour and K is the units of capital.

Calculate the approximate percentage change in Q if L is increased by 5% and K decreased by 3%.

SOLUTION

$Q = 10L^{0.7}K^{0.3}$

The small changes formula (7.6) is $\Delta Q \cong \frac{\partial Q}{\partial L}\Delta L + \frac{\partial Q}{\partial K}\Delta K$

Then $\Delta L = \frac{5L}{100}; \Delta K = -\frac{2K}{100}$.

The first-order derivatives are $\frac{\partial Q}{\partial L} = 7L^{-0.3}K^{0.3}$ and $\frac{\partial Q}{\partial K} = 3L^{0.7}K^{-0.7}$ (see Worked Example 7.3(b)).

Substituting ΔL and ΔK and the first-order derivatives into the incremental changes formula (7.6) we have

$$\Delta Q \cong \frac{\partial Q}{\partial L}\Delta L + \frac{\partial Q}{\partial K}\Delta K = \frac{10\left(0.7L^{-0.3}\right)K^{0.3}}{1}\frac{5L}{100} + \frac{10L^{0.7}\left(0.3K^{-0.7}\right)}{1}\left(\frac{-2K}{100}\right)$$

$$= \frac{10L^{0.7}K^{0.3}}{1}\left(\frac{(0.7)5}{100}\right) + \frac{10L^{0.7}K^{0.3}}{1}\left(\frac{(0.3)(-2)}{100}\right)$$

factoring out the original formula for Q $(Q = 10L^{0.7}K^{0.3})$

$$= \frac{10L^{0.7}K^{0.3}}{1}\left(\frac{(0.7)5}{100} - \frac{(0.3)(2)}{100}\right)$$

$$= \frac{Q}{1}\left(\frac{3.5}{100} - \frac{0.6}{100}\right)$$

$$= \frac{Q}{1}\left(\frac{2.9}{100}\right)$$

Hence $\Delta Q \cong Q\frac{2.9}{100}$; production increases by approximately 2.9%.

Summary

- **Functions of one variable:** $y = f(x)$

  The **total derivative** is dy/dx.

  The **differential** of y is

$$dy = \left(\frac{dy}{dx}\right)dx$$

where   dy: differential of y,   derivative of y w.r.t. x,   dx: differential of x
and dy and dx are infinitesimally small.

- **Functions of two variables:** $z = f(x, y)$

  The first-order partial derivatives are:

  $\partial z / \partial x$ is the partial derivative of z w.r.t. x, y treated as a constant (abbreviated to z_x).

  $\partial z / \partial y$ is the partial derivative of z w.r.t. y, x treated as a constant (abbreviated to z_y).

  The second-order partial derivatives are:

  $$\frac{\partial^2 z}{\partial x^2} \equiv z_{xx}, \quad \frac{\partial^2 z}{\partial y^2} \equiv z_{yy}, \quad \frac{\partial^2 z}{\partial y \partial x} \equiv z_{yx}, \quad \frac{\partial^2 z}{\partial x \partial y} \equiv z_{xy}$$

  The total differential of z is given by

  $$dz = \left(\frac{\partial f}{\partial x} \right) dx + \left(\frac{\partial f}{\partial y} \right) dy$$

  or for small changes $dx \to \Delta x$, etc., the **small (incremental) changes formula** is given by

  $$\Delta z \cong \left(\frac{\partial f}{\partial x} \right) \Delta x + \left(\frac{\partial f}{\partial y} \right) \Delta y$$

PROGRESS EXERCISES 7.3

Differentials and Small Changes

Find the differential, dz, for questions 1–5:

1. $z = x^2$
2. $z = x^2 + y^2$
3. $z = 3e^x + y$
4. $z = 10L^{0.2}K^{0.6}$
5. $z = \ln x + 5 \ln y$

In questions 6–11, find the approximate percentage change which results from the indicated changes in the other variables.

6. $Q = L^{0.5}K^{0.5}$, L and K both increase by 3%.
7. $U = 5xy$, x and y both increase by 2%.
8. $TR = 50P^{0.6}A^{1.2}$, P decreases by 3% and A increases by 2%.
9. $TR = 125P^{0.5}A^{1.5}Q^{0.2}$, P decreases by 3%, A increases by 2%, Q is unchanged.
10. $TR = 125P^{0.5}A^{1.5}Q^{0.2}$, P, A and Q increase by 2%.
11. $Q = 200 - 2P + 0.02Y + 0.3P_s$, Y and P_s increase by 3%, P does not change, where P_s is the price of substitute products.
12. If $z = e^{x+y}$ show that $z = 0.5(z_x + z_y)$.
13. Given the production function $Q = 15L^{\frac{1}{3}}K^{\frac{2}{3}}$
    (a) Show that $L\frac{\partial Q}{\partial L} + K\frac{\partial Q}{\partial K} = Q$.

    (b) If L is increased by 3% and K is increased by 6%, calculate the approximate percentage change in Q using partial derivatives. Why is this result only approximate?
14. (a) Determine the first and second partial derivatives of

    $$z = x^2 + 5xy + y^2$$

(b) A firm's profit (π) is related to the number of sales people (S) and the price of the product (P) according to the equation $\pi = 150S^{1.2}P^{-1.4}$.

Use partial derivatives to calculate the percentage advertising expenditure when the number of sales people is reduced by 5% and the price is increases by 7%.

15. A utility function is given by the equation $U = 250x^{0.3}y^{0.7}$
   (a) Calculate all the first- and second-order derivatives of U.
   (b) Calculate the percentage change in U when x decreases by 5% and y increases by 3%.

16. The utility function is given by the equation $U = 20Q_1^{\frac{4}{5}}Q_2^{\frac{1}{5}}$ where Q_1 and Q_2 are the number of hours spent at work and sleeping per day.

Find the approximate percentage change in U if hours at work (Q_1) is increased by 5% and hours sleeping (Q_2) is decreased by 3.8%.

7.2 Applications of Partial Differentiation

In this section partial differentiation is applied to:

- Production functions, Cobb–Douglas production function in general form
- Returns to scale
- Utility functions
- Partial elasticities
- National income multipliers.

We shall sketch each function, find marginal functions, rates of change and elasticities.

7.2.1 Production functions

At the end of this section you should be able to:

- Derive expressions for the marginal functions, MP_L and MP_K.
- Outline the relationship between marginal and average functions.
- Derive the equation of and sketch the graph of an isoquant.
- Derive the equation for the slope of an isoquant.

In the previous section the general production function described output as a function of labour (L) and capital (K), that is,

$$Q = f(L, K)$$

A production function that is widely used in economic analysis is the Cobb–Douglas production function which is expressed in general form as

$$Q = AL^{\alpha}K^{\beta} \qquad\qquad (7.7)$$

where A is a constant and $0 < \alpha < 1, 0 < \beta < 1, L > 0, K > 0$. For example, a Cobb–Douglas production function in specific form could be written as

$$Q = 50L^{0.4}K^{0.6}$$

Table 7.4   Summary of marginal functions for $Q = AL^\alpha K^\beta$: $0 < \alpha < 1, 0 < \beta < 1$

Name	Partial derivative	Interpretation
Marginal product of labour	$Q_L = MP_L$	$MP_L > 0$
	$MP_L = \dfrac{\partial Q}{\partial L} = A\alpha L^{\alpha-1} K^\beta$	With capital held constant, Q increases as labour input, L, increases.
	$Q_{LL} = \dfrac{\partial(MP_L)}{\partial L}$	$Q_{LL} < 0$
	$\dfrac{\partial^2 Q}{\partial L^2} = (\alpha - 1)\dfrac{A\alpha L^\alpha K^\beta}{L^2}$ $= (\alpha - 1)\dfrac{\alpha Q}{L^2}$	With capital held constant, Q increases, but at a decreasing rate; therefore MP_L decreases (its rate of change is negative) as labour input, L, increases.
	Since $Q_{LL} < 0 \rightarrow$ curve concave to origin	**Law of diminishing returns to labour**
Marginal product of capital	$Q_K = MP_K$	$MP_K > 0$
	$MP_K = \dfrac{\partial Q}{\partial K} = A\beta L^\alpha K^{\beta-1}$	With labour held constant, Q increases as capital input, K, increases.
	$Q_{KK} = \dfrac{\partial(MP_K)}{\partial K} =$ $\dfrac{\partial^2 Q}{\partial K^2} = A\beta L^\alpha(\beta - 1)K^{\beta-2}$ $= (\beta - 1)\dfrac{\beta Q}{K^2}$	$Q_{KK} < 0$ With labour held constant, Q increases but at a decreasing rate; therefore MP_K decreases (its rate of change is negative) as capital input, K, increases.
	Since $Q_{KK} < 0 \rightarrow$ curve concave to origin.	**Law of diminishing returns to capital**
	$Q_{KL} = Q_{LK} =$ $\dfrac{\partial^2 Q}{\partial K \partial L} = A\alpha\beta L^{\alpha-1} K^{\beta-1}$	$Q_{KL} > 0$ and $Q_{LK} > 0$ MP_L increases (its rate of change is positive) as capital input, K, increases and MP_K increases as L increases.

Marginal functions in general

The partial derivatives for the Cobb–Douglas production function are summarised in Table 7.4. The first derivatives are the marginal functions, referred to as marginal products. The second derivatives (the rate of change of the first derivatives) indicate whether the marginal product (MP_L, MP_K) is increasing or decreasing.

In Section 6.5, the law of diminishing returns to labour was outlined. The law of diminishing returns to capital reflects the change in output resulting from a change in capital, keeping labour constant.

A Cobb–Douglas production function exhibits diminishing returns to each factor. This is confirmed by the negative second derivatives:

$$\text{Law of diminishing returns to labour:} \quad Q_{LL} < 0$$
$$\text{Law of diminishing returns to capital:} \quad Q_{KK} < 0$$

Remember

Negative second derivatives also indicate that the curves are concave towards the origin.

WORKED EXAMPLE 7.7

MP_L AND MP_K: INCREASING OR DECREASING?

Find the first and second derivatives for the production function, $Q = 10L^{0.5}K^{0.5}$. Use the second-order partial derivatives to determine whether the marginal functions (marginal products) are increasing or decreasing.

SOLUTION

The second derivatives for the production function are calculated by differentiating the first derivatives again, as follows:

Q_{KK}	Q_{LL}	Q_{LK}
$Q = 10L^{0.5}K^{0.5}$	$Q = 10L^{0.5}K^{0.5}$	$Q = 10L^{0.5}K^{0.5}$
$Q_K = 5L^{0.5}K^{-0.5}$	$Q_L = 5L^{-0.5}K^{0.5}$	$Q_K = 5L^{0.5}K^{-0.5}$
$Q_{KK} =$ $5(-0.5)L^{0.5}K^{-0.5-1}$	$Q_{LL} =$ $5(-0.5)L^{-0.5-1}K^{0.5}$	$Q_{LK} =$ $5(0.5)L^{0.5-1}K^{-0.5}$
$= -2.5L^{0.5}K^{-1.5}$	$= -2.5L^{-1.5}K^{0.5}$	$= 2.5L^{-0.5}K^{-0.5}$
Negative: MP_K (or Q_K) decreases as K input increases. Law of diminishing returns to capital (short-run law of production)	Negative: MP_L (or Q_L) decreases as L input increases. Law of diminishing returns to labour (short-run law of production)	Positive: MP_K increases as L input increases
Note: MP_K is positive.	**Note:** MP_L is positive.	

The relationship between marginal and average functions

In Chapter 6, the APL (average product of labour) was defined as total output divided by the number of units of labour employed. Similarly, the APK (the average product of capital) is defined as total output divided by the number of units of capital. The relationships between the APL and the MP_L and between the APK and the MP_K are summarised in Table 7.5 for the Cobb–Douglas production function, $Q = AL^{\alpha}K^{\beta}$.

Production conditions

Normally, a producer would desire productivity to increase as the amount of each input increases, indicated by positive marginal functions. However, the rate of increase usually slows down as the amounts of inputs become progressively larger, indicated by negative second derivatives.

Table 7.5   Average and marginal functions for the production function, $Q = AL^\alpha K^\beta$

	Average function	Marginal function	Comment
Labour	$Q \quad = AL^\alpha K^\beta$	$Q \quad = AL^\alpha K^\beta$	$MP_L = \alpha(AL^{\alpha-1}K^\beta)$
	$APL = \dfrac{Q}{L}$	$MP_L = \dfrac{\partial Q}{\partial L}$	$= \alpha(APL)$
	$= \dfrac{AL^\alpha K^\beta}{L}$	$= A\alpha L^{\alpha-1}K^\beta$	$\Rightarrow MP_L < APL$
	$= AL^{\alpha-1}K^\beta$		since $0 < \alpha < 1$
Capital	$APK = \dfrac{Q}{K}$	$MP_K = \dfrac{\partial Q}{\partial K}$	$MP_K = \beta(AL^\alpha K^{\beta-1})$
	$= \dfrac{AL^\alpha K^\beta}{K}$	$= A\beta L^\alpha K^{\beta-1}$	$= \beta(APK)$
	$= AL^\alpha K^{\beta-1}$		$\Rightarrow MP_K < APK$
			since $0 < \beta < 1$

In practice, a firm produces output over a certain range of the production function as outlined by the following production conditions.

Conditions for using labour

$MP_L \quad = \dfrac{\partial Q}{\partial L}$ is positive

$\dfrac{d(MP_L)}{dL} = \dfrac{\partial^2 Q}{\partial L^2} < 0$

$MP_L \quad < APL$

Conditions for using capital

$MP_K \quad = \dfrac{\partial Q}{\partial K}$ is positive

$\dfrac{d(MP_K)}{dK} = \dfrac{\partial^2 Q}{\partial K^2} < 0$

$MP_K \quad < APK$

Graphical representation of production functions: isoquants

In Section 7.1 a production function $Q = f(L, K)$ was graphically represented by a two-dimensional graph known as an isoquant, $K = f(L)$. To plot the isoquant as a two-dimensional graph, the quantity was fixed at a constant value, and then K was expressed in terms of L. The isoquant gives all the combinations of L and K for which Q (production) has the same fixed value. The graphs of four isoquants were plotted in Figures 7.2 and 7.3.

The slope of an isoquant (MRTS)

The slope dK/dL may be derived directly from the equation of the isoquant, as in Worked Example 7.1. This slope is called the **marginal rate of technical substitution** (MRTS).

At any point $(L = L_0, K = K_0)$ on an isoquant, the value of the slope is a measure of the decrease in capital for each unit increase in labour, that is, the number of units of capital which would be replaced (substituted) when labour increases by one unit, while still maintaining the same output, Q.

$$\left.\dfrac{dK}{dL}\right|_{L_0 K_0} \quad \text{is a short way of denoting the value of the slope at } (L_0, K_0).$$

The isoquant in Figure 7.5 exhibits a **diminishing marginal rate of technical substitution**, that is, $\Delta K/\Delta L$, or the rate at which the amount of capital decreases for each unit increase in labour gets smaller (diminishes) as L increases.

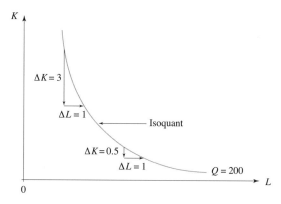

Figure 7.5   MRTS = slope of isoquant, slope is diminishing

 Remember

Do not mix up the marginal products: MP_L, MP_K and the $MRTS = dK/dL$.

The slope of an isoquant is the ratio of the marginal products

The slope of an isoquant may be expressed in terms of the marginal products of labour and capital by means of the total differential

$$dQ = \left(\frac{\partial Q}{\partial L}\right) dL + \left(\frac{\partial Q}{\partial K}\right) dK \qquad (7.8)$$

Along any isoquant, $dQ = 0$, therefore substituting $dQ = 0$ into equation (7.8), an expression for dK/dL may be written as

$$0 = \left(\frac{\partial Q}{\partial L}\right) dL + \left(\frac{\partial Q}{\partial K}\right) dK$$

$$-\left(\frac{\partial Q}{\partial K}\right) dK = \left(\frac{\partial Q}{\partial L}\right) dL$$

$$-Q_K \, dK = Q_L dL \qquad (7.9)$$

$$\frac{-Q_K \, dK}{dL} = \frac{Q_L}{1}$$

$$\frac{dK}{dL} = -\frac{Q_L}{Q_K}$$

As the partial derivatives Q_L and Q_K are the marginal product of labour and marginal product of capital, respectively, usually denoted by MP_L and MP_K, dK/dL may also be written as

$$\frac{dK}{dL} = -\frac{MP_L}{MP_K} \qquad (7.10)$$

WORKED EXAMPLE 7.8
SLOPE OF AN ISOQUANT IN TERMS OF MP_L, MP_K

A production function is given by the equation $Q = 10L^{0.5}K^{0.5}$.

(a) Graph the isoquants for $Q = 50$ and $Q = 70$.
(b) Derive expressions for the marginal product of labour and the marginal product of capital.
(c) Deduce an expression for the slope of the isoquants, dK/dL. For $Q = 50$, calculate the MRTS at $L = 1$ and give a verbal interpretation of your answer.

SOLUTION

(a) Substitute $Q = 50$ into the equation for the production function and solve for $K = f(L)$.

$$50 = 10L^{0.5}K^{0.5}$$
$$5 = L^{0.5}K^{0.5}$$
$$\frac{5}{L^{0.5}} = K^{0.5}$$
$$\frac{25}{L} = K \quad \text{square each side of the equation}$$
$$\text{or } K = \frac{25}{L}$$

Similarly, the equation of the isoquant for $Q = 70$ is $K = 49/L$.
A series of points are calculated for each isoquant, as shown in Table 7.6. The graphs are plotted in Figure 7.6.

Table 7.6   Points for plotting the isoquants $K = 25/L$ and $K = 49/L$

L	0.10	0.20	0.30	0.40	0.50	0.60	0.70	0.80	0.90	1.00	1.10	1.20	1.30	1.40
$K = 25/L$	250	125	83	63	50	42	36	31	28	25	23	21	19	18
$K = 49/L$	490	245	163	123	98	82	70	61	54	49	45	41	38	35

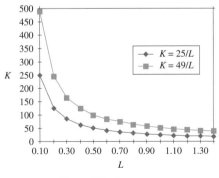

Figure 7.6   Isoquants

(b) The marginal product of labour:

$$MP_L = \frac{\partial Q}{\partial L} = 10(0.5L^{0.5-1})K^{0.5} = 5L^{-0.5}K^{0.5}$$

The marginal product of capital:

$$MP_K = \frac{\partial Q}{\partial K} = 10L^{0.5}(0.5K^{0.5-1}) = 5L^{0.5}K^{-0.5}$$

(c) To derive an expression for dK/dL or the MRTS, start by writing out an expression for the total differential as given in equation (7.8) which is rewritten here:

$$dQ = \left(\frac{\partial Q}{\partial L}\right)dL + \left(\frac{\partial Q}{\partial K}\right)dK$$

As Q is constant along a given isoquant, then $dQ = 0$. Substitute $dQ = 0$ and the partial derivatives MP_L and MP_K into equation (7.8), then solve for dK/dL as follows:

$$0 = (5L^{-0.5}K^{0.5})dL + (5L^{0.5}K^{-0.5})dK$$

$$-(5L^{0.5}K^{-0.5})dK = (5L^{-0.5}K^{0.5})dL$$

$$\frac{-(5L^{0.5}K^{-0.5})dK}{dL} = \frac{(5L^{-0.5}K^{0.5})}{1} \quad \text{divide both sides by dL}$$

$$\frac{dK}{dL} = -\frac{5L^{-0.5}K^{0.5}}{5L^{0.5}K^{-0.5}} \quad \text{divide both sides by } -(5L^{0.5}K^{-0.5})$$

$$\frac{dK}{dL} = -\frac{\cancel{5}L^{-0.5-0.5}K^{0.5-(-0.5)}}{\cancel{5}}$$

$$\frac{dK}{dL} = -\frac{L^{-1}K^{1}}{1} = -\frac{K}{L} \quad \text{after simplifying using the rules for indices}$$

Note: Exactly the same answer is obtained by using equation (7.10) directly. (However, formulae are easily forgotten, so you should always be able to work from basics.)

$$\frac{dK}{dL} = -\frac{MP_L}{MP_K} = -\frac{5L^{-05}K^{0.5}}{5L^{0.5}K^{-0.5}} = -\frac{K}{L}$$

At $L = 1$, the MRTS or

$$\frac{dK}{dL} = -\frac{K}{L} = -\frac{25}{1} = -25$$

This means that the producer is willing to give up 25 units of capital to receive one more unit of labour and still maintain production, $Q = 50$.

The MRTS in reduced form for a Cobb–Douglas production function

Given the general Cobb–Douglas production function, $Q = AL^\alpha K^\beta$, then

$$\text{MRTS:} \frac{dK}{dL} = -\frac{\partial Q/\partial L}{\partial Q/\partial K} = -\frac{MP_L}{MP_K} = -\frac{A\alpha L^{\alpha-1}K^\beta}{A\beta L^\alpha K^{\beta-1}}$$

$$= -\frac{\alpha K^\beta K^{-\beta+1}}{\beta L^\alpha L^{-\alpha+1}} \quad \text{using rules for indices}$$

$$= -\frac{\alpha K}{\beta L}$$

The MRTS in reduced form is expressed as

$$\text{MRTS} = -\frac{\alpha}{\beta}\frac{K}{L} \tag{7.11}$$

PROGRESS EXERCISES 7.4

Partial Differentiation and Production Functions

1. A firm has a production function $Q = 10L^{0.5}K^{0.5}$.
   (a) Derive expressions for MP_L and MP_K.
   (b) Prove that

   $$Q = L\frac{\partial Q}{\partial L} + K\frac{\partial Q}{\partial K}$$

2. Find all the first and second partial derivatives for the production function

   $$Q = 80L^{0.5}K^{0.2}$$

   (a) Give a verbal description of each derivative.
   (b) Are the marginal functions increasing or decreasing? Give reasons for your answer.
   (c) Evaluate MP_L and MP_K when $Q = 80$ and $L = 2, 5, 10$.
   Do these calculations confirm your answer in part (b)?
3. A firm's production function is given by the equation $Q = 80L^{0.5}K^{0.2}$.
   (a) Derive expressions for MP_L and APL. Show that $MP_L < APL$.
   (b) Derive expressions for MP_K and APK. Show that $MP_K < APK$.
4. A firm producing keyboards has a production function $Q = 10LK^{0.5}$ keyboards per hour.
   (a) If production is set at 80 keyboards $(Q = 80)$, write down the equation of the corresponding isoquant in the form $K = f(L)$. Calculate K when $L = 2, 4, 6$. Graph the isoquant.
   (b) Derive expressions for MP_L and MP_K. Evaluate MP_L and MP_K when $L = 2, 4, 6$.
   (c) Use the results in (b) to evaluate $MRTS = dK/dL$ when $L = 2, 4, 6$. Comment on the results.
5. (See question 2. $Q = 80L^{0.5}K^{0.2}$.)
   (a) Derive an expression for $MRTS = dK/dL$ in terms of MP_L and MP_K.

(b) Write down the equation of the isoquant $Q = 320$.
Evaluate the MRTS on the isoquant $Q = 320$ at $L = 4, 9, 16$. Do your calculations confirm the law of diminishing marginal rate of technical substitution?

6. Confirm that

$$MRTS = -\frac{\alpha K}{\beta L}$$

for the production function $Q = 25L^{0.6}K^{0.5}$.

7. A manager proposes that a firm increases capital by 5% and decreases labour by 3%. If the firm's production function is $Q = 80L^{0.5}K^{0.2}$, use partial differentiation to find the approximate change in output, Q.

8. Given the Cobb–Douglas production function $Q = 50L^{0.3}K^{0.5}$, calculate and comment on the values of the APL and the MP_L at $L = 2, 4, 6$ when K remains constant at $K = 10$.

7.2.2 Returns to scale

If both inputs (L, K) in the Cobb–Douglas production function are changed by the same proportion (for example, each input is doubled or each input is reduced by one-fifth, etc.) we can easily determine the proportionate change in output, Q. For any constant, λ, if L is replaced by λL and K is replaced by λK, then the proportionate change in Q may be determined from the equation of the production function.

Given the Cobb–Douglas production function

$$Q_1 = AL^{\alpha}K^{\beta} \tag{7.12}$$

then, replacing L by λL and K by λK in equation (7.12), the new output, Q_2, is given by

$$Q_2 = A(\lambda L)^{\alpha}(\lambda K)^{\beta}$$

$$= A\lambda^{\alpha}L^{\alpha}\lambda^{\beta}K^{\beta}$$

$$= \lambda^{\alpha+\beta}AL^{\alpha}K^{\beta}$$

$$= \lambda^{\alpha+\beta}Q_1 \tag{7.13}$$

- If $\alpha + \beta = 1$, then the proportionate change in output is the same as the proportionate change in each input. This is described as **constant returns to scale**. For example, if $\lambda = 2$, then $Q_2 = 2^1 Q_1 = 2Q_1$.
- If $\alpha + \beta < 1$, then the proportionate change in output is less than the proportionate change in each input. This is described as **decreasing returns to scale**. For example, if $\lambda = 2$, $\alpha + \beta = 0.8$, then $Q_2 = 2^{0.8}Q_1 = 1.7Q_1$.
- If $\alpha + \beta > 1$, then the proportionate change in output is greater than the proportionate change in each input. This is described as **increasing returns to scale**. For example, if $\lambda = 2$, $\alpha + \beta = 1.3$, then $Q_2 = 2^{1.3}Q_1 = 2.5Q_1$.

Note: Returns to scale are known as the long-run laws of production, whereas the law of diminishing returns to labour or capital is known as the short-run law of production.

Homogeneous functions of degree *r*

In general, a Cobb–Douglas production function is described as homogeneous, order r, if

$$f(\lambda L, \lambda K) = \lambda^r f(L, K)$$

where $r = (\alpha + \beta)$.

WORKED EXAMPLE 7.9
CONSTANT, INCREASING AND DECREASING RETURNS TO SCALE

Given the Cobb–Douglas production functions

$$Q = 50L^{0.3}K^{0.5} \tag{7.14}$$
$$Q = 50L^{0.4}K^{0.6} \tag{7.15}$$
$$Q = 50L^{0.5}K^{0.6} \tag{7.16}$$

(i) Calculate the level of output when $L = 10$, $K = 15$.
(ii) Calculate the level of output when both inputs double, $L = 20$, $K = 30$.

Comment on the returns to scale.

SOLUTION
The level of output is calculated by substituting the appropriate values of L and K into each production function. The calculations are:

Production function	(i) $L = 10$, $K = 15$	(ii) $L = 20$, $K = 30$ (inputs doubled)	Comment
$Q = 50L^{0.3}K^{0.5}$	$\begin{aligned} Q_1 &= 50L^{0.3}K^{0.5} \\ &= 50(10)^{0.3}(15)^{0.5} \\ &= 50(2.0)(3.9) \\ &= 390 \end{aligned}$	$\begin{aligned} Q_2 &= 50(L)^{0.3}(K)^{0.5} \\ &= 50(20)^{0.3}(30)^{0.5} \\ &= 50(2.46)(5.48) \\ &= 674 \\ &< 390 \times 2, \text{ i.e.,} \\ &< Q_1 \times 2 \end{aligned}$	Production function exhibits **decreasing returns to scale.** $(\alpha + \beta) < 1$
$Q = 50L^{0.4}K^{0.6}$	$\begin{aligned} Q_1 &= 50L^{0.4}K^{0.6} \\ &= 50(10)^{0.4}(15)^{0.6} \\ &= 50(2.51)(5.08) \\ &= 637 \end{aligned}$	$\begin{aligned} Q_2 &= 50(L)^{0.4}(K)^{0.6} \\ &= 50(20)^{0.4}(30)^{0.6} \\ &= 50(3.31)(7.70) \\ &= 1274 \\ &= 637 \times 2, \text{ i.e.,} \\ &= Q_1 \times 2 \end{aligned}$	Production function exhibits **constant returns to scale.** $(\alpha + \beta) = 1$
$Q = 50L^{0.5}K^{0.6}$	$\begin{aligned} Q_1 &= 50L^{0.5}K^{0.6} \\ &= 50(10)^{0.5}(15)^{0.6} \\ &= 50(3.2)(5.1) \\ &= 816 \end{aligned}$	$\begin{aligned} Q_2 &= 50(L)^{0.5}(K)^{0.6} \\ &= 50(20)^{0.5}(30)^{0.6} \\ &= 50(4.47)(7.69) \\ &= 1719 \\ &= 816 \times 2, \text{ i.e.,} \\ &= Q_1 \times 2 \end{aligned}$	Production function exhibits **increasing returns to scale.** $(\alpha + \beta) > 1$

Incremental changes

If both labour and capital change simultaneously by small but different proportions or amounts, then the total change in output can be determined by the 'small changes' formula which was expressed as equation (7.6). In this case, with $Q = f(L, K)$, equation (7.6) is modified to become

$$\Delta Q \cong \left(\frac{\partial Q}{\partial L}\right) \Delta L + \left(\frac{\partial Q}{\partial L}\right) \Delta K \qquad \textbf{(7.17)}$$

7.2.3 Utility functions

At the end of this section you should be able to:

- Define marginal utilities.
- Derive the equation for an indifference curve and sketch its graph.
- Derive the slope of an indifference curve in terms of marginal utilities.

The analysis of utility functions parallels that of production functions in the previous section. A utility function describes utility as a function of the goods consumed and may be written in general form as

$$U = f(x, y)$$

A utility function that is widely used in economic analysis is the Cobb–Douglas utility function which is expressed in general form as

$$U = Ax^\alpha y^\beta \qquad \textbf{(7.18)}$$

where A is a constant and $0 < \alpha < 1, 0 < \beta < 1, x > 0, y > 0$. x and y are the quantities of goods X and Y consumed. For example, a Cobb–Douglas utility function in specific form could be written as

$$U = 10x^{0.3}y^{0.5}$$

Marginal utility

In Chapter 6, utility was expressed as a function of one variable, x. Hence, marginal utility was defined as the derivative of total utility w.r.t. x,

$$U = f(x), \quad \text{then} \quad \frac{dU}{dx} = U_x = MU_x$$

In this section, where $U = f(x,y)$, the marginal utilities are defined as the partial derivatives:

$$\frac{\partial U}{\partial x} = U_x = MU_x \quad \text{is the marginal utility w.r.t. good X, and}$$

$$\frac{\partial U}{\partial y} = U_y = MU_y \quad \text{is the marginal utility w.r.t. good Y}$$

Graphical representation of utility functions

The graphical representation of utility functions is similar to that of production functions. A utility function $U = f(x, y)$ can be represented by a series of two-dimensional graphs known as **indifference curves** (equal utility), $y = f(x)$. To plot an indifference curve, utility is fixed at a constant value, then y is expressed in terms of x. An indifference curve gives all combinations of x and y for which utility, U, has the same value.

> An isoquant is a combination of inputs L and K which, when used, give a firm the same level of output.
>
> An indifference curve is a combination of goods X and Y which, when consumed, give the consumer the same level of utility.

Note on notation: X and Y represent the names of good X and good Y.
P_X = price per unit of good X and x = number of units of good X.
P_Y = price per unit of good Y and y = number of units of good Y.

Slope of an indifference curve

The slope of an indifference curve dy/dx is called the **marginal rate of substitution** (MRS), and may be expressed in terms of the marginal utilities of good X and good Y. The expression is derived from the total differential,

$$dU = \left(\frac{\partial U}{\partial x}\right) dx + \left(\frac{\partial U}{\partial y}\right) dy \tag{7.19}$$

Since U is constant along a given indifference curve, then $dU = 0$. Substituting $dU = 0$ into equation (7.19), an expression for dy/dx may be written as

$$0 = \left(\tfrac{\partial U}{\partial x}\right) dx + \left(\tfrac{\partial U}{\partial y}\right) dy$$

$$-\left(\tfrac{\partial U}{\partial y}\right) dy = \left(\tfrac{\partial U}{\partial x}\right) dx$$

$$-U_y dy = U_x dx$$

$$\frac{-U_y dy}{dx} = \frac{U_x}{1}$$

$$\frac{dy}{dx} = -\frac{U_x}{U_y} \tag{7.20}$$

Conventionally, $U_x \equiv MU_x$ and $U_y \equiv MU_y$; hence, equation (7.20) can be rewritten as

$$\frac{dy}{dx} = -\frac{MU_x}{MU_y} \tag{7.21}$$

Thus, the slope of an indifference curve, MRS, is equal to the ratio of the marginal utility of good X to the marginal utility of good Y.

The value of the slope of an indifference curve at any given point $(x = x_0, y = y_0)$, $dy/dx|_{\text{at } x_0, y_0}$, is the number of units by which good Y decreases when good X increases by one unit, while maintaining the same level of utility. The MRS is similar to the MRTS for isoquants. The MRS is illustrated in Figure 7.7. The indifference curve in Figure 7.7 exhibits a **diminishing marginal rate of substitution,**

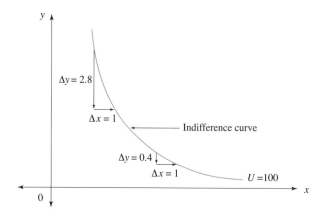

Figure 7.7   MRS = slope of an indifference curve, slope is diminishing

that is, $(\Delta y/\Delta x)$, or the rate at which the amount of good Y decreases for each unit increase in good X becomes smaller (diminishes) as x increases.

Remember

Do not mix up marginal utilities, $dU/dx = MU_x$, $dU/dy = MU_y$ and the MRS, dy/dx.

WORKED EXAMPLE 7.10
INDIFFERENCE CURVES AND SLOPE

A utility function is given by the equation $U = 8x^{0.5}y^{0.5}$.

(a) Graph the indifference curves for $U = 40$, $U = 48$ and $U = 56$.
(b) Derive expressions for the marginal utility of good X and good Y.
(c) Deduce an expression for the slope of the indifference curves, dy/dx.

SOLUTION

(a) Substitute $U = 40$ into the equation for the utility function and solve for $y = f(x)$:

$$40 = 8x^{0.5}y^{0.5}$$

$$5 = x^{0.5}y^{0.5}$$

$$\frac{5}{x^{0.5}} = y^{0.5}$$

$$\frac{25}{x} = y \quad \text{square each side of the equation}$$

$$\text{or } y = \frac{25}{x}$$

Similarly, the equations of the indifference curves for $U = 48$ are $y = 36/x$ and for $U = 56$ are $y = 49/x$. A series of points are calculated for each indifference curve, as shown in Table 7.7. The graphs are plotted in Figure 7.8.

Table 7.7   Points for plotting indifference curves

x	1	2	3	4	5	6
$y = 25/x$	25	12.5	8.3	6.3	5	4.2
$y = 36/x$	36	18	12	9	7.2	6
$y = 49/x$	49	24.5	16.3	12.3	9.8	8.2

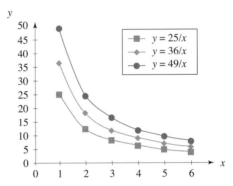

Figure 7.8   Indifference curves, $U = f(x,y)$

(b) The marginal utility of good X is: $MU_x = \partial U/\partial x = 8(0.5x^{0.5-1})Y^{0.5} = 4x^{-0.5}y^{0.5}$. The marginal utility of good Y is: $MU_y = \partial U/\partial y = 8x^{0.5}(0.5y^{0.5-1}) = 4x^{0.5}y^{-0.5}$.

(c) To find an expression for the MRS, substitute the expressions for the marginal utility of each good into equation (7.21):

$$\frac{dy}{dx} = -\frac{MU_x}{MU_y} = -\frac{4x^{-0.5}y^{0.5}}{4x^{0.5}y^{-0.5}} = -x^{-1}y = -\frac{y}{x}$$

Incremental changes

If the quantities of both products X and Y are changed by small but different amounts or proportions, then the resulting overall change in utility may be calculated by means of the incremental changes formula (7.6). Economic theory and the analysis of consumer behaviour argues that consumers will always prefer more to less, thereby increasing their overall utility.

PROGRESS EXERCISES 7.5

Returns to Scale: Utility and Production Functions

1. A consumer has a utility function $U = 10x^{0.2}y^{0.8}$, where x is the number of cups of tea consumed, y is the number of glasses of brandy.

(a) Derive expressions for U_x and U_y.

(b) Prove that

$$U = x\frac{\partial U}{\partial x} + y\frac{\partial U}{\partial y}$$

2. A utility function is given as

$$U = \frac{500}{y} - 10\ln x$$

(a) Write down the equation of the indifference curve $U = 20$ in the form $y = f(x)$.

(b) Evaluate y when $x = 2, 10, 25$. Plot the indifference curve.

(c) Derive an expression for the slope of the indifference curve $U = 20$ by differentiating the equation of the indifference curve directly.

Calculate the slope of the indifference curve. Hence evaluate the slope at $x = 2, 10, 25$. Do your calculations confirm the law of diminishing marginal rate of substitution?

3. (See question 2)

$$U = \frac{500}{y} - 10\ln x$$

(a) Derive expressions for U_x and U_y.

(b) Derive an expression for the MRS, dy/dx, the slope of an indifference curve, in terms of the marginal utilities.

Your answers should agree with those in question 2.

4. Given the Cobb–Douglas utility function $U = 25x^{0.5}y^{0.2}$,

(a) Derive expressions for the marginal utilities.

(b) Derive an expression for the slope of the indifference curve in terms of the marginal utilities.

(c) Confirm that the slope of the indifference curve satisfies the equation

$$\frac{dy}{dx} = -\frac{\alpha y}{\beta x}$$

5. The consumer in question 1 plans to decrease her consumption of cups of tea by 5% and decrease consumption of glasses of brandy by 8%. Use partial differentiation to find the approximate change in utility, U.

6. Determine whether each of the following production functions has increasing, decreasing or constant returns to scale.

(a) $Q = 10K^{0.5}L^{0.5}$ (b) $Q = 10K^{0.7}L^{0.6}$ (c) $Q = 10K^{0.3}L^{0.5}$.

7. Given the Cobb–Douglas production function, $Q = L^{0.7}K^{0.6}$,

(a) Calculate Q when $L = 50$ and $K = 200$.

(b) Calculate Q when both inputs are increased by 10%. Compare your answer when Q in part (a) is increased by 10%. What do these calculations imply about the level of returns to scale?

(c) If L increases by 5% and K increases by 10%, use the incremental changes formula to find the approximate increase in Q.

7.2.4   Partial elasticities

Partial elasticities of demand

In Section 6.7 price elasticity of demand was defined as the responsiveness of quantity demanded to changes in the price of the good itself. For the demand function $Q = f(P)$, price elasticity of demand was expressed as

$$\varepsilon_d = \frac{dQ}{dP} \cdot \frac{P}{Q}$$

We now, however, want to examine price elasticity of demand and other demand elasticities for a more general demand function of the form

$$Q_A = f(P_A, Y, P_B)$$

where Q_A = quantity demanded of good A

$\quad$ P_A = price of good A
$\quad$ Y = consumer income
$\quad$ P_B = price of another good B.

The responsiveness of the quantity demanded of good A to a change in any of these three variables is now examined.

Price elasticity of demand

This measures the percentage change in Q_A w.r.t. P_A keeping Y and P_B constant. It is given by

$$\varepsilon_d = \frac{\partial Q_A}{\partial P_A} \frac{P_A}{Q_A} \tag{7.22}$$

$\varepsilon_d = (\partial Q_A / \partial P_A)(P_A / Q_A)$ and $\varepsilon_d = dQ/dP \cdot P/Q$ are similar except that 'd' is replaced by '∂' since demand is a function of several variables.

Income elasticity of demand: This measures the percentage change in Q_A w.r.t. Y keeping P_A and P_B constant. It is given by

$$\varepsilon_Y = \frac{\partial Q_A}{\partial Y} \frac{Y}{Q_A} \tag{7.23}$$

Cross-price elasticity of demand: This measures the percentage change in Q_A w.r.t. P_B keeping P_A and Y constant. It is given by

$$\varepsilon_c = \frac{\partial Q_A}{\partial P_B} \frac{P_B}{Q_A} \tag{7.24}$$

WORKED EXAMPLE 7.11
PARTIAL ELASTICITIES OF DEMAND

The demand function for good A is given by

$$Q_A = 100 - 2P_A + 0.2Y + 0.3P_B$$

Find the price, income and cross-price elasticities of demand at $P_A = 6$, $Y = 500$, $P_B = 10$.

SOLUTION
The respective elasticity values are:

Price elasticity of demand
Differentiate the demand function partially w.r.t. P_A.

$$\frac{\partial Q_A}{\partial P_A} = -2$$

then use equation (7.22)

$$\varepsilon_d = \frac{\partial Q_A}{\partial P_A} \frac{P_A}{Q_A^*}$$

$$= -2\frac{6}{191} = -0.06$$

Income elasticity of demand
Differentiate the demand function partially w.r.t. Y.

$$\frac{\partial Q_A}{\partial Y} = 0.2$$

then use equation (7.23)

$$\varepsilon_Y = \frac{\partial Q_A}{\partial Y} \frac{Y}{Q_A}$$

$$= 0.2\frac{500}{191} = 0.52$$

Cross-price elasticity of demand
Differentiate the function partially w.r.t. P_B.

$$\frac{\partial Q_A}{\partial P_B} = 0.3$$

then use equation (7.24)

$$\varepsilon_c = \frac{\partial Q_A}{\partial P_B} \frac{P_B}{Q_A}$$

$$= 0.3\frac{10}{191} = 0.02$$

* Since at $P_A = 6$, $Y = 500$, $P_B = 10$, $Q_A = 100 - 2(6) + 0.2(500) + 0.3(10) = 191$

Partial elasticity of labour and capital

Partial elasticity w.r.t. labour is defined as the proportionate change in output (Q) resulting from a proportionate change in labour input (L) when capital is held constant:

$$\varepsilon_{QL} = \frac{\partial Q}{\partial L} \frac{L}{Q} \tag{7.25}$$

Note: The partial elasticity w.r.t. labour may also be expressed as the ratio of the marginal product of labour to the average product of labour:

$$\varepsilon_{QL} = \frac{\partial Q}{\partial L} \frac{L}{Q} = \frac{(\partial Q/\partial L)}{(Q/L)} = \frac{MP_L}{AP_L} \tag{7.26}$$

For the Cobb–Douglas production function, $Q = AL^\alpha K^\beta$,

$$\varepsilon_{QL} = \frac{\alpha AL^{\alpha-1}K^\beta}{1} \frac{L}{Q} = \frac{\alpha AL^\alpha K^\beta}{Q} = \frac{\alpha Q}{Q} = \alpha \tag{7.27}$$

Similarly, **the partial elasticity w.r.t. capital** is defined as the proportionate change in output (Q)

resulting from a proportionate change in capital input (K) when labour is held constant:

$$\varepsilon_{QK} = \frac{\partial Q}{\partial K}\frac{K}{Q} \tag{7.28}$$

Note: The partial elasticity w.r.t. capital may also be expressed as the ratio of the marginal product of capital to the average product of capital:

$$\varepsilon_{QK} = \frac{\partial Q}{\partial K}\frac{K}{Q} = \frac{(\partial Q/\partial K)}{(Q/K)} = \frac{MP_K}{AP\,K} \tag{7.29}$$

For the Cobb–Douglas production function, $Q = AL^{\alpha}K^{\beta}$,

$$\varepsilon_{QK} = \frac{\beta AL^{\alpha}K^{\beta-1}}{1}\frac{K}{Q} = \frac{\beta AL^{\alpha}K^{\beta}}{Q} = \frac{\beta Q}{Q} = \beta \tag{7.30}$$

WORKED EXAMPLE 7.12
PARTIAL ELASTICITIES OF LABOUR AND CAPITAL

Derive the partial elasticities w.r.t. labour and capital for the production function $Q = 10L^{0.5}K^{0.5}$.

SOLUTION
The partial elasticities are derived as follows:

Partial elasticity w.r.t. labour	Partial elasticity w.r.t. capital
$$\varepsilon_{QL} = \frac{\partial Q}{\partial L}\frac{L}{Q}$$	$$\varepsilon_{QK} = \frac{\partial Q}{\partial K}\frac{K}{Q}$$
Substitute the partial derivative,	Substitute the partial derivative,
$$\varepsilon_{QL} = \frac{10(0.5L^{-0.5})K^{0.5}}{1}\frac{L}{Q}$$	$$\varepsilon_{QK} = \frac{10L^{0.5}(0.5K^{-0.5})}{1}\frac{K}{Q}$$
$$= \frac{0.5(10L^{0.5}K^{0.5})}{Q} = \frac{0.5(Q)}{Q}$$	$$= \frac{0.5(10L^{0.5}K^{0.5})}{Q} = \frac{0.5(Q)}{Q}$$
$$= 0.5$$	$$= 0.5$$
$$= \alpha$$	$$= \beta$$
The answer agrees with equation (7.27).	The answer agrees with equation (7.30).

7.2.5 The multipliers for the linear national income model

In Chapter 3 the linear national income model for a three-sector economy was given as

$$Y = C + I + G$$
$$C = C_0 + bY_d$$
$$T = tY$$

The equilibrium level of income in its reduced form was derived as

$$Y_e \text{ equilibrium level of income in its reduced form} = \frac{C_0 + I_0 + G_0}{1 - b(1 - t)}$$

Partial differentiation may be used to determine the effect on the equilibrium level of income of very small changes in any one of the variables: C, I, equilibrium level of income in its reduced form G or T.

WORKED EXAMPLE 7.13
USE PARTIAL DERIVATIVES TO DERIVE EXPRESSIONS FOR VARIOUS MULTIPLIERS

The national income model for a three-sector economy is given as

$$Y = C + I + G, \quad C = C_0 + bY_d, \quad T = tY$$

and equilibrium national income is given by the equation

$$Y_e = \frac{C_0 + I_0 + G_0}{1 - b(1 - t)}$$

Use partial derivatives to derive expressions for the changes in Y_e which result from very small changes in (a) investment, I, only; (b) government expenditure, G, only; and (c) taxation rate, t, only.

SOLUTION

(a) The rate of change of Y_e w.r.t. I is the first partial derivative of Y_e w.r.t. I.

   Note: Since the level of private investment is changing, write I instead of I_0.

$$Y_e = \frac{C_0 + I + G_0}{1 - b(1 - t)}$$

$$\frac{\partial Y_e}{\partial I} = \frac{1}{1 - b(1 - t)}$$

   This partial derivative is known as the investment multiplier.

(b) The rate of change of Y_e w.r.t. G is the first partial derivative of Y_e w.r.t. G.

   Note: Since the level of government expenditure is changing, write G instead of G_0.

$$Y_e = \frac{C_0 + I_0 + G}{1 - b(1 - t)}$$

$$\frac{\partial Y_e}{\partial G} = \frac{1}{1 - b(1 - t)}$$

   This partial derivative is known as the government expenditure multiplier.

(c) The rate of change of Y_e w.r.t. the tax rate, t, is the first partial derivative of Y_e w.r.t. t.

$$Y_e = \frac{C_0 + I_0 + G_0}{1 - b(1 - t)} = (C_0 + I_0 + G_0)(1 - b + bt)^{-1}$$

$$\frac{\partial Y_e}{\partial t} = (C_0 + I_0 + G_0)[(-1)(1 - b + bt)^{-2}(b)]$$

differentiating by the chain rule

$$= \frac{-b(C_0 + I_0 + G_0)}{(1 - b(1 - t))^2}$$

$$= -\frac{bY_e}{1 - b(1 - t)}$$

This partial derivative is known as the income tax rate multiplier.

The multipliers are summarised:

$$\frac{\partial Y_e}{\partial I} = \frac{1}{1 - b(1 - t)} \quad \text{the investment multiplier}$$

$$\frac{\partial Y_e}{\partial G} = \frac{1}{1 - b(1 - t)} \quad \text{the government expenditure multiplier}$$

$$\frac{\partial Y_e}{\partial t} = \frac{bY_e}{1 - b(1 - t)} \quad \text{the income tax rate multiplier}$$

PROGRESS EXERCISES 7.6

Partial Elasticities, National Income Model Multipliers

1. The demand function for good X is given by

$$Q_x = 520 - 20P_x + 0.6Y + 2.9P_y$$

Find the price, income and cross-price elasticities of demand at $P_X = 10$, $Y = 700$, $P_Y = 21$.

2. Given a Cobb–Douglas production function $Q = 250L^{0.6}K^{0.2}$,
   (a) Derive expressions for the average and marginal products of labour.
   (b) Derive the partial elasticity w.r.t. labour.
   (c) Confirm that the partial elasticity is equal to the ratio of the marginal to average product.

3. The national income model for a three-sector economy is given by $Y = C + I + G$, where $C = 120 + 0.6Y_d$, $T = 0.25Y_d$, $G = 200$ and $I = 120$.
   (a) Calculate the equilibrium level of national income.
   (b) Use partial derivatives to evaluate the following multipliers:
       (i) investment, (ii) government expenditure, (iii) income tax rate.

4. Derive the partial elasticities w.r.t. labour and capital for the production functions I: $Q = 10L^{0.6}K^{0.4}$, II: $Q = 180L^{0.4}K^{0.3}$.

(a) Use the definition of elasticity to find the approximate change in output Q in (i) I, when labour increases by 3%, (ii) II, when capital increases by 5%.

(b) Use partial derivatives to calculate the approximate percentage change in Q in (i) I, when L increases by 3%, (ii) II, when K increases by 5%.

Compare the answers in (a) and (b).

5. (See question 2) For the Cobb–Douglas production function $Q = 250L^{0.6}K^{0.2}$,

(a) Derive expressions for the average and marginal product of capital.

(b) Derive the partial elasticity w.r.t. capital.

(c) Confirm that the partial elasticity is equal to the ratio of the marginal to average product.

7.3 Unconstrained Optimisation

At the end of this section you should be able to:

- Locate the optimum points for functions of two variables
- Calculate maximum profit for a firm producing two goods
- Show how firms use price discrimination in order to maximise profits.

7.3.1 Find the optimum points for functions of two variables

Optimisation of functions of one variable revisited

The methods for finding optimum points (maxima, minima), or points of inflection, for functions of two variables are simply extensions of those used for finding the maxima, minima and points of inflection for functions of one variable which are covered in Chapter 6. If you recall, we used a three-step method; that is, for the function, $y = f(x)$, the steps are:

Step 1: Find the first and second derivatives: dy/dx, d^2y/dx^2.

Step 2: Equate the first derivative to zero, $dy/dx = 0$, and solve this equation to find the x-coordinate of the potential turning point(s).

Step 2a: If required, substitute the x-coordinate of each turning point(s) into the equation of the curve to find the corresponding y-coordinate.

Step 3: Use the second derivatives to determine the nature of the turning point(s). Evaluate d^2y/dx^2 at each turning point:

$$\text{if } \left.\frac{d^2y}{dx^2}\right|_{\text{at turning point}} > 0 \quad \text{the point is a minimum}$$

$$\text{if } \left.\frac{d^2y}{dx^2}\right|_{\text{at turning point}} < 0 \quad \text{the point is a maximum}$$

$$\text{if } \left.\frac{d^2y}{dx^2}\right|_{\text{at turning point}} = 0 \quad \text{the point may be a point of inflection}$$

Optimisation of functions of two variables: method

In order to optimise functions of two variables, such as $z = f(x,y)$, each of the above three steps is extended. Given the function, $z = f(x,y)$, the method is as follows:

Step 1: Find the first and second derivatives:

$$\frac{\partial z}{\partial x}, \frac{\partial z}{\partial y}, \frac{\partial^2 z}{\partial x^2}, \frac{\partial^2 z}{\partial y^2}, \frac{\partial^2 z}{\partial x \partial y}$$

Step 2: Equate the first derivatives to zero, then solve these equations to find the x-coordinate and the y-coordinate of the turning point(s):

$$\frac{\partial z}{\partial x} = 0, \quad \frac{\partial z}{\partial y} = 0$$

Step 2a: If required, find z for the x- and y-coordinates of the turning point(s).

Step 3: Use the second derivatives to determine the nature of the turning points. Evaluate all second derivatives at the x- and y-coordinates of the turning points.

Extension: As the number of dimensions increases (and consequently the number of derivatives), so also do the number of conditions which must be satisfied to confirm the nature of the optimum point. We define a new term, Δ, where

$$\Delta = \left(\frac{\partial^2 z}{\partial x^2}\right)\left(\frac{\partial^2 z}{\partial y^2}\right) - \left(\frac{\partial^2 z}{\partial x \partial y}\right)^2 \tag{7.31}$$

The point is a minimum if $\partial^2 z/\partial x^2 > 0$ and $\partial^2 z/\partial y^2 > 0$ and provided $\Delta > 0$.

The point is a maximum if $\partial^2 z/\partial x^2 < 0$ and $\partial^2 z/\partial y^2 < 0$ and provided $\Delta > 0$.

The point is a point of inflection if both second derivatives have the same sign but $\Delta < 0$.

The point is a saddle point if the second derivatives have different signs and $\Delta < 0$.

If $\Delta = 0$ then there is no conclusion.

A saddle point is a point which is a maximum when viewed along one axis but a minimum when viewed along the other. In fact, it looks like a saddle; hence its name. Figure 7.9 shows (a) a maximum, (b) a minimum and (c) a saddle point in three dimensions.

At this stage it might seem that there is a confusing number of conditions to remember; however, if you view these conditions as extensions of the methods of Chapter 6 they will soon fall into place, especially with practice when working through the examples.

Note: In this text it is possible to classify all given functions according to the second-order conditions outlined above. However, for those functions for which these conditions fail, then higher-order derivatives are required, but this is beyond the scope of this text.

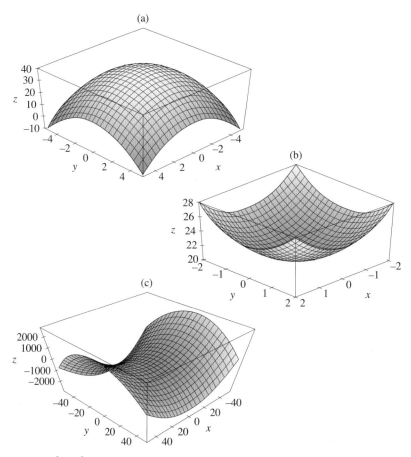

Figure 7.9 (a) $z = -x^2 - y^2 + 40$ has a maximum point at $x = 0$, $y = 0$, $z = 40$. (b) $z = x^2 + y^2 + 20$ has a minimum point at $x = 0$, $y = 0$, $z = 20$. (c) $z = x^2 - y^2 - 4x + 4y$: a saddle point at $x = 2$, $y = 2$, $z = 0$.

WORKED EXAMPLE 7.14
OPTIMUM POINTS FOR FUNCTIONS OF TWO VARIABLES

Given that $z = x^2 + y^2 + 20$,
Find the optimum point.
Determine the nature of the optimum point.

SOLUTION
Work through the method for finding the optimum point for functions of two variables.

Step 1: Find the derivatives. The first and second derivatives are $z_x = 2x$, $z_y = 2y$, $z_{xx} = 2$, $z_{yy} = 2$, $z_{xy} = 0$.

Step 2: Equate the first derivatives to zero, and solve for the x- and y-coordinates of the optimum point:

$$z_x = 2x, \ z_y = 2y$$
$$0 = 2x, \ 0 = 2y$$

The only solution to each of these equations is $x = 0$ and $y = 0$.

Step 2a: Find the z-coordinate of the optimum point (we need three coordinates for three dimensions). Substituting $x = 0$ and $y = 0$ into the equation for z gives $z = 0^2 + 0^2 + 20 = 20$; therefore, there is an optimum point at $x = 0$, $y = 0$, $z = 20$.

Step 3: Use the second derivatives to determine the nature of the turning point. The two straight second derivatives are constants and positive:

$$z_{xx} = 2 > 0 \quad \text{and} \quad z_{yy} = 2 > 0$$

Therefore this point is a minimum provided $\Delta > 0$, which is calculated as

$$\Delta = (z_{xx})(z_{yy}) - (z_{xy})^2$$
$$= (2)(2) - (0)^2 = 4 > 0$$

Therefore a minimum point exists at $x = 0$, $y = 0$, $z = 20$. This is illustrated in Figure 7.9(b).

PROGRESS EXERCISES 7.7

Unconstrained Optimisation

Find and classify the stationary points for the functions given in questions 1 to 6.

1. $z = x^2 + y^2 - 10x - 12y$

2. $z = 20x - x^2 - y^2 + 8y$

3. $F(x, y) = 2y^2 + 2xy + x^2 - 16x - 20y$

4. $C = 5x^2 + 2y^2 - 4xy - 36x$

5. $Q = 5\ln(L) + 2\ln(K) - 0.1L - 0.4K$

6. $U = x^3 + y^3 - 27x - 12y + 210$

7.3.2 Total revenue maximisation and profit maximisation

The method of optimisation of functions of two variables is now used to examine more practical problems in economics and business. For example, we might want to calculate the maximum total revenue or maximum profit for a multi-product monopolist or a perfectly competitive firm.

WORKED EXAMPLE 7.15
MONOPOLIST MAXIMISING TOTAL REVENUE FOR TWO GOODS

A monopolist produces two goods, X and Y; the demand function for each good is given as

$$P_X = 36 - 3x \quad \text{and} \quad P_Y = 56 - 4y$$

(a) Write down the equation for the monopolist's total revenue.
(b) Determine the quantities of each good which must be sold so that total revenue is maximised. What is the maximum revenue?

SOLUTION

(a) Total revenue = price × quantity for each good produced, that is,

$$TR = P_X x + P_Y y$$
$$= (36 - 3x)x + (56 - 4y)y$$
$$= 36x - 3x^2 + 56y - 4y^2$$

(b) Work through each step of the optimisation method:

Step 1: Find derivatives

$$\frac{\partial TR}{\partial x} \text{ or } TR_x = 36 - 6x$$
$$TR_y = 56 - 8y$$
$$TR_{xx} = -6$$
$$TR_{yy} = -8$$
$$TR_{xy} = 0$$

Step 2: Solve

$TR_x = 0$, therefore
$$0 = 36 - 6x$$
$$6x = 36$$
$$x = 6$$
and $TR_y = 0$, therefore
$$0 = 56 - 8y$$
$$8y = 56$$
$$y = 7$$

Step 2a: Find TR when $x = 6$ and $y = 7$.
$$TR = 36(6) - 3(6)^2$$
$$+ 56(7) - 4(7)^2$$
$$= 304$$

Step 3: Use second-order derivatives to decide the nature of the turning point.

$$TR_{xx} = -6$$
$$TR_{yy} = -8$$
$$TR_{xy} = 0$$

$$\left. \begin{array}{l} TR_{xx} = -6 \\ TR_{yy} = -8 \end{array} \right\} \text{both negative so}$$

possible maximum, but need to calculate Δ, so
$$\Delta = (TR_{xx})(TR_{yy}) - (TR_{xy})^2$$
$$= (-6)(-8) - (0)^2$$
$$= 48 > 0$$
So a maximum is confirmed.

Maximum revenue is £304 when $x = 6$ and $y = 7$.

When we are given the firm's total cost function along with the demand functions, we may use unconstrained optimisation to find the quantities of each good which must be produced and sold in order to maximise profit.

WORKED EXAMPLE 7.16
MAXIMISE PROFIT FOR A MULTI-PRODUCT FIRM

 Find animated worked examples at **www.wiley.com/college/bradley**

A perfectly competitive firm produces two goods, X and Y, which are sold at £54 and £52 per unit, respectively. The firm has a total cost function given by

$$TC = 3x^2 + 3xy + 2y^2 - 100$$

Find the quantities of each good which must be produced and sold in order to maximise profits. What is the maximum profit?

SOLUTION
A perfectly competitive firm charges the same price irrespective of the number of units sold; therefore

$$\text{Total revenue} = \text{revenue from good X} + \text{revenue from good Y}$$
$$= x P_X + y P_Y$$
$$= 54x + 52y$$

Profit = total revenue − total cost, therefore

$$\pi = TR - TC$$
$$= 54x + 52y - 3x^2 - 3xy - 2y^2 + 100$$

Now work through the three steps for optimisation.

Step 1: Find the first- and second-order partial derivatives:

$$\pi_x = 54 - 6x - 3y, \quad \pi_y = 52 - 3x - 4y,$$
$$\pi_{xx} = -6, \quad \pi_{yy} = -4, \quad \pi_{xy} = -3$$

Step 2: Equate the first-order partial derivatives to zero and solve.

$$(1) \quad \pi_x = 0 \rightarrow 0 = 54 - 6x - 3y$$
$$(2) \quad \pi_y = 0 \rightarrow 0 = 52 - 3x - 4y$$

Solve (1) and (2) for x and y:

(1) $\times 1$ $0 = 54 - 6x - 3y$

(2) $\times 2$ $0 = 104 - 6x - 8y$

Subtract $0 = -50 + 5y$ therefore $y = 10$

Solve for x by substituting $y = 10$ into either (1) or (2); therefore $x = 4$.

Step 3: Use second-order conditions to determine the type of optimum point at $x = 4$, $y = 10$:

$$\pi_{xx} = -6, \quad \pi_{yy} = -4, \quad \pi_{xy} = -3$$

$$\Delta = (\pi_{xx})(\pi_{yy}) - (\pi_{xy})^2 = (-6)(-4) - (-3)^2 = 15$$

Both straight second-order partial derivatives are negative, indicating a possible maximum. Since Δ is positive, then this confirms that the point is a maximum. Maximum profits are £468 when $y = 10$ and $x = 4$.

7.3.3 Price discrimination

Price discrimination for a monopolist producing a single good, X, which is sold in two separate markets, market 1 and market 2 (each with a different demand function), was introduced in Worked Example 6.24. In this section the analysis of profit and revenue for such a monopolist is continued with more general methods.

To find the prices which should be charged in each market to maximise profit

Start by writing down the equation for the profit function:

$\pi = TR - TC$ where

$TR = (TR)_1 + (TR)_2$ TR is the sum of the total revenues; for each market

     $= P_1 x_1 + P_2 x_2$ P_1 and x_1 are the price and quantity in market 1; P_2 and x_2 are the price and quantity in market 2

$x = x_1 + x_2$ the total quantity sold is the sum of the quantities sold in each market

(In this section refer to the quantities sold in each market as x_1 and x_2 instead of Q_1 and Q_2.)

To find the conditions for maximum profit, work through the steps for optimisation.

Step 1: Find the first- and second-order partial derivatives:

$$\frac{\partial \pi}{\partial x_1} = \frac{\partial TR_1}{\partial x_1} - \frac{\partial TC}{\partial x_1} \quad \rightarrow \quad \pi_{x_1} = TR_{x_1} - TC_{x_1} = MR_1 - MC_1$$

$$\frac{\partial \pi}{\partial x_2} = \frac{\partial TR_2}{\partial x_2} - \frac{\partial TC}{\partial x_2} \quad \rightarrow \quad \pi_{x_2} = TR_{x_2} - TC_{x_2} = MR_2 - MC_2$$

$$\pi_{x_1 x_1} = TR_{x_1 x_1} - TC_{x_1 x_1}; \ \pi_{x_2 x_2} = TR_{x_2 x_2} - TC_{x_2 x_2}; \ \pi_{x_1 x_2} = TR_{x_1 x_2} - TC_{x_1 x_2}$$

Step 2: Equate the first-order partial derivatives to zero and solve:

$$\pi_{x_1} = 0 \rightarrow TR_{x_1} - TC_{x_1} = 0 \rightarrow MR_1 - MC_1 = 0 \rightarrow MR_1 = MC_1$$
$$\pi_{x_2} = 0 \rightarrow TR_{x_2} - TC_{x_2} = 0 \rightarrow MR_2 - MC_2 = 0 \rightarrow MR_2 = MC_2$$

In other words, $MR = MC$ for maximum profit for each market.

However, since the single good X is produced by one firm, the cost function is the same for each market; consequently the marginal cost function is the same for each market. Therefore

$$MC_1 = MC_2 = MC$$

Thus the first-order condition for maximum profit is

$$MR_1 = MR_2 = MC \tag{7.32}$$

However, the second-order conditions must be used to classify the point as a maximum, minimum or otherwise.

WORKED EXAMPLE 7.17
MONOPOLIST: PRICE AND NON-PRICE DISCRIMINATION

A monopolist produces a single good X but sells it in two separate markets. The demand function for each market is

$$P_1 = 50 - 4x_1 \quad \text{and} \quad P_2 = 80 - 3x_2$$

where P_1 and x_1, P_2 and x_2 are the price and quantity in markets 1 and 2, respectively. The cost function is $TC = 120 + 8x$, where $x = x_1 + x_2$.

(a) Find the price and quantity of the good in each market which maximises profit.
(b) Determine the price elasticity of demand for each market.
(c) Find the price and quantity of the good in each market which maximises profit when the monopolist does not use price discrimination, that is, $P_1 = P_2$.

SOLUTION
It is now left as an exercise to the reader to use the more general methods outlined above to obtain the same answers for part (a) as those obtained in Worked Example 6.24(a). Part (c) is the same as in Worked Example 6.24(b).
(b) Elasticity in each market:

$$\varepsilon_1 = \frac{dx_1}{dP_1} \frac{P_1}{x_1} \quad \text{and} \quad \varepsilon_2 = \frac{dx_2}{dP_2} \frac{P_2}{x_2}$$

(You may find it convenient to use $\varepsilon = P/(P - a)$, the elasticity for linear functions, $P = a - bQ$, derived in Chapter 6.) The derivatives are obtained from the equations of the respective demand functions:

Market 1	**Market 2**
$\varepsilon_1 = \dfrac{dx_1}{dP_1}\dfrac{P_1}{x_1}$	$\varepsilon_2 = \dfrac{dx_2}{dP_2}\dfrac{P_2}{x_2}$
$= \dfrac{1}{(dP_1/dx_1)}\dfrac{P_1}{x_1}$	$= \dfrac{1}{(dP_2/dx_2)}\dfrac{P_2}{x_2}$
$= \dfrac{1}{-4}\dfrac{P_1}{x_1}$	$= \dfrac{1}{-3}\dfrac{P_2}{x_2}$
$= -\dfrac{1}{4}\left(\dfrac{29}{5.25}\right) = -1.4$	$= -\dfrac{1}{3}\left(\dfrac{44}{12}\right) = -1.2$

When $P_2 > P_1$, the demand elasticities $|\varepsilon_2| < |\varepsilon_1|$. Therefore, the higher price is charged in the market with the most inelastic demand (where demand is less responsive to price changes).

PROGRESS EXERCISES 7.8

Unconstrained Optimisation: Applications

1. A utility function is given by the equation $U = x - 2x^2 + xy + 40y - y^2$, where x is the number of units of good X and y is the number of units of good Y consumed. Determine the values for x and y which maximise U. (Confirm that the point is a maximum.)
2. Given the demand functions for two goods $P_X = 320 - 4x + 2y$ and $P_Y = 106 + 2x - 20y$:
   (a) Write down the equation for overall total revenue.
   (b) Find the values of P_X, x, P_Y, y for which revenue is optimised. Hence calculate the maximum total revenue.
3. (See question 2) A firm producing two goods has demand functions as given in question 2 and a total cost function $TC = 4xy + 40x + 26y$.
   (a) Write down the equation of the profit function.
   (b) Determine the values of x and y for which profit is optimised. Hence calculate the maximum profit. Also calculate total revenue and the prices charged when profit is maximised.
4. A firm sells two related products whose demand functions are given as $Q_1 = 190 - 4P_1, -P_2$ and $Q_2 = 120 - 2P_1 - 3P_2$, where P_1, Q_1, P_2, Q_2 are the prices and quantities of goods 1 and 2, respectively.
   (a) Write down the equation of the overall total revenue in terms of P_1 and P_2 only.
   (b) Determine the values of P_1, Q_1, P_2, Q_2 for which revenue is maximised.
   (c) Calculate the maximum revenue.
5. Show that the total cost function $TC = 4Q_1Q_2 + 10Q_1 + 15Q_2$ has no economically meaningful optimum point.

6. A firm has two separate plants, each producing the same product. The cost functions for each plant are given by the equations

$$C_1(x) = 400 + \frac{x^2}{20} - 10x \quad \text{and} \quad C_2(y) = 400 - y + \frac{y^3}{147}$$

where x and y are the number of units produced in each plant. If overall costs are $C = C_1(x) + C_2(y)$, determine the number of units which should be manufactured in each plant to minimise costs.

7. The profit function for two goods is given by the equation

$$\pi = 527 - 2x^2 - 3y^2 - 2xy + 20x + 20y$$

Determine the values of x and y for which profit is maximised.

8. The total cost and demand functions for two goods are given by the equations

$$TC = 10 + 4x + 6y, \quad P_x = 8 - 2x, \quad P_y = 20 - 0.5y$$

where x is the number of units of the first good, y is the number of units of the second good.
   (a) Write down the equations for (i) total revenue, and (ii) profit.
   (b) Determine the number of units of each good which should be sold to maximise revenue. Calculate the maximum revenue.
   (c) Determine the number of units of each good which should be bought and sold to maximise profits. Calculate the maximum profit.
   (d) Calculate the price elasticity of demand at maximum revenue and maximum profit for each good. Comment.

9. A monopolist can charge different prices in each of two markets whose demand and total cost functions are given as

$$P_X = 80 - 2.5x, \quad P_Y = 125 - 10y, \quad TC = 200 + 5(x + y)$$

   (a) Calculate the maximum profit with price discrimination.
   (b) Calculate the maximum profit with no price discrimination.
   (c) Calculate the price elasticity of demand in each market when profits are maximised with price discrimination.

10. The total cost and the demand functions for two goods are given by the equations

$$TC = 25 + 4x + 10y, \quad P_X = 12 - 2x, \quad P_Y = 24 - y$$

where P_X and x, P_Y and y represent the prices and quantities of goods X and Y, respectively.

Write down the expression for profit and hence determine the values of x and y for which profit is maximised.

11. A firm produces two related goods whose demand functions are given by the equations

$$P_X = 200 - x + 4y \quad \text{and} \quad P_Y = 500 + 6x - 75y$$

where P_X, x, P_Y, y are the prices and quantities for goods X and Y, respectively:
(a) Write down the equation for the firm's total revenue.
(b) Calculate the price and quantity of each good which must be sold to maximise total revenue.

7.4 Constrained Optimisation and Lagrange Multipliers

At the end of this section you should be able to:

* Understand the meaning of a constrained maximum or minimum
* Use Lagrange multipliers to find the maximum or minimum values subject to a constraint
* Find maximum utility subject to a budget constraint
* Find maximum output subject to a cost constraint
* Find minimum cost subject to a production constraint.

In the previous section, unconstrained optimum values were calculated for different economic variables such as total revenue and profit. However, in most real-life situations, consumers and firms face financial and production constraints as there are limits on the availability of labour, capital and raw materials and, therefore, on the total level of output and income produced within an economy. We are still interested in finding optimum values, but now these optimum values will be subject to constraints.

7.4.1 What is a constrained maximum or minimum?

In this text we deal with linear constraints (see Section 2.7). For example, a budget constraint for two goods is usually given by an equation of the form

$$\begin{array}{lll} \text{expenditure} = \text{limit} & \text{or expenditure} \leq \text{limit} & \text{or expenditure} \geq \text{limit} \\[1em] x P_X + y P_Y = M & \text{or} \begin{array}{l} x P_X + y P_Y < M \\ x P_X + y P_Y \leq M \end{array} & \text{or} \begin{array}{l} x P_X + y P_Y > M \\ x P_X + y P_Y \geq M \end{array} \end{array}$$

where M is the monetary limit. In this text only equality constraints are used.

In Worked Example 7.15, total revenue for goods X and Y was maximised. It was found that the maximum revenue was £304 when $x = 6$ and $y = 7$. However, there was no limit on the number of units of each good which could be sold. Suppose a constraint is imposed that only allows £80 worth of goods to be produced and sold, where each unit of X costs £5 while each unit of Y costs £10. The problem is then reduced to finding how many units of each good must be produced and sold so that revenue is maximised. This means that the firm faces a budget constraint which is given by the equation

$$P_X \times x + P_Y \times y = 80$$

$$5x + 10y = 80$$

Without resorting to further mathematics, we can see that the number of units of each good ($x = 6$ and $y = 7$), for which revenue is a maximum when no constraints were imposed, can no longer be used, since $TR = 5(6) + 10(7) = 100$ exceeds the limit of 80. The problem we are now faced with is how to find the combination of x and y which will yield the maximun possible revenue, subject to a given constraint.

7.4.2 Finding the constrained extrema with Lagrange multipliers

To find the optimum values of a function $z = (x, y)$ subject to a constraint, $ax + by = M$, we define the **Lagrangian function**, L, where

$$L = L(x, y, \lambda) = f(x, y) + \lambda(M - ax - by) \qquad \textbf{(7.33)}$$

where λ is called a **Lagrange multiplier** and is treated as a variable. L is the sum of the original function to be optimised and $\lambda \times$ (constraint $= 0$). The optimum value of λ must be determined in addition to the optimum values of the variables x and y.

The method for finding the optimum value(s) of L is the same as in the previous section, but extended slightly (again!), since the problem consists of three independent variables, x, y and λ and, therefore, three first-order derivatives, L_x, L_y, L_λ. The method of Lagrange multipliers is demonstrated in the following worked examples.

WORKED EXAMPLE 7.18
MAXIMISING TOTAL REVENUE SUBJECT TO A BUDGET CONSTRAINT

The total revenue function for two goods is given by the equation

$$TR = 36x - 3x^2 + 56y - 4y^2$$

Find the number of units of each good which must be sold if profit is to be maximised when the firm is subject to a budget constraint, $5x + 10y = 80$.

SOLUTION
Define the Lagrangian: $L = 36x - 3x^2 + 56y - 4y^2 + \lambda(80 - 5x - 10y)$

Now use the methods in Section 7.3 to find the values of x, y and λ for which revenue is maximised.

Step 1: Find the first-order partial derivatives:

$$L_x = 36 - 6x - 5\lambda, \quad L_y = 56 - 8y - 10\lambda, \quad L_\lambda = 80 - 5x - 10y$$

Step 2: Equate the first-order partial derivatives to zero and solve for x, y and λ. It is usually more convenient to eliminate λ from the first two equations and then solve for x and y.

(1) $36 - 6x - 5\lambda = 0$
(2) $56 - 8y - 10\lambda = 0$
(3) $80 - 5x - 10y = 0$

Solve as follows:

(1) $x - 2 \rightarrow$ $-72 + 12x + 10\lambda = 0$
(2) $\rightarrow$ $\underline{56 - 8y - 10\lambda = 0}$
(4) add $\rightarrow$ $-16 + 12x - 8y = 0$

Now use equations (3) and (4) to solve for x and y:

(5) $\rightarrow$ (3)/5 $\rightarrow$ $16 - x - 2y = 0$
(6) $\rightarrow$ (4)/-4 $\rightarrow$ $\underline{4 - 3x + 2y = 0}$
        and $20 - 4x + 0 = 0 \rightarrow x = 20/4 = 5$

Substitute $x = 5$ into any equation containing x and y such as (5), and solve for y:

$$y = (16 - x)/2 = 11/2 = 5.5$$

Therefore $y = 5.5$ and $x = 5$ are the quantities of goods X and Y for which revenue is maximised. Maximum total revenue is

$$TR = 36(5) - 3(5)^2 + 56(5.5) - 4(5.5)^2 = 292$$

which is less than the value for maximum revenue (£304) for the unconstrained maximum calculated in Worked Example 7.15. Therefore, the constraint is a limitation. See Figure 7.10 and Table 7.8.

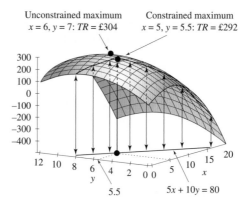

Figure 7.10 Constrained maximum: $TR = 36x - 3x^2 + 56y - 4y^2$, subject to $5x + 10y = 80$

Table 7.8 TR evaluated at selected points on the constraint: $80 = 5x + 8y (y = 8 - 0.5x)$

x	0	2	4	5	6	8	10	12	16
$y = 8 - 0.5x$	8	7	6	5.5	5	4	3	2	0
TR	192	256	288	292	288	256	192	96	−192
$5x + 10y =$	80	80	80	80	80	80	80	80	80

Note 1: The values of x and y which optimise L are also the values of x and y which optimise the original function, $f(x, y)$, subject to the constraint. A formal mathematical proof is beyond the scope of this text.

Note 2: The second-order conditions are more complex when three independent variables x, y, λ are involved. So, yet again, formal proofs are beyond the scope of this text. However, for the usual economic/business problems presented in this text, it is quite safe to assume that the method of Lagrange multipliers yields a maximum for a maximisation problem and a minimum for a minimisation problem.

In the remaining worked examples, the methods will be the same as those used in Worked Example 7.18; however, the equations, $L_x = 0$, $L_y = 0$, $L_\lambda = 0$, may not be three linear equations in three unknowns. The following worked examples will suggest various methods for solving three non-linear equations in three unknowns.

7.4.3   Maximum utility subject to a budget constraint

In this section, maximum utility subject to a constraint is found by using the standard method of Lagrange multipliers. We then demonstrate the general condition

$$-\frac{U_x}{U_y} = -\frac{P_X}{P_Y}$$

which is stated in economic textbooks for finding maximum utility subject to a budget constraint. This condition is derived by Lagrange multipliers and the standard graphical methods of economics.

WORKED EXAMPLE 7.19
LAGRANGE MULTIPLIERS AND UTILITY MAXIMISATION

Use Lagrange multipliers to find the maximum utility for the utility function $U = 5xy$, when subject to a budget of £30, where the price of each unit of X is £5 and each unit of Y is £1.

SOLUTION
The equation of the budget constraint is

$$x P_X + y P_Y = M$$
$$x(5) + y(1) = 30$$
$$5x + y = 30$$

The equation of the Lagrangian is therefore

$$L = 5xy + \lambda(30 - 5x - y)$$

Step 1: Find the first-order partial derivatives:

$$L_x = 5y - 5\lambda, \quad L_y = 5x - \lambda, \quad L_\lambda = 30 - 5x - y$$

Step 2: Equate the first-order derivatives to zero and solve

(1) $5y - 5\lambda = 0$ or $5y = 5\lambda$
(2) $5x - \lambda = 0$ or $5x = \lambda$
(3) $30 - 5x - y = 0$

To solve, first eliminate λ from equations (1) and (2):

(1) $\rightarrow$ $5y = 5\lambda$
(2) $\times 5$ $\rightarrow$ $\underline{25x = 5\lambda}$

(4) subtract $\rightarrow$ $5y - 25x = 0$

Now use equations (3) and (4) to solve for x and y:

(5) $\rightarrow$ (4)/5 $\rightarrow$ $y - 5x = 0$
(6) $\rightarrow$ (3) rearranged $\rightarrow$ $\underline{y + 5x = 30}$

add $2y + 0 = 30$ $\rightarrow$ $y = 15$

Substituting $y = 15$ into equation (5) gives $x = 3$; therefore, maximum utility occurs when $x = 3$, $y = 15$. The level of maximum utility is $U = 5(3)(15) = 225$. From equation (1), $\lambda = 15$.

Graphical analysis for locating maximum utility, $-(U_x/U_y) = -(P_X/P_Y)$

Figure 7.11 illustrates three indifference curves, each representing three different levels of utility. If there is no limit on the available quantities of goods X and Y that can be consumed, then utility will increase indefinitely. There will be no limiting maximum value. If, however, a budget constraint is imposed, then the quantities consumed of each good, x and y, are limited. The budget constraint is also illustrated in Figure 7.11.

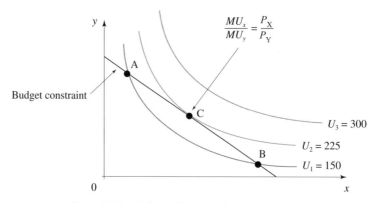

Figure 7.11   Utility subject to a budget constraint

The equation of the budget constraint is

$$x P_X + y P_Y = M$$

$$y = \frac{M}{P_Y} - \frac{P_X}{P_Y} x \tag{7.34}$$

The point of intersection of any indifference curve with the budget line gives (x, y), the quantities of goods X and Y, respectively, which satisfy the budget constraint. However, utility increases as the indifference curves move further from the origin. Therefore the maximum possible utility occurs at the point where the 'highest' indifference curve is tangent to the budget line. Point C in Figure 7.11 is such a point.

Therefore, the quantities of x and y for which utility is a maximum are given by the coordinates of point C at which the indifference curve is tangent to the budget constraint. From Chapter 6, the slope of the curve at a point was shown to be the same as the slope of the tangent at that point. Hence, equating the slope of the indifference curve and the slope of the budget constraint gives

$$-\frac{U_x}{U_y} = -\frac{P_X}{P_Y} \tag{7.35}$$

where the slope of the indifference curve is $-(U_x/U_y)$ (this was derived in Section 7.2) and the slope of the budget line is $-(P_X/P_Y)$ (read slope directly from equation 7.34). Equation (7.35) may also be written as

$$\frac{MU_x}{MU_y} = \frac{P_X}{P_Y} \tag{7.36}$$

where the minus signs are cancelled and U_x is replaced by the alternative representation, MU_x, etc.

WORKED EXAMPLE 7.20
USE LAGRANGE MULTIPLIERS TO DERIVE THE IDENTITY
$U_x/U_y = P_X/P_Y$

Use Lagrange multipliers to show that a utility function $U(x, y)$ which is subject to a budget constraint $x P_X + y P_Y = M$ is maximised when $U_x/U_y = P_X/P_Y$ (ignore negative signs).

SOLUTION
The Lagrangian is $L = U(x, y) + \lambda(M - x P_X - y P_Y)$. To find the maximum (or any optimum point!) equate the first-order partial derivatives to zero:

(1) $L_x = U_x - \lambda P_X \quad \rightarrow \quad 0 = U_x - \lambda P_X \quad$ therefore $\quad U_x = \lambda P_X$
(2) $L_y = U_y - \lambda P_Y \quad \rightarrow \quad 0 = U_y - \lambda P_Y \quad$ therefore $\quad U_y = \lambda P_Y$
(3) $L_\lambda = M - x P_X - y P_Y \quad\quad\quad\quad\quad\quad\quad\quad\quad$ therefore $\quad M = x P_X + y P_Y$

If we divide U_x by U_y, then from equations (1) and (2) we may write

$$\frac{U_x}{U_y} = \frac{\lambda P_X}{\lambda P_Y} = \frac{P_X}{P_Y}$$

This is the same as equation (7.35) as required. The budget equation, expressed as equation (3) above, must also be satisfied. So the Lagrange multiplier method produces exactly the same results as the graphical analysis.

Interpretation of the Lagrange multiplier λ

The Lagrange multiplier method is a more systematic method for finding constrained maximum or minimum points. In addition, the value of the Lagrange multiplier, λ, has many applications. For example, the value of λ gives the change in the maximum or minimum value of the function being optimised for each unit change in the constraint. To demonstrate, Worked Example 7.19 is repeated in Worked Example 7.21, but with the budget constraint being increased from £30 to £31, that is, by one unit.

WORKED EXAMPLE 7.21
MEANING OF λ

Given the utility function $U = 5xy$, find the change in the maximum level of utility when the budget constraint $5x + y = 30$ is increased by one unit.

SOLUTION

This utility function and budget constraint were given in Worked Example 7.19 where utility was maximised at $x = 3, y = 15$ and maximum utility, $U = 225$.

When the budget constraint is increased by one unit its equation becomes

$$5x + y = 31$$

Therefore the equation of the Lagrangian is

$$L = 5xy + \lambda(31 - 5x - y)$$

Step 1: Find the first-order partial derivatives:

$$L_x = 5y - 5\lambda, \quad L_y = 5x - \lambda, \quad L_\lambda = 31 - 5x - y$$

Step 2: Equate the first-order partial derivatives to zero and solve:

(1) $0 = 5y - 5\lambda$ or $5y = 5\lambda$
(2) $0 = 5x - \lambda$ or $5x = \lambda$
(3) $0 = 31 - 5x - y$

Now, to solve, first eliminate λ from equations (1) and (2):

(1) rewritten $\quad\rightarrow\quad$ $\qquad 5y = 5\lambda$
(2) $\times 5$ $\qquad\rightarrow\quad$ $\qquad \underline{25x = 5\lambda}$
(4) $\quad$ subtract $\qquad 5y - 25x = 0$

Now use equations (3) and (4) to solve for x and y:

(5) $\quad\rightarrow\quad$ (4)/5 $\qquad\rightarrow\quad$ $y - 5x = 0$
(6) $\quad\rightarrow\quad$ (3) rewritten $\quad\rightarrow\quad$ $\underline{y + 5x = 31}$
$\qquad$ add $\qquad\qquad\qquad 2y + 0 = 31 \quad\rightarrow\quad y = 15.5$

Substituting $y = 15.5$ into equation (5) gives $x = 3.1$; therefore utility is at a maximum when $y = 15.5$ and $x = 3.1$. The maximum utility is $U = 5(3.1)(15.5) = 240.25$.

From equation (1), $\lambda = 15.5$. U has increased by $(240.25 - 225) = 15.25$. The increase in U is approximately equal to λ.

If $\lambda > 0$, the optimum value of utility will increase by λ for each unit increase in the constraint and decrease by λ for each unit reduction in the constraint, $xP_X + yP_Y = M$.

	Change in M	**Change in U**
$\lambda > 0$	+1 unit	λ approximately
	−1 unit	−λ approximately
$\lambda < 0$	+1 unit	−λ approximately
	−1 unit	λ approximately

7.4.4 Production functions

The method of Lagrange multipliers may also be used to find the maximum level of output subject to constraints on the availability of labour, L, capital, K, and fixed cost expenditure (C).

WORKED EXAMPLE 7.22
MAXIMISE OUTPUT SUBJECT TO A COST CONSTRAINT

 Find animated worked examples at **www.wiley.com/college/bradley**

Given a Cobb–Douglas production function $Q = L^{0.3} K^{0.7}$, subject to a cost constraint of £150 with the price of labour $w = 3$ and the price of capital $r = 15$:

(a) Find the values of L and K for which production is maximised.
(b) What is the maximum level of production possible subject to the constraint?
(c) What is the maximum level of production possible with no constraint?

SOLUTION

(a) First, write down the equation for the cost constraint:

$$wL + rK = C$$

$$3L + 15K = 150$$

Therefore the Lagrangian is given by the equation

$$L = L^{0.3}K^{0.7} + \lambda(150 - 3L - 15K)$$

Note: In this application both the Lagrangian and the units of labour are represented by the symbol L. Be careful to distinguish between them: you could use La to represent the Lagrangian, if you wish.

Now use the usual method for finding the optimum point (steps 1 and 2 only).
Step 1: Find the first-order partial derivatives:

$$L_L = 0.3L^{-0.7}K^{0.7} - 3\lambda$$

$$L_K = 0.7L^{0.3}K^{-0.3} - 15\lambda$$

$$L_\lambda = 150 - 3L - 15K$$

Step 2: Equate the first-order partial derivatives to zero and solve:
(1) $\quad 0 = 0.3L^{-0.7}K^{0.7} - 3\lambda$
(2) $\quad 0 = 0.7L^{0.3}K^{-0.3} - 15\lambda$
(3) $\quad 0 = 150 - 3L - 15K$
Again, we suggest that you eliminate λ from equations (1) and (2):

(1) $\times 5 \qquad \rightarrow \qquad 0 = 1.5L^{-0.7}K^{0.7} - 15\lambda$
(2) $\times 1 \qquad \rightarrow \qquad \underline{0 = 0.7L^{0.3}K^{-0.3} - 15\lambda}$
$\qquad$ subtract $\qquad 0 = 1.5L^{-0.7}K^{0.7} - 0.7L^{0.3}K^{-0.3}$

and now simplify. There are several ways to simplify, and the following is just one way.

$$1.5L^{-0.7}K^{0.7} = 0.7L^{0.3}K^{-0.3}$$

$$\frac{1.5K^{0.7}L^{0.7}}{0.7L^{0.3}K^{-0.3}} = 1 \quad \text{dividing both sides by } 0.7L^{0.3}K^{-0.3}$$

$$\frac{1.5K^{0.7-(-0.3)}}{0.7L^{0.3-(-0.7)}} = 1 \quad \begin{array}{l}\text{dividing numbers with the same base, subtract} \\ \text{the indices}\end{array}$$

$$\frac{1.5K}{0.7L} = 1 \quad \text{simplify the indices}$$

(4) $1.5K = 0.7L \quad$ or $\quad 15K = 7L \quad$ multiplying both sides by 10 to avoid decimals
We now have two equations in L and K, hence solve for L and K.

(5) $\rightarrow$ (4) rearranged $\quad 7L - 15K = 0$

(6) $\rightarrow$ (3) rearranged $\quad \underline{3L + 15K = 150}$

add $\qquad\qquad\qquad 10L = 150 \rightarrow L = 15$

Substituting $L = 15$ into equation (5) gives $7(15) - 15K = 0$ or $K = 7$. Therefore, production is maximised when $L = 15$ and $K = 7$.
Check that these values satisfy the cost constraint:

$$3L + 15K = 150 \quad \rightarrow \quad 3(15) + 15(7) = 150 \quad \rightarrow \quad 45 + 105 = 150 \text{ TRUE}$$

(b) The maximum level of production is found by substituting these values back into the equation for the production function:

$$Q = L^{0.3}K^{0.7} = 15^{0.3}7^{0.7} = (2.2533)(3.9045) = 8.798 \quad \text{using the calculator}$$

(c) If there were no limits on L and K, then there would be no limit on the level of production, since $L^{0.3}$ increases indefinitely as L increases, and so also does $K^{0.7}$ as K increases. (Try some values on the calculator!)

7.4.5   Minimising cost subject to a production constraint

Again, the Lagrange multiplier method is used to find the minimum cost subject to a production constraint.

WORKED EXAMPLE 7.23
MINIMISE COSTS SUBJECT TO A PRODUCTION CONSTRAINT

A production function is $Q = 12L^{0.5}K^{0.5}$. Find the values of L and K which minimise costs when labour costs £25 per unit (wage rate), capital costs £50 per unit (rent) and when the production constraint is 240 units of output.

SOLUTION
The equation of the cost function is

$$wL + rK = C$$

$$25L + 50K = C$$

This is the function which is to be minimised subject to the production constraint. The Lagrangian is

$$L = 25L + 50K + \lambda(240 - 12L^{0.5}K^{0.5})$$

Now work through the method for finding optimum points.

Step 1: Find the first-order partial derivatives:

$$L_L = 25 - \lambda 12(0.5L^{0.5-1}K^{0.5}) = 25 - \lambda 6L^{-0.5}K^{0.5}$$

$$L_K = 50 - \lambda 12(0.5L^{0.5}K^{0.5-1}) = 50 - \lambda 6L^{0.5}K^{-0.5}$$

$$L_\lambda = 240 - 12L^{0.5}K^{0.5}$$

Step 2: Equate the first-order partial derivatives to zero and solve:

(1) $0 = 25 - \lambda 6L^{-0.5}K^{0.5}$ → $25 = \lambda 6L^{-0.5}K^{0.5}$
(2) $0 = 50 - \lambda 6L^{0.5}K^{-0.5}$ → $50 = \lambda 6L^{0.5}K^{-0.5}$
(3) $0 = 240 - 12L^{0.5}K^{0.5}$ → $240 = 12L^{0.5}K^{0.5}$

As usual, eliminate λ from equations (1) and (2). Divide the corresponding sides of each equation:

$$\frac{(1)}{(2)} \rightarrow \frac{25}{50} = \frac{\lambda 6L^{-0.5}K^{0.5}}{\lambda 6L^{0.5}K^{-0.5}}$$

$$\frac{1}{2} = \frac{L^{-0.5}K^{0.5}}{L^{0.5}K^{-0.5}}$$

$$\frac{1}{2} = \frac{K}{L}$$

$$\frac{L}{2} = K$$

Substitute $K = L/2$ into equation (3):

$$240 = 12L^{0.5}\left(\frac{L}{2}\right)^{0.5}$$

$$240 = \frac{12}{\sqrt{2}}L$$

$$L = \frac{\sqrt{2}(240)}{12}$$

$$L = 28.28$$

and $K = 14.14$. Therefore $C = 25L + 50K = 25(28.28) + 50(14.14) = 1414$.

PROGRESS EXERCISES 7.9

Constrained Optimisation in General

1. Find the values of x and y for which the utility function $U = x^{0.2}y^{0.8}$ is maximised, subject to the budget constraint, $180 = 4x + 2y$ (assume second-order conditions are satisfied).

2. (See question 1) Given $U = x^{0.2} y^{0.8}$:
   (a) Write down the equations for the marginal utilities.
   (b) Show that $U_x/U_y = P_X/P_Y$ when $U = x^{0.2} y^{0.8}$ is maximised, subject to the budget constraint, $x P_X + y P_Y = M$.
3. A student is restricted to a monthly budget constraint of £200 which is allocated to two activities: bridge and skating. His utility function is given by the equation $U = 5x^{0.2}y^{0.8}$, where x and y represent the number of hours spent on skating and bridge, respectively. Skating costs £5 per hour, while bridge costs £2 per hour.
   (a) Derive expressions for MU_x and MU_y.
   (b) Solve the simultaneous equations

$$\frac{MU_x}{MU_y} = \frac{P_X}{P_Y} \quad \text{and} \quad x P_X + y P_Y = M$$

   (c) Use Lagrange multipliers to determine the number of hours that should be spent on each activity to maximise utility.
   Comment on the answers in (b) and (c).
4. The production function for a garage which services cars and trucks is $Q = 15L^{2/3}K^{1/3}$. If each labour unit used costs £5, while each unit of capital costs £3, find the number of units of capital and labour which should be used to maximise productivity subject to the constraint $LP_L + KP_K = 450$.
5. Determine the values of K and L which minimise the cost function $C = 3K + 5L$ subject to the production constraint, $250 = 10K^{0.5}L^{0.5}$.

   Note: This is rather different from the usual 'maximise the production function subject to the constraint'.
6. A firm has a production function $Q = 15L^{1/5}K^{1/3}$, where L and K are the number of units of labour and capital. Find the maximum level of production, subject to the constraint $K + 4L = 20$.
7. A monopolist has a total cost function $TC = 10x + 6$, where $x = x_1 + x_2$, the sum of the number of units of goods 1 and 2 produced, respectively. The product may be sold in two separate markets, in which the demand functions are given by the equations

$$P_1 = 50 - 5x_1 \quad \text{and} \quad P_2 = 30 - 2x_2$$

   (a) Write down the equation of the profit function. Hence determine the price which should be charged in each market to maximise profits. Calculate the maximum profit.
   (b) Determine the price elasticity of demand in each market. Comment.
8. A school has a utility function $U = 5 \ln(x_A) + 10 \ln(x_B)$, for two services, A (sports) and B (remedial teaching). Find the maximum level of utility if sports (A) costs £200 per unit, while remedial teaching (B) costs £400 per unit. The school's budget limit is £3600.
9. A firm's production function is $Q = 50L^{0.6}K^{0.4}$, subject to the cost constraint $8L + 4K = 400$.
   (a) Deduce the price of a unit of labour and capital from the constraint equation.
   (b) Determine the levels of L and K at which production is maximised.
   (c) Show that $MP_L/MP_K = P_L/P_K$ when production is maximised.

10. A firm manufactures a good J (jam) from raw materials X (strawberries) and Y (blackberries). The quantity of jam (in jars) manufactured per day is $Q = 20x^{0.2}y^{0.8}$. If each unit of X (strawberries) costs £20 and each unit of Y (blackberries) costs £2, but total daily expenditure is restricted to £1450,
   (a) Write down the equation of the constraint.
   (b) Use the method of Lagrange to determine the number of units of strawberries and blackberries which should be purchased to maximise production subject to the constraint. Calculate the maximum production.
   (c) Show that $MP_x/MP_y = P_X/P_Y$ when production is maximised.
11. The demand functions and total cost function for two goods are given as follows:

$$P_1 = 106 - 23Q_1 + 5Q_2, \quad P_2 = 546 - 5Q_2 + 10Q_1, \quad TC = 5Q_1Q_2 + 4Q_1 + 504Q_2$$

   (a) Show that the profit function (π) is

$$\pi = 102Q_1 + 42Q_2 - 23Q_1^2 + 10Q_1Q_2 - 5Q_2^2$$

   Hence determine the values of Q_1 and Q_2 for which profit is a maximum. Calculate the maximum profit.
   (b) A production requirement is that

$$Q_2 + Q_1 = 20$$

   Find the maximum profit subject to this constraint.
12. A utility function is given by the equation

$$U = 120\, x^{0.7}\, y^{0.3}$$

   where x is the number of hours spent playing golf and y is the hours spent playing cards per month.

   Calculate the values of x and y for which utility is maximised, subject to the constraint

$$21x + 3y = 180$$

7.5   Summary

Function of one variable $y = f(x)$

The **derivative** is dy/dx.

The **differential** of y is

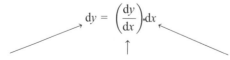

$$dy = \left(\frac{dy}{dx}\right)\cdot dx$$

where   dy: differential of y   derivative of y w.r.t. x   dx: differential of x
and dy and dx are infinitesimally small.

Function of two variables: $z = f(x, y)$

The first-order partial derivatives are

$\partial z/\partial x$ is the partial derivative of z w.r.t. x, y treated as a constant (abbreviated to z_x)

$\partial z/\partial y$ is the partial derivative of z w.r.t. y, x treated as a constant (abbreviated to z_y)

The second-order partial derivatives are

$$\frac{\partial^2 z}{\partial x^2} \equiv z_{xx}, \quad \frac{\partial^2 z}{\partial y^2} \equiv z_{yy}, \quad \frac{\partial^2 z}{\partial y \partial x} \equiv z_{yx}, \quad \frac{\partial^2 z}{\partial x \partial y} \equiv z_{xy}$$

The total differential of z is given by

$$dz = \left(\frac{\partial f}{\partial x}\right) dx + \left(\frac{\partial f}{\partial y}\right) dy$$

or for small changes $dx \to \Delta x$, etc., the **small (incremental) changes formula** is given by

$$\Delta z \cong \left(\frac{\partial f}{\partial x}\right) \Delta x + \left(\frac{\partial f}{\partial y}\right) \Delta y$$

Unconstrained optimisation

Step 1: Find the first and second derivatives

$$\frac{\partial z}{\partial x}, \quad \frac{\partial z}{\partial y}, \quad \frac{\partial^2 z}{\partial x^2}, \quad \frac{\partial^2 z}{\partial y^2}, \quad \frac{\partial^2 z}{\partial x \partial y}$$

Step 2: First-order conditions:
At stationary point $\partial z/\partial x = 0$, $\partial z/\partial y = 0$
Solve these two equations for the x- and y-coordinates of the turning point(s).
Step 2a: If required, find z for the x- and y-coordinates of the turning point(s).
Step 3: Second-order conditions:

The point is a minimum if $\partial^2 z/\partial x^2 > 0$ and $\partial^2 z/\partial y^2 > 0$ and provided $\Delta > 0$.

The point is a maximum if $\partial^2 z/\partial x^2 < 0$ and $\partial^2 z/\partial y^2 < 0$ and provided $\Delta > 0$.

The point is a point of inflection if both second derivatives have the same sign but $\Delta < 0$.

The point is a saddle point if the second derivatives have different signs and $\Delta < 0$, where

$$\Delta = \left(\frac{\partial^2 z}{\partial x^2}\right)\left(\frac{\partial^2 z}{\partial y^2}\right) - \left(\frac{\partial^2 z}{\partial x \partial y}\right)^2$$

Constrained optimisation: Lagrange multipliers

Given a function $z = f(x,y)$ to be optimised, subject to a constraint, $M - ax - by = 0$, the Lagrangian (L) is defined as $L = f(x, y) + \lambda(M - ax - by)$. Solve the equations

$$\frac{\partial L}{\partial x} = 0, \quad \frac{\partial L}{\partial y} = 0, \quad \frac{\partial L}{\partial \lambda} = 0$$

for the values of x, y and λ which optimise L.
   Second-order conditions are not required in this text.

Applications

Production: Cobb–Douglas production: $Q = AL^\alpha K^\beta$:

1. Represent production functions in two-dimensional diagrams (isoquants) by fixing the value of Q, then express $K = f(L)$.
2. The slope dK/dL is called the **marginal rate of technical substitution** (MRTS).
3. Slope is also expressed as

$$(i) \ \frac{dK}{dL} = -\frac{MP_L}{MP_K}, \quad (ii) \ MRTS = \frac{\alpha}{\beta} \frac{K}{L}$$

4. Returns to scale. When L and K are each replaced by λL and λK, respectively, in

$$Q_1 = AL^\alpha K^\beta \text{ then } Q_2 = A(\lambda L)^\alpha (\lambda K)^\beta = \lambda^{\alpha+\beta} AL^\alpha K^\beta = \lambda^{\alpha+\beta} Q_1$$

If $\alpha + \beta = 1$: constant returns to scale, i.e., $Q_2 = \lambda Q_1$
If $\alpha + \beta < 1$: decreasing returns to scale, i.e., $Q_2 < \lambda Q_1$
If $\alpha + \beta > 1$: increasing returns to scale, i.e., $Q_2 > \lambda Q_1$
(all these rules apply to a Cobb–Douglas function).

Partial elasticity

1. Price elasticity of demand:

$$\varepsilon_d = \frac{\partial Q_A}{\partial P_A} \frac{P_A}{Q_A}$$

2. Income elasticity of demand:

$$\varepsilon_Y = \frac{\partial Q_A}{\partial Y} \frac{Y}{Q_A}$$

3. Cross-price elasticity of demand:

$$\varepsilon_c = \frac{\partial Q_A}{\partial P_B} \frac{P_B}{Q_A}$$

4. Elasticity w.r.t. labour:

$$\varepsilon_{QL} = \frac{\partial Q}{\partial L} \frac{L}{Q}$$

Note:

$$\varepsilon_{QL} = \frac{\partial Q}{\partial L} \frac{L}{Q} = \frac{\partial Q/\partial L}{Q/L} = \frac{MP_L}{AP\,L}$$

5. Elasticity w.r.t. capital:

$$\varepsilon_{QK} = \frac{\partial Q}{\partial K} \frac{K}{Q} \quad \text{and} \quad \varepsilon_{QK} = \frac{\partial Q}{\partial K} \frac{K}{Q} = \frac{\partial Q/\partial K}{Q/K} = \frac{MP_K}{AP\,K}$$

National income model multipliers

$$\frac{\partial Y_e}{\partial I} = \frac{1}{1 - b(1 - t)} : \text{the investment multiplier}$$

$$\frac{\partial Y_e}{\partial G} = \frac{1}{1 - b(1 - t)} : \text{the government expenditure multiplier}$$

$$\frac{\partial Y_e}{\partial t} = \frac{bY_e}{1 - b(1 - t)} : \text{the income tax rate multiplier}$$

www.wiley.com/college/bradley

Go to the website for Problems in Context

TEST EXERCISES 7

1. Determine the first and second partial derivatives of the following:
   (a) $U = x^3 + 2y^3$ (b) $Z = x^2y + 10xy$ (c) $Q = 100L^{0.4}K^{0.6}$
2. A production function $Q = 100L^{0.4}K^{0.6}$ relates output, Q, to the number of labour units, L, and capital units K.
   (a) Derive the equation for the marginal and the average products of labour and capital.
   (b) Prove the identity

$$Q = L\frac{\partial Q}{\partial L} + K\frac{\partial Q}{\partial K}$$

(c) Use partial derivatives to calculate the approximate percentage change in Q when L increases by 6% while K decreases by 4%.

3. The satisfaction derived from activities X (watching TV) and Y (playing golf) is given by the utility function: $U = 4x^{0.5}y^{0.5}$.
   (a) Show that, when $U = 40$, the equation of the corresponding indifference curve is $y = 100/x$.
   (b) Plot the indifference curve for $0 < x < 5$ and derive the equation for the slope of the indifference curve. Use the second derivative to determine the curvature algebraically.

4. (See question 3) $U = 4x^{0.5}y^{0.5}$.
   Derive the equations for
   (a) the marginal utility w.r.t. x ($MU_x = \partial U/\partial x$) and (b) the marginal utility w.r.t. y ($MU_y = \partial U/\partial y$).
   Show that the slope of the indifference curve may be expressed as

$$\frac{dy}{dx} = -\frac{MU_x}{MU_y}$$

5. (See question 3) $U = 4x^{0.5}\,y^{0.5}$.
   (a) Explain the terms (i) constant, (ii) increasing and (iii) decreasing returns to scale.
   (b) Calculate the level of utility when the hours spent at each activity are
       (i) 4 hours of TV and 6 hours of golf, (ii) 2 hours of TV and 3 hours of golf.
       Do the results indicate that the function exhibits increasing returns to scale?
   (c) Use the definitions of returns to scale to show that the utility function exhibits constant returns to scale.

6. Given a Cobb–Douglas production function $Q = AL^\alpha K^\beta$, define MP_L (the marginal product of labour), APL (the average product of labour), ε_{QL} (the partial elasticity w.r.t. labour). Show that $\varepsilon_{QL} = MP_L/APL$.

7. Given a Cobb–Douglas production function $Q = 200L^{0.4}K^{0.8}$,
   (a) Determine the price elasticity w.r.t. labour: the price elasticity w.r.t. capital.
   (b) Determine the maximum level of production subject to the constraint $L + 4K = 300$.

8. A utility function $U = 100x^{0.8}y^{0.2}$ relates level of benefit to the number of glasses of lemon tea and wine consumed, respectively.
   (a) Use partial derivatives to calculate the overall increase in utility when consumption of wine increases by 5% and consumption of tea increases by 8%.
   (b) If consumption of both beverages increases by 50% calculate the overall increase in utility. What type of returns to scale is exhibited by this function? Give reasons.

9. (See question 8) A budget constraint of £40 is imposed, hence the equation of the constraint is $x + 4y = 40$. Find the number of units of each beverage which should be consumed to maximise utility.

INTEGRATION AND APPLICATIONS

8

8.1   Integration as the Reverse of Differentiation

8.2   The Power Rule for Integration

8.3   Integration of the Natural Exponential Function

8.4   Integration by Algebraic Substitution

8.5   The Definite Integral and the Area under a Curve

8.6   Consumer and Producer Surplus

8.7   First-order Differential Equations and Applications

8.8   Differential Equations for Limited and Unlimited Growth

8.9   *Integration by Substitution and Integration by Parts* (website only)

8.10   Summary

Chapter Objectives

Integration is the process that reverses differentiation. In this chapter applications include the calculation of consumer and producer surplus and the solution of first-order differential equations. At the end of this chapter you should be able to:

- Integrate standard functions and evaluate definite integrals of standard functions
- Integrate functions of linear functions and evaluate their definite integrals
- Determine the net area enclosed between a curve and the lines $x = a$ and $x = b$
- Calculate the consumer and producer surplus and illustrate these graphically
- Integrate marginal functions to obtain the corresponding total function
- Solve differential equations of the type $dy/dt = f(t); dy/dt = ky; dy/dt = f(t)g(t)$
- Determine whether the solution of a differential equation exhibits limited or unlimited growth
- Determine whether a function exhibits a constant proportional rate of growth.

8.1   Integration as the Reverse of Differentiation

You have already encountered many examples of mathematical operations which reverse each other. One example is multiplication then division by the same number, or vice versa:

$$N \times x \div x = N \quad \text{such as } 345 \times 20 \div 20 = 345$$

Another example is taking the antilog of the log of any number (to the same base) or vice versa:

$$\text{antilog}\,[\log(N)] = N \quad \text{such as antilog}[\log(50)] = 50 \,\text{or}\, 10^{\log(50)} = 50$$

A third example is to integrate the derivative (w.r.t. the same variable), or vice versa; in other words, when a function is differentiated then integrated, you will end up with the same function back again. This statement is written

$$\int [d(\text{function})] = \text{function} \tag{8.1}$$

$$d\left[\int (\text{function})\right] = \text{function} \tag{8.2}$$

where $\int$ indicates integration and d indicates differentiation (Figure 8.1).

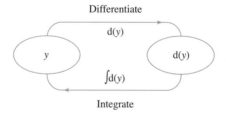

Figure 8.1   Integration reverses differentiation

Or, when the independent variable is stated explicitly, e.g., $y = f(x) + c$, the operation 'integration w.r.t. x' is denoted by enclosing the function to be integrated between the integration symbol $\int$ and dx. In general, the function obtained by integrating $f(x)$ is symbolised by $F(x)$ in this text (Figure 8.2).

Note the inclusion of the constant c. The derivative of a constant term is zero, so when integrating (thereby reversing differentiation), allowance must be made for the possibility that the function (which was differentiated) may have contained a non-zero constant term. This definition of integration as the operation which reverses differentiation allows us to deduce the rules for integration, such as the power rule.

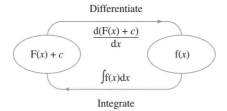

Figure 8.2 Integration reverses differentiation

Note: As always, rigorous mathematical analysis and proofs are beyond the scope of the text. The examples and exercises in this chapter deal with functions which are continuous over the interval of integration. A rigorous treatment of this topic is beyond the scope of this book.

8.2 The Power Rule for Integration

At the end of this section you should be able to:

- Integrate power functions, x^n
- Integrate the sum or difference of several power functions
- Integrate any linear function raised to a power.

Deduce the power rule for integration

Following from Section 8.1, the integral of $f(x) = x$ is the function which is differentiated to give x. The derivative of $\frac{1}{2}x^2 + c$ is x since

$$\frac{d}{dx}\left(\frac{x^2}{2} + c\right) = \frac{2x}{2} + 0 = x \quad \text{therefore} \quad \int x\,dx = \frac{x^2}{2} + c$$

See Figure 8.3.

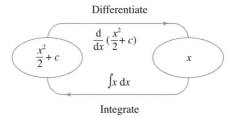

Figure 8.3 Integration reverses differentiation

This example and the following examples demonstrate how the power rule for integration is deduced.

	$y = F(x) + c$	$\dfrac{d[F(x) + c]}{dx} = f(x)$	$\int f(x)dx = F(x) + c$
Example (a)	$y = \dfrac{x^2}{2} + c$	$\dfrac{dy}{dx} = \dfrac{2x}{2} = x$	$\int x\,dx = \dfrac{x^2}{2} + c$
Example (b)	$y = \dfrac{x^3}{3} + c$	$\dfrac{dy}{dx} = x^2$	$\int x^2 dx = \dfrac{x^3}{3} + c$
Example (c)	$y = \dfrac{x^4}{4} + c$	$\dfrac{dy}{dx} = x^3$	$\int x^3 dx = \dfrac{x^4}{4} + c$
Example (d)	$y = \dfrac{x^{n+1}}{n+1} + c$	$\dfrac{dy}{dx} = x^n$	$\int x^n dx = \dfrac{x^{n+1}}{n+1} + c$

Example (d) is the power rule for integration:

$$\int x^n dx = \frac{x^{n+1}}{n+1} + c \tag{8.3}$$

The power rule is very simple to apply:

Step 1: Always start by simplifying where possible. If you do not simplify, you will give yourself a lot of extra work or, worse still, it may not be possible to integrate the unsimplified function at all.

Step 2: Write the power function 'on line', so that you may state the power n clearly.

Step 3: Apply the power rule for integration:

$$\int x^n dx = \frac{x^{n+1}}{n+1} + c$$

WORKED EXAMPLE 8.1
USING THE POWER RULE FOR INTEGRATION

Integrate the following functions:

(a) x^4 (b) $\dfrac{x^4}{x^2}$ (c) $\dfrac{1}{x^4}$ (d) $\sqrt{x}$ (e) $\dfrac{1}{\sqrt{x^3}}$

SOLUTION

	Step 1 Simplify, if possible	Step 2 State n and $n + 1$	Step 3 Apply power rule $\int x^n dx = \dfrac{x^{n+1}}{n+1} + c$	Simplify the answer if possible
(a) x^4	x^4	$n = 4$ $n + 1 = 5$	$\int x^4 dx = \dfrac{x^5}{5} + c$	Not necessary
(b) $\dfrac{x^4}{x^2}$	$\dfrac{x^4}{x^2} = x^{4-2} = \dfrac{x^4}{x^2}$	$n = 2$ $n + 1 = 3$	$\int x^2 dx = \dfrac{x^3}{3} + c$	Not necessary
(c) $\dfrac{1}{x^4}$	$\dfrac{1}{x^4} = x^{-4}$	$n = -4$ $n + 1 = -3$	$\int x^{-4} dx = \dfrac{x^{-3}}{-3} + c$	$\dfrac{x^{-3}}{-3} = -\dfrac{x^{-3}}{3}$ $= \dfrac{1}{3x^3} + c$
(d) $\sqrt{x}$	$\sqrt{x} = x^{0.5}$	$n = 0.5$ $n + 1 = 1.5$	$\int x^{0.5} dx = \dfrac{x^{1.5}}{-3} + c$	Not necessary
(e) $\dfrac{1}{\sqrt{x^3}}$	$\dfrac{1}{\sqrt{x^3}} = \dfrac{1}{(x^3)^{0.5}}$ $= x^{-1.5}$	$n = -1.5$ $n + 1 = -0.5$	$\int x^{-1.5} dx = \dfrac{x^{-0.5}}{-0.5} + c$	$-\dfrac{x^{-0.5}}{0.5} = -2x^{0.5}$ $= -\dfrac{2}{x^{0.5}} + c$

The minus one exception to the power rule

The power rule applies for any value of n except $n = -1$. If $n = -1$ then $n + 1 = -1 + 1 = 0$, therefore, according to the power rule, we have

$$\int x^{-1} dx = \frac{x^{-1+1}}{-1+1} = \frac{x^0}{0} = ?$$

Division by zero is not defined, so the power rule cannot be used in this case. Therefore to integrate x^{-1} look for the function whose derivative is x^{-1}. See Figure 8.4.

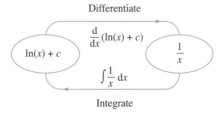

Differentiate

$\dfrac{d}{dx}(\ln(x) + c)$

$\ln(x) + c$ $\dfrac{1}{x}$

$\int \dfrac{1}{x} dx$

Integrate

Figure 8.4 $\int \frac{1}{x} dx = \ln(x) + c$ since integration reverses differentiation

Hence the rule for integrating x^{-1} is

$$\boxed{\int \frac{1}{x}\,dx = \int x^{-1}\,dx = \ln|x| + c}$$ (8.4)

where $|x|$ means the absolute value of x. The use of $|x|$ will be important in the section on definite integration. In this section, on indefinite integration, we simply write

$$\int \frac{1}{x}\,dx = \ln(x) + c$$

The integral of a constant term

The integral of a constant term (K) is deduced as follows:

$$\int K\,dx = \int Kx^0\,dx \quad \text{since } x^0 = 1$$

Now apply the power rule. When $n = 0 \rightarrow n + 1 = 0 + 1 = 1$, therefore

$$\int Kx^0\,dx = K\int x^0\,dx = K\frac{x^1}{1} + c = Kx + c$$

$$\boxed{\int K\,dx = Kx + c}$$ (8.5)

In other words, the integral of a constant term is the constant term multiplied by the independent variable.

Working rules for integration

The following working rules (similar to those used in differentiation) apply when integrating sums and differences of several functions.

- As in differentiation, the sum or difference of several power functions is integrated by integrating each function separately:

$$\int \{f(x) + g(x)\}\,dx = \int f(x)\,dx + \int g(x)\,dx$$ (8.6)

- As in differentiation, the integral of a constant multiplied by a variable term is the constant multiplied by the integral of the variable term:

$$\int Kf(x)\,dx = K\int f(x)\,dx$$ (8.7)

WORKED EXAMPLE 8.2
INTEGRATING SUMS AND DIFFERENCES, CONSTANT MULTIPLIED BY VARIABLE TERM

Integrate (a) $x^2 + \dfrac{1}{x^2}$ and (b) $5x - 3x^2$.

SOLUTION

(a)

Step 1: Simplify, $x^2 + \dfrac{1}{x^2} = x^2 + x^{-2}$

Step 2: State n, hence $n + 1$ for each term

$$\text{for } x^2 \text{ we have } n = 2 \rightarrow n + 1 = 3$$
$$\text{for } x^{-2} \text{ we have } n = -2 \rightarrow n + 1 = -2 + 1 = -1$$

Step 3: Apply the power rule to obtain

$$\int (x^2 + x^{-2})dx = \frac{x^3}{3} + \left(\frac{x^{-1}}{-1}\right) = \frac{x^3}{3} - \frac{x^{-1}}{1} = \frac{x^3}{3} - \frac{1}{x} + c$$

(b) The steps are worked through very briefly (explain each step to yourself):

$$\int (5x - 3x^2)dx = \int 5x\,dx - \int 3x^2\,dx$$

$$= 5\int x\,dx - 3\int x^2\,dx$$

$$= 5\frac{x^2}{2} - 3\frac{x^3}{3} = 2.5x^2 - x^3 + c$$

Note 1: You can easily check that your integration is correct. According to equation (8.2), $d\left[\int (\text{function})\right] = \text{function}$. So differentiate your answer and you should end up with the function you originally set out to integrate.

Note 2: In the examples so far we have integrated functions of x w.r.t. x. The same rules follow for any other variable; integrate functions of Q w.r.t. Q, functions of P w.r.t. P, etc. Now for some more general worked examples.

WORKED EXAMPLE 8.3

INTEGRATING MORE GENERAL FUNCTIONS

Integrate the following: (a) $\dfrac{x + 5x^2}{x}$ and (b) $(Q + 3)\sqrt{Q}$.

SOLUTION

(a)

Step 1: Simplify, $\dfrac{x + 5x^2}{x^2} = \dfrac{x}{x^2} + \dfrac{5x^2}{x^2} = \dfrac{1}{x} + 5$

Step 2: State n for each term: $1/x = x^{-1} \to n - 1$, so this is the exception to the power rule. Integration of $1/x$ is given by equation (8.4). The term 5 is a constant; from equation (8.5) we have that integrating a constant gives the constant multiplied by the variable.

Step 3: Write out the integration, $\int (1/x + 5)dx = \ln|x| + 5x + c$. This is the answer; it cannot be further simplified.

(b) Since this is a function of Q, integration is w.r.t. Q. Determine $\int (Q + 3)\sqrt{Q}dQ$.

Step 1: Simplify before attempting to integrate:

$$(Q + 3)\sqrt{Q} = (Q + 3)(Q)^{0.5}$$

$$= (Q + 3)Q^{0.5}$$

$$= Q \times Q^{0.5} + 3 \times Q^{0.5}$$

$$= Q^{1.5} + 3Q^{0.5} \quad \text{since} \quad Q^1 Q^{0.5} = Q^{1.5}$$

Step 2: State n, hence $n + 1$ for each term:

$$\text{for } Q^{1.5} \text{ we have } n = 1.5 \quad \to \quad n + 1 = 2.5$$

$$\text{for } Q^{0.5} \text{ we have } n = 0.5 \quad \to \quad n + 1 = 1.5$$

Step 3: Apply the power rule to obtain

$$\int (Q + 3)\sqrt{Q}dQ = \int (Q^{1.5} + 3 \times Q^{0.5})dQ = \dfrac{Q^{2.5}}{2.5} + 3 \times \dfrac{Q^{1.5}}{1.5} + c$$

Simplify the answer; in this case the numbers:

$$\dfrac{Q^{2.5}}{2.5} + 3 \times \dfrac{Q^{1.5}}{1.5} = 0.4Q^{2.5} + 2Q^{1.5} + c$$

PROGRESS EXERCISES 8.1

Integration by the Power Rule

Integrate each of the following.

1. $x + x^3 + x^{3.5}$

2. $\dfrac{1}{\sqrt{x^5}}$

3. $\dfrac{x}{\sqrt{x^5}}$

4. $\dfrac{x + \sqrt{x}}{x}$

5. $2x + \dfrac{2}{x}$

6. $2x + \dfrac{1}{2x}$

7. $x\sqrt{x} + 3$

8. $5x + \sqrt{5x}$

9. $x(x - 3)^2$

10. $\dfrac{x + 7}{x}$

11. $x^2\left(1 + \dfrac{1}{x^2}\right)$

12. $Q(Q - \sqrt{Q})$

13. $t\left(1 + \dfrac{2}{t}\right)$

14. $4 + \dfrac{4}{t}$

15. $5 + \dfrac{1}{Q}$

16. $\dfrac{2Q - 5}{Q}$

17. $Q(20 - 0.5Q)$

18. $\dfrac{Q + 4}{Q^2}$

19. $\dfrac{1}{5Q} - 5$

20. $\dfrac{1}{0.2}\left(Y - \dfrac{Y}{2}\right)$

8.3   Integration of the Natural Exponential Function

In Chapter 4 it was stated that the base e in the exponential function e^x is the natural base for describing growth and decay for all types of system. The number e is also a natural number in calculus. e^x is the only function which does not change when differentiated or integrated (Figure 8.5).

Therefore

$$\int e^x \mathrm{d}x = e^x + c \tag{8.8}$$

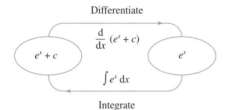

Figure 8.5   Integration of e^x

WORKED EXAMPLE 8.4

INTEGRATING FUNCTIONS CONTAINING e^x

Integrate (a) $3e^x + 5$ and (b) $\dfrac{e^t + e^{2t}}{e^t}$

SOLUTION

Follow the usual procedure for integration, simplifying first if necessary. State and apply the appropriate rule for integration. The solutions are as follows:

(a)

$3e^x$ is 3 multiplied by the standard function, which is integrated according to equation (8.8); 5 is a constant term, which is integrated according to equation (8.5) to give $5x$

Therefore

$$\int (3e^x + 5)\, dx$$

$$= 3 \int e^x dx + \int 5\, dx$$

$$= 3e^x + 5x + c$$

(b)

Here we must simplify first:

$$\frac{e^t + e^{2t}}{e^t} = \frac{e^t}{e^t} + \frac{e^{2t}}{e^t} \quad \text{dividing each by } e^t$$

$$= 1 + e^{2t}e^{-t}$$

$$= 1 + e^t \quad \text{rule 1, indices}$$

This simplifies to the sum of two terms: 1 is a constant term, integrated according equation (8.5) and e^t is a standard function; use equation (8.8) to obtain

$$\int \frac{e^t + e^{2t}}{e^t} dt = \int (1 + e^t) dt = 1 \times t + e^t + c$$

$$= t + e^t + c$$

The independent variable is x in part (a), so the integration must be w.r.t. x; the independent variable is t in part (b), so the integration must be w.r.t. t.

8.4   Integration by Algebraic Substitution

At the end of this section you should be able to:

- Integrate functions of linear functions by substitution
- Integrate certain products by substitution.

8.4.1   Using substitution to integrate functions of linear functions

Compare the functions on the left of the table with those on the right.

Remember

A function is a rule for operating on x.

Function of x, $f(x)$	Function of a linear function, $f(5x - 2)$
x^{10}, x raised to the power 10	$(5x - 2)^{10}$, $(5x - 2)$ raised to the power 10
$\dfrac{1}{x}$, 1 divided by x	$\dfrac{1}{(5x - 2)}$, 1 divided by $(5x - 2)$
e^x, e to the power x	$e^{(5x-2)}$, e to the power $(5x - 2)$

- The functions of x on the left are standard functions whose integrals are known.
- The functions on the right are of exactly the same type as those on the left, except that each x has been replaced by $(5x - 2)$, a linear function of x. These are called functions of linear functions; here they are functions of $(5x - 2)$.

To integrate these functions of linear functions, consider the effect of replacing the linear function by a single variable, say u. The substitution of u for $(5x - 2)$ would reduce each to a standard function in u. However, there is just one further adjustment necessary before integration of the standard function of u may proceed, as demonstrated in the following worked example.

WORKED EXAMPLE 8.5
INTEGRATING FUNCTIONS OF LINEAR FUNCTIONS BY SUBSTITUTION

Use the method of substitution to integrate the following:

(a) $\displaystyle\int (5x - 2)^{10}dx$ (b) $\displaystyle\int \frac{1}{5x - 2}dx$ (c) $\displaystyle\int e^{5x-2}dx$

SOLUTION
(a) Substitute u for the linear function $(5x - 2)$:

$$\int (5x - 2)^{10}dx \rightarrow \int (u)^{10} \underbrace{dx}_{\downarrow}$$

a du is required here in order to integrate functions of u

☺ This symbol indicates integration $\rightarrow \int (\ldots)du$

du indicates that only functions of u will be integrated

To deduce an expression for dx in terms of du, use the substitution equation $u = (5x - 2)$ as outlined below.

Method

Substitution equation $u = 5x - 2$

Differentiate w.r.t. x $\dfrac{du}{dx} = 5 \rightarrow du = 5\,dx$

Multiply both sides by $\frac{1}{5}$ $\dfrac{1}{5}du = \dfrac{1}{\not{5}}\dfrac{\not{5}dx}{1} = dx \rightarrow dx = \dfrac{1}{5}du$

Now, replace dx by $\frac{1}{5}du$ $\displaystyle\int u^{10}dx \rightarrow \int u^{10}\left(\dfrac{1}{5}\right)du$

Factor out the constant $\frac{1}{5}$ and integrate $\dfrac{1}{5}\int u^{10}du = \dfrac{1}{5}\dfrac{u^{11}}{11} + c$

Write u back in terms of x $= \dfrac{(5x-2)^{11}}{55} + c$

The solutions to parts (a), (b) and (c) are written briefly as follows.

(a) $\displaystyle\int (5x-2)^{10}dx$	(b) $\displaystyle\int \dfrac{1}{(5x-2)}dx$	(c) $\displaystyle\int e^{5x-2}dx$
Step 1 Let $u = 5x - 2$, so the integral becomes $\int (u)^{10}dx$, and dx needs to be expressed in terms of du	**Step 1** Let $u = 5x - 2$, so the integral becomes $\int u^{-1}dx$, and dx needs to be expressed in terms of du	**Step 1** Let $u = 5x - 2$, so the integral becomes $\int e^{u}\,dx$, and dx needs to be expressed in terms of du
Step 2 Use the substitution equation to deduce an expression for dx in terms of du by differentiating w.r.t. x, then solving for dx. In this problem $dx = \frac{1}{5}du$	**Step 2** Use the substitution equation to deduce an expression for dx in terms of du by differentiating w.r.t. x, then solving for dx. In this problem $dx = \frac{1}{5}du$	**Step 2** Use the substitution equation to deduce an expression for dx in terms of du by differentiating w.r.t. x, then solving for dx. In this problem $dx = \frac{1}{5}du$
Step 3 Integrate function f(u): $\displaystyle\int u^{10}dx \rightarrow \int u^{10}\left(\frac{1}{5}\right)du$ $\displaystyle\int u^{10}\left(\frac{1}{5}\right)du = \frac{1}{5}\int u^{10}du$ $= \dfrac{1}{5}\dfrac{u^{11}}{11} + c$	**Step 3** Integrate function f(u): $\displaystyle\int \frac{1}{u}dx \rightarrow \int \frac{1}{u}\left(\frac{1}{5}\right)du$ $\displaystyle\int \frac{1}{u}\left(\frac{1}{5}\right)du = \frac{1}{5}\int \frac{1}{u}du$ $= \dfrac{1}{5}\ln|u| + c$	**Step 3** Integrate function f(u): $\displaystyle\int e^{u}dx \rightarrow \int e^{u}\left(\frac{1}{5}\right)du$ $\displaystyle\int e^{u}\left(\frac{1}{5}\right)du = \frac{1}{5}\int e^{u}du$ $= \dfrac{1}{5}e^{u} + c$
Step 4 u is written in terms of x:	**Step 4** u is written in terms of x:	**Step 4** u is written in terms of x:
$\dfrac{1}{5}\dfrac{u^{11}}{11} + c = \dfrac{(5x-2)^{11}}{55} + c$	$\dfrac{1}{5}\ln|u| + c = \dfrac{1}{5}\ln|5x-2| + c$	$\dfrac{1}{5}e^{u} + c = \dfrac{1}{5}e^{5x-2} + c$

8.4.2 General functions of linear functions

In each problem above, the substitution $u = (5x - 2)$ transformed the function to be integrated to a standard function in u and dx to $\frac{1}{5}du$:

(a) $\dfrac{1}{5}\displaystyle\int u^{10}du$ (b) $\dfrac{1}{5}\displaystyle\int \dfrac{1}{u}du$ (c) $\dfrac{1}{5}\displaystyle\int e^{u}du$

Integration of the standard functions of u was then possible. If the above worked example were repeated for a *general* linear function, the substitution $u = mx + c$ would transform the function to be integrated to a standard function of u and dx to $1/m\, du$, that is,

$$\text{(a)} \quad \frac{1}{m} \int u^n du \qquad \text{(b)} \quad \frac{1}{m} \int \frac{1}{u} du \qquad \text{(c)} \quad \frac{1}{m} \int e^u du \quad (m \neq 0)$$

These standard functions may then be integrated w.r.t. u. When integration is complete and simplified, write u back in terms of x. This allows us to write down a general rule for integrating functions of linear functions: integrate the function **then** multiply by the derivative of the linear function inverted. For example,

$$\int (mx + c)^n dx = \frac{1}{m} \frac{(mx + c)^{n+1}}{n + 1} + c \quad (m \neq 0, n \neq -1) \tag{8.9}$$

$$\int \frac{1}{(mx + c)} dx = \frac{1}{m} \ln |mx + c| + c \quad (m \neq 0) \tag{8.10}$$

$$\int e^{mx+c} dx = \frac{1}{m} e^{mx+c} + c \quad (m \neq 0) \tag{8.11}$$

In the following worked examples, these rules are used instead of working through the substitution each time.

WORKED EXAMPLE 8.6
INTEGRATING LINEAR FUNCTIONS RAISED TO A POWER

Integrate the following:

$$\text{(a)} \quad \int \sqrt{3x + 4}\, dx \qquad \text{(b)} \quad \int \frac{5}{(6x + 3)^4}\, dx \qquad \text{(c)} \quad \int (1 - 3Q)^5 dQ$$

SOLUTION
In each of these questions, do the usual simplifications first by (i) writing roots as powers, and (ii) bringing all the variable terms above the line in fractions so that the value of n for the power rule may be stated.

(a)	(b)	(c)
$\int \sqrt{3x + 4}\, dx$	$\int \frac{5}{(6x + 3)^4}\, dx$	$\int (1 - 3Q)^5 dQ$
$= \int (3x + 4)^{1/2} dx$	$= \int 5(6x + 3)^{-4} dx$	

$$= \frac{1}{3} \frac{(3x+4)^{1/2+1}}{1/2+1}$$

$$= \frac{1}{3} \times \frac{(3x+4)^{3/2}}{3/2}$$

(multiple by $\frac{3}{2}$ inverted)

$$= \frac{1}{3} \frac{2}{3} \frac{(3x+4)^{3/2}}{1}$$

$$= \frac{2(3x+4)^{3/2}}{9} + c$$

$$= 5 \times \frac{1}{6} \frac{(6x+3)^{-4+1}}{-4+1}$$

$$= \frac{5}{1} \times \frac{1}{6} \frac{(6x+3)^{-3}}{-3}$$

$$= -\frac{5}{6} \frac{1}{3} \frac{(6x+3)^{-3}}{1}$$

$$= -\frac{5(6x+3)^{-3}}{18} + c$$

$$= -\frac{5}{18(6x+3)^3} + c$$

$$= \frac{1}{(-3)} \times \frac{(1-3Q)^{5+1}}{5+1}$$

$$= -\frac{1}{3} \times \frac{(1-3Q)^6}{6}$$

$$= -\frac{(1-3Q)^6}{18} + c$$

WORKED EXAMPLE 8.7
MORE EXAMPLES ON INTEGRATING FUNCTIONS OF LINEAR FUNCTIONS

Integrate each of the following:

(a) $\displaystyle\int \frac{1}{8Q+1} dQ$ (b) $\displaystyle\int 5(1 - e^{-0.8Y}) dY$ (c) $\displaystyle\int \frac{2}{5-8Q} dQ$

SOLUTION

(a)

$$\int \frac{1}{8Q+1} dQ$$

$$= \int (8Q+1)^{-1} dQ$$

$n = -1$, this is the exception to the general power rule, therefore

$$\int (8Q+1)^{-1} dQ$$

$$= \frac{1}{8} \ln(8Q+1) + c$$

(b)

$$\int 5(1 - e^{-0.8Y}) dY$$

No simplification is necessary; factor out the constant 5 and proceed with integration:

$$\int 5(1 - e^{-0.8Y}) dY$$

$$= 5 \times \int (1 - e^{-0.8Y}) dY$$

$$= 5 \left(Y - \frac{1}{(-0.8)} e^{-0.8Y} \right)$$

$$= 5 \left(Y + \frac{1}{0.8} e^{-0.8Y} \right) + c$$

(c)

$$\int \frac{2}{5-8Q} dQ$$

$$= \int 2(5-8Q)^{-1} dQ$$

$n = -1$, therefore

$$\int 2(5-8Q)^{-1} dQ$$

$$= 2 \times \frac{1}{(-8)} \times \ln(5-8Q) + c$$

$$= -\frac{2}{8} \ln(5 - Q) + c$$

$$= -\frac{1}{4} \ln(5 - 8Q) + c$$

PROGRESS EXERCISES 8.2

Integration of Functions of Linear Functions

Integrate the following:

1. $\displaystyle\int (x - 5)^{0.2} dx$
2. $\displaystyle\int (2x - 5)^{0.2} dx$
3. $\displaystyle\int (3x - 5)^{0.2} dx$

4. $\displaystyle\int \sqrt{9x - 5}\, dx$
5. $\displaystyle\int \sqrt[3]{9x - 5}\, dx$
6. $\displaystyle\int \frac{10}{x - 1} dx$

7. $\displaystyle\int \frac{1}{10x - 1} dx$
8. $\displaystyle\int \frac{10}{x^2} dx$
9. $\displaystyle\int \frac{10}{(x - 1)^5} dx$

10. $\displaystyle\int (20 - 2Q)^{0.45} dQ$
11. $\displaystyle\int \frac{1}{2 + 7Q} dQ$
12. $\displaystyle\int \frac{4}{2 + 7Q} dQ$

13. $\displaystyle\int \frac{1}{2 - 7Q} dQ$
14. $\displaystyle\int \frac{15}{Q^7} dQ$
15. $\displaystyle\int \frac{15}{(Q - 5)^7} dQ$

16. $\displaystyle\int \frac{15}{(2Q - 5)^7} dQ$
17. $\displaystyle\int \left(\sqrt{Q + 1} + \frac{1}{\sqrt{Q}}\right) dQ$
18. $\displaystyle\int \frac{5\sqrt{4 + Y}}{3} dY$

19. $\displaystyle\int 5\frac{\sqrt{4 - Y}}{3} dY$
20. $\displaystyle\int (1 + 0.8t)^{1.12} dt$
21. $\displaystyle\int (1 - 0.8t)^{1.12} dt$

22. $\displaystyle\int e^{12x} dx$
23. $\displaystyle\int \frac{4}{e^{2x}} dx$
24. $\displaystyle\int 50e^{-0.85Q} dQ$

25. $\displaystyle\int 100(1 - e^{-0.5Y}) dY$
26. $\displaystyle\int \frac{1 + 2e^x}{e^{2x}} dx$
27. $\displaystyle\int 3e^{1+Y} dY$

28. $\displaystyle\int 10e^{2-5Q} dQ$
29. $\displaystyle\int \left(\frac{1}{e^{2t}} + 2e^{t+2} + 3\right) dt$

8.5 The Definite Integral and the Area under a Curve

At the end of this section you should be able to:

- State that the area under any given curve is approximately equal to $\sum_{x=a}^{x=b} y_i \Delta x$.
- State that as $\Delta x \to 0$, the area under the curve is determined by the definite integral $\int_{x=a}^{x=b} y\, dx$.

The approximate area under a curve

In the last section, integration was defined as the operation which reverses differentiation. Integration is now used to calculate the area under a curve. Figure 8.6 shows a curve $y = f(x)$.

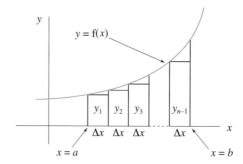

Figure 8.6   Area under the curve $\simeq$ sum of areas of rectangles

The area under the curve between $x = a$ and $x = b$ is approximately equal to the sum of the areas of the narrow rectangles, each of width Δx, but various heights, $y_1, y_2, \ldots, y_{n-1}$:

$$\text{area} \approx y_1 \Delta x + y_2 \Delta x + y_3 \Delta x + \cdots + y_{n-1} \Delta x$$

$$\text{area} \approx \sum_{i=a}^{i=b} y_i \Delta x$$

The height, y_i, of each rectangle is calculated by substituting each x-value into the equation of the curve, starting with $x = a$; but as you can see in Figure 8.6, the rectangles underestimate the area beneath the curve. The situation may be improved by dividing the interval a to b into rectangles with smaller widths (Figure 8.7).

Ultimately, by allowing the width of the rectangles to become as small as possible, but not equal to zero, the sum of the areas of the rectangles will be exactly equal to the area under the curve. The height of the first rectangle is the value of y when $x = a$ and the height of the last rectangle is the value of y when $x = b$.

When infinitely small changes are dealt with, the notation changes (Figure 8.8). As in differentiation, replace Δx by dx to indicate that Δx is approaching zero. Also replace the summation sign $\sum$ by the integral sign $\int$:

$$\sum_{x=a}^{x=b} y \Delta x \rightarrow \int_{x=a}^{x=b} y dx \quad \text{as } \Delta x \rightarrow 0$$

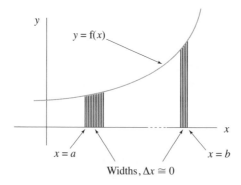

Figure 8.7   Decreasing the size of Δx gives a better approximation to area

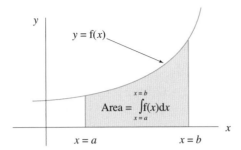

Figure 8.8   Area under the curve is determined exactly by integration

Evaluating the integral between the limits $x = a$ and $x = b$ is carried out as follows.

Step 1: Integrate as normal, $\int f(x)\,dx = F(x) + c$, where $F(x)$ denotes the result of integrating $f(x)$ with respect to x.

Step 2: The area under the curve between $x = a$ and $x = b$ is given by

$$\text{Net area} = \int_{x=a}^{x=b} f(x)\,dx = F(x)|_{x=a}^{x=b} = F(b) - F(a) \qquad \textbf{(8.12)}$$

where $F(b)$ is found by substituting $x = b$ into $F(x)$. $F(a)$ is found by substituting $x = a$ into $F(x)$.

WORKED EXAMPLE 8.8
EVALUATING THE DEFINITE INTEGRAL

Evaluate (a) $\int_{x=1}^{x=3} (x + 2)\,dx$ and (b) $\int_{x=0}^{x=4} x^2\,dx$. In (a) calculate the area (i) by integration and (ii) geometrically.

SOLUTION

(a) $\int_{x=1}^{x=3} (x + 2)\,dx$

Step 1

$\int_{x=1}^{x=3} (x + 2)\,dx = \dfrac{x^2}{2} + 2x + c = F(x)$

Step 2

$F(x) = \dfrac{x^2}{2} + 2x + c$, therefore

at $x = 3$, $F(3) = \dfrac{(3)^2}{2} + 2(3) + c = 10.5 + c$

at $x = 1$, $F(1) = \dfrac{(1)^2}{2} + 2(1) + c = 2.5 + c$

(b) $\int_{x=0}^{x=4} x^2\,dx$

Step 1

$\int_{x=1}^{x=3} x^2\,dx = \dfrac{x^3}{3} + c = F(x)$

Step 2

$F(x) = \dfrac{x^3}{3} + c$, therefore

at $x = 4$, $F(4) = \dfrac{(4)^3}{3} + c = \dfrac{64}{3} + c$

at $x = 0$, $F(0) = \dfrac{(0)^3}{3} + c = c$

Therefore

$$\int_{x=1}^{x=3} (x+2)\,dx = \frac{x^2}{2} + 2x + c$$

$$= F(3) - F(1)$$

$$= (10.5 + c) - (2.5 + c)$$

$$= 10.5 + c - 2.5 - c$$

$$= 8$$

Therefore

$$\int_{x=0}^{x=4} x^2\,dx = \frac{x^3}{3} + c$$

$$= F(4) - F(0)$$

$$= \frac{64}{3} + c - c$$

$$= \frac{64}{3} = 21.33$$

Note: The constants $+c$ and $-c$ cancel. Therefore it is not necessary to include c in definite integration.

To find the area geometrically for part (a), first sketch the function $f(x) = x + 2$. This is a straight line with slope 1 and intercept 2 (Figure 8.9). The area under the line between $x = 1$ and $x = 2$ is the area of the rectangle plus the area of the triangle:

$$\text{Area of rectangle} = \text{width} \times \text{height} = 2 \times 3 = 6$$

$$\text{Area of triangle} = \tfrac{1}{2} \times \text{width} \times \text{height} = \tfrac{1}{2} \times 2 \times 2 = 2$$

$$\text{Total area} = 8$$

This is the same as we obtained by integration.

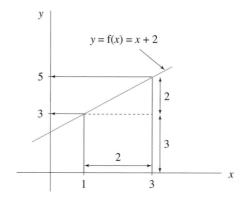

Figure 8.9   Area under $f(x) = x + 2$

WORKED EXAMPLE 8.9
DEFINITE INTEGRAL AND e^x

Evaluate (a) $\displaystyle\int_{x=0}^{x=2} 40e^{2x-3}\,dx$ and (b) $\displaystyle\int_{Y=0}^{Y=3} (1 - e^{-0.5Y})\,dY$

SOLUTION

(a) $\displaystyle\int_{x=0}^{x=2} 40e^{2x-3}\,dx$

(b) $\displaystyle\int_{Y=0}^{Y=3} (1 - e^{-0.5Y})\,dY$

Step 1: Integrate as normal

$$\int_{x=0}^{x=2} 40e^{2x-3}\,dx = 40\frac{1}{2}e^{2x-3}\Big|_{x=0}^{x=2}$$

Step 1: Integrate as normal

$$\int_{Y=0}^{Y=3} (1 - e^{-0.5Y})\,dY = Y - \frac{1}{-0.5}e^{-0.5Y}$$

Step 2: Evaluate F(x) at the limits of integration and subtract

$$\int_{x=0}^{x=2} 40e^{2x-3}\,dx = 40\frac{1}{2}e^{2x-3}\Big|_{x=0}^{x=2}$$

$$= \frac{40}{2}[e^{2(2)-3} - e^{2(0)-3}]$$

$$= 20(e^1 - e^{-3})$$

$$= 20(2.7183 - 0.0498)$$

$$= 53.37$$

Step 2: Evaluate F(x) at the limits of integration and subtract

$$\int_{Y=0}^{Y=3} (1 - e^{-0.5Y})\,dY = Y + 2e^{-0.5Y}\Big|_{Y=0}^{Y=3}$$

$$= [(3 + 2e^{-0.5(3)})$$

$$-(0 + 2e^{-0.5(0)})]$$

$$= [(3 + 0.4462) - (0 + 2)]$$

$$= 3.4462 - 2 = 1.4462$$

Definite integration gives the net area contained between the curve and the x-axis

In Worked Example 8.8 the area calculated was above the x-axis, but when integration is used to calculate the area, the following cases arise:

Case 1: Areas above the x-axis are evaluated with a positive sign.
Case 2: Areas below the x-axis are evaluated with a negative sign.

It follows that when integration is used to calculate an area which is partially above the x-axis and partially below the x-axis, the numerical answer is the net area, the sum of the positive and negative areas.

WORKED EXAMPLE 8.10
DEFINITE INTEGRATION AND NET AREA
BETWEEN CURVE AND x-AXIS

Sketch the function $f(x) = x^2 - 1$. Show that the graph crosses the x-axis at $x = 1$. Use integration to determine (a) the area between $x = 0$ and $x = 1$, (b) the area between $x = 1$ and $x = 2$, (c) the area between $x = 0$ and $x = 2$.

SOLUTION

The graph is sketched in Figure 8.10. The graph crosses the x-axis when $y = 0$; that is, at $0 = x^2 - 1 \rightarrow x^2 = 1 \rightarrow x = \pm 1$.

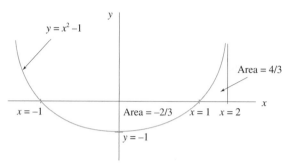

Figure 8.10   Definite integration calculates the net enclosed area

The areas for parts (a), (b) and (c) are now determined by integration:

(a)	(b)	(c)
$\text{Area} = \displaystyle\int_{x=0}^{x=1} (x^2 - 1)\,dx$	$\text{Area} = \displaystyle\int_{x=1}^{x=2} (x^2 - 1)\ dx$	$\text{Area} = \displaystyle\int_{x=0}^{x=2} (x^2 - 1)\ dx$
$= \left(\dfrac{x^3}{3} - x\right)\Big|_{x=0}^{x=1}$	$= \left(\dfrac{x^3}{3} - x\right)\Big|_{x=1}^{x=2}$	$= \left(\dfrac{x^3}{x} - x\right)\Big|_{x=0}^{x=2}$
$= F(1) - F(0)$	$= F(2) - F(1)$	$= F(2) - F(0)$
$= \left(\dfrac{1}{3} - 1\right) - \left(\dfrac{0}{3} - 0\right)$	$= \left(\dfrac{8}{3} - 2\right) - \left(\dfrac{1}{3} - 1\right)$	$= \left(\dfrac{8}{3} - 2\right) - \left(\dfrac{0}{3} - 0\right)$
$= -\dfrac{2}{3} - 0$	$= \dfrac{2}{3} - \left(-\dfrac{2}{3}\right)$	$= \dfrac{2}{3} - 0$
$= -\dfrac{2}{3}$	$= \dfrac{4}{3}$	$= \dfrac{2}{3}$

These results demonstrate that (a) + (b) = (c); that is, the area calculated by integration is the net area contained between the curve and the x-axis over the interval $x = 0$ to $x = 2$.

Evaluation of the definite integral when F(x) = ln |x|

Earlier in this chapter it was stated that, strictly speaking,

$$\int \frac{1}{x}dx = \ln |x| + c$$

where $|x|$ is the absolute value of x, for example, $\ln | -2| = \ln(2) = 0.3010$. In definite integration, if the log of a negative number arises, the negative sign is ignored. See Worked Example 8.11(a).

Note: Division by zero must not occur for any value of x within the limits of integration. If this occurs there will be a vertical asymptote (Chapter 4), so the area cannot be determined.

WORKED EXAMPLE 8.11
DEFINITE INTEGRATION AND LOGS

Evaluate (a) $\int_{x=-2}^{x=-1} \dfrac{5}{x}\,dx$ and (b) $\int_{x=0.2}^{x=1} \left(x^2 + \dfrac{1}{x}\right)dx$.

SOLUTION

(a)

$$Area = \int_{x=-2}^{x=-1} \dfrac{5}{x}\,dx = \int_{x=-2}^{x=-1} \dfrac{5}{1}\dfrac{1}{x}\,dx$$

$$= 5\ln|x|\Big|_{x=-2}^{x=-1}$$

$$= F(-1) - F(-2)$$

$$= 5\ln|-1| - 5|-2|$$

Use $|-2| = 2, |-1| = 1$, hence

$$Area = 5\ln(1) - 5\ln(2)$$

$$= 0 - 3.4657$$

$$= -3.4657$$

(b)

$$Area = \int_{x=0.2}^{x=1} \left(x^2 + \dfrac{1}{x}\right)dx$$

$$= \dfrac{x^3}{3} + \ln|x|\Big|_{x=0.2}^{x=1}$$

$$= F(1) - F(0.2)$$

$$= \left(\dfrac{(1)^3}{3} + \ln|1|\right) - \left(\dfrac{(0.2)^3}{3} + \ln|0.2|\right)$$

$$= \left(\dfrac{1}{3} + 0\right) - \left(\dfrac{0.008}{3} + (-1.609\,44)\right)$$

(recall: logs of numbers less than 1 are negative)

$$Area = \dfrac{1}{3} - (-1.6068)$$

$$= 0.3333 + 1.6068 = 1.9401$$

PROGRESS EXERCISES 8.3

Definite Integration

Evaluate the following definite integrals, correct to two decimal places:

1. $\int_{x=1}^{x=3} (x + 5)\,dx$

2. $\int_{x=0}^{x=4} (x^3 + 2x)\,dx$

3. $\int_{x=1}^{x=10} \dfrac{x+5}{2}\,dx$

4. $\int_{x=-2}^{x=2} (x^2 - 3)\,dx$

5. $\int_{x=1}^{x=5} \left(4x + \dfrac{4}{x}\right)dx$

6. $\int_{x=3}^{x=10} \dfrac{x^2 + 8x}{x^3}\,dx$

7. $\int_{x=0}^{x=5} \sqrt{Q}(Q^2 + Q)\,dQ$

8. $\int_{Q=0}^{Q=20} Q(100 - 2Q)\,dQ$

9. $\int_{Q=0.75}^{Q=12.25} (5Q + 2)\,dQ$

10. $\int_{x=5}^{x=9} \dfrac{10}{6+x}\,dx$

11. $\int_{Q=0.3}^{Q=1.3} \left(\dfrac{10}{Q}\right)dQ$

12. $\int_{Q=1}^{Q=12} \dfrac{Q^2 + 20Q + 12}{Q}\,dQ$

13. $\displaystyle\int_{Q=5}^{Q=9} \frac{4}{10-Q}\,dQ$

14. $\displaystyle\int_{Q=0}^{Q=5} \left(10 + \frac{1}{Q+1}\right)dQ$

15. $\displaystyle\int_{Q=1}^{Q=8} \sqrt{Q+8}\,dQ$

16. $\displaystyle\int_{x=0}^{x=3} e^{2x}\,dx$

17. $\displaystyle\int_{t=0}^{t=1} 500e^{0.4t}\,dt$

18. $\displaystyle\int_{Q=0}^{Q=5} \frac{e^{2Q}+4}{e^{2Q}}\,dQ$

19. $\displaystyle\int_{x=1}^{x=3} 4(1 - e^{0.9x})\,dQ$

In questions 20 to 25:

(a) Sketch f(Q), the function to be integrated, over the interval of integration.
(b) Calculate the net area enclosed over the interval of integration.
(c) Calculate the areas above and below the axis separately.

20. $\displaystyle\int_{Q=0}^{Q=10} (10 - Q)\,dQ$

21. $\displaystyle\int_{Q=0}^{Q=4} (Q^2 - 4)\,dQ$

22. $\displaystyle\int_{Q=0}^{Q=10} (16 - Q^2)\,dQ$

23. $\displaystyle\int_{Q=0}^{Q=10} (50 + 5Q - Q^2)\,dQ$

24. $\displaystyle\int_{Q=0}^{Q=5} (18 + 8Q - Q^2)\,dQ$

25. $\displaystyle\int_{Q=0.5}^{Q=4} \left(4 + \frac{2}{Q}\right)dQ$

8.6 Consumer and Producer Surplus

At the end of this section you should be able to:

• Calculate consumer surplus
• Calculate producer surplus
• Calculate net profit and loss.

Net profit and loss are found by calculating the area between the total revenue curve and the total cost curve. In Chapter 3, consumer surplus (CS) and producer surplus (PS) were introduced and calculated geometrically. It might be an idea to review Section 3.3 before proceeding. In this section the definite integral is used to determine consumer surplus and producer surplus for non-linear functions.

Consumer surplus (CS)

Consumer surplus was described as the difference between the expenditure a consumer is willing to make on successive units of a good, from $Q = 0$ to $Q = Q_0$, and the actual amount spent on Q_0 units of the good at the market price of P_0 per unit:

$$CS = \left(\begin{array}{c}\text{revenue consumer was willing}\\\text{to pay at higher prices}\end{array}\right) - \left(\begin{array}{c}\text{actual expenditure}\\\text{at } P = P_0\end{array}\right)$$

$$= \left(\begin{array}{c}\text{the area enclosed between the demand function}\\\text{over the interval } Q = 0 \text{ to } Q = Q_0\end{array}\right) - P_0 Q_0$$

- For a linear demand function, such as Figure 8.11, CS is given by the area of the triangle (shaded).
- For a non-linear function, such as Figure 8.12, integration is used to calculate the total area under the curve.

In general, consumer surplus at $P = P_0$ (corresponding quantity Q_0) is

$$\text{Area under the curve} - \text{area of rectangle} = \int_{Q=0}^{Q=Q_0} (\text{demand function}) \, dQ - P_0 Q_0 \qquad \textbf{(8.13)}$$

WORKED EXAMPLE 8.12
USING THE DEFINITE INTEGRAL TO CALCULATE CONSUMER SURPLUS

 Find animated worked examples at **www.wiley.com/college/bradley**

Calculate the consumer surplus for (a) the demand function $P = 60 - 2Q$ when the market price is $P_0 = 12$, (i) geometrically and (ii) by integration; (b) the demand function $P = 100/(Q + 2)$ when the market price is $P_0 = 20$.

In each case graph the demand function and shade in the consumer surplus.

SOLUTION

(a) The demand function is linear with a slope of –2 and an intercept of 60 (Figure 8.11).

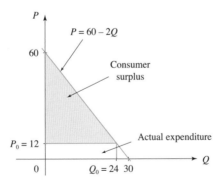

Figure 8.11   Consumer surplus for Worked Example 8.12(a)

Substitute $P_0 = 12$ into the equation for the demand curve to find the corresponding value of Q_0:

$$12 = 60 - 2Q \quad \rightarrow \quad 2Q = 60 - 12 \quad \rightarrow \quad Q = 24$$

Therefore $P_0 Q_0 = 12 \times 24 = 288$.

To calculate CS, the shaded area in Figure 8.11, (i) use the area of the triangle definition and (ii) apply equation (8.13).

(i) CS = area of triangle = $1/2 \times$ base $\times$ perpendicular height = $0.5 \times 24 \times 48 = 576$.

(ii) The following method is longer and is applied here to show that both methods yield the same answer.

Area under the curve − area of rectangle

$$= \int_{Q=0}^{Q=Q_0} (\text{demand function})\, dQ - P_0 Q_0$$

$$= \int_{Q=0}^{Q=Q_0} (60 - 2Q)\, dQ - P_0 Q_0$$

$$= \left(60Q - 2\frac{Q^2}{2}\right)\Bigg|_{Q=0}^{Q=24} - (12 \times 24)$$

$$= [60(24) - (24)^2] - [60(0) - (0)^2] - (12 \times 24)$$

$$= 864 - 0 - 288$$

$$= 576$$

Consumer surplus is 576.

(b) $P = 100/(Q + 2)$ is a hyperbolic demand function (Chapter 4). See Figure 8.12.

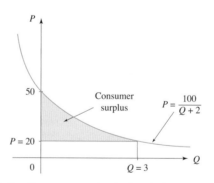

Figure 8.12   Consumer surplus for Worked Example 8.12(b)

Substitute $P_0 = 20$ into the equation for the demand curve to find the corresponding value of Q_0:

$$20 = \frac{100}{Q+2} \quad \rightarrow \quad 20(Q+2) = 100 \quad \rightarrow \quad (Q+2) = \frac{100}{20}$$

$$\rightarrow \quad (Q+2) = 5 \quad \rightarrow \quad Q = 5 - 2 \quad \rightarrow \quad Q = 3$$

Therefore $P_0Q_0 = 20 \times 3 = 60$. To calculate CS (shaded area), apply equation (8.13):

$$\text{Area under the curve} - \text{area of rectangle} = \int_{Q=0}^{Q=Q_0} (\text{demand function}) \, dQ - P_0Q_0$$

$$= \int_{Q=0}^{Q=3} \frac{100}{Q+2} \, dQ - P_0Q_0$$

$$= 100 \, \ln(Q+2) \Big|_{Q=0}^{Q=3} - (20 \times 3)$$

$$= 100[\ln(3+2) - \ln(0+2)] - (20 \times 3)$$

$$= 100[\ln(5) - \ln(2)] - 60$$

$$= (100 \times 0.9163) - 60$$

$$= 91.6291 - 60 = 31.6291$$

Consumer surplus is 31.6291.

Producer surplus (PS)

In Chapter 3 the producer surplus was defined as the difference between the revenue the producer receives for Q_0 units of a good when the market price is P_0 per unit and the revenue that she was willing to accept for successive units of the good from $Q = 0$ to $Q = Q_0$:

$$PS = (\text{actual revenue at price } P = P_0) - (\text{acceptable revenue at lower prices})$$

$$= P_0 \, Q_0 - \left(\begin{array}{l} \text{the area under the supply function} \\ \text{over the interval } Q = 0 \text{ to } Q = Q_0 \end{array} \right)$$

- For a linear demand function, such as in Worked Example 8.13(a), PS is given by the area of the triangle (shaded).
- For a non-linear function, the area under the curve may be calculated by integration.

In general, producer surplus at $P = P_0$ (corresponding quantity Q_0) is

$$\boxed{\text{Area of rectangle} - \text{area under curve} = P_0Q_0 - \int_{Q=0}^{Q=Q_0} (\text{supply function}) \, dQ} \qquad \textbf{(8.14)}$$

WORKED EXAMPLE 8.13
USING THE DEFINITE INTEGRAL TO CALCULATE PRODUCER SURPLUS

Sketch the supply functions (a) $P = 20 + 4Q$ and (b) $P = Q^2 + 6Q$. In each case calculate the producer surplus at $Q = 4$. Shade the producer surplus on each sketch.

SOLUTION

(a) This is a linear supply function; it is sketched in Figure 8.13.

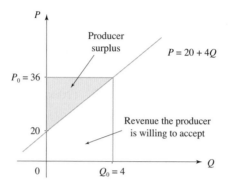

Figure 8.13   Producer surplus for Worked Example 8.13(a)

At $Q_0 = 4$, $P_0 = 36$, to calculate the producer surplus, (i) calculate the area of the shaded triangle in Figure 8.13, and (ii) apply equation (8.14).
(i) CS = area of triangle = 1/2 × base × perpendicular height = 0.5 × 4 × 16 = 32.
(ii) In this example PS for this linear function is calculated by integration to show that the same result is obtained.

Area of rectangle − area under the curve

$$= P_0 Q_0 - \int_{Q=0}^{Q=Q_0} (\text{supply function}) \, dQ$$

$$= (36 \times 4) - \int_{Q=0}^{Q=Q_0} (20 + 4Q) \, dQ$$

$$= 144 - \left(20Q + 4\frac{Q^2}{2} \right) \Big|_{Q=0}^{Q=4}$$

$$= 144 - \{ [20(4) + 2(4)^2] - [20(0) + 2(0)^2] \}$$

$$= 144 - (112 - 0) = 32$$

Producer surplus is 32.
(b) Integration is required to calculate the area under the supply curve since the supply function is non-linear: $P = Q^2 + 6Q$ is a quadratic (Figure 8.14).
When $Q_0 = 4$, $P_0 = 40$, since $P_0 = Q_0^2 + 6Q_0 = (4)^2 + 6(4) = 40$. To calculate the producer surplus, apply equation (8.14):

Area of rectangle − area under the curve

$$= P_0 Q_0 - \int_{Q=0}^{Q=Q_0} (\text{supply function}) \, dQ$$

$$= (40 \times 4) - \int_{Q=0}^{Q=4} (Q^2 + 6Q) \, dQ$$

$$= 160 - \left(\frac{Q^3}{3} + 6\frac{Q^2}{2} \right) \Bigg|_{Q=0}^{Q=4}$$

$$= 160 - \{[(4)^3/3 + 3(4)^2] - [(0)^3/3 + 3(0)^2]\}$$

$$= 160 - (21.3333 + 48 - 0) = 90.6667$$

Producer surplus is 90.6667.

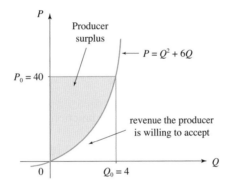

Figure 8.14   Producer surplus for Worked Example 8.13(b)

When demand and supply are modelled by exponential functions, consumer surplus and producer surplus are calculated in the same way, but the integration involves exponential functions instead of power functions.

WORKED EXAMPLE 8.14
CONSUMER AND PRODUCER SURPLUS:
EXPONENTIAL FUNCTIONS

The demand and supply curves for a good are given respectively by the equations

$$P = 300e^{-0.2Q} \quad \text{and} \quad P = 2e^{0.8Q}$$

(a) Find the equilibrium price and quantity.
(b) Calculate consumer surplus and producer surplus at equilibrium.

SOLUTION

(a) At equilibrium $P_d = P_s$, so equating the demand and supply prices gives

$$300e^{-0.2Q} = 2e^{0.8Q}$$

$$150e^{-0.2Q} = e^{0.8Q}$$

$$150 = \frac{e^{0.8Q}}{e^{-0.2Q}} \qquad \text{dividing across by } e^{-0.2Q}$$

$$150 = e^{0.8Q}e^{0.2Q} \qquad \text{using the rules for indices}$$

$$150 = e^{1Q} = e^{Q} \qquad \text{rule 1 for indices}$$

$$\ln(150) = Q \qquad \text{taking logs of both sides}$$

$$5.0106 = Q$$

Substitute the equilibrium quantity into the demand function or the supply function to find the equilibrium price; for example, substituting into the supply function gives

$$P = 2e^{0.8Q} = 2e^{0.8(5.0106)} = 2e^{4.0085} = 110.1284$$

(b) **Consumer surplus** is $\int_{Q=0}^{Q=Q_0} (\text{demand function}) \, dQ - P_0 Q_0$

Consumer surplus at $Q_0 = 5.0106$, $P_0 = 110.1284$ is

$$\int_{Q=0}^{Q=5.0106} 300e^{-0.2Q} dQ - (110.1284 \times 5.0106)$$

$$= 300 \left(\frac{1}{-0.2} e^{-0.2Q} \right) \Bigg|_{Q=0}^{Q=5.0106} - 551.8094$$

$$= \frac{300}{-0.2}[0.3671 - 1] - 551.8094 = 397.5406$$

Producer surplus is $P_0 Q_0 - \int_{Q=0}^{Q=Q_0} (\text{supply function}) \, dQ$

Producer surplus at $Q_0 = 5.0106$, $P_0 = 110.1284$ is

$$(110.1284 \times 5.0106) - \int_{Q=0}^{Q=5.0106} 2e^{0.8Q} dQ$$

$$= 551.8094 - \left(2\frac{e^{0.8Q}}{0.8} \right) \Bigg|_{Q=0}^{Q=5.0106}$$

$$= 551.8094 - \frac{2}{0.8} \left[\left(e^{0.8(5.0106)} - e^{0.8(0)} \right) \right]$$

$$= 551.8094 - 2.5(55.0631 - 1) = 416.6516$$

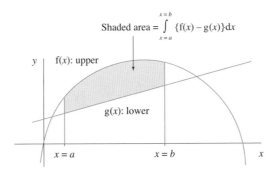

Figure 8.15   Shaded area = area under upper curve – area under lower curve

Area between two curves and other applications of definite integration

The area between two curves over an interval $a \leq x \leq b$ is the difference between the area under the upper curve and the area under the lower curve, as shown by the shaded area in Figure 8.15. This area may be calculated as follows:

$$
\begin{pmatrix} \text{shaded} \\ \text{area} \end{pmatrix} = \begin{pmatrix} \text{area under} \\ \text{upper curve} \end{pmatrix} - \begin{pmatrix} \text{area under} \\ \text{lower curve} \end{pmatrix}
$$

$$
= \int_{x=a}^{x=b} f(x)\, dx - \int_{x=a}^{x=b} g(x)\, dx
$$

$$
= \int_{x=a}^{x=b} [f(x) - g(x)]\, dx
$$

In fact, we have already calculated the TR implicitly by this method when determining CS and PS, since area $= TR = P \times Q$.

PROGRESS EXERCISES 8.4

Consumer and Producer Surplus and TR

In questions 1 to 6 you should: (a) sketch the demand function then shade the consumer surplus at the given value of P or Q; and (b) calculate the consumer surplus correct to two decimal places.

1. $P = 50 - 4Q$ at $P = 10$

2. $P = \dfrac{200}{Q+1}$ at $Q = 9$

3. $P = 100 - Q^2$ at $Q = 8$

4. $Q = 30 - 0.5P$ at $P = 20$

5. $P = \dfrac{100}{Q+1}$ at $Q = 9$

6. $Q = \dfrac{200}{P} - 5$ at $P = 10$

In questions 7 to 10 you should: (a) sketch the demand function then shade the consumer surplus at the given value of P or Q; and (b) calculate the producer surplus correct to two decimal places.

7. $P = 5 + 6Q$ at $P = 29$ 8. $P = 10 - \dfrac{1}{Q+1}$ at $Q = 4$

9. $Q = -8 + 2P$ at $Q = 4$ 10. $P = Q^2 + 4$ at $Q = 5$

11. The demand and supply functions for a good are $P = 50 - 2Q$ and $P = 14 + 4Q$, respectively.
    (a) Calculate the equilibrium price and quantity; confirm your answer graphically.
    (b) Calculate the consumer surplus (CS) and the producer surplus (PS) at equilibrium, correct to two decimal places; shade them on the graph in (a).
12. The demand and supply functions for a good are $P = 100 - 0.5Q$ and $P = 10 + 0.5Q$, respectively.
    (a) Calculate the equilibrium price and quantity; confirm your answer graphically.
    (b) Calculate consumer and producer surplus at equilibrium.
13. The demand and supply functions for a good are $P = 50 - Q^2$ and $P = 10 + 6Q$, respectively.
    (a) Calculate the equilibrium price and quantity.
    (b) Calculate the consumer and producer surplus at equilibrium.
14. A farmer may sell his produce in two different markets; the demand functions for the two markets are $P = 100/(Q + 1)$ and $P = 100 - Q^2$.
    (a) Graph both demand functions over the interval $Q = 0$ to $Q = 14$; comment on the revenue from the two markets.
    (b) Calculate the total revenue received from (i) the sale of the first 10 units in each market and (ii) the sale of units 8 to 12 inclusive.

For each of the following demand functions, calculate the consumer surplus at the indicated value of Q or P.

15. $P = 5 + 15e^{-0.05Q}$ at $Q = 9$ 16. $P = 100e^{-Q}$ at $Q = 2.5$
17. $4 - 0.1Q = \ln(P)$ at $P = 20$ 18. $P = 101 - e^Q$ at $Q = 3$

For each of the following supply functions, calculate the producer surplus at the indicated value of Q or P.

19. $P = 15 + e^{0.05Q}$ at $Q = 20$ 20. $P = 50e^{1.2Q}$ at $Q = 2$

8.7 First-order Differential Equations and Applications

At the end of this section you should be able to:

- Solve differential equations of the type $dy/dx = f(x)$
- Calculate the total function from the equation for the rate of change
- Solve differential equations of the type $dy/dx = ky$
- Solve differential equations of the type $y/x = (x)(y)$.

A differential equation is an equation which contains derivatives, for example,

$$\frac{dy}{dx} = 10x \qquad\qquad \textbf{(8.15)}$$

Equation (8.15) is called a first-order differential equation, since the highest derivative is a first-order derivative.

General and particular solutions of differential equations

Unlike most equations encountered so far, a differential equation, such as equation (8.15), cannot be used to determine the value of y for a given value of x. (Try to evaluate y when $x = 2$!) To evaluate y, we need to deduce an equation that does not contain derivatives from the differential equation. The equation deduced from the differential equation is called the solution of the differential equation. So, to find the solution of a differential equation, we integrate to reverse the differentiation.

For equation (8.15), the solution is obtained by integrating both sides of the differential equation w.r.t. x:

$$\frac{dy}{dx} = 10x$$

$$\int \left(\frac{dy}{dx}\right) dx = \int 10x \, dx \quad \text{integrating both sides w.r.t. } x$$

$$\int d(y) = \int 10x \, dx \quad \text{since integration reverses differentiation} \rightarrow \int d(y) = y$$

$$y = 10\frac{x^2}{2} + c \quad \text{where } c, \text{ the constant of integration, is called the arbitrary constant}$$

A general solution is a 'family' of related curves. In this example, the family of related curves is a series of quadratics which cut the y-axis at every possible value of the constant c, some of which are shown in Figure 8.16.

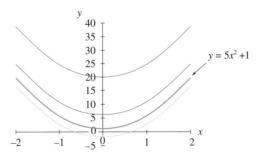

$$y = 5x^2 + 1$$

Figure 8.16 The general solution represented by a 'family' of related curves and a particular solution

The value of c may be calculated if one particular point or condition on the solution curve is given. For example, suppose we were told that the solution must pass through the point $x = 1$, $y = 6$. Then substitute the values given for x and y into the general solution and solve for c:

$$y = 5x^2 + c \qquad \text{this is the general solution}$$
$$x = 1, \; y = 6 \qquad \text{this is the given condition}$$
$$6 = 5(1)^2 + c \qquad \text{substituting the given condition into the general solution}$$
$$6 - 5 = 1 = c \qquad \text{we now have the value of } c, \text{ subject to one particular condition}$$
$$y = 5x^2 + 1 \qquad \text{substituting the value of } c \text{ into the general solution}$$

This solution, which contains no arbitrary constant, is called a **particular solution**, as shown in Figure 8.16.

Solution of differential equations of the form $dy/dx = f(x)$

The method used to solve the differential equation $dy/dx = f(x)$ is outlined as follows:

Step 1: Write the differential equation in the form $dy/dx = \text{RHS}$.
Step 2: Integrate both sides w.r.t. x. This gives the general solution.
Step 3: If conditions are given for x and y, substitute these values into the general solution and solve for the arbitrary constant, c.
Step 4: Substitute this value of c into the general solution to find the particular solution.

WORKED EXAMPLE 8.15
SOLUTION OF DIFFERENTIAL EQUATIONS: $dy/dx = f(x)$

Find the general and particular solutions for the differential equation

$$\frac{dy}{dx} - 6x + 2 = 0 \qquad \text{given that } x = 3 \text{ when } y = 0$$

SOLUTION
Work through the method as outlined above.

Step 1:

$$\text{Write } \frac{dy}{dx} = \text{RHS} \rightarrow \frac{dy}{dx} = 6x - 2$$

Step 2: Integrate both sides w.r.t. x:

$$\int \left(\frac{dy}{dx} \right) dx = \int 6x - 2 \, dx$$
$$y = 6\frac{x^2}{2} - 2x + c$$

This is the general solution. Now find the particular solution.

Step 3: Given that $x = 3$ when $y = 0$, substitute these values into the general solution

$$y = 3x^2 - 2x + c \rightarrow 0 = 3(3)^2 - 2(3) + c \rightarrow 0 = 27 - 6 + c \rightarrow c = -21$$

Step 4: The particular solution is $y = 3x^2 - 2x - 21$.

PROGRESS EXERCISES 8.5

Differential Equations of the Form $dy/dx = f(x)$

Determine the general solution for the following differential equations:

1. $\dfrac{dy}{dx} = 15x$
2. $\dfrac{dy}{dt} - 10t + 1 = 0$
3. $12 - 8x\dfrac{dy}{dx} = 4$

4. $Q\dfrac{dC}{dQ} = 10Q^2$
5. $\dfrac{P}{Q}\dfrac{dP}{dQ} - QP = 0$
6. $\dfrac{dY}{dt} = e^{0.5t}$

Find the particular solution for each of the following differential equations:

7. $\dfrac{dy}{dx} = 5x$, given $x = 2$ when $y = 4$
8. $\dfrac{dy}{dt} - 10t - 2 = 0$, given $y = 15$ when $t = 0$

9. $12 - 8x\dfrac{dy}{dx} = 4x$, given $y = 6$ when $x = 1$

10. $Q\dfrac{dC}{dQ} = 10 - Q^2$, given $C = 50$ when $Q = 1$

11. $\dfrac{P}{Q}\dfrac{dP}{dQ} - (Q+5)P = 0$, given $P = 10$ when $Q = 0$

12. $\dfrac{dY}{dt} = 12e^{0.6t}$ given $Y = 80$ when $t = 0$

Differential equations in economics

In Chapter 6 the marginal function was defined as the derivative of the total function. The equation for a marginal function may be written as a differential equation, which may then be solved to obtain the equation of the total function.

WORKED EXAMPLE 8.16
FIND TOTAL COST FROM MARGINAL COST

The marginal cost for a product is given by the equation $MC = 10/Q$.

(a) Write down the differential equation for total cost in terms of Q.
(b) Find the equation of the total cost function if total costs are 500 when $Q = 10$.

SOLUTION

(a) $MC = 10/Q$ may be written as

$$MC = \frac{d(TC)}{dQ} = \frac{10}{Q}$$

(b) This differential equation is solved to obtain the equation for the total cost.

Step 1: MC is

$$\frac{d(TC)}{dQ} = \frac{10}{Q}$$

Step 2:

$$TC = \int \frac{d(TC)}{dQ} dQ = \int (MC) dQ = \int \frac{10}{Q} dQ$$

$$TC = 10 \ln(Q) + c$$

where the arbitrary constant c is fixed cost. The general solution gives the general form of the total cost function.

Step 3: The condition that $TC = 500$ when $Q = 10$ is given. Substitute these conditions into the general solution to evaluate c:

$$500 = 10 \ln(10) + c \rightarrow 500 - 23.0259 = 476.9741 = c$$

Step 4: Therefore $TC = 10 \ln(Q) + 476.9741$.

This is the particular solution. It is the equation of the total cost function which satisfies a particular condition $TC = 500$ when $Q = 10$.

Differential equations and rates of change

Many practical applications are described in terms of rates of change (dy/dt) such as the rate at which the quantity of a good demanded, Q, changes over a period of time; the rate at which natural resources are depleted; the rate at which pollutants invade the environment; the rate at which the volume of sales changes; the rate at which a drug is absorbed into the bloodstream; the rate at which an epidemic spreads; the rate at which random information is absorbed, etc. These rates are described mathematically as differential equations. For example, given the rate of change of quantity Q, the total quantity accumulated is simply the definite integral of the rate of change from $t = a$ to $t = b$. The explanation is briefly outlined as follows.

$$\Delta Q = \frac{\Delta Q}{\Delta t} \times \Delta t \quad \text{(change in } Q\text{)} = \text{rate} \times \text{(time interval)}$$

The total accumulated value of Q between $t = a$ and $t = b$ is approximately equal to the sum of the quantities accumulated over each successive small interval of time:

$$Q = \Delta Q_1 + \Delta Q_2 + \cdots + \Delta Q_n = \sum \Delta Q_i = \sum_{\Delta t_i} \frac{\Delta Q}{\Delta t_i} \times \Delta t_i$$

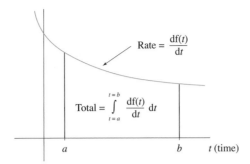

Figure 8.17   The definite integral of the rate = total accumulation

The exact quantity accumulated is obtained by letting the time intervals become infinitesimally small. Consequently, the summation sign is replaced by the integral sign (as in the introduction to the definite integral):

$$\text{Total } Q = \int_{t=a}^{t=b} (\text{rate of change } Q)\, dt = \int_{t=a}^{t=b} \frac{dQ}{dt} dt$$

The total quantity accumulated between two points in time is calculated by integrating the equation for the rate of change between the two given time limits. This idea is illustrated graphically in Figure 8.17. The area under the curve between any two points in time $t = a$ and $t = b$ gives the total amount accumulated (or consumed) during that time interval.

Note: Integrating marginal functions to obtain total functions is, in effect, integrating the rate to obtain the total.

WORKED EXAMPLE 8.17
DIFFERENTIAL EQUATIONS AND RATES OF CHANGE

The present rate at which heating oil (in thousands of litres) is consumed annually by a large department store is given by the equation

$$\text{Rate of consumption} = \frac{dQ}{dt} = 1560e^{0.012t} \text{ where } t \text{ is in years}$$

(a) Plot the rate of consumption over the interval $0 \le t \le 20$. Derive the equation of the consumption function, given $Q = 0$ at $t = 0$.

　　Assuming that present rates of consumption continue,
(b) Calculate the total amount consumed during the first 10 years.
(c) Calculate the total amount consumed in the years, $10 < t \le 20$.

SOLUTION

(a) The rate of consumption is an exponential, as sketched in Figure 8.18.

$$\frac{dQ}{dt} = 1560e^{0.012t} \rightarrow \int \frac{dQ}{dt} dt = \int 1560e^{0.012t} dt \rightarrow Q = 130\,000e^{0.012t} + C$$

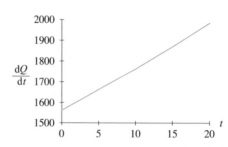

Figure 8.18   Consumption of oil for years 0 to 20

Substitute $Q = 0$ at $t = 0$ into the equation $Q = 130\,000e^{0.012t} + C$ and solve for C:

$$0 = 130\,000 + C, \text{ therefore } C = -130\,000$$

Hence, the particular solution: $Q = 130\,000e^{0.012t} + C \rightarrow Q = 130\,000(e^{0.012t} - 1)$.

(b) The total amount consumed during the first 10 years is given by the equation

$$\text{Total consumed} = \int_{t=0}^{t=10} \frac{dQ}{dt} dt$$

$$= \int_{t=0}^{t=10} 1560e^{0.012t} dt$$

$$= 1560 \left(\frac{e^{0.012t}}{0.012} \right) \Bigg|_{t=0}^{t=10}$$

$$= \frac{1560}{0.012} (e^{0.012(10)} - e^{0.012(0)}) = \frac{1560}{0.012} (0.1275)$$

$$= 16\,574.6 \text{ thousand litres}$$

(c) The total amount consumed between the years 11 (starts at $t = 10$) to 20 inclusive is calculated as in part (b) above, with the limits of integration changed to $t = 10$ to $t = 20$:

$$\text{Total consumed} = \int_{t=10}^{t=20} 1560e^{0.012t} dt$$

$$= 1560 \left(\frac{e^{0.012t}}{0.012} \right) \Bigg|_{t=10}^{t=20}$$

$$= \frac{1560}{0.012} (e^{0.012(20)} - e^{0.012(10)}) = \frac{1560}{0.012} (0.1438)$$

$$= 18\,694 \text{ thousand litres}$$

PROGRESS EXERCISES 8.6

Integrating Marginals / Rates to Obtain Totals

1. The marginal cost function for a good is given by the equation $MC = -Q^2 + 80Q$.
   (a) Write down the differential equation which relates total cost to marginal costs.
   (b) Find the equation of the total cost function if fixed costs are 500.
   (c) Calculate the cost of producing successive units from $Q = 3$ to $Q = 12$.
2. The marginal propensity to consume is given by the equation $MPC = e^{-0.8Y}$.
   (a) Write down the differential equation which relates MPC and total consumption.
   (b) Solve for the consumption function if it is known that consumption is 25 at $Y = 0$.
3. The marginal revenue function for a firm is given by the equation $MR = 120 - 2Q$.
   (a) Write down the differential equation relating marginal revenue and total revenue.
   (b) Find the equation of the total revenue function if total revenue is zero when $Q = 0$.
4. A firm's marginal cost is given by the equation $MC = Q^2 - 42Q + 200$.
   (a) Find the equation of the total cost function if fixed costs are 900.
   (b) Calculate the total cost of producing 12 units.
5. The marginal labour cost function is given by the equation $MLC = 3 + 4L$.
   (a) Write down the differential equation which relates MLC and total labour cost.
   (b) Find the equation of the total labour cost function if total labour cost is zero when $L = 0$.
   (c) Calculate the cost of employing successive labour units from $L = 1$ to $L = 7$.
6. The marginal productivity of labour is given by the equation $MP_L = 10 - 2L$.
   (a) Write down the differential equation which relates output Q and MP_L.
   (b) Find the equation of the production function if $Q = 0$ when $L = 0$.
   (c) Calculate the total output of employing successive labour units from $L = 2$ to $L = 5$.
7. A firm expects its annual income, (Y), and expenditure to change according to the equations $Y(t) = 250 + 0.5t$ and $E(t) = t^2 - 17t + 287.5$ (t is time in years).
   (a) Sketch the graphs of income and expenditure on the same diagram over the interval $0 \le t \le 20$. Confirm the break-even points algebraically.
   (b) Use integration to calculate the net savings (total income – total expenditure) over the intervals

$$(i) \quad 0 \le t \le 2.5 \qquad (ii) \quad 2.5 \le t \le 15$$

8. An intensive pig-fattening unit is adding pollutant (in hundreds of gallons) to a river at the rate of $r(t) = \sqrt{t}$, where t is the number of months that the unit has been in operation.
   (a) Graph the rate of pollution for $t = 0$ to 12 months. Comment.
   (b) Calculate the volume of pollutant in the river during the first six months.
9. The rate at which a certain individual can memorise a sequence of items is given by the equation

$$n(t) = \frac{10}{(1+t)} \qquad \text{where } t \text{ is in minutes}$$

(a) Graph and hence describe the rate of memory-retention curve.
(b) Derive an expression for the total number of items memorised over an interval of time, given that no items have been memorised at $t = 0$. Graph the number of items retained in memory for $t = 0$ to $t = 30$.
(c) Calculate the number of items memorised during:
(i) the first 10 minutes, (ii) the second 10 minutes, (iii) the third 10 minutes.

10. A tool-hire company has estimated the rate of increase in maintenance cost for power drills which is given by the equation $C(t) = 2 + t^{1.5}$, where costs are in £, time in weeks.
(a) Plot the rate of increase in maintenance costs for $t = 0$ to 30.
(b) Calculate the total maintenance costs during (i) the first five weeks, and (ii) the next five weeks.
(c) What is the average weekly cost of maintenance during:
(i) The first five weeks?
(ii) The second five weeks?

11. (a) The marginal cost for a good is given by the equation

$$MC = 80$$

Determine the equation of total cost when fixed costs are 800.
(b) Marginal revenue for the same good is

$$MR = 200 - 20Q$$

Determine the equation of total revenue, assuming $TR = 0$ when $Q = 0$.
(c) Calculate the value of Q at which profit is maximised.

12. A manufacturer of a good has a demand function $20Q + 5P = 440$ and a marginal cost function $MC = 1 - 0.5Q$.

If the manufacturer decides to maximise revenue instead of its profit, calculate the effect on the consumer surplus.

(Calculate the consumer surplus, by integration, in each case.)

Solution of differential equations of the form dy/dx = ky

Equations of this type are solved by the same method as outlined above for $dy/dx = f(x)$, with the inclusion of one further step before step 2.

Step 1: Write the differential equation in the form $dy/dx = $ RHS, i.e., $dy/dx = ky$. In this case, it is not possible to integrate both sides w.r.t. x immediately since the RHS contains y:

$$\int \left(\frac{dy}{dx}\right) dx = \int ky \, dx$$

We can only integrate functions of x w.r.t. x and functions of y w.r.t. y, etc. Therefore, separate the y-terms with dy and the x-terms with dx before integrating. This is called separating the variables.

Step 2: **Separate the variables:** separate the x-terms with dx, and the y-terms with dy:

$$\frac{1}{y}\frac{dy}{dx} = k \rightarrow \frac{1}{y}dy = k\,dx$$

Step 3: Integrate both sides. This gives the general solution.

Step 4: If conditions are given for x and y, substitute these values into the general solution and solve for the arbitrary constant, c.

Step 5: Substitute this value of c into the general solution to find the particular solution.

WORKED EXAMPLE 8.18
SOLVING DIFFERENTIAL EQUATIONS OF
THE FORM dy/dx = *ky*

Solve the differential equations:

(a) $\dfrac{dy}{dx} - 0.5y = 0$, given $y = 10$ when $x = 2$

(b) $\dfrac{dI}{dt} - 0.07I$, given $I = 500$ when $x = 0$.

SOLUTION

(a) **Step 1:** $\dfrac{dy}{dx} = 0.5y$

Step 2: $\dfrac{1}{y}\dfrac{dy}{dx} = 0.5 \rightarrow \dfrac{1}{y}dy = 0.5\,dx$

Step 3: $\displaystyle\int \frac{1}{y}dy = 0.5\,dx$

$$\ln(y) = 0.5x + c \qquad \text{integrating}$$
$$y = e^{0.5x+c} = e^{0.5x}e^c \quad \text{going from log form to index form}$$
$$y = Ae^{0.5x} \qquad \text{since } e^c \text{ is a constant term, call it } A$$

This is the general solution. The general solution is a family of growth curves which cut the vertical axis at A.

Step 4: The condition that $y = 10$ when $x = 2$ is given, so substitute these values into the general solution, and you may solve for A:

$$y = Ae^{0.5x} \quad \text{this is the general solution}$$

Substituting $y = 10$, $x = 2$ into the general solution gives the equation

$$10 = Ae^{0.5(2)} \rightarrow 10 = A(2.7183) \rightarrow \frac{10}{2.7183} = 3.6788 = A$$

Step 5: Substitute the value calculated for A into the general solution to obtain the particular solution $y = 3.6788e^{0.5x}$.

(b) $\dfrac{dI}{dt} = -0.07I$, given $I = 500$ when $t = 0$

Step 1: $\dfrac{dI}{dt} = -0.07I$

Step 2: $\dfrac{1}{I}\dfrac{dI}{dt} = -0.07 \rightarrow \dfrac{1}{I}dI = -0.07\,dt$

Step 3: $\displaystyle\int \dfrac{1}{I}dI = \int -0.07\,dt$

$$
\begin{aligned}
\ln(I) &= -0.07t + c && \text{integrating} \\
I^t &= e^{-0.07t+c} = e^{-0.07t}\,e^{c} && \text{going from log form to index form} \\
I &= Ae^{-0.07t} && \text{since } e^{c} \text{ is a constant term, call it } A
\end{aligned}
$$

This is the general solution. The general solution is a family of growth curves which cut the vertical axis at A. See Figure 8.19.

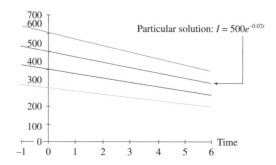

Figure 8.19   General solution, with particular solution indicated

Step 4: The condition that $I = 500$ when $t = 0$ is given, so substitute these values into the general solution, and solve for A:

$$I = Ae^{-0.07t} \quad \text{this is the general solution}$$

Substituting $I = 500$, $t = 0$ into the general solution gives the equation

$$500 = Ae^{-0.07(0)} \quad \rightarrow \quad 500 = A(1) \quad \rightarrow \quad 500 = A$$

Step 5: Substitute the value calculated for A into the general solution to obtain the particular solution $I = 500e^{-0.07t}$.

Solution of differential equations of the form $dy/dx = f(x)g(y)$

In this case, the RHS of the differential equation may be factored into the product of functions of x, $f(x)$, and functions of y, $g(y)$. Such equations are solved by separating the variables as outlined above.

WORKED EXAMPLE 8.19
SOLVING DIFFERENTIAL EQUATIONS
OF THE FORM $dy/dx = f(x)g(y)$

Solve the differential equation $dP/dQ = P(Q + 1)$.

SOLUTION

Step 1: $\dfrac{dP}{dQ} = P(Q + 1)$

Step 2: $dP = P(Q+1)\,dQ \quad \rightarrow \quad \dfrac{1}{P}dP = (Q+1)\,dQ$

Step 3: $\displaystyle\int \frac{1}{P}dP = \int (Q+1)\,dQ$

$$\ln(P) = \frac{Q^2}{2} + Q + c$$

$$P = e^{0.5Q^2 + Q + c} = e^{0.5Q^2 + Q}e^c \quad \text{going from log form to index form}$$
$$P = Ae^{0.5Q^2 + Q} \qquad\qquad \text{since } e^c \text{ is a constant term, call it } A$$

This is the general solution. No conditions were given, so a particular solution cannot be obtained.

PROGRESS EXERCISES 8.7

Differential Equations of the Form $dy/dx = f(y)$; $dy/dx = f(y)g(y)$

Find the general solution and particular solutions of the following differential equations.

1. $\dfrac{dy}{dt} = 2y$, given $y = 50$ at $t = 0$

2. $\dfrac{dy}{dt} + \dfrac{y}{10} = 0$, given $y = 800$ at $t = 0$

3. $0.4N + \dfrac{dN}{dt} = 0$, given $N = 812$ at $t = 2$

4. $\dfrac{dy}{dx} = 2\dfrac{x}{y}$, given $y = 10$ when $x = 10$

5. $\dfrac{dy}{dt} = 3t^2y$, given $y = 50$ at $t = 0$

6. $3\dfrac{dP}{dt} + \dfrac{t}{2} = 60$, given $P = 21$ at $t = 0$

7. $\dfrac{dy}{dx} = xy$, given $y = 20$ when $x = 0$

8. $0.05I + \dfrac{dI}{dt} = 0$, given $I = 900$ at $t = 0$

9. $\dfrac{dQ}{dt}\dfrac{P}{Q} = -1$, given $P = 0.25$ when $Q = 4$

10. $\dfrac{dQ}{dP}\dfrac{P}{Q} = -\dfrac{1}{Q}$, given $P = 120$ when $Q = 0$

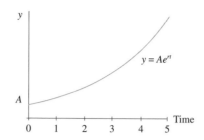

Figure 8.20   Unlimited growth $y = Ae^{rt}$

8.8   Differential Equations for Limited and Unlimited Growth

In this section you will be introduced to:

- Differential equations of the form $dy/dt = ry$, which describe unlimited growth
- Differential equations of the form $dy/dt = r(A - y)$, which describe limited growth.

Law of unlimited growth

Differential equations of the type $dy/dt = ry$ describe systems in which the proportional rate of growth, r, is constant. The solution of this type of differential equation is $y = Ae^{rt}$, where A is the arbitrary constant, describing unlimited growth (or decay, if the index is negative). See Figure 8.20. The solution to this type of equation has already been outlined in Worked Example 8.18.

Law of limited growth

Differential equations of the type $dy/dt = r(A - y)$ describe systems which grow fast initially, but level off just short of a maximum limit, $y = A$, as illustrated in Figure 8.21.

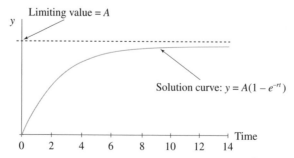

Figure 8.21   The solution of the differential equation: $dy/dt = r(A - y)$ models limited growth

WORKED EXAMPLE 8.20

LIMITED GROWTH

 Find animated worked examples at **www.wiley.com/college/bradley**

The rate at which the volume of sales (Q) for a new type of printer increases after an advertising campaign is given by the equation $dQ/dt = 0.04(700 - Q)$, given that $Q = 0$ at $t = 0$. Q is the number of printers sold, t is time in years.

(a) Solve the differential equation to obtain an expression for Q in terms of t.
(b) Sketch the solution curve, hence describe verbally how the volume of sales should increase in time.

SOLUTION

(a) To solve this equation the variables must be separated:

Step 1: $\dfrac{dQ}{dt} = 0.04(700 - Q)$

Step 2: $dQ = 0.04(700 - Q)\,dt$

$\dfrac{1}{700 - Q}dQ = 0.04\,dt$

Step 3: $\displaystyle\int \dfrac{1}{700 - Q}dQ = \int 0.04\,dt$

$\dfrac{\ln(700 - Q)}{-1} = 0.04t + c$

$\ln(700 - Q) = -0.04t - c$

$700 - Q = e^{-0.04t-c} = e^{-0.04t}e^{-c} = Be^{-0.04t}$

$Q = 700 - Be^{-0.04t}$

This is the general solution. Substituting the given condition that $Q = 0$ at $t = 0$ gives an equation which may be solved for the arbitrary constant B:

$$0 = 700 - Be^{(0)} = 700 - B \rightarrow B = 700$$

Therefore, the particular solution is $Q = 700(1 - e^{-0.04t})$.

(b) The sketch of the solution curve is given in Figure 8.22, demonstrating that the limit or 'ceiling' towards which the volume of sales increases is 700, at a rate of 0.04 per year.

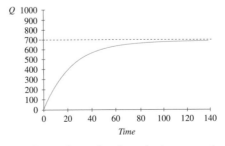

Figure 8.22 Limited growth, where the limiting value is 700

Constant proportional rates of growth

In Chapter 4 the growth in populations, investment, etc., was modelled by the exponential function of the form given in equation (8.16):

$$P_t = P_0 e^{rt} \qquad\qquad (8.16)$$

where P_t, P_0 are the population sizes at time t and at some base time $t = 0$, respectively. The constant r is called the **constant proportional rate of growth**, since the proportional rate, or the relative rate, of growth is defined in general as

$$\frac{\text{rate of change of } P \text{ at time } t}{\text{size of } P \text{ at time } t} = \frac{\mathrm{d}P/\mathrm{d}t}{P_t} \qquad\qquad (8.17)$$

To show that r is a constant proportional rate of growth, take equation (8.16) and calculate the rate of change:

$$P_t = P_0 e^{rt}$$

$$\frac{\mathrm{d}P_t}{\mathrm{d}t} = P_0(r e^{rt}) = r P_0 e^{rt}$$

$$\frac{\mathrm{d}P}{\mathrm{d}t} = r P_t \quad (\text{since } P_t = P_0 e^{rt})$$

That is, the rate of change of the size of P at time $t = r \times$ the size of P at time t. Therefore

$$\frac{\mathrm{d}P_t/\mathrm{d}t}{P_t} = r$$

This means that the proportional rate of growth (or decay), r, for functions of the form $P_t = P_0 e^{rt}$ is the same at any point in time. In other growth (and decay) models, the proportional rate of growth may have different values at different points in time, for example,

$$P_t = \frac{12}{t}, \quad P_t = \frac{P_0}{500 - t}, \quad L_t = 200 \ln(t + 1)$$

WORKED EXAMPLE 8.21
DETERMINING THE PROPORTIONAL RATES OF GROWTH

For each of the following functions:

(i) $P_t = 5312 e^{0.015t}$ (ii) $P_t = 47 e^{-0.02t}$ (iii) $P_t = 12/t$

(a) Derive a general expression for the proportional rate of growth.
(b) Calculate the proportional rates of growth at $t = 5$.

SOLUTION

The proportional rate of growth is given by

$$r = \frac{dP_t/dt}{P_t}$$

For each of the functions, the calculations are as follows.

(i)

(a) $P_t = 5312e^{0.015t}$

$$\frac{dP_t}{dt} = 5312(0.015e^{0.015t})$$

Therefore

$$r = \frac{dP_t/dt}{P_t}$$

$$= \frac{5312(0.015e^{0.015t})}{5312e^{0.015t}}$$

$$= 0.015$$

r does not depend on time, therefore r is constant, as demonstrated in part (b) below:

(b)

At $t = 5, r = 0.015$

At $t = 10, r = 0.015$

(ii)

(a) $P_t = 47e^{-0.02t}$

$$\frac{dP_t}{dt} = 47(-0.02e^{-0.02t})$$

Therefore

$$r = \frac{dP_t/dt}{P_t}$$

$$= \frac{47(-0.02e^{0.02t})}{47e^{-0.02t}}$$

$$= -0.02$$

r does not depend on time, therefore r is constant, as demonstrated in part (b) below:

(b)

At $t = 5, r = -0.02$

At $t = 10, r = -0.02$

(iii)

(a) $P_t = \dfrac{12}{t} = 12t^{-1}$

$$\frac{dP_t}{dt} = 12(-1)t^{-2}$$

Therefore

$$r = \frac{dP_t/dt}{P_t}$$

$$= -\frac{12t^{-2}}{12t^{-1}} = -t^{-1}$$

$$= -\frac{1}{t}$$

r depends on time, therefore r is not constant, as demonstrated in part (b) below:

(b)

At $t = 5$

$$r = -\frac{1}{t} = -\frac{1}{5} = -0.02$$

At $t = 10$

$$r = -\frac{1}{t} = -\frac{1}{10} = -0.1$$

PROGRESS EXERCISES 8.8

Growth, Decay and Other Applications

1. The rate at which an ore is mined is given by the differential equation $dA/dt = 0.02A$.
   (a) Solve the differential equation and graph the general solution.
   (b) If the total reserves of ore in 1989 were 400 million tons, find the particular solution.
   (c) Calculate the total amount of ore mined from the beginning of 1995 to the beginning of 2000 if mining continues at the present rate.

2. The differential equation $dP/dt = 0.01P$ models the relationship between the numbers in a population at t. What is the proportional rate of population growth?
   (a) Deduce the equation which describes the total population at any time t.
   (b) Calculate the numbers in the population in 2200, given that $P = 58.6$ million in 1998 (let $t = 0$ at the start of 1998).
   (c) Calculate the number of years taken for the population to rise to 70 million.

3. The amount of sulphur dioxide (in tons) present in the atmosphere is given by the equation

$$\frac{d(\text{pollutant})}{dt} = -0.2P \quad (t \text{ is in hours})$$

   (a) If 50 tons is released into the atmosphere (i.e., pollutant $= 50$ at $t = 0$), derive an expression for the amount of sulphur dioxide present at any time t.
   (b) Use your expression from (a) to determine the amount of sulphur dioxide (i) after four hours and (ii) after eight hours.
   (c) Use your expression from (a) to calculate the time taken for sulphur dioxide to reduce to five tons.

4. A capital investment (I) depreciates continuously at the rate of 5% annually,

$$\frac{dI}{dt} = -0.05I$$

   (a) If the investment is valued at £12 000 initially (i.e., at $t = 0$), calculate the value of the investment after 5.5 years.
   (b) How many years will it take for the value of the investment to fall to £5000?

5. The price elasticity of demand for a good is constant at $\varepsilon = -1$.
   (a) Write down the differential equation that relates the price of the good to the quantity demanded. (**Hint:** Use the definition of point elasticity.)
   (b) Deduce an expression for the demand function, given that $P = 12$ when $Q = 48$.

6. The rate at which the fish stocks increase in an inland lake is given by the equation $dF/dt = 0.12(80 - F)$, where F is in thousands, t is in years.
   (a) Derive an expression for the total number of fish present at any time t given $F = 0$ at $t = 0$.
   (b) Graph the total fish stock as a function of time from $t = 0$ to $t = 30$. Does the lake appear to have a maximum carrying capacity?
   (c) Calculate the number of years taken for the fish stocks to reach 40 000.

7. A hospital is considering the installation of two alternative heating systems. The rate of increase in costs for each system is given by the equations

$$\frac{dC_1}{dt} = 0.01C_1 \quad \text{and} \quad \frac{dC_2}{dt} = t^{0.25}$$

   (a) Derive expressions for the total cost of each system at any time t given that cost $= £50\,000$ at $t = 0$.
   (b) Graph both cost functions from $t = 0$ to $t = 10$ years. Which system is more cost effective in the long run? Give reasons.

8. The price elasticity of demand is given by $\varepsilon = -10/Q$.
   (a) Write down the differential equation in terms of P and Q.
   (b) Find the equation of the demand function if $P = 20$ when $Q = 0$.

   In questions 9 to 12, determine the proportional rate of growth or decay, as defined in equation (8.18), for each of the following functions at $t = 5$ and $t = 10$. State whether the functions have a constant proportional rate of growth.

9. $P_t = 500e^{0.2t}$ 10. $P_t = 685e^{-0.1t}$ 11. $P_t = 2/t + 2$ 12. $I_t = 20t^{2.5}$

 www.wiley.com/college/bradley

Go to the website for the following additional material that accompanies Chapter 8:

8.9   Integration by Substitution and Integration by Parts

8.9.1 Integration of certain products by algebraic substitution

- Certain products 1
  Worked Example 8.22 Integration of certain products by substitution
  Worked Example 8.23 Integration of certain products by substitution and/or quotients by substitution
- Certain products 2
  Worked Example 8.24 A further example of substitution to integrate certain products
  Progress Exercises 8.9 Integration of certain products by substitution

8.9.2 Integration by parts

- Integration by parts
  Worked Example 8.25 Integration by parts 1
- Choice of u in integration by parts
  Worked Example 8.26 Integration by parts: choosing u
  Progress Exercises 8.10 Integration by parts and applications

8.9.3 Further first-order differential equation: separating the variables

Worked Example 8.27 First-order differential equations $\dfrac{dy}{dx} = f(x)g(y)$

Progress Exercises 8.11

Test Exercises 8: questions 6, 7, 8, 9, and 10

Solutions to PE

8.10 Summary

Rules for integration

Standard functions		Functions of linear functions	
$f(x)$	$F(x) = \int f(x)\, dx$	$f(mx + c)(m \neq 0)$	$F(mx + c) = \int f(mx + c)\, dx$
K	$Kx + c$	N/A	N/A
$x^n, n \neq -1$	$\dfrac{x^{n+1}}{n+1} + c$	$(mx + c)^n, n \neq -1$	$\dfrac{1}{m}\dfrac{(mx + c)^{n+1}}{n+1} + c$
x^{-1} or $\dfrac{1}{x}$	$\ln(x) + c$	$\dfrac{1}{mx + c}$	$\dfrac{1}{m}\ln(mx + c) + c$
e^x	$e^x + c$	$e^{(mx+c)}$	$\dfrac{1}{m}e^{(mx+c)} + c$

The definite integral

$$\int_{x=a}^{x=b} f(x)\, dx = F(b) - F(a)$$

This is the net area enclosed between the curve and the x-axis over the interval $x = a$ and $x = b$.

First-order differential equations

Use integration to solve first-order differential equations of the form

$$\frac{dy}{dx} = f(x) \qquad \text{solve by direct integration}$$

$$\frac{dy}{dx} = Ky \left(\frac{dy}{dt} = ry\right), \qquad \frac{dy}{dt} = r(A - y) \qquad \text{solve by separating the variables}$$

(Solutions model unlimited and limited growth)

$$\frac{dy}{dx} = f(x)\, g(y) \qquad \text{solve by separating the variables}$$

Consumer and producer surplus

At $P = P_0, Q = Q_0$

(a) Consumer surplus is defined as the

$$\text{area under the curve} - \text{area of rectangle} = \int_{Q=0}^{Q=Q_0} (\text{demand function})\, dQ - P_0 Q_0$$

(b) Producer surplus is defined as the

$$\text{area of rectangle} - \text{area under the curve} = P_0 Q_0 - \int_{Q=0}^{Q=Q_0} (\text{supply function}) \, dQ$$

Integrate marginal functions to obtain total functions

$$TC = \int \frac{d(TC)}{dQ} \, dQ = \int (MC) \, dQ$$

$$TR = \int \frac{d(TR)}{dQ} \, dQ = \int (MR) \, dQ$$

Integrate rates (w.r.t. time) to obtain the total amount accumulated over a given time interval

Example: Given the rate at which a resource is consumed, $dQ/dt = Ae^{bt}$, then the total amount consumed between $t = t_0$ and t_1 is given by the integral

$$\int_{t=t_0}^{t=t_1} \frac{dQ}{dt} \, dt = \int_{t=t_0}^{t=t_1} Ae^{bt} \, dt$$

Solution of certain first-order differential equations

Differential equations are used to model situations which involve change, for example, limited and unlimited growth:

(a) $dy/dt = ry$ has a solution of the form $y = Ae^{rt}$, which describes unlimited growth.
(b) $dy/dt = r(A - y)$ has a solution of the form $Q = A - Be^{-rt}$, which describes limited growth.

www.wiley.com/college/bradley

Go to the website for Problems in Context

TEST EXERCISES 8

1. (a) Integrate the following:
   (i) $\int x^4 + 2x \, dx$ (ii) $\int \sqrt{x} + 2 \, dx$ (iii) $\int \frac{2}{x} + 2x \, dx$
   (b) Given the following functions,
   $y = 4 - 2x, \quad y = 4 - x^2, \quad y = x^2 - x + 4$
     (i) Graph each function for $0 \le x \le 4$.
     (ii) Use integration to determine the net area over the interval $0 \le x \le 4$.
     (iii) Use integration to determine the total area enclosed between the graph and the axis over the interval $0 \le x \le 4$.

2. Solve the differential equations

$$\text{(a)} \quad \frac{dy}{dx} = 0.2y \quad \text{(b)} \quad \frac{dy}{dx} = 120(1 - 0.2y)$$

Which solution models limited growth? What is the limit?

3. A farm releases 500 tons of effluent into a river. The amount of effluent (E) present is given by the differential equation $dE/dt = 0.1E$, where t is in hours:
   (a) Deduce an expression for E in terms of t.
   (b) Calculate the amount of effluent present after five hours.
   (c) Calculate the time taken for the amount of effluent to reduce to 10.50 tons.

4. (a) Evaluate the following:

$$\int_{t=0}^{t=5} 40e^{-0.3t}\,dt$$

   (b) The demand and supply functions for a good are given by the equations

$$P = 4(Q - 16)^2 \qquad P = (Q + 2)^2 \quad \text{respectively}$$

   (i) Determine the equilibrium price and quantity.
   (ii) Calculate the consumer and producer surplus at equilibrium.

5. (a) Solve the differential equation $dy/dt = (1 - y)$.
   (b) The rate at which an infection spreads in a poultry house is given by the equation $dP/dt = 0.75(2500 - P)$, where t is time in days. If $P = 0$ at $t = 0$:
   (i) Solve the differential equation to determine an expression for the number of poultry (P) infected at any time t.
   (ii) Graph the solution for $t = 0$ to 12. Hence describe the time path.
   (iii) Calculate the time taken for 1500 poultry to become infected.

LINEAR ALGEBRA AND APPLICATIONS

9.1   Linear Programming

9.2   Matrices

9.3   Solution of Equations: Elimination Methods

9.4   Determinants

9.5   The Inverse Matrix and Input/Output Analysis

9.6   Excel for Linear Algebra

9.7   Summary

Chapter Objectives

In this chapter definitions and properties of matrices and determinants are introduced along with the rules for their arithmetic manipulation. Applications include the solution of systems of linear equations, determination of equilibrium and input/output analysis. At the end of this chapter you should be able to:

- Solve linear programming problems and illustrate your results graphically
- Define a matrix and carry out arithmetic operations involving matrices
- Solve a system of linear equations by Gaussian elimination and Gauss–Jordan elimination
- Define a determinant and evaluate 2×2 and 3×3 determinants
- Solve a system of linear equations using Cramer's rule
- Calculate the inverse of a 2×2 and 3×3 matrix
- Solve a system of linear equations by the inverse matrix method
- Solve 3×3 input/output problems
- Use Excel to carry out elementary row operations to solve a system of linear equations and to calculate the inverse of a matrix.

9.1 Linear Programming

In Chapter 7 the optimal value(s) of non-linear functions (utility, production, cost, etc.) were found, subject to one constraint only. In this section the optimal value of linear functions (cost, profit, output, etc.), subject to several constraints, will be found. If both the function to be optimised and the constraints are restricted to two variables, the optimum value may be found graphically as well as algebraically. This is the subject of this section. When the number of variables exceeds two, the graphical approach is no longer possible. The optimal value is found by more general linear programming methods, such as the simplex method, which is beyond the scope of this text. The graphical and mathematical methods used in solving a typical linear programming problem are described in Worked Example 9.1.

Remember

Cost, budget and income constraints were presented in Chapter 2. These will be required in the graphical methods which follow.

WORKED EXAMPLE 9.1
FIND THE MINIMUM COST SUBJECT TO CONSTRAINTS

A daily diet requires a minimum of 600 mg of vitamin C, 360 mg of vitamin D, 40 mg of vitamin E. The contents of these vitamins (per portion) in two food mixes, X and Y, are given in Table 9.1.

Table 9.1   Vitamin content of X and Y

	Vitamin C (mg)	Vitamin D (mg)	Vitamin E (mg)	Cost per 1 mg
One portion of X	20	10	4	5
One portion of Y	30	20	1	4
Minimum daily requirement	600	360	40	

(a) Express the information given on vitamins in terms of inequality constraints.
(b) Graph the inequality constraints and shade in the feasible region.
(c) Write down the equation of the cost function. Plot the cost function for costs = 200.
(d) From the graph determine the number of portions of mixes X and Y which fulfil the daily requirements at a minimum cost. Confirm your answer algebraically.

SOLUTION

Let the optimum number of portions of mixes X and Y be x and y, respectively. The problem sets out to find the values of x and y which satisfy the minimum dietary requirements at a minimum cost.

(a) **The expressions for the inequality constraints**

The constraints are the minimum daily requirements of each vitamin. The constraints are called **inequality constraints** since the requirements are not exact values but minimum values, so the variables may be greater than minimum. To express the inequality constraints mathematically proceed as follows.

The quantities of vitamins C, D and E provided by one portion of mix X are given in the first row of Table 9.1. The quantities of vitamins C, D and E provided by x portions of mix X are calculated by multiplying the quantities provided by one portion (given in the first row of Table 9.1) by x. The quantities of vitamins C, D and E provided by y portions of mix Y are calculated by multiplying the quantities provided by one portion of mix Y by y. These calculations are summarised in Table 9.2.

Table 9.2   Vitamin content of x and y portions of mix X and Y, respectively

	Vitamin C	Vitamin D	Vitamin E	Cost per portion
x portions of mix X	$20 \times x$	$10 \times x$	$4 \times x$	$5 \times x$
y portions of mix Y	$30 \times y$	$20 \times y$	$1 \times y$	$4 \times y$
Minimum daily requirement	600	360	40	

Therefore, to satisfy the minimum daily requirements we must have

Total vitamin $C \geq 600$	that is, $20x + 30y \geq 600$	**(9.1)**
Total vitamin $D \geq 360$	that is, $10x + 20y \geq 360$	**(9.2)**
Total vitamin $E \geq 40$	that is, $4x + y \geq 40$	**(9.3)**

Since the number of portions is always greater than or equal to zero, $x \geq 0$ and $y \geq 0$ are two further constraints. These are the **inequality constraints**.

(b) **To graph the inequalities**, treat each inequality as an equation, then plot the equation line. The inequality is represented graphically by every point on or above the plotted line. The easiest way to plot these lines is to find the points of intersection with the axes. The points of intersection with the axes are summarised for all three lines, equations (9.1), (9.2), (9.3), as follows:

	Equation of line	Cuts x-axis when $y = 0$	Cuts y-axis when $x = 0$
Vitamin C	(1) $20x + 30y = 600$	$20x = 600 \rightarrow x = 30$	$30y = 600 \rightarrow y = 20$
Vitamin D	(2) $10x + 20y = 360$	$10x = 360 \rightarrow x = 36$	$20y = 360 \rightarrow y = 18$
Vitamin E	(3) $4x + y = 40$	$4x = 40 \rightarrow x = 10$	$y = 40$
	$x \geq 0, y \geq 0$	First quadrant	

The unknowns are x and y. Therefore, when plotting graphs, either one of the variables may be plotted on the horizontal axis, the other on the vertical axis. In this example, x is plotted on the horizontal and y on the vertical. The three inequality constraints are plotted in Figure 9.1.

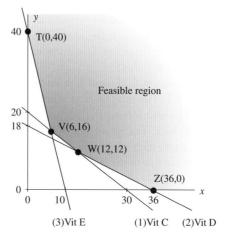

Figure 9.1   Inequality constraints and the feasible region

All three inequality constraints are satisfied simultaneously by the overlap region in the plane, as shown in Figure 9.1. This overlap region is called the **feasible region**.

(c) **To find the equation of the cost function**

The function to be minimised (or maximised) is called the **objective function**. In this example, the objective function is cost, where

$$Cost = (units\ of\ X \times price\ per\ unit) + (units\ of\ Y \times price\ per\ unit)$$

Since the price of each unit of X and Y is given, therefore the cost of x units of mix X and y units of mix Y is: $cost = (x \times 5) + (y \times 4)$, that is,

$$Cost = 5x + 4y \qquad\qquad \textbf{(9.4)}$$

This cost function is in fact an **isocost** function: it represents the combinations of all quantities of products X and Y which result in the same overall cost.

To plot the cost function (isocost line)

Since the value of cost is unknown, begin by assuming any value for cost, calculate the points of intersection with the axes, then plot the line. For example, when cost = 200, the isocost line is given by $cost = 5x + 4y \rightarrow 200 = 5x + 4y$.

This isocost line cuts the x-axis when $y = 0$, that is, $200 = 5x \rightarrow x = 40$.

This isocost line cuts the y-axis when $x = 0$, that is, $200 = 4y \rightarrow y = 50$.

When one isocost line is drawn, any number of isocost lines may be drawn by drawing lines parallel to this isocost line (see Chapter 2). As cost decreases, the cost line moves downwards towards the origin as shown in Figure 9.2.

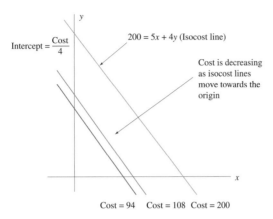

Figure 9.2   Cost decreases as isocost lines move towards the origin

Alternatively, write the isocost equation in the form $y = mx + c$; read off the slope, m; plot any line with slope m:

$$y = -\frac{5}{4}x + \frac{\text{cost}}{4} : \quad m = -\frac{5}{4} = \frac{\Delta y}{\Delta x} : \quad \begin{array}{l} \text{All isocost lines will be parallel to the line} \\ \text{that cuts the } y\text{-axis at 5 and the } x\text{-axis at 4} \end{array}$$

(d) **To find the minimum cost graphically**
The minimum cost is the point in the feasible region which is on the 'lowest' isocost line. To find this point graphically, move the isocost line downwards towards the origin until it contains just one point in the feasible region. From Figure 9.3, this point is at V, the point which has coordinates $x = 6$, $y = 16$.

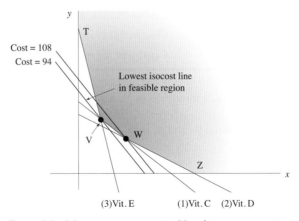

Figure 9.3   Minimum cost at point V, subject to constraints

Therefore, $x = 6$ portions of mix X and $y = 16$ portions of mix Y will satisfy the minimum daily vitamin requirements at the lowest possible cost. Since cost $= 5x + 4y$, the minimum cost $= 5x + 4y = 5(6) + 4(16) = 94$.

The minimum cost by mathematical methods

The extreme point theorem: This theorem states that, if an optimal solution exists, it is found at one of the corner points of the feasible region. In Figure 9.3, the corner points, T, V, W and Z, are found by solving each pair of equations bounding the feasible region for their point of intersection. For example, point V is the point of intersection of lines (1), vitamin C, and (3), vitamin E: $20x + 30y = 600$ and $4x + y = 40$. So solving these two equations gives point V, as follows:

$$\begin{aligned}
&\text{(1)} && 20x + 30y = 600 \\
&\text{(4)} \rightarrow \text{(3)} \times 5 && 20x + 5y = 200 \\
&\text{subtract (1)} \rightarrow \text{(4)} && 25y = 400, \text{ therefore, } y = 16
\end{aligned}$$

Substituting $y = 16$ into any of these three equations gives $x = 6$. This answer agrees with that obtained graphically.

To demonstrate the extreme point theorem

That is, show that the cost is lower at point V than at any other corner point of the feasible region. The points at the corners of the feasible region are calculated by solving the appropriate pair of simultaneous equations, as was done for point V above. The results for all corner points should be: T: (0, 40), V: (6, 16), W: (12, 12), Z: (36, 0).

The total cost is evaluated at each of these points and the results summarised as:

Point (x, y)	Cost $= 5x + 4y$
T(0, 40)	Cost $= 5(0) + 4(40) = 160$
V(6, 16)	Cost $= 5(6) + 4(16) = 94$
W(12, 12)	Cost $= 5(12) + 4(12) = 108$
Z(36, 0)	Cost $= 5(36) + 4(0) = 180$

Therefore the mathematical result, a minimum at V, agrees with the graphical result.

Note: If a constraint is not required to find at least one corner point in the feasible region, then this constraint is not a limitation.

Maximisation

Worked Example 9.2 demonstrates how to find the maximum values of a function subject to several constraints.

WORKED EXAMPLE 9.2
PROFIT MAXIMISATION SUBJECT TO CONSTRAINTS

A company manufactures two types of wrought iron gates. The number of labour-hours required to produce each type of gate, along with the maximum number of hours available, are given in Table 9.3.

Table 9.3   Requirements for gates type I and II

	Welding	Finishing	Admin.	Selling price ($£$)	Profit ($£$)
Type I gate	6	2	1	120	55
Type II gate	2	1	1	95	25
Max. hours available	840	300	250		

(a) Write down expressions for:
  (i) The constraints.
  (ii) Total revenue.
  (iii) Profit.
(b) Plot the constraints and shade in the feasible region.
(c) Plot the revenue and profit lines, indicating the direction of increasing revenue and profit.
(d) Determine, graphically, the number of gates which should be produced and sold to maximise:
  (i) Revenue.
  (ii) Profit.
  Confirm these answers algebraically, using the extreme point theorem.
(e) Calculate the number of labour-hours which are not used when:
  (i) Revenue is maximised.
  (ii) Profit is maximised.

SOLUTION

(a) For x type I gates and y type II gates, the number of labour-hours, revenue and profit are calculated by multiplying the requirements for one gate of each type, as given in Table 9.3, by x and y, respectively. This is given in Table 9.4.

Table 9.4   Requirements for x type I gates and y type II gates

	Welding	Finishing	Admin.	Selling price ($£$)	Profit ($£$)
x type I gates	$6x$	$2x$	x	$120x$	$55x$
y type II gates	$2y$	$1y$	y	$95y$	$25y$
Max. hours available	840	300	250		

The constraints are

Welding	$6x + 2y \leq 840$	(9.5)
Finishing	$2x + y \leq 300$	(9.6)
Administration	$x + y \leq 250 \quad x \geq 0, \ y \geq 0$	(9.7)

The equations for total revenue and profit are

Total revenue:	$TR = 120x + 95y$	(9.8)
Profit:	$\pi = 55x + 25y$	(9.9)

(b) Plot the constraints, by calculating the points of intersection with the axes.

Activity	Equation	Cuts x-axis when $y = 0$	Cuts y-axis when $x = 0$
Welding	(1) $6x + 2y = 840$	$6x = 840 \rightarrow x = 140$	$2y = 840 \rightarrow y = 420$
Finishing	(2) $2x + y = 300$	$2x = 300 \rightarrow x = 150$	$y = 300$
Administration	(3) $x + y = 250$	$x = 250$	$y = 250$
	$x \geq 0, y \geq 0$	First quadrant	

These lines are sketched in Figure 9.4. Since the inequalities are all less than or equal to a limit, the feasible region is the overlap region of all three constraints, (c) $TR = 120x + 95y$.

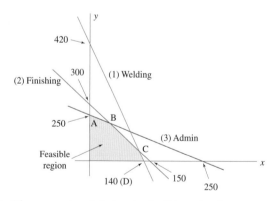

Figure 9.4   The constraints defining the feasible region for gate manufacturing

To plot one revenue line, choose any value of TR, for example, $TR = 9500$. This line (called an isorevenue line) cuts the y-axis at $y = 100.00$ (when x is zero) and the x-axis at $x = 79.167$ (when $y = 0$). See Figure 9.5. As revenue increases, the isorevenue lines move upwards away from the origin.
Profit: $\pi = 55x + 55y$.
To plot one isoprofit line, choose any value of π, for example, $\pi = 1100$. This line cuts the y-axis at $y = 44$ (when x is zero) and the x-axis at $x = 20$ (when $y = 0$). As profit increases, the isoprofit lines move upwards away from the origin. See Figure 9.5.

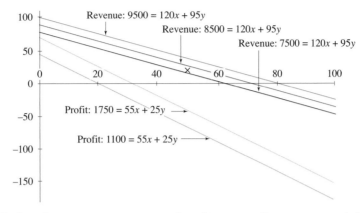

Figure 9.5   Profit and revenue increases as isoprofit and isorevenue lines move upwards from the origin

(d) For maximum revenue, it is required that the isorevenue line be as 'high' as possible, but contain one point in the feasible region. From Figure 9.6(a), this point is at B, where $x = 50$, $y = 200$. For maximum profit, it is required that the isoprofit line be as 'high' as possible, but still contain one point in the feasible region. From Figure 9.6(b), this point is at C, where $x = 120$, $y = 60$.

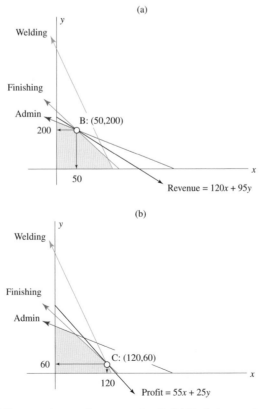

Figure 9.6   (a) Revenue is a maximum at point B. (b) Profit is a maximum at point C

The answer is confirmed algebraically by solving the appropriate pairs of equations at points B and C.

Point B: Maximum revenue is at the intersection of equations (2), finishing, and (3), administration.

(2) $2x + y = 300$
(3) $x + y = 250$
The solution of these equations is $x = 50$, $y = 200$

Point C: The maximum profit is at the intersection of equations (1), welding, and (2), finishing.

(1) $6x + 2y = 840$
(2) $2x + y = 300$
The solution of these equations is $x = 120$, $y = 60$

To demonstrate the extreme point theorem, calculate profit and revenue at all corner points (just to check the answers above).

Point (x, y)	$TR = 120x + 95y$	Profit $= 55x + 25y$
A $(x = 0, y = 250)$	$TR = 120(0) + 95(250) = 23\ 750$	Profit $= 55(0) + 25(250) = 6250$
B $(x = 50, y = 200)$	$TR = 120(50) + 95(200) = 25\ 000^*$	Profit $= 55(50) + 25(200) = 7750$
C $(x = 120, y = 60)$	$TR = 120(120) + 95(60) = 20\ 100$	Profit $= 55(120) + 25(60) = 8100^*$
D $(x = 140, y = 0)$	$TR = 120(140) + 95(0) = 16\ 800$	Profit $= 55(140) + 25(0) = 7700$

*Maximum value, subject to constraints

(e) When revenue is maximised ($x = 50$, $y = 200$), there are 140 unused welding hours:

	Welding	Finishing	Admin.
50 type I gates	6 (50)	2 (50)	1 (50)
200 type II gates	2 (200)	1 (200)	1 (200)
Hours used	700	300	250
Max. hours available	840	300	250
Unused hours	140	0	0

When profit is maximised ($x = 120$, $y = 60$) there are 70 unused administration hours:

	Welding	Finishing	Admin.
120 type I gates	6(120)	2(120)	1(120)
60 type II gates	2(60)	1(60)	1(60)
Hours used	840	300	180
Max. hours available	840	300	250
Unused hours	0	0	70

LINEAR ALGEBRA AND APPLICATIONS

PROGRESS EXERCISES 9.1

Linear Programming

Assume the non-negativity constraint $(x \geq 0, y \geq 0)$ applies in each of the following problems. In questions 1 to 4:

(a) Graph the inequality constraints.
(b) Shade in the feasible region.
(c) Calculate the corner points of the feasible region.

1. (i) $\begin{aligned} 3x + 2y &\geq 15 \\ 6x + 9y &\geq 36 \end{aligned}$ (ii) $\begin{aligned} 6x + 2y &\leq 30 \\ 2x + 6y &\leq 26 \end{aligned}$

2. (i) $4y + 5x \leq 20$ (ii) $y + 3x \leq 6$ (iii) $3y + 7x \leq 21$

3. (i) $4y + 5x \leq 20$ (ii) $8y + 15x \leq 48$ (iii) $3y + 7x \leq 21$

4. (i) $y + 5x \leq 10$ (ii) $y + 2x \leq 6$ (iii) $2y + 7x \geq 14$

5. Draw two isofunction lines for each of the following:
   (a) $TC = 12x + 24y$ (b) $TR = 9x + 3y$ (c) $\pi = 5.5x + 2.85y$

6. (See question 3)
   Find the maximum value of $W = 3x + 2y$, subject to the constraints
   (i) $4y + 5x \leq 20$ (ii) $8y + 15x \leq 48$ (iii) $3y + 7x \leq 21$
   (a) algebraically and (b) graphically.

7. (See question 6)
   Find the minimum value of $C = 3x + 2y$, subject to the constraints
   (i) $4y + 5x \leq 20$ (ii) $8y + 15x \leq 48$ (iii) $3y + 7x \leq 21$
   (a) algebraically and (b) graphically.

8. For the following constraints
   (i) $12x + 36y \leq 720$ (ii) $60x + 70y \leq 4200$ (iii) $24x + 40y \leq 1200$
   (a) Graph the inequalities.
   (b) Shade in the feasible region. Hence state whether each constraint is acting as a limitation.
   (c) Calculate the corner points of the feasible region.

9. For the following constraints
   (i) $25x + 80y \geq 800$ (ii) $22x + 5y \leq 220$ (iii) $x + y \geq 10$
   (a) Graph the inequalities.
   (b) Shade in the feasible region. Hence state whether each constraint is acting as a limitation.
   (c) Calculate the corner points of the feasible region.

10. A company can use two types of machine A and B, in a manufacturing plant. The number of operators required and the running cost per day are given as

	Cost per day	Available operators	Floor area (m²)	Profit per machine
Machine A	6	2	2	20
Machine B	3	4	2	30
Maximum available	360	280	160	

(a) Write down the inequality constraints and the profit function.
(b) Graph the inequalities. From the graph, determine the number of machines A and B which should be used to maximise profits.
(c) Confirm your answer in (b) algebraically.

11. A gardener requires fertiliser to have a minimum of 80 units of nitrogen, 15 units of potassium and 10 units of iron. Two fertiliser mixes are available.
Mix X contains 20, 2 and 1 units of the minerals per kg, while
Mix Y contains 4, 1, 2 units of the minerals per kg, respectively.
Mix X costs £18 per kg while Mix Y costs £6 per kg.
(a) Write down the inequality constraints and the equation of the cost function.
(b) Graph the inequality constraints. Hence determine the number of kg of each mix which provides the minimum mineral requirements at minimum cost.
(c) Confirm your answer in (b) algebraically.

9.2 Matrices

At the end of this section you should be able to:

- Define matrices
- Add, subtract and transpose matrices
- Multiply matrices
- Apply matrix arithmetic
- Use matrix arithmetic to simplify calculations for large arrays of data.

Matrices are rectangular arrays of numbers or symbols. The usual arithmetic operations – addition, subtraction and multiplication – are performed with these arrays. However, dealing with arrays of numbers, and not just single numbers as in ordinary arithmetic, means that there will be restrictions on all these arithmetic operations. Since matrices and determinants (see next section) are both arrays of numbers, much of the terminology used to describe matrices is the same as that used in determinants such as minors, cofactor, dimension.

9.2.1 Matrices: definition

Matrices are rectangular arrays of numbers or symbols.
The dimension of a matrix is stated as the number of rows by the number of columns. The following are examples of matrices, with their dimensions:

$$A = \begin{pmatrix} 1 & 2 \\ -2 & 4 \end{pmatrix} \quad B = \begin{pmatrix} 0 & 2 \\ 1 & 0 \\ 2 & 5 \end{pmatrix} \quad D = (-2 \quad 1 \quad 4)$$

Dimension $\qquad$ 2×2 $\qquad$ 3×2 $\qquad$ 1×3

Special matrices

The null matrix is a matrix of any dimension in which every element is zero, such as

$$\begin{pmatrix} 0 & 0 \\ 0 & 0 \end{pmatrix}, \quad \begin{pmatrix} 0 \\ 0 \\ 0 \end{pmatrix}, \quad (0 \ 0 \ 0 \ 0)$$

dimension: $\quad 2 \times 2 \qquad 3 \times 1 \qquad 1 \times 4$

The unit matrix or identity matrix is any square matrix in which every element is zero, except the elements on the main diagonal, each of which has the value 1, such as

$$I = \begin{pmatrix} 1 & 0 \\ 0 & 1 \end{pmatrix}, \qquad I = \begin{pmatrix} 1 & 0 & 0 \\ 0 & 1 & 0 \\ 0 & 0 & 1 \end{pmatrix}$$

The unit or identity matrix is represented by the symbol I.

Matrices are equal if they are of the same dimension and the corresponding elements are identical:

$$\begin{pmatrix} 1 & -1 \\ 0 & 3 \end{pmatrix}, \quad \begin{pmatrix} 1 & -1 \\ 0 & 3 \end{pmatrix} \qquad \text{are equal, but the matrices}$$

$$\begin{pmatrix} 1 & 1 & 1 \\ 1 & 1 & 1 \end{pmatrix}, \quad \begin{pmatrix} 1 & 1 \\ 1 & 1 \\ 1 & 1 \end{pmatrix} \qquad \begin{array}{l} \text{are not equal. The elements are identical,} \\ \text{but the dimensions are not.} \end{array}$$

The transpose of a matrix is the matrix obtained by writing the rows of any matrix as columns or vice versa, as follows:

$$\begin{array}{cc} & \text{row 1} \quad \text{row 2} \\ & \downarrow \quad \downarrow \\ \begin{array}{l} \text{row 1} \rightarrow \\ \text{row 2} \rightarrow \end{array} \begin{pmatrix} a & b \\ c & d \end{pmatrix}^{T} = & \begin{pmatrix} a & c \\ b & d \end{pmatrix} \end{array}$$

where the superscript T indicates that the matrix is to be transposed.

9.2.2 Matrix addition and subtraction

To add matrices, add the corresponding elements, for example,

$$\begin{pmatrix} a & b \\ c & d \end{pmatrix} + \begin{pmatrix} e & f \\ g & h \end{pmatrix} = \begin{pmatrix} a+e & b+f \\ c+g & d+h \end{pmatrix} \tag{9.10}$$

To **subtract** matrices, subtract the corresponding elements, for example,

$$\begin{pmatrix} a & b \\ c & d \end{pmatrix} - \begin{pmatrix} e & f \\ g & h \end{pmatrix} = \begin{pmatrix} a-e & b-f \\ c-g & d-h \end{pmatrix}$$

(9.11)

WORKED EXAMPLE 9.3
ADDING AND SUBTRACTING MATRICES

Given the following matrices:

$$A = \begin{pmatrix} 1 & 2 \\ -2 & 4 \end{pmatrix} \qquad B = \begin{pmatrix} 0 & 2 & 2 \\ 1 & 0 & 5 \end{pmatrix} \qquad C = \begin{pmatrix} 3 & -2 \\ 5 & 0 \end{pmatrix}$$

(9.12)

(a) Calculate
   (i) $A + C$
   (ii) $A - C$
(b) Why is it not possible to calculate $A + B$ or $A - B$?
   Hence, state the restrictions on matrix addition and subtraction.

SOLUTION

(a) (i) $\begin{pmatrix} 1 & 2 \\ -2 & 4 \end{pmatrix} + \begin{pmatrix} 3 & -2 \\ 5 & 0 \end{pmatrix} = \begin{pmatrix} 1+3 & 2+(-2) \\ -2+5 & 4+0 \end{pmatrix} = \begin{pmatrix} 4 & 0 \\ 3 & 4 \end{pmatrix}$

(ii) $\begin{pmatrix} 1 & 2 \\ -2 & 4 \end{pmatrix} - \begin{pmatrix} 3 & -2 \\ 5 & 0 \end{pmatrix} = \begin{pmatrix} 1-3 & 2-(-2) \\ -2-5 & 4-0 \end{pmatrix} = \begin{pmatrix} -2 & 4 \\ -7 & 4 \end{pmatrix}$

(b) When attempting to add the corresponding elements of the matrices A and B it is found that there is no third column in matrix A, therefore it is not possible to add pairs of corresponding elements in the two matrices. Matrix addition is not possible.

$$\begin{pmatrix} 1 & 2 \\ -2 & 4 \end{pmatrix} + \begin{pmatrix} 0 & 2 & 2 \\ 1 & 0 & 5 \end{pmatrix} = \begin{pmatrix} 1+0 & 2+2 & ?+2 \\ -2+1 & 4+0 & ?+5 \end{pmatrix}$$

No corresponding elements to add!

The same problem arises when attempting to subtract matrices A and B: there is no third column in matrix A from which to subtract the elements in the third column of matrix B.

To add or subtract two or more matrices, all matrices must have exactly the same dimensions.

Scalar multiplication

A scalar is an ordinary number: 2, 5, –8, etc. When a matrix is multiplied by a scalar, each element in the matrix is multiplied by the scalar:

$$k \begin{pmatrix} a & b \\ c & d \end{pmatrix} = \begin{pmatrix} ka & kb \\ kc & kd \end{pmatrix} \tag{9.13}$$

WORKED EXAMPLE 9.4
MULTIPLICATION OF A MATRIX BY A SCALAR

(a) Given the matrix

$$A = \begin{pmatrix} 1 & 2 \\ -2 & 4 \end{pmatrix}$$

calculate 5A.

(b) Calculate 3I, where I is the 3 × 3 unit matrix.

SOLUTION

(a) $5A = \begin{pmatrix} 5 \times 1 & 5 \times 2 \\ 5 \times (-2) & 5 \times 4 \end{pmatrix} = \begin{pmatrix} 5 & 10 \\ -10 & 20 \end{pmatrix}$

(b) $3I = 3 \begin{pmatrix} 1 & 0 & 0 \\ 0 & 1 & 0 \\ 0 & 0 & 1 \end{pmatrix} = \begin{pmatrix} 3 & 0 & 0 \\ 0 & 3 & 0 \\ 0 & 0 & 3 \end{pmatrix}$

9.2.3 Matrix multiplication

Two matrices A and B are multiplied as follows:

$$\begin{pmatrix} a & b \\ c & d \end{pmatrix} \times \begin{pmatrix} e & f \\ g & h \end{pmatrix} = \begin{pmatrix} \text{row 1} \times \text{col. 1} & \text{row 1} \times \text{col. 2} \\ \text{row 2} \times \text{col. 1} & \text{row 2} \times \text{col. 2} \end{pmatrix} \tag{9.14}$$

where row 1, row 2 are from matrix A, col. 1, col. 2 from matrix B.

To multiply a row by a column, for example (row 1 × col. 1), proceed as follows:

$$\begin{pmatrix} a & b \\ - & - \end{pmatrix} \times \begin{pmatrix} e & - \\ h & - \end{pmatrix} = \begin{pmatrix} (a \times e) + (b \times g) & - \\ - & - \end{pmatrix} \tag{9.15}$$

that is, from row 1 and column 1, multiply the first pair of elements in each, the second pair of elements in each, then add the products of these pairs.

The product of the two matrices written out in full is

$$\begin{pmatrix} a & b \\ c & d \end{pmatrix} \times \begin{pmatrix} e & f \\ g & h \end{pmatrix} = \begin{pmatrix} ae+bg & af+bh \\ ce+dg & cf+dh \end{pmatrix} \qquad (9.16)$$

dimension: $\quad (2 \times 2) \times (2 \times 2) \quad = (2 \times 2)$

WORKED EXAMPLE 9.5
MATRIX MULTIPLICATION

Given the matrices

$$A = \begin{pmatrix} 1 & 2 \\ -2 & 4 \end{pmatrix} \qquad B = \begin{pmatrix} 0 & 2 & 2 \\ 1 & 0 & 5 \end{pmatrix} \qquad C = \begin{pmatrix} 3 & -2 \\ 5 & 0 \end{pmatrix}$$

evaluate: (a) AC (b) CA (c) AB (d) BA.
Compare your answers for (a) and (b).

SOLUTION

(a) The product of the matrices A and C is given in general as

$$AC = \begin{pmatrix} 1 & 2 \\ -2 & 4 \end{pmatrix} \times \begin{pmatrix} 3 & -2 \\ 5 & 0 \end{pmatrix} = \begin{pmatrix} \text{row 1} \times \text{col. 1} & \text{row 1} \times \text{col. 2} \\ \text{row 2} \times \text{col. 1} & \text{row 2} \times \text{col. 2} \end{pmatrix}$$

$$= \begin{pmatrix} (1)(3)+(2)(5) & (1)(-2)+(2)(0) \\ (-2)(3)+(4)(5) & (-2)(-2)+(4)(0) \end{pmatrix}$$

$$= \begin{pmatrix} 13 & -2 \\ 14 & 4 \end{pmatrix}$$

(b) Similarly, CA is calculated as follows:

$$CA = \begin{pmatrix} 3 & -2 \\ 5 & 0 \end{pmatrix} \times \begin{pmatrix} 1 & 2 \\ -2 & 4 \end{pmatrix} = \begin{pmatrix} \text{row 1} \times \text{col. 1} & \text{row 1} \times \text{col. 2} \\ \text{row 2} \times \text{col. 1} & \text{row 2} \times \text{col. 2} \end{pmatrix}$$

$$= \begin{pmatrix} (3)(1)+(-2)(-2) & (3)(2)+(-2)(4) \\ (5)(1)+(0)(-2) & (5)(2)+(0)(4) \end{pmatrix}$$

$$= \begin{pmatrix} 7 & -2 \\ 5 & 10 \end{pmatrix}$$

In the above example

$$AC = \begin{pmatrix} 13 & -2 \\ 14 & 4 \end{pmatrix}$$

and

$$CA = \begin{pmatrix} 7 & -2 \\ 5 & 10 \end{pmatrix}$$

therefore $AC \neq CA$. (Since two matrices are equal only if all corresponding elements are identical.)

So in matrix multiplication, the order of multiplication is important. In general, in matrix multiplication, $AC \neq CA$.

(c) The product of the matrices A and B is given in general as

$$AB = \begin{pmatrix} 1 & 2 \\ -2 & 4 \end{pmatrix} \times \begin{pmatrix} 0 & 2 & 2 \\ 1 & 0 & 5 \end{pmatrix}$$

$$= \begin{pmatrix} \text{row 1} \times \text{col. 1} & \text{row 1} \times \text{col. 2} & \text{row 1} \times \text{col. 3} \\ \text{row 2} \times \text{col. 1} & \text{row 2} \times \text{col. 2} & \text{row 2} \times \text{col. 3} \end{pmatrix}$$

dimension: $(2 \times 2) \times (2 \times 3) = (2 \times 3)$

Multiplying the rows by the columns,

$$AB = \begin{pmatrix} 1 & 2 \\ -2 & 4 \end{pmatrix} \times \begin{pmatrix} 0 & 2 & 2 \\ 1 & 0 & 5 \end{pmatrix}$$

$$= \begin{pmatrix} (1)(0) + (2)(1) & (1)(2) + (2)(0) & (1)(2) + (2)(5) \\ (-2)(0) + (4)(1) & (-2)(2) + (4)(0) & (-2)(2) + (4)(5) \end{pmatrix}$$

$$= \begin{pmatrix} 2 & 2 & 12 \\ 4 & -4 & 16 \end{pmatrix}$$

(d) To evaluate BA, proceed as usual:

$$BA = \begin{pmatrix} 0 & 2 & 2 \\ 1 & 0 & 5 \end{pmatrix} \times \begin{pmatrix} 1 & 2 \\ -2 & 4 \end{pmatrix} = \begin{pmatrix} \text{row 1} \times \text{col. 1} & \text{row 1} \times \text{col. 2} \\ \text{row 2} \times \text{col. 1} & \text{row 2} \times \text{col. 2} \end{pmatrix}$$

dimension: $(2 \times 3) \times (2 \times 2)$

Then multiplying the rows in the first matrix by the columns in the second matrix, it is found that:

$$BA = \begin{pmatrix} 0 & 2 & 2 \\ 1 & 0 & 5 \end{pmatrix} \times \begin{pmatrix} 1 & 2 \\ -2 & 4 \end{pmatrix}$$

no third element in the columns of A to multiply the third elements in the rows of B

$$= \begin{pmatrix} (0)(1) + (2)(-2) + (2)(?) & (0)(2) + (2)(4) + (2)(?) \\ (1)(1) + (0)(-2) + (5)(?) & (1)(2) + (0)(4) + (5)(?) \end{pmatrix}$$

Since the rows in B contain three elements and the columns in A only two, matrix multiplication is not possible.

Matrix multiplication BA is possible if the number of elements in the rows of the first matrix (B) is the same as the number of elements in the columns of the second matrix (A).

This condition for matrix multiplication can be established quickly by writing down the dimensions of the matrices to be multiplied, in order:

$$A \times B = \text{product}$$

dimension of product: 2×3

dimensions: $(2 \times 2) \times (2 \times 3) = (2 \times 3)$

the same, so multiplication is possible

The 'inside' numbers are the same, therefore multiplication is possible. The 'outside' numbers give us the dimension of the product.

9.2.4 Applications of matrix arithmetic

WORKED EXAMPLE 9.6
APPLICATIONS OF MATRIX ARITHMETIC

A distributor records the weekly sales of personal computers (PCs) in three retail outlets in different parts of the country (see Table 9.5). The cost price of each model is:

Table 9.5   Number of computers sold in each shop

	Basic	Extra	Latest
Shop A	150	320	180
Shop B	170	420	190
Shop C	201	63	58

Basic £480, Extra £600, Latest £1020.
The retail price of each model in each of the three shops is given in Table 9.6.

Table 9.6   Selling price of computers in each shop

	Basic	Extra	Latest
Shop A	560	750	1580
Shop B	520	690	1390
Shop C	590	720	1780

Use matrix multiplication to calculate:

(a) The total weekly cost of computers to each shop.
(b) The total weekly revenue for each model for each shop.
(c) The total weekly profit for each shop.
   Which shop makes the greatest overall profit?

SOLUTION

(a) The numbers sold from Table 9.5 may be written as a matrix, Q:

$$\begin{array}{cc} & \begin{array}{ccc} \text{PC} \rightarrow & \text{Basic} & \text{Extra} \quad \text{Latest} \end{array} \\ Q = \begin{array}{c} \text{shop A} \\ \text{shop B} \\ \text{shop C} \end{array} & \begin{pmatrix} 150 & 320 & 180 \\ 170 & 420 & 190 \\ 201 & 63 & 58 \end{pmatrix} \end{array}$$

Write the cost of each type of computer as a column matrix:

$$C = \begin{pmatrix} 480 \\ 600 \\ 1020 \end{pmatrix}$$

If this cost matrix C is premultiplied by the numbers sold matrix, Q, the product will be a 3×1 matrix, in which the elements in each row give the total cost of computers to each shop:

$$Q \cdot C = \text{total cost}$$

$$\text{dimension: } (3 \times 3)(3 \times 1) = (3 \times 1)$$

$$\begin{pmatrix} 150 & 320 & 180 \\ 170 & 420 & 190 \\ 201 & 63 & 58 \end{pmatrix} \begin{pmatrix} 480 \\ 600 \\ 1020 \end{pmatrix} = \begin{pmatrix} 150(480) + 320(600) + 180(1020) \\ 170(480) + 420(600) + 190(1020) \\ 201(480) + 63(600) + 58(1020) \end{pmatrix}$$

$$= \begin{pmatrix} 447\,600 \\ 527\,400 \\ 193\,440 \end{pmatrix}$$

Cost to: shop A $= £447\,600$, shop B $= £527\,400$, shop C $= £193\,440$.

(b) The total revenue = price × quantity. The quantities are given by the matrix, Q, for the data in Table 9.5. The prices are obtained from the data in Table 9.6. Matrix multiplication, however, is carried out by multiplying rows by columns; therefore in order to multiply quantity × price for each PC, the rows in Table 9.6 must be written as columns in the prices matrix. That is, the prices matrix must be transposed before premultiplying by the quantities matrix, Q.

$$P = \begin{pmatrix} 560 & 750 & 1580 \\ 520 & 690 & 1390 \\ 590 & 720 & 1780 \end{pmatrix}^T \quad \begin{array}{l} \text{Price of Basic PC} \rightarrow \\ \text{Price of Extra PC} \rightarrow \\ \text{Price of Latest PC} \rightarrow \end{array} \begin{pmatrix} 560 & 520 & 590 \\ 750 & 690 & 720 \\ 1580 & 1390 & 1780 \end{pmatrix}$$

Now multiply Q (quantities matrix) by P (prices matrix from data in Table 9.6):

PC type Basic Extra Latest
$$Q \times P = \begin{array}{l} \text{shop A} \\ \text{shop B} \\ \text{shop C} \end{array} \begin{pmatrix} 150 & 320 & 180 \\ 170 & 420 & 190 \\ 201 & 63 & 58 \end{pmatrix}$$

$$\begin{array}{l} \text{price of Basic} \rightarrow \\ \times \text{ price of Extra} \rightarrow \\ \text{price of Latest} \rightarrow \end{array} \begin{pmatrix} 560 & 520 & 590 \\ 750 & 690 & 720 \\ 1580 & 1390 & 1780 \end{pmatrix}$$

$Q \times P$ shop A   shop B

$$= \begin{pmatrix} 150(560) + 320(750) + 180(1580) & N/A \\ N/A & 170(520) + 420(690) + 190(1390) \\ N/A & N/A \end{pmatrix}$$

shop C

$$\times \begin{pmatrix} N/A \\ N/A \\ 201(590) + 63(720) + 58(1780) \end{pmatrix}$$

The total revenue for each shop is given by the elements on the main diagonal of the product matrix. Total revenue for shops A, B and C is summarised in a column matrix, TR:

Basic PC   Extra PC   Latest PC

$$\text{Revenue matrix, } TR = \begin{pmatrix} 84\,000 + 240\,000 + 284\,400 \\ 88\,400 + 289\,800 + 264\,100 \\ 118\,590 + 45\,360 + 103\,240 \end{pmatrix} \begin{array}{l} \leftarrow \text{shop A} \\ \leftarrow \text{shop B} \\ \leftarrow \text{shop C} \end{array} = \begin{pmatrix} 604\,400 \\ 642\,300 \\ 267\,190 \end{pmatrix}$$

(c) Profit $= TR - TC$

$$= \begin{pmatrix} 604\,400 \\ 642\,300 \\ 267\,190 \end{pmatrix} - \begin{pmatrix} 447\,600 \\ 527\,400 \\ 193\,440 \end{pmatrix} = \begin{pmatrix} 156\,800 \\ 114\,900 \\ 73\,750 \end{pmatrix} \begin{matrix} \leftarrow \text{shop A} \\ \leftarrow \text{shop B} \\ \leftarrow \text{shop C} \end{matrix}$$

So shop A makes the highest profit.

PROGRESS EXERCISES 9.2

Introductory Matrix Algebra and Applications

1. Given the matrices: $A = \begin{pmatrix} 1 & -4 \\ 0 & -9 \end{pmatrix}$, $B = \begin{pmatrix} 4 & 3 \\ -7 & 0 \end{pmatrix}$, $C = \begin{pmatrix} 5 & -1 & -1 \\ 12 & 0 & 2 \end{pmatrix}$

   calculate, if possible:

   (a) $A + B$ (b) $A - B$ (c) $A + 4B$ (d) $A + I$ (e) AI (f) $A + C$
   (g) $A + B^{\mathrm{T}}$ (h) BC (i) CB (j) CB^{T} (k) $(AB)^{\mathrm{T}}$ (l) $C + 5I$
   (m) $C^{\mathrm{T}} A$ (n) $(BC)^{\mathrm{T}}$ (o) $AC + B$

2. Given the matrices: $A = \begin{pmatrix} 1 \\ 2 \\ -2 \end{pmatrix}$, $B = \begin{pmatrix} 1 & 2 & 0 \\ 0 & 1 & -1 \end{pmatrix}$, $C = \begin{pmatrix} 1 & 0 \\ 2 & -5 \\ -1 & 1 \end{pmatrix}$

   Determine each of the following, if possible:

   (a) $A + C$ (b) $B^{\mathrm{T}} + C$ (c) BC (d) $A^{\mathrm{T}}B^{\mathrm{T}}$ (e) $B^{\mathrm{T}}A^{\mathrm{T}}$

3. A fast-food chain has three shops, A, B and C. The average daily sales and profit in each shop are given in the following table:

	Units sold			Units profit		
	Shop A	Shop B	Shop C	Shop A	Shop B	Shop C
Burgers	800	400	500	20p	40p	33p
Chips	950	600	700	50p	45p	60p
Drinks	500	1200	900	30p	35p	20p

   Use matrix multiplication to determine:
   (a) The profit for each product.
   (b) The profit for each shop.

4. The percentage of voters who will vote for party candidates A, B and C is given in the following table:

	A	B	C	No. of voters
Area 1	60%	20%	20%	25 000
Area 2	45%	30%	25%	60 000
Area 3	38%	30%	32%	98 000

   Use matrix multiplication to calculate the total number of votes for each candidate.

5. Given the following matrices:

$$A = \begin{pmatrix} 2 & -1 \\ 4 & 3 \end{pmatrix}, \quad B = \begin{pmatrix} 0 & 1 \\ -1 & 2 \end{pmatrix}, \quad C = \begin{pmatrix} 1 \\ 5 \\ 3 \end{pmatrix}, \quad D = \begin{pmatrix} 3 & 1 & 2 \\ 0 & 1 & 1 \end{pmatrix}$$

(a) Show that $AB \neq BA$.
(b) Determine the following if possible:
   (i) AC (ii) AD (iii) DC (iv) DCC^{T}

9.3   Solution of Equations: Elimination Methods

In Chapter 3, simultaneous equations were solved by adding multiples of equations to other equations until one equation in one variable was obtained. With one variable known, the remaining variables were easily found. See Worked Examples 3.1, 3.2, 3.3, 3.6. Worked Example 3.6 will be solved here by the method of Gaussian elimination.

9.3.1   Gaussian elimination

In its simplest form, Gaussian elimination is a technique for solving a system of n linear equations in n unknowns by systematically adding multiples of equations to other equations in such a way that we end up with a series of n equations, each containing one less unknown than the previous equation, with the last equation containing just one unknown. In other words, a set of equations such as

$$\begin{aligned} x + y - z &= 3 \\ 2x + y - z &= 4 \\ 2x + 2y + z &= 12 \end{aligned}$$

(9.17)

may be reduced systematically to

$$\begin{aligned} x + y - z &= 3 \\ -y + z &= -2 \\ 2z &= 4 \end{aligned}$$

(9.18)

The solutions to both sets of equations (9.17) and (9.18) are identical, but the reduced set of equations is easily solvable. The last equation contains just one unknown, z, so solve this first. Substitute this value of z into the middle equation to solve for y. Finally, with y and z known, solve the first equation for x. This is called solving by back substitution. The following worked example will demonstrate the method of Gaussian elimination and solving by back substitution.

WORKED EXAMPLE 9.7
SOLUTION OF A SYSTEM OF EQUATIONS: GAUSSIAN ELIMINATION

Solve the following equations by Gaussian elimination:

$$x + y - z = 3 \tag{1}$$
$$2x + y - z = 4 \tag{2}$$
$$2x + 2y + z = 12 \tag{3}$$

SOLUTION

All three equations must be written in the same format: variables x, y, z (in the same order) on the LHS and constants on the RHS. With the equations written in order, the variables may be dropped in the calculations which follow.

$$
\begin{array}{ll}
x + y - z = 3 & (1) \\
2x + y - z = 4 & (2) \\
2x + 2y + z = 12 & (3)
\end{array}
\rightarrow
\begin{array}{cccc}
x & y & z & \text{RHS} \\
\begin{pmatrix}
1 & 1 & -1 & 3 \\
2 & 1 & -1 & 4 \\
2 & 2 & 1 & 12
\end{pmatrix}
\end{array}
\begin{array}{l}
(1) \\
(2) \\
(3)
\end{array}
$$

The 3×4 matrix is called the augmented matrix, A. The aim of Gaussian elimination is to systematically reduce this array of numbers to upper triangular form, similar in form to the reduced set of equations (9.18). A matrix of upper triangular form has zeros beneath the main diagonal (boxed background):

$$
\begin{pmatrix}
\boxed{a_{1,1}} & a_{1,2} & a_{1,3} & a_{1,4} \\
0 & \boxed{a_{2,2}} & a_{2,3} & a_{2,4} \\
0 & 0 & \boxed{a_{3,3}} & a_{3,4}
\end{pmatrix}
$$

Therefore, in column 1, we require zeros beneath the first element, and in column 2 we require zeros beneath the second element.

Action	Augmented matrix	Calculations
To get the required 0s, you must add $(-2 \times$ row 1) to row 2 and $(-2 \times$ row 1) to row 3	$\begin{array}{cccc} x & y & z & \text{RHS} \\ \downarrow & \downarrow & \downarrow & \downarrow \\ \begin{pmatrix} 1 & 1 & -1 & 3 \\ (2) & 1 & -1 & 4 \\ (2) & 2 & 1 & 12 \end{pmatrix} \end{array} \begin{array}{l} (1) \\ (2) \\ (3) \end{array}$	

Action	Augmented matrix	Calculations
$\boxed{\text{row } 2 + (-2 \times \text{row } 1)}$ ⟶	$\begin{pmatrix} 1 & 1 & -1 & 3 \\ 0 & -1 & 1 & -2 \\ 0 & 0 & 1 & 2 \end{pmatrix}$ $\begin{matrix}(1)\\(2)^1\\(3)\end{matrix}$	$\left\{\begin{matrix} -2 & -2 & 2 & -6 \\ 2 & 1 & -1 & 4 \\ \hline 0 & -1 & 1 & -2 \end{matrix}\right.$ $\begin{matrix}(1)\times-2\\(2)\\ \\ \text{adding}\end{matrix}$
$\boxed{\text{row } 3 + (-2 \times \text{row } 1)}$ ⟶	$\begin{pmatrix} 1 & 1 & -1 & 3 \\ 0 & -1 & 1 & -2 \\ 0 & 0 & 3 & 6 \end{pmatrix}$ $\begin{matrix}(1)\\(2)^1\\(3)^1\end{matrix}$	$\left\{\begin{matrix} -2 & -2 & 2 & -6 \\ 2 & 2 & 1 & 12 \\ \hline 0 & 0 & 3 & 6 \end{matrix}\right.$ $\begin{matrix}(1)\times-2\\(2)\\ \\ \text{adding}\end{matrix}$

[1]Updated once. [2]Updated twice.

No further elimination is necessary, since there is a 0 beneath the second element in column 2. However, divide row 3 by 3 to simplify (optional)

$$\begin{pmatrix} 1 & 1 & -1 & 3 \\ 0 & -1 & 1 & -2 \\ 0 & 0 & 1 & 2 \end{pmatrix} \begin{matrix}(1)\\(2)^1\\(3)^2\end{matrix}$$ $\boxed{\text{divide row } 3^1 \text{ by } 3}$

Rewrite the equations from the augmented matrix:

$$\begin{aligned} x + y - z &= 3 && (1) \\ -y + z &= -2 && (2)^1 \\ z &= 2 && (3)^2 \end{aligned}$$ **(9.19)**

Solve by back substitution

Start with equation $(3)^2$

$(3)^2 \rightarrow z = 2$

Substitute $z = 2$ into equation $(2)^1$

$(2)^1 \rightarrow -y + z = -2 \quad$ hence $\quad -y + 2 = -2 \rightarrow y = 4$

Substitute $z = 2$ and $y = 4$ into equation (1)

$(1)^2 \rightarrow -x + y - z = 3 \quad$ hence $\quad x + 4 - 2 = 3 \rightarrow x = 1$

Solution: $x = 1, y = 4, z = 2$

In Worked Example 9.7 the elimination worked out quickly and easily; the numbers in the augmented matrix made the arithmetic easy. However, when the numbers are not so convenient, divide the row that is being used in the elimination by the value of the eliminating element, as illustrated in Worked Example 9.8.

WORKED EXAMPLE 9.8
MORE GAUSSIAN ELIMINATION

Solve the following equations by Gaussian elimination:

$$\begin{aligned} 2x + y + z &= 12 \\ 6x + 5y - 3z &= 6 \\ 4x - y + 3z &= 5 \end{aligned}$$

SOLUTION

The equations are already written in the required form: variables in order on the LHS and constants on the RHS. Therefore, write down the augmented matrix and proceed with the Gaussian elimination:

$$
\begin{array}{cccc}
x & y & z & \text{RHS}
\end{array}
$$

$$
\begin{array}{c}
2x + y + z = 12 \\
6x + 5y - 3z = 6 \\
4x - y + 3z = 5
\end{array}
\quad \rightarrow \quad
\begin{pmatrix}
2 & 1 & 1 & 12 \\
6 & 5 & -3 & 6 \\
4 & -1 & 3 & 5
\end{pmatrix}
$$

Now proceed with the elimination on the augmented matrix.

Action	Augmented matrix	Calculations
Make the boxed 2 into a 1 by dividing row 1 by 2	$\begin{array}{cccc} x & y & z & \text{RHS} \end{array}$ $\begin{pmatrix} \boxed{2} & 1 & 1 & 12 \\ 6 & 5 & -3 & 6 \\ 4 & -1 & 3 & 5 \end{pmatrix} \begin{array}{c} (1) \\ (2) \\ (3) \end{array}$ $\begin{pmatrix} 1 & 0.5 & 0.5 & 6 \\ 6 & 5 & -3 & 6 \\ 4 & -1 & 3 & 5 \end{pmatrix} \begin{array}{c} (1)^1 \\ (2) \\ (3) \end{array}$	
row 2 + (6 × row 1¹)	$\begin{pmatrix} 1 & 0.5 & 0.5 & 6 \\ 0 & 2 & -6 & -30 \\ 4 & -1 & 3 & 5 \end{pmatrix} \begin{array}{c} (1)^1 \\ (2)^1 \\ (3) \end{array}$	$\begin{array}{ccccl} -6 & -3 & -3 & -36 & (1)^1 \times -6 \\ 6 & 5 & -3 & 6 & (2) \\ \hline 0 & 2 & -6 & 30 & \text{adding} \end{array}$
row 3 + (−4 × row 1¹)	$\begin{pmatrix} 1 & 0.5 & 0.5 & 6 \\ 0 & 2 & -6 & -30 \\ 0 & -3 & 1 & -19 \end{pmatrix} \begin{array}{c} (1)^1 \\ (2)^1 \\ (3)^1 \end{array}$	$\begin{array}{ccccl} -4 & -2 & -2 & -24 & (1)^1 \times -4 \\ 4 & -1 & -3 & 5 & (3) \\ \hline 0 & -3 & 1 & -19 & \text{adding} \end{array}$
Make the boxed 2 into a 1 by dividing row 2¹ by 2	$\begin{pmatrix} 1 & 0.5 & 0.5 & 6 \\ 0 & \boxed{2} & -6 & -30 \\ 0 & -3 & 1 & -19 \end{pmatrix} \begin{array}{c} (1)^1 \\ (2)^1 \\ (3)^1 \end{array}$ $\begin{pmatrix} 1 & 0.5 & 0.5 & 6 \\ 0 & 1 & -3 & -15 \\ 0 & -3 & 1 & -19 \end{pmatrix} \begin{array}{c} (1)^1 \\ (2)^2 \\ (3)^1 \end{array}$	
row 3¹ + (−3 × row 2²)	$\begin{pmatrix} 1 & 0.5 & 0.5 & 6 \\ 0 & 1 & -3 & -15 \\ 0 & 0 & -8 & -64 \end{pmatrix} \begin{array}{c} (1)^1 \\ (2)^2 \\ (3)^1 \end{array}$	$\begin{array}{ccccl} 0 & -3 & -9 & -45 & (2)^2 \times 3 \\ 0 & -3 & 1 & -19 & (3)^1 \\ \hline 0 & -0 & -8 & -64 & \text{adding} \end{array}$

The elimination is now complete; solve by back substitution. From equation $(3)^2$

$$-8z = -64 \quad \text{hence} \quad z = \frac{-64}{-8} = 8$$

Substitute $z = 8$ into equation $(2)^2$

$$y - (3 \times 8) = -15 \quad \text{hence} \quad y = -15 + (3 \times 8) = 9$$

Substitute $z = 8$ and $y = 9$ into equation $(1)^1$

$$x + (0.5 \times 9) + (0.5 \times 8) = 6 \quad \text{hence} \quad x = 6 - 4.5 - 4 = -2.5$$

Solution: $x = -2.5$, $y = 9$, $z = 8$

9.3.2   Gauss–Jordan elimination

Gauss–Jordan elimination goes further than Gaussian elimination, producing an augmented matrix with a main diagonal of ones:

$$\begin{pmatrix} \boxed{1} & 0 & 0 & b_{1,4} \\ 0 & \boxed{1} & 0 & b_{2,4} \\ 0 & 0 & \boxed{1} & b_{3,4} \end{pmatrix}$$

In this form the solutions may be read off immediately, as in Worked Example 9.9.

WORKED EXAMPLE 9.9
GAUSS–JORDAN ELIMINATION

 Find animated worked examples at **www.wiley.com/college/bradley**

Solve the following equations by Gauss–Jordan elimination:

$$2x + y + z = 12$$
$$6x + 5y - 3z = 6$$
$$4x - y + 3z = 5$$

SOLUTION
Rearrange the equations to have variables on the LHS and constants on the RHS, as for Gaussian elimination. Then write them as an augmented matrix. Start by carrying out the Gaussian elimination, i.e., reducing the augmented matrix to upper triangular form. Since this is the same set of equations as those in Worked Example 9.8, we will continue from this point.

Action	Augmented matrix	Calculations
Make the boxed –8 into a 1 by dividing row 3^2 by –8	$\begin{pmatrix} 1 & 0.5 & 0.5 & 6 \\ 0 & 1 & -3 & -15 \\ 0 & 0 & \boxed{-8} & -64 \end{pmatrix} \begin{array}{l} (1)^1 \\ (2)^2 \\ (3)^2 \end{array}$	
Add multiples of row 3 to rows 1 and 2 to generate 0s in column 3	$\begin{pmatrix} 1 & 0.5 & 0.5 & 6 \\ 0 & 1 & -3 & -15 \\ 0 & 0 & 1 & 8 \end{pmatrix} \begin{array}{l} (1)^1 \\ (2)^2 \\ (3)^3 \end{array}$	
$(-0.5 \times \text{row } 3^3) + \text{row } 1^1$ $(3 \times \text{row } 3^3) + \text{row } 2^2$	$\begin{pmatrix} 1 & 0.5 & 0 & 2 \\ 0 & 1 & 0 & 9 \\ 0 & 0 & 1 & 8 \end{pmatrix} \begin{array}{l} (1)^2 \\ (2)^3 \\ (3)^3 \end{array}$	Calculate $(-0.5 \times \text{row } 3^3)$ $\rightarrow (0 \ \ 0 \ \ -0.5 \ \ -4)$ Calculate $(3 \times \text{row } 3^3)$ $\rightarrow (0 \ \ 0 \ \ -3 \ \ 24)$
Add multiples of row 2 to row 1 to generate 0s in column 2	$\begin{pmatrix} 1 & 0 & 0 & -2.5 \\ 0 & 1 & 0 & 9 \\ 0 & 0 & 1 & 8 \end{pmatrix} \begin{array}{l} (1)^3 \\ (2)^3 \\ (3)^3 \end{array}$	Calculate $(-0.5 \times \text{row } 2^3)$ $\rightarrow (0 \ \ -0.5 \ \ 0 \ \ -4.5)$

Write down the equations from the augmented matrix:

$$x + 0y + 0z = -2.5$$
$$0x + y + 0z = 9$$
$$0x + 0y + z = 8$$

or read off the solution directly: $x = -2.5, y = 9, z = 8$

Note: If the elimination process produces fewer equations than unknowns, then there is no unique solution.

Gauss–Jordan elimination will be used in Section 9.5 to calculate the inverse of a matrix.

PROGRESS EXERCISES 9.3

Gaussian and Gauss–Jordan Elimination

1. Write the following system of equations as an augmented matrix:

$$3x + 3y + 6z = 12$$
$$x - 3y + 5z = 5$$
$$2x + 10y - 3z = 0$$

(a) Reduce the augmented matrix to upper triangular form, then solve by back substitution.
(b) Solve by Gauss–Jordan elimination.

2. Solve the following systems of equations by (i) Gaussian elimination, and (ii) Gauss–Jordan elimination:

(a)
$$x + y = 12$$
$$2x + 5y + 2z = 20$$
$$6x + 3y + 6z = 0$$

(b)
$$x + y = 12$$
$$2x - 5y + 2z = 20$$
$$6x + 3y + 6z = 0$$

(c)
$$x + y = 12$$
$$2x + 2y = 20$$
$$6x + 3y + 6z = 0$$

(d)
$$x + y - 2z = 12$$
$$x - 5y + 4z = 20$$
$$-6x + 3y - 15z = 0$$

Solve the equations in questions 3 to 7 and leave your answers as fractions.

3.
$$3x + 4y - 9z = -2$$
$$6x + 15y - 21 = 0$$
$$5x - 4y - 9 = 0$$

4.
$$P_1 + 4P_2 + 8P_3 = 26$$
$$5P_1 + 7P_2 = 38$$
$$8P_1 + 12P_2 + 2P_3 = 66$$

5.
$$2Y - 5C + 0.8T = 580$$
$$-Y + C + 0.6T + 340 = 0$$
$$0.4Y - T = 100$$

6.
$$5x + 5y - 4z = 15$$
$$7x - 3y + 4z = 15$$
$$z = 2x - 8y$$

7.
$$5x + 1.7z = 21.6$$
$$y - z = 0$$
$$2.5y - 2.5x = 1.35x - 10.8$$

8. Use an elimination method to find the equilibrium prices and quantities where the supply and demand functions for each good are as follows:

$$Q_{d1} = 50 - 2P_1 + 5P_2 - 3P_3 \qquad Q_{s1} = 8P_1 - 5$$
$$Q_{d2} = 22 + 7P_1 - 2P_2 + 5P_3 \qquad Q_{s2} = 12P_2 - 5$$
$$Q_{d3} = 17 + P_1 + 5P_2 - 3P_3 \qquad Q_{s3} = 4P_3 - 1$$

9.4 Determinants

At the end of this section you should be able to:

- Evaluate 2×2 and 3×3 determinants
- State and use Cramer's rule to solve two and three simultaneous equations in the same number of unknowns
- Find equilibrium values for the national income model and other applications.

9.4.1 Evaluate 2×2 determinants

Determinants: definitions

- A determinant is a square array of numbers or symbols, for example,

$$A = \begin{vmatrix} a & b \\ c & d \end{vmatrix} \qquad B = \begin{vmatrix} 2 & 5 \\ 3 & -4 \end{vmatrix} \qquad D = \begin{vmatrix} 1 & 0 & -2 \\ 2 & 2 & 3 \\ 1 & 3 & 2 \end{vmatrix}$$

- The dimensions of a determinant are stated as (number of rows, r) $\times$ (number of columns, c). Therefore, the dimensions of the determinants A, B and D are

$$A : 2 \times 2, \qquad B : 2 \times 2, \qquad D : 3 \times 3$$

The elements within a determinant (or matrix) are referred to by the row and column in which the element occurs, for example,

row 1, col. 1, etc.
$\downarrow$

$$B = \begin{vmatrix} b_{1,1} = 2 & b_{1,2} = 5 \\ b_{2,1} = 3 & b_{2,2} = -4 \end{vmatrix}$$

How to evaluate a 2 × 2 determinant

A 2 × 2 determinant is evaluated as follows:

$$\begin{vmatrix} a & b \\ c & d \end{vmatrix} = (a)(d) - (c)(b) \tag{9.20}$$

Hence the value of determinant B is

$$\begin{vmatrix} 2 & 5 \\ 3 & -4 \end{vmatrix} = (2)(-4) - (3)(5) = -8 - (15) = -23$$

 Warning

Most mistakes made in evaluating determinants arise from signs. So use brackets for negatives.

9.4.2   Use determinants to solve equations: Cramer's rule

You might wonder 'how can determinants be used to solve equations?' The use of determinants is demonstrated in Worked Example 9.10.

WORKED EXAMPLE 9.10
USING DETERMINANTS TO SOLVE SIMULTANEOUS EQUATIONS

Solve the simultaneous equations by eliminating one variable, then solve for the other variable.

(a) $\begin{aligned} 2x + 5y &= 10 \\ 3x + 4y &= 5.8 \end{aligned}$ (b) $\begin{aligned} a_1 x + b_1 y &= d_1 \\ a_2 x + b_2 y &= d_2 \end{aligned}$

Describe the method of solution in general terms using the equations given in (b). Hence, deduce a method of solution which uses determinants.

SOLUTION

To describe the method of solution in general we shall solve (a) and (b) in parallel.

(a)

(1) $2x + 5y = 10$
(2) $3x + 4y = 5.8$

To eliminate y, multiply equation (1) by 4 and equation (2) by 5.

(1) × 4 $8x + 20y = 40$
(2) × 5 $15x + 20y = 29$

Subtract

$$8x - 15x + 0y = 40 - 29$$

Gathering the x-terms

$$-7x = 11$$

Divide across by 7

$$x = \frac{11}{-7} = -1.57$$

(b)

(1) $a_1x + b_1y = d_1$
(2) $a_2x + b_2y = d_2$

To eliminate y, multiply equation (1) by b_2 and equation (2) by b_1.

(1) × b_2 $a_1b_2x + b_1b_2y = d_1b_2$
(2) × b_1 $a_2b_2x + b_2b_2y = d_2b_2$

Subtract

$$a_1b_2x - a_2b_1x = d_1b_2 - d_2b_1$$

Gathering the x-terms

$$(a_1b_2 - a_2b_1)x = d_1b_2 - d_2b_1$$

Divide across by $(a_1b_2 - a_2b_1)$

$$x = \frac{d_1b_2 - d_2b_1}{(a_1b_2 - a_2b_1)} \qquad \textbf{(9.21)}$$

Look carefully at the numerator and the denominator of equation (9.21) and notice that in each case there is (product of two values) − (product of two values). This is exactly the same format that was used when evaluating a 2 × 2 determinant. Therefore, the numerator and denominator in equation (9.21) may be written as determinants:

$$x = \frac{d_1b_2 - d_2b_1}{(a_1b_2 - a_2b_1)} = \frac{\begin{vmatrix} d_1 & b_1 \\ d_2 & b_2 \end{vmatrix}}{\begin{vmatrix} a_1 & b_1 \\ a_2 & b_2 \end{vmatrix}} \qquad \textbf{(9.22)}$$

Similarly, if x were eliminated in equations (b) and solved for y, it would be found that

$$y = \frac{a_1d_2 - a_2d_1}{(a_1b_2 - a_2b_1)} = \frac{\begin{vmatrix} a_1 & d_1 \\ a_2 & d_2 \end{vmatrix}}{\begin{vmatrix} a_1 & b_1 \\ a_2 & b_2 \end{vmatrix}} \qquad \textbf{(9.23)}$$

The formulae for x and y are now examined in detail. A general rule is deduced, called Cramer's rule, which uses determinants to solve simultaneous equations.

Cramer's rule

The solution of the simultaneous equations

$$a_1 x + b_1 y = d_1$$
$$a_2 x + b_2 y = d_2$$

is given by the formulae

$$x = \frac{\begin{vmatrix} d_1 & b_1 \\ d_2 & b_2 \end{vmatrix}}{\begin{vmatrix} a_1 & b_1 \\ a_2 & b_2 \end{vmatrix}}, \qquad y = \frac{\begin{vmatrix} a_1 & d_1 \\ a_2 & d_2 \end{vmatrix}}{\begin{vmatrix} a_1 & b_1 \\ a_2 & b_2 \end{vmatrix}}$$

When these formulae are examined in detail, it is noted that:

The denominator is the same in each:

$$x = \frac{\begin{vmatrix} d_1 & b_1 \\ d_2 & b_2 \end{vmatrix}}{\begin{vmatrix} a_1 & b_1 \\ a_2 & b_2 \end{vmatrix}}, \qquad y = \frac{\begin{vmatrix} a_1 & d_1 \\ a_2 & d_2 \end{vmatrix}}{\begin{vmatrix} a_1 & b_1 \\ a_2 & b_2 \end{vmatrix}} \qquad \boxed{\text{same}}$$

(9.24)

The denominator in each case is:

$$\begin{vmatrix} a_1 & b_1 \\ a_2 & b_2 \end{vmatrix}$$

The columns in this determinant are related to the original equations as follows:

$$a_1 x + b_1 y = d_1$$
$$a_2 x + b_2 y = d_2$$
$$\downarrow \quad \downarrow$$
$$\begin{vmatrix} a_1 & b_1 \\ a_2 & b_2 \end{vmatrix}$$

Column 1 consists of the coefficients of the x variables from the original set of equations. Column 2 consists of the coefficients of the y variables from the original set of equations. The determinant of the coefficients will be referred to as Δ, that is,

$$\Delta = \begin{vmatrix} a_1 & b_1 \\ a_2 & b_2 \end{vmatrix}$$

Looking again at equation (9.24), a rule for writing down the numerators may be established:

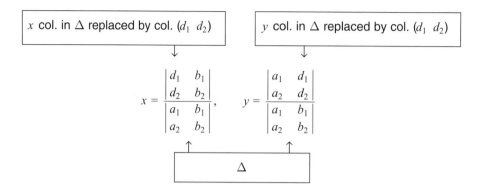

The determinant in which the column of x-coefficients is replaced by the column constants is referred to as Δ_x:

$$\Delta_x = \begin{vmatrix} d_1 & b_1 \\ d_2 & b_2 \end{vmatrix}$$

The determinant in which the column of y-coefficients is replaced by the column constants is referred to as Δ_y:

$$\Delta_y = \begin{vmatrix} a_1 & d_1 \\ a_2 & d_2 \end{vmatrix}$$

These observations are summarised next.

The solution of two simultaneous equations in two unknowns is given by

$$x = \frac{\Delta_x}{\Delta} \qquad y = \frac{\Delta_y}{\Delta} \tag{9.25}$$

In fact, this general formula may be extended to any number of equations in the same number of unknowns. This general rule is called Cramer's rule.

Cramer's rule:

The solution of three linear equations in three unknowns is given by

$$x = \frac{\Delta_x}{\Delta}, \qquad y = \frac{\Delta_y}{\Delta}, \qquad z = \frac{\Delta_z}{\Delta} \tag{9.26}$$

Note: If $\Delta = 0$, there is division by zero when applying Cramer's rule formulae. $\Delta = 0$ means that the set of equations has no unique solution (see Chapter 3).

WORKED EXAMPLE 9.11
USING CRAMER'S RULE TO SOLVE SIMULTANEOUS EQUATIONS

Use Cramer's rule to solve the equations:

$$\text{(a)} \begin{array}{l} y = 10x + 12 \\ 4x + 2y = 36 \end{array} \qquad \text{(b)} \begin{array}{l} P = 50 - 2Q \\ P = 5 + 3Q \end{array}$$

SOLUTION

In each case, Cramer's rule is used:
The solution of two simultaneous equations in two unknowns is

$$x = \frac{\Delta_x}{\Delta}, \qquad y = \frac{\Delta_y}{\Delta}$$

The procedure is as follows:

Step 1: Write the equations in order: LHS (variable terms in the same order) = RHS (constants).

Step 2: Write down the determinants Δ, Δ_x, Δ_y. Evaluate these determinants.

Step 3: Use Cramer's rule to solve for the unknowns.

(a) $\begin{array}{l} y = 10x + 12 \\ 4x + 2y = 36 \end{array}$

Step 1

$$-10x + y = 12$$
$$4x + 2y = 36$$

Step 2

$$\Delta = \begin{vmatrix} -10 & 1 \\ 4 & 2 \end{vmatrix} = (-10)(2) - (4)(1)$$

$$= -20 - 4 = -24$$

$$\Delta_x = \begin{vmatrix} 12 & 1 \\ 36 & 2 \end{vmatrix} = (12)(2) - (36)(1)$$

$$= 24 - 36 = -12$$

$$\Delta_y = \begin{vmatrix} -10 & 12 \\ 4 & 36 \end{vmatrix} = (-10)(36)$$

$$-(4)(12) = -408$$

(b) $\begin{array}{l} P = 50 - 2Q \\ P = 5 + 3Q \end{array}$

Step 1

$$P + 2Q = 50$$
$$P - 3Q = 5$$

Step 2

$$\Delta = \begin{vmatrix} 1 & 2 \\ 1 & -3 \end{vmatrix} = (1)(-3) - (1)(2)$$

$$= -3 - 2 = -5$$

$$\Delta_P = \begin{vmatrix} 50 & 2 \\ 5 & -3 \end{vmatrix} = (50)(-3) - (5)(2)$$

$$= -160$$

$$\Delta_Q = \begin{vmatrix} 1 & 50 \\ 1 & 5 \end{vmatrix} = (1)(5) - (1)(50)$$

$$= -45$$

Step 3 Use Cramer's rule:

$$x = \frac{\Delta_x}{\Delta} = \frac{-12}{-24} = 0.5$$

$$y = \frac{\Delta_y}{\Delta} = \frac{-408}{-24} = 17$$

Step 3 Use Cramer's rule:

$$P = \frac{\Delta_P}{\Delta} = \frac{-160}{-5} = 32$$

$$Q = \frac{\Delta_Q}{\Delta} = \frac{-45}{-5} = 9$$

WORKED EXAMPLE 9.12
FIND THE MARKET EQUILIBRIUM USING CRAMER'S RULE

Given the supply and demand functions for two related goods, A and B,

Good A : $\begin{cases} Q_{da} = 30 - 8P_a + 2P_b \\ Q_{sa} = -15 + 7P_a \end{cases}$ Good B : $\begin{cases} Q_{db} = 28 + 4P_a - 6P_b \\ Q_{sb} = 12 + 2P_b \end{cases}$

(a) Write down the equilibrium condition for each good. Hence, deduce two equations in P_a and P_b.

(b) Use Cramer's rule to find the equilibrium prices and quantities for goods A and B.

SOLUTION

(a) The equilibrium condition for each good is that $Q_s = Q_d$.

For good A:
$$-15 + 7P_a = 30 - 8P_a + 2P_a$$
$$15P_a - 2P_b = 45$$

For good B:
$$12 + 2P_b = 28 + 4P_a - 6P_b$$
$$-4P_a + 8P_b = 16$$

Therefore, the simultaneous equations are

(1) $15P_a - 2P_b = 45$
(2) $-4P_a + 8P_b = 16$

(b) Applying Cramer's rule:

$$P_a = \frac{\Delta_{P_a}}{\Delta} = \frac{\begin{vmatrix} 45 & -2 \\ 16 & 8 \end{vmatrix}}{\begin{vmatrix} 15 & -2 \\ -4 & 8 \end{vmatrix}} = \frac{(45)(8) - (16)(-2)}{(15)(8) - (-4)(-2)} = \frac{360 - (-32)}{120 - 8} = \frac{392}{112} = 3.5$$

$$P_b = \frac{\Delta_{P_b}}{\Delta} = \frac{\begin{vmatrix} 15 & 45 \\ -4 & 16 \end{vmatrix}}{\begin{vmatrix} 15 & -2 \\ -4 & 8 \end{vmatrix}} = \frac{(15)(16) - (-4)(45)}{(15)(8) - (-4)(-2)} = \frac{240 - (-180)}{120 - (8)} = \frac{420}{112} = 3.75$$

Substituting these values of P_a and P_b into any of the original equations, solve for Q_a and Q_b.
Solution: $Q_a = 9.5$, $Q_b = 19.5$.

General expressions for equilibrium in the income-determination model

Cramer's rule may be used to find general expressions for the equilibrium level of income, consumption, investment, etc., in the income-determination model.

WORKED EXAMPLE 9.13
USE CRAMER'S RULE FOR THE INCOME-DETERMINATION MODEL

Given the general income-determination model (no government sector):

$$Y = C + I \qquad \text{(9.27)}$$
$$C = C_0 + bY \qquad \text{(9.28)}$$

where $I = I_0, 0 < b < 1$ (b, I_0 and C_0 are constants).

(a) Write equations (9.24) and (9.25) in the form

$$a_1 Y + a_2 C = a_3 \qquad a_1, a_2, a_3 \text{ are constants}$$

(b) Hence, use Cramer's rule to express the equilibrium levels of income (Y_e) and consumption (C_e) in terms of the constants b, I_0 and C_0.

SOLUTION

(a) Rearranging equations (9.27) and (9.28) in the required form, variable terms on LHS and constant terms on RHS, and substituting I_0 for I:

$$Y - C = I_0 \qquad \text{(9.29)}$$
$$-bY + C = C_0 \qquad \text{(9.30)}$$

(b) Solve equations (9.29) and (9.30) for Y and C by Cramer's rule:

$$Y_e = \frac{\Delta_Y}{\Delta} = \frac{I_0 + C_0}{1 - b} \quad \text{and} \quad C_e = \frac{\Delta_C}{\Delta} = \frac{C_0 + bI_0}{1 - b}$$

Since:
$$\Delta = \begin{vmatrix} 1 & -1 \\ -b & 1 \end{vmatrix} = 1 - (-b)(-1) = 1 - b$$

$$\Delta_Y = \begin{vmatrix} I_0 & -1 \\ C_0 & 1 \end{vmatrix} = (I_0) - (-C_0) = I_0 + C_0$$

$$\Delta_C = \begin{vmatrix} 1 & I_0 \\ -b & C_0 \end{vmatrix} = C_0 - (-b)(I_0) = C_0 + bI_0$$

PROGRESS EXERCISES 9.4

2 × 2 Determinants, with Applications

1. Evaluate the following determinants:

(a) $\begin{vmatrix} 4 & 1 \\ 9 & 3 \end{vmatrix}$ (b) $\begin{vmatrix} -2 & 0 \\ 2 & 1 \end{vmatrix}$ (c) $\begin{vmatrix} 4 & -3 \\ 1 & 8 \end{vmatrix}$ (d) $\begin{vmatrix} 2 & 2 \\ 5 & 5 \end{vmatrix}$ (e) $\begin{vmatrix} 6 & -1 \\ -6 & 2 \end{vmatrix}$

(f) $\begin{vmatrix} 6 & -1 \\ -6 & 1 \end{vmatrix}$ (g) $\begin{vmatrix} c & 2c \\ 1 & c \end{vmatrix}$ (h) $\begin{vmatrix} a & (1-a) \\ -a & a \end{vmatrix}$ (i) $\begin{vmatrix} \frac{1}{a} & -\frac{1}{1-a} \\ -a & a \end{vmatrix}$

Use Cramer's rule to solve the simultaneous equations (correct to two decimal places) in questions 2 to 7.

2. $x + 2y = 10$, $5x + 8y = 40$ 3. $2x - y = 12$, $21x + 12y = 63$

4. $5P_1 + 9P_2 = 36$, $12P_1 + 2P_2 = 16$ 5. $P = 100 - 8Q$, $P = 30 + 5Q$

6. $50P + 8Q - 190 = 0$, $P = 25 + 7Q$ 7. $Y = 0.2r + 20$, $Y = -0.05r + 42$

8. The demand and supply functions for two related products (1: pens, 2: paper) are given by the equations

$$Q_{d1} = 30 - P_1 + 4P_2, \qquad Q_{s1} = 3P_1 - 6$$
$$Q_{d2} = 36 + 3P_1 - 2P_2, \qquad Q_{s2} = 12P_2 - 4$$

Use Cramer's rule to find the equilibrium prices and quantities.

9. Given the general income-determination model:
   (1) $Y = C + I_0$ and (2) $C = C_0 + bY$ where $0 < b < 1$, and I_0 and C_0 are constants.
   (a) Write equations (1) and (2) in the form $a_1Y + a_2C = a_3$ where a_1, a_2, a_3 are constants.
   (b) Hence use Cramer's rule to express the equilibrium condition for income (Y) and consumption (C) in terms of the constants b, I_0 and C_0.

10. If the equilibrium condition in,
    (i) the goods market, is given by the equation

$$Y = C + I \quad \text{where } C = 237.8 + 0.2Y \text{ and } I = 10 - 0.4r$$

    (ii) the money market, is given by the equation

$$M_d = M_s \quad \text{where } M_d = 100 + 0.1Y - 0.3r \text{ and } M_s = 129.225$$

    (a) Write the equilibrium equations for each market in the form $aY + cr = c$ where a, b and c are constants
    (b) Use Cramer's rule to solve for the equilibrium levels of income (Y) and interest rate (r), for which the product and money markets are simultaneously in equilibrium.

11. Given the demand function for two goods,

$$Q_1 = 5 - P_1 + 4P_2$$
$$Q_2 = -3P_2 + 2P_1$$

use Cramer's rule to derive expressions for P_1 and P_2 in terms of Q_1 and Q_2.

9.4.3 Evaluate 3 × 3 determinants

Using Cramer's rule to solve three equations in three unknowns involves the evaluation of 3×3 determinants. This is quite a different procedure from evaluating 2×2 determinants. To evaluate the general 3×3 determinant A, proceed as follows:

$$A = \begin{vmatrix} a_{1,1} & a_{1,2} & a_{1,3} \\ a_{2,1} & a_{2,2} & a_{2,3} \\ a_{3,1} & a_{3,2} & a_{3,3} \end{vmatrix} = (a_{1,1}) \begin{vmatrix} a_{2,2} & a_{2,3} \\ a_{3,2} & a_{3,3} \end{vmatrix} - (a_{1,2}) \begin{vmatrix} a_{2,1} & a_{2,3} \\ a_{3,1} & a_{3,3} \end{vmatrix} + (a_{1,3}) \begin{vmatrix} a_{2,1} & a_{2,2} \\ a_{3,1} & a_{3,2} \end{vmatrix} \qquad \textbf{(9.31)}$$

Note: Brackets are used to avoid errors when substituting negative numbers.

WORKED EXAMPLE 9.14
EVALUATION OF A 3 × 3 DETERMINANT

Evaluate the 3×3 determinant, D, where

$$D = \begin{vmatrix} 1 & 0 & -2 \\ 2 & 2 & 3 \\ 1 & 3 & 2 \end{vmatrix}$$

SOLUTION

Applying the method given in equation (9.31), the value of determinant D is calculated as

$$D = 1 \begin{vmatrix} 2 & 3 \\ 3 & 2 \end{vmatrix} - (0) \begin{vmatrix} 2 & 3 \\ 1 & 2 \end{vmatrix} + (-2) \begin{vmatrix} 2 & 2 \\ 1 & 3 \end{vmatrix}$$

$$= 1[(2)(2) - (3)(3)] - 0[(2)(2) - (1)(3)] + (-2)[(2)(3) - (1)(2)]$$

$$= 1(-5) + (0)(1) + (-2)(4) = -5 + 0 - 8 = -13$$

WORKED EXAMPLE 9.15
SOLVE THREE SIMULTANEOUS EQUATIONS BY CRAMER'S RULE

The equilibrium condition for three related products simplifies to the following equation:

$$15P_1 - 4P_2 - 7P_3 = 14$$
$$-4P_1 + 6P_2 - 2P_3 = 34$$
$$-3P_1 - 2P_2 + 12P_3 = 1$$

Use Cramer's rule to solve for the equilibrium prices P_1, P_2 and P_3.

SOLUTION

Go through the usual steps for Cramer's rule:

Step 1: The equations are already arranged in the required form
LHS (variable terms, in order) = RHS (constants).

Step 2: Write out and evaluate the four determinants: Δ, Δ_{P_1}, Δ_{P_2}, Δ_{P_3}. The determinants are:

$$\Delta = \begin{vmatrix} 15 & -4 & -7 \\ -4 & 6 & -2 \\ -3 & -2 & 12 \end{vmatrix} \quad \Delta_{P_1} = \begin{vmatrix} 14 & -4 & -7 \\ 34 & 6 & -2 \\ 1 & -2 & 12 \end{vmatrix} \quad \Delta_{P_2} = \begin{vmatrix} 15 & 14 & -7 \\ -4 & 34 & -2 \\ -3 & 1 & 12 \end{vmatrix}$$

$$\Delta_{P_3} = \begin{vmatrix} 15 & -4 & 14 \\ -4 & 6 & 34 \\ -3 & -2 & 1 \end{vmatrix}$$

Evaluate each determinant as follows:

$$\Delta = \begin{vmatrix} 15 & -4 & -7 \\ -4 & 6 & -2 \\ -3 & -2 & 12 \end{vmatrix} = 15 \begin{vmatrix} 6 & -2 \\ -2 & 12 \end{vmatrix} - (-4) \begin{vmatrix} -4 & -2 \\ -3 & 12 \end{vmatrix} + (-7) \begin{vmatrix} -4 & 6 \\ -3 & -2 \end{vmatrix}$$

$$= 15[(6)(12) - (-2)(-2)] + 4[(-4)(12) - (-3)(-2)] - 7[(-4)(-2) - (-3)(6)]$$

$$= (15)(68) + (4)(-54) + (-7)(26)$$

$$= 1020 - 216 - 182 = 622$$

$$\Delta_{P_1} = \begin{vmatrix} 14 & -4 & -7 \\ 34 & 6 & -2 \\ 1 & -2 & 12 \end{vmatrix} = 14 \begin{vmatrix} 6 & -2 \\ -2 & 12 \end{vmatrix} - (-4) \begin{vmatrix} 34 & -2 \\ 1 & 12 \end{vmatrix} + (-7) \begin{vmatrix} 34 & 6 \\ 1 & -2 \end{vmatrix}$$

$$= 14[(6)(12) - (-2)(-2)] + 4[(34)(12) - (1)(-2)] - 7[(34)(-2) - (1)(6)]$$

$$= (14)[68] + (4)[410] - 7[-74]$$

$$= 952 + 1640 + 518 = 3110$$

Similarly, Δ_{P_2} is evaluated as follows:

$$\Delta_{P_2} = \begin{vmatrix} 15 & 14 & -7 \\ -4 & 34 & -2 \\ -3 & 1 & 12 \end{vmatrix} = 15 \begin{vmatrix} 34 & -2 \\ 1 & 12 \end{vmatrix} - 14 \begin{vmatrix} -4 & -2 \\ -3 & 12 \end{vmatrix} + (-7) \begin{vmatrix} -4 & 34 \\ -3 & 1 \end{vmatrix}$$

Now evaluate each 2×2 determinant:

$$\Delta_{P_2} = 15[(34)(12) - (1)(-2)] - 14[(-4)(12) - (-3)(-2)] - 7[(-4)(1) - (-3)(34)]$$

$$= (15)[410] - 14[-54] - 7[98]$$

$$= 6150 + 756 - 686 = 6220$$

Similarly, Δ_{P_3} is evaluated as

$$\Delta_{P_3} = \begin{vmatrix} 15 & -4 & 14 \\ -4 & 6 & 34 \\ -3 & -2 & 1 \end{vmatrix} = 15 \begin{vmatrix} 6 & 34 \\ -2 & 1 \end{vmatrix} - (-4) \begin{vmatrix} -4 & 34 \\ -3 & 1 \end{vmatrix} + (14) \begin{vmatrix} -4 & 6 \\ -3 & -2 \end{vmatrix}$$

Evaluate the 2×2 determinants,

$$\Delta_{P_3} = 15[(6)(1) - (-2)(34)] + 4[(-4)(1) - (-3)(34)] + 14[(-4)(-2) - (-3)(6)]$$

$$= 15[74] + 4[98] + 14[26]$$

$$= 1110 + 392 + 364 = 1866$$

Step 3: Use Cramer's rule to solve for P_1, P_2 and P_3:

$$P_1 = \frac{\Delta P_1}{\Delta} = \frac{3110}{622} = 5, \quad P_2 = \frac{\Delta P_2}{\Delta} = \frac{6220}{622} = 10, \quad P_3 = \frac{\Delta P_3}{\Delta} = \frac{1866}{622} = 3$$

Applications

Cramer's rule may be used in any application which involves the solution of three or more simultaneous linear equations, such as in equilibrium, break-even, etc. In Worked Example 9.12 the equilibrium for a two-product market was calculated. The equilibrium for markets with three products or more may now be calculated. Worked Example 9.13, the income-determination model, may also be extended to three or more variables.

WORKED EXAMPLE 9.16
EQUILIBRIUM LEVELS IN THE NATIONAL INCOME MODEL

Given the following national income model:

$$Y = C + I_0 + G_0$$

$$C = C_0 + b(Y - T), \quad \text{and} \quad T = T_0 + tY$$

where Y = income, C = consumption, T = taxation, C_0, b, t, I_0, G_0 and T_0 (autonomous taxation) are constants and $C_0 > 0$; $0 < b < 1$; $0 < t < 1$:

(a) Write this model as three equations in terms of the variables Y, C and T.
(b) Use Cramer's rule to derive expressions for the equilibrium level of income, consumption and taxation.

SOLUTION

(a) The variable terms containing Y, C and T are arranged on the LHS of each equation, the constants on the RHS:

$$Y - C = I_0 + G_0$$
$$-bY + C + bT = C_0$$
$$-tY + T = T_0$$

(b) Using Cramer's rule, solve for Y, C, T:

$$Y = \frac{\Delta_Y}{\Delta} = \frac{I_0 + G_0 + C_0 - bT_0}{1 - b + bt} = \text{the equilibrium level of income}$$

$$C = \frac{\Delta_C}{\Delta} = \frac{C_0 + bT_0 + b(1 - t)(I_0 + G_0)}{1 - b + bt} = \text{equilibrium level of consumption}$$

$$T = \frac{\Delta_T}{\Delta} = \frac{T_0(1 - b) + t(C_0 + I_0 + G_0)}{1 - b + bt} = \text{equilibrium level of taxation}$$

Since:

$$\Delta = \begin{vmatrix} 1 & -1 & 0 \\ -b & 1 & b \\ -t & 0 & 1 \end{vmatrix} = (1) \begin{vmatrix} 1 & b \\ 0 & 1 \end{vmatrix} - (-1) \begin{vmatrix} -b & b \\ -t & 1 \end{vmatrix} + (0) = 1 + (-b - (-bt))$$

$$= 1 - b + bt$$

$$\Delta_Y = \begin{vmatrix} I_0 + G_0 & -1 & 0 \\ C_0 & 1 & b \\ T_0 & 0 & 1 \end{vmatrix} = (I_0 + G_0) \begin{vmatrix} 1 & b \\ 0 & 1 \end{vmatrix} - (-1) \begin{vmatrix} C_0 & b \\ T_0 & 1 \end{vmatrix} + (0)$$

$$= (I_0 + G_0) + (C_0 - bT_0)$$

$$\Delta_C = \begin{vmatrix} 1 & I_0 + G_0 & 0 \\ -b & C_0 & b \\ -t & T_0 & 1 \end{vmatrix} = (1) \begin{vmatrix} C_0 & b \\ T_0 & 1 \end{vmatrix} - (I_0 + G_0) \begin{vmatrix} -b & b \\ -t & 1 \end{vmatrix} + (0)$$

$$= (C_0 - bT_0) - (I_0 + G_0)(-b - (-bt))$$

$$= C_0 - bT_0 + b(I_0 + G_0)(1 - t)$$

$$\Delta_T = \begin{vmatrix} 1 & -1 & I_0 + G_0 \\ -b & 1 & C_0 \\ -t & 0 & T_0 \end{vmatrix} = (1) \begin{vmatrix} 1 & C_0 \\ 0 & T_0 \end{vmatrix} - (-1) \begin{vmatrix} -b & C_0 \\ -t & T_0 \end{vmatrix} + (I_0 + G_0) \begin{vmatrix} -b & 1 \\ -t & 0 \end{vmatrix}$$

$$= T_0 - bT_0 + tC_0 + (I_0 + G_0)(0 - (-t))$$

$$= T_0(1 - b) + t(C_0 + I_0 + G_0)$$

LINEAR ALGEBRA AND APPLICATIONS

PROGRESS EXERCISES 9.5

3×3 Determinants, with Applications

1. Evaluate each of the following determinants:

(a) $\begin{vmatrix} 1 & 2 & -5 \\ 0 & 6 & 5 \\ -1 & 2 & 7 \end{vmatrix}$ (b) $\begin{vmatrix} -3 & 0 & 3 \\ 3 & 2 & 6 \\ 4 & 0 & 9 \end{vmatrix}$ (c) $\begin{vmatrix} -2 & 3 & 2 \\ 5 & 12 & 2 \\ 3 & -4.5 & 3 \end{vmatrix}$

2. Solve the following system of equations by Cramer's rule:

(a) $x + y = 12$
$2x + 5y + 2z = 20$
$6x + 3y + 6z = 0$

(b) $x + y = 12$
$2x - 5y + 2z = 20$
$6x + 3y + 6z = 0$

(c) $x + y = 12$
$2x + 2y = 20$
$6x + 3y + 6z = 0$

(d) $x + y - 2z = 12$
$x - 5y + 4z = 20$
$-6x + 3y - 15z = 0$

Use Cramer's rule to solve the following equations in questions 3 to 7.

3. $3x + 4y - 9z = -2$
$6x + 15z - 21 = 0$
$5x + 4y - 9 = 0$

4. $P_1 + 4P_2 + 8P_3 = 26$
$5P_1 + 7P_2 = 38$
$8P_1 + 12P_2 + 2P_3 = 66$

5. $2Y - 5C + 0.8T = 580$
$-Y + C + 0.6T + 340 = 0$
$0.4Y - T = 100$

6. $5x + 5y - 4z = 15$
$7x - 3y + 4z = 15$
$z = 2x - 8y$

7. $5x + 1.7z = 21.6$
$y - z = 0$
$2.5y - 2.5x = 1.35x - 10.8$

8. Use Cramer's rule to find the equilibrium prices and quantities where the supply and demand functions for each good are:

$$Q_{d1} = 50 - 2P_1 + 5P_2 - 3P_3, \quad Q_{s1} = 8P_1 - 5$$

$$Q_{d2} = 22 + 7P_1 - 2P_2 + 5P_3, \quad Q_{s2} = 12P_2 - 5$$

$$Q_{d3} = 17 + P_1 + 5P_2 - 3P_3, \quad Q_{s3} = 4P_3 - 1$$

9. (a) State Cramer's rule.
Show graphically how a system of two equations in two unknowns has:
(i) no solution, (ii) a unique solution, (iii) infinitely many solutions.
Is it possible to solve all three systems above by Cramer's rule?
(b) Use Cramer's rule to solve for P_1, P_2 and P_3

$$0.5P_1 + 11.6P_2 - 8P_3 = -47.39$$

$$1.9P_2 + 4.5P_3 = 59.94$$

$$0.8P_1 + 3.5P_3 = 49.7$$

9.5 The Inverse Matrix and Input/Output Analysis

At the end of this section you should be able to:

- Calculate the inverse of a matrix
- Use the inverse matrix to solve equations
- Solve problems related to input/output analysis.

Section 9.4 used determinants to solve three equations in three unknowns. This section uses the inverse matrix for the same purpose. It also introduces one other application of the inverse matrix, input/output analysis.

The inverse matrix

Section 9.2 introduced matrix arithmetic: addition, subtraction and multiplication. For each arithmetic operation, the restrictions imposed were noted because matrices were used instead of single numbers. In matrix multiplication the order of multiplication generally produced different answers. When a matrix B is divided by a matrix A, B is multiplied by A^{-1}, the inverse of A, but there are two possibilities:

$$\frac{B}{A} = \begin{cases} B A^{-1} \\ A^{-1} B \end{cases}$$

each of which will generally give a different answer. Therefore matrix division is not used directly; instead we multiply by the inverse of the matrix.

9.5.1 To find the inverse of a matrix: elimination method

Write the matrix A and a unit matrix of the same dimension as an augmented matrix

$$A|I$$

Perform a series of row operations on the augmented matrix to convert the matrix A to the unit matrix: the unit matrix, I, is simultaneously converted to A^{-1}.

The explanation is as follows. Multiply each matrix by A^{-1}:

$$A|I = AA^{-1}|I A^{-1} = I|A^{-1}$$

This method is illustrated in Worked Example 9.17.

WORKED EXAMPLE 9.17
THE INVERSE OF A MATRIX: ELIMINATION METHOD

(a) Find the inverse of the matrix

$$D = \begin{pmatrix} 1 & 0 & -2 \\ 2 & 2 & 3 \\ 1 & 3 & 2 \end{pmatrix}$$

by Gauss–Jordan elimination.

(b) Show that $DD^{-1} = I$.

SOLUTION

(a) Write out the augmented matrix consisting of the matrix D and the unit matrix of the same dimension:

$$\begin{pmatrix} 1 & 0 & -2 & 1 & 0 & 0 \\ 2 & 2 & 3 & 0 & 1 & 0 \\ 1 & 3 & 2 & 0 & 0 & 1 \end{pmatrix}$$

Carry out Gauss–Jordan elimination on the matrix D by the method given in Worked Example 9.9.

Action	Augmented matrix	Calculations
	$\begin{pmatrix} 1 & 0 & -2 & 1 & 0 & 0 \\ 2 & 2 & 3 & 0 & 1 & 0 \\ 1 & 3 & 2 & 0 & 0 & 1 \end{pmatrix}\begin{matrix}(1)\\(2)\\(3)\end{matrix}$	
row 2 + (−2 × row 1) row 3 + (−1 × row 1)	$\begin{pmatrix} 1 & 0 & -2 & 1 & 0 & 0 \\ 0 & 2 & 7 & -2 & 1 & 0 \\ 0 & 3 & 4 & -1 & 0 & 1 \end{pmatrix}\begin{matrix}(1)\\(2)^1\\(3)^1\end{matrix}$	Calculate (−2 × row 1) $\rightarrow (-2 \ \ 0 \ \ 4 \ \ -2 \ \ 0 \ \ 0)$ Calculate (−1 × row 1) $\rightarrow (-1 \ \ 0 \ \ 2 \ \ -1 \ \ 0 \ \ 0)$
row $2^1 \times \frac{1}{2}$	$\begin{pmatrix} 1 & 0 & -2 & 1 & 0 & 0 \\ 0 & 1 & \frac{7}{2} & -1 & \frac{1}{2} & 0 \\ 0 & 3 & 4 & 0 & 0 & 1 \end{pmatrix}\begin{matrix}(1)\\(2)^2\\(3)^1\end{matrix}$	
row 3^1 + (−1 × row 2^2)	$\begin{pmatrix} 1 & 0 & -2 & 1 & 0 & 0 \\ 0 & 1 & \frac{7}{2} & -1 & \frac{1}{2} & 0 \\ 0 & 0 & -\frac{13}{2} & 2 & -\frac{3}{2} & 1 \end{pmatrix}\begin{matrix}(1)\\(2)^2\\(3)^2\end{matrix}$	Calculate (−3 × row 2^2) $\rightarrow (0 \ \ -3 \ \ -\frac{21}{2} \ \ 3 \ \ -\frac{3}{2} \ \ 0)$
row $3^2 \times \frac{2}{13}$	$\begin{pmatrix} 1 & 0 & -2 & 1 & 0 & 0 \\ 0 & 1 & \frac{7}{2} & -1 & \frac{1}{2} & 0 \\ 0 & 0 & 1 & -\frac{4}{13} & \frac{3}{13} & -\frac{2}{13} \end{pmatrix}\begin{matrix}(1)\\(2)^2\\(3)^3\end{matrix}$	
row 1 + (2 × row 3^3) row 2^2 + ($-\frac{7}{2}$ × row 3^3)	$\begin{pmatrix} 1 & 0 & 0 & \frac{5}{13} & \frac{6}{13} & -\frac{4}{13} \\ 0 & 1 & 0 & \frac{1}{13} & -\frac{4}{13} & \frac{7}{13} \\ 0 & 0 & 1 & -\frac{4}{13} & \frac{3}{13} & -\frac{2}{13} \end{pmatrix}\begin{matrix}(1)^1\\(2)^2\\(3)^3\end{matrix}$	Calculate (−2 × row 3^3) $\rightarrow (0 \ \ 0 \ \ 2 \ \ -\frac{8}{13} \ \ \frac{6}{13} \ \ -\frac{4}{13})$ Calculate ($-\frac{7}{2}$ × row 3^3) $\rightarrow (0 \ \ 0 \ \ -\frac{7}{2} \ \ \frac{14}{13} \ \ \frac{10.5}{13} \ \ \frac{7}{13})$

The original matrix, D, is now reduced to the unit matrix. The inverse of D is given by the transformed unit matrix: the 4th, 5th and 6th columns of the augmented matrix.

$$D^{-1} = \begin{pmatrix} \frac{5}{13} & \frac{6}{13} & -\frac{4}{13} \\ \frac{1}{13} & -\frac{4}{13} & \frac{7}{13} \\ -\frac{4}{13} & \frac{3}{13} & -\frac{2}{13} \end{pmatrix} = \frac{1}{13}\begin{pmatrix} 5 & 6 & -4 \\ 1 & -4 & 7 \\ -4 & 3 & -2 \end{pmatrix}$$

Note: If division by zero arises at any stage, then D does not have an inverse.

(b) Every element in D^{-1} is multiplied by $1/13$. Before multiplying D^{-1} by D, factor out this scalar to simplify the arithmetic involved:

$$DD^{-1} = \begin{pmatrix} 1 & 0 & -2 \\ 2 & 2 & 3 \\ 1 & 3 & 2 \end{pmatrix} \frac{1}{13} \begin{pmatrix} 5 & 6 & -4 \\ 1 & -4 & 7 \\ -4 & 3 & -2 \end{pmatrix} = \frac{1}{13} \begin{pmatrix} 13 & 0 & 0 \\ 0 & 13 & 0 \\ 0 & 0 & 13 \end{pmatrix} = \begin{pmatrix} 1 & 0 & 0 \\ 0 & 1 & 0 \\ 0 & 0 & 1 \end{pmatrix}$$

9.5.2 To find the inverse of a matrix: cofactor method

Initially, some of the terminology associated with determinants and matrices is explained. The **minor** of an element is the determinant of what is left when the row and column containing that element are crossed out. For example, in determinant

$$D = \begin{vmatrix} 1 & 0 & -2 \\ 2 & 2 & 3 \\ 1 & 3 & 2 \end{vmatrix}$$

the minor of the first element in the first row, $d_{1,1}$, is

$$D = \begin{vmatrix} \boxed{1} & 0 & -2 \\ 2 & 2 & 3 \\ 1 & 3 & 2 \end{vmatrix} = \begin{vmatrix} 2 & 3 \\ 3 & 2 \end{vmatrix} = (2)(2) - (3)(3) = -5$$

The **cofactor** of a given element is the minor of that element multiplied by either $+1$ or -1. If the given element is in the same position as $+1$ in the determinant of ±1, given below, multiply the minor by $+1$, otherwise multiply the minor by -1. Here is the determinant of ±1:

$$\begin{vmatrix} +1 & -1 & +1 \\ -1 & +1 & -1 \\ +1 & -1 & +1 \end{vmatrix}$$

The cofactor of an element is given the symbol C subscripted with the location (row, column) of that element. For example, $C_{1,1}$ is the cofactor of $d_{1,1}$ in determinant D above; it is calculated as follows:

$$C_{1,1} = (\text{minor of } d_{1,1}) \times (+1) = \begin{vmatrix} 2 & 3 \\ 3 & 2 \end{vmatrix} \times (1) = (-5)(1) = -5$$

The minor, -5, was evaluated above.

The value of a 3×3 determinant is calculated as follows:

$$|D| = (d_{1,1} \times C_{1,1}) + (d_{1,2} \times C_{1,2}) + (d_{1,3} \times C_{1,3}) \tag{9.32}$$

that is, the value of $|D|$ is the sum of the products of each element in row 1 and its **cofactor**. (In fact, $|D|$ may also be evaluated by summing the products of each element $\times$ cofactor from **any** one row or column. This is particularly useful if a row or column contains several zeros.) This method of evaluation is called Laplace expansion or the cofactor method.

Given a matrix A, the inverse of A is defined as follows:

$$A^{-1} = \frac{(C)^{T}}{|A|}$$

where C^{T} is the matrix in which every element is replaced by its cofactor:

$$A^{-1} = \frac{1}{|A|} \begin{pmatrix} C_{1,1} & C_{1,2} & C_{1,3} \\ C_{2,1} & C_{2,2} & C_{2,3} \\ C_{3,1} & C_{3,2} & C_{3,3} \end{pmatrix}^{T} \qquad \text{(9.33)}$$

Note: If

$$|A| = 0. \qquad \frac{1}{|A|} \quad \rightarrow \quad \frac{1}{0}$$

A **has no inverse**, since division by zero is not defined. Therefore, the inverse of A is determined by:

Step 1: Evaluating $|A|$. If $|A| = 0$, there is no inverse.
Step 2: Calculating the cofactor of each element.
Step 3: Replacing each element by its cofactor, then transposing the matrix of cofactors (this matrix is called the adjoint of A).
Step 4: Multiplying the transposed matrix of cofactors by $1/|A|$.

The inverse of a 2 × 2 matrix

If

$$A = \begin{pmatrix} a & b \\ c & d \end{pmatrix}$$

then

$$A^{-1} = \frac{1}{ad - bc} \begin{pmatrix} d & -b \\ -c & a \end{pmatrix} \quad (ad - bc \neq 0)$$

The proof is left as an exercise.

WORKED EXAMPLE 9.18
THE INVERSE OF A 3 × 3 MATRIX

(a) Find the inverse of the matrix

$$D = \begin{pmatrix} 1 & 0 & -2 \\ 2 & 2 & 3 \\ 1 & 3 & 2 \end{pmatrix}$$

(b) Show that $DD^{-1} = I$.

SOLUTION

(a) Use the definition of the inverse given in equation (9.30) to determine the inverse:

Step 1: Evaluate $|D|$. This is calculated by multiplying each element in row 1 by its cofactor. If the table for calculating cofactors is set up, the required cofactors are the first three cofactors in the table. So step 1 is deferred to step 2.

Step 2: Set up the table to calculate all the cofactors.

Element
(by rows)   Minor × (sign) = cofactor

1   $\begin{vmatrix} 1 & 0 & -2 \\ 2 & 2 & 3 \\ 1 & 3 & 2 \end{vmatrix} \rightarrow \begin{vmatrix} 2 & 3 \\ 3 & 2 \end{vmatrix}(1) = 4 - (9) = -5$

0   $\begin{vmatrix} 1 & 0 & -2 \\ 2 & 2 & 3 \\ 1 & 3 & 2 \end{vmatrix} \rightarrow \begin{vmatrix} 2 & 3 \\ 1 & 2 \end{vmatrix}(-1) = [4 - (3)](-1) = -1$

-2 $\begin{vmatrix} 1 & 0 & -2 \\ 2 & 2 & 3 \\ 1 & 3 & 2 \end{vmatrix} \rightarrow \begin{vmatrix} 2 & 2 \\ 1 & 3 \end{vmatrix}(1) = [6 - (2)](1) = 4$

> $|D|$ by row 1
> $= (1)(-5) + 0(-1) + (-2)(4)$
> $= -13$

2   $\begin{vmatrix} 1 & 0 & -2 \\ 2 & 2 & 3 \\ 1 & 3 & 2 \end{vmatrix} \rightarrow \begin{vmatrix} 0 & -2 \\ 3 & 2 \end{vmatrix}(-1) = [0 - (-6)](-1) = -6$

2   $\begin{vmatrix} 1 & 0 & -2 \\ 2 & 2 & 3 \\ 1 & 3 & 2 \end{vmatrix} \rightarrow \begin{vmatrix} 1 & -2 \\ 1 & 2 \end{vmatrix}(1) = [2 - (-2)](1) = 4$

3   $\begin{vmatrix} 1 & 0 & -2 \\ 2 & 2 & 3 \\ 1 & 3 & 2 \end{vmatrix} \rightarrow \begin{vmatrix} 1 & 0 \\ 1 & 3 \end{vmatrix}(-1) = [3 - (0)](-1) = -3$

1   $\begin{vmatrix} 1 & 0 & -2 \\ 2 & 2 & 3 \\ 1 & 3 & 2 \end{vmatrix} \rightarrow \begin{vmatrix} 0 & -2 \\ 2 & 3 \end{vmatrix}(1) = [0 - (-4)](1) = 4$

3   $\begin{vmatrix} 1 & 0 & -2 \\ 2 & 2 & 3 \\ 1 & 3 & 2 \end{vmatrix} \rightarrow \begin{vmatrix} 1 & -2 \\ 2 & 3 \end{vmatrix}(-1) = [3 - (-4)](-1) = -7$

2   $\begin{vmatrix} 1 & 0 & -2 \\ 2 & 2 & 3 \\ 1 & 3 & 2 \end{vmatrix} \rightarrow \begin{vmatrix} 1 & 0 \\ 2 & 2 \end{vmatrix}(1) = [2 - (0)](1) = 2$

Step 3: Replace each element in D by its cofactor. Then transpose.

$$C^T = \begin{pmatrix} -5 & -1 & 4 \\ -6 & 4 & -3 \\ 4 & -7 & 2 \end{pmatrix}^T = \begin{pmatrix} -5 & -6 & 4 \\ -1 & 4 & -7 \\ 4 & -3 & 2 \end{pmatrix} = (\text{adjoint of } D)$$

Step 4: Multiply the adjoint matrix by

$$\frac{1}{|D|} = \frac{1}{-13}$$

$$D^{-1} = \frac{1}{-13} \begin{pmatrix} -5 & -6 & 4 \\ -1 & 4 & -7 \\ 4 & -3 & 2 \end{pmatrix}$$

Every element in this matrix could be multiplied by $-(1/13)$, but this will introduce awkward fractions, so the scalar multiplication is usually left until a final single matrix is required.

$$\text{(b)}\ DD^{-1} = \begin{pmatrix} 1 & 0 & -2 \\ 2 & 2 & 3 \\ 1 & 3 & 2 \end{pmatrix} \times \frac{1}{-13} \begin{pmatrix} -5 & -6 & 4 \\ -1 & 4 & -7 \\ 4 & -3 & 2 \end{pmatrix}$$

$$= -\frac{1}{13} \begin{pmatrix} -13 & 0 & 0 \\ 0 & -13 & 0 \\ 0 & 0 & -13 \end{pmatrix} = \begin{pmatrix} 1 & 0 & 0 \\ 0 & 1 & 0 \\ 0 & 0 & 1 \end{pmatrix}$$

To write a system of equations in matrix form

In general, a system of equations, all written in the same format, may be expressed concisely in terms of matrices as follows:

$$\begin{array}{l} (1)\ a_1x + b_1y + c_1z = d_1 \\ (2)\ a_2x + b_2y + c_2z = d_2 \\ (3)\ a_3x + b_3y + c_3z = d_3 \end{array} \to \begin{pmatrix} a_1 & b_1 & c_1 \\ a_2 & b_2 & c_2 \\ a_3 & b_3 & c_3 \end{pmatrix} \begin{pmatrix} x \\ y \\ z \end{pmatrix} = \begin{pmatrix} d_1 \\ d_2 \\ d_3 \end{pmatrix} \qquad \textbf{(9.34)}$$

This statement is written concisely as $\qquad AX = B$

Dimension: $\qquad (3 \times 3)(3 \times 1) = (3 \times 1) \qquad \textbf{(9.35)}$

If the matrix A is multiplied by the column matrix X, the result is a 3×1 matrix, in which each element is the LHS of equations (1), (2) and (3) above:

$$\begin{pmatrix} a_1 & b_1 & c_1 \\ a_2 & b_2 & c_2 \\ a_3 & b_3 & c_3 \end{pmatrix} \begin{pmatrix} x \\ y \\ z \end{pmatrix} = \begin{pmatrix} d_1 \\ d_2 \\ d_3 \end{pmatrix} \to \begin{pmatrix} a_1x + b_1y + c_1z \\ a_2x + b_2y + c_2z \\ a_3x + b_3y + c_3z \end{pmatrix} = \begin{pmatrix} d_1 \\ d_2 \\ d_3 \end{pmatrix}$$

Equating the corresponding elements of these 3×1 matrices, AX and B, reproduces equations (1), (2) and (3).

Note: The columns of matrix A consist of the coefficients of x, y, z from equations (1), (2) and (3).

To solve a set of equations using the inverse matrix

Premultiply both sides of equation (9.35), $AX = B$, by the inverse of matrix A:

$A^{-1} AX = A^{-1} B$ but $AA^{-1} = I$, the unit matrix

$IX = A^{-1} B$ but $IX = X$, since I in matrix multiplication behaves like a 1 in ordinary multiplication, therefore

$$X = A^{-1}B \tag{9.36}$$

where X is the column of unknowns, A is the matrix of coefficients, B is the column of constants from the RHS of the equations, taken in order. So, if the column matrix B is premultiplied by the inverse of the matrix A, the resulting column matrix is the solution for x, y, z.

WORKED EXAMPLE 9.19
SOLVE A SYSTEM OF EQUATIONS BY THE INVERSE MATRIX

Use the inverse matrix to solve the following simultaneous equations:

$$x - 2z = 4$$
$$2x + 2y + 3z = 15$$
$$x + 3y + 2z = 12$$

SOLUTION

Step 1: Write all the equations in the same order: variables (in order) = RHS. Write down the matrices A and B. The matrix A is the matrix of coefficients of the three equations, all arranged in the same format; B is the column matrix consisting of the constants from the RHS of the equations.

$$A = \begin{pmatrix} 1 & 0 & -2 \\ 2 & 2 & 3 \\ 1 & 3 & 2 \end{pmatrix}, \qquad B = \begin{pmatrix} 4 \\ 15 \\ 12 \end{pmatrix}$$

Step 2: Since $X = A^{-1} B$, determine the inverse of A. However, if you look carefully at A, you will see that this matrix is identical to the matrix D in Worked Examples 9.17 and 9.18 in which the inverse of D was determined; therefore

$$A^{-1} = \frac{1}{-13} \begin{pmatrix} -5 & -6 & 4 \\ -1 & 4 & -7 \\ 4 & -3 & 2 \end{pmatrix}$$

So proceed straight away to the next step.

Step 3: Premultiply the matrix B by the inverse of A:

$$X = A^{-1}B$$

$$\begin{pmatrix} x \\ y \\ z \end{pmatrix} = \frac{1}{-13} \begin{pmatrix} -5 & -6 & 4 \\ -1 & 4 & -7 \\ 4 & -3 & 2 \end{pmatrix} \begin{pmatrix} 4 \\ 15 \\ 12 \end{pmatrix} = -\frac{1}{13} \begin{pmatrix} -62 \\ -28 \\ -5 \end{pmatrix} = \begin{pmatrix} 4.77 \\ 2.15 \\ 0.38 \end{pmatrix}$$

$$\begin{pmatrix} x \\ y \\ z \end{pmatrix} = \begin{pmatrix} 4.77 \\ 2.15 \\ 0.38 \end{pmatrix}$$

The solutions may simply be read off (correct to two decimal places), $x = 4.77$, $y = 2.15$, $z = 0.38$.

Input/output analysis

Consider a three-sector economy, such as industry, agriculture and financial services. The output from any sector, such as agriculture, may be required by:

(i) The other sectors.
(ii) The same sector.
(iii) External demands, such as sales, exports, etc.

For example, the output from all three sectors (£million) in this three-sector economy could be distributed as follows:

		Input to				
		Agric.	Industry	Services	Other demands	Total output
Output from	Agric. →	150	225	125	100	600
	Industry →	210	250	140	300	900
	Services →	170	0	30	100	300

So the total output required from each sector must satisfy the final demand (sales, exports, etc.), as well as the demands from other sectors which require this as basic raw material or input. For example, the output from the agriculture sector is required as raw material by agriculture itself (calves, foodstuffs, etc.); industry requires agricultural output for processing; government and other agencies provide financial and other services. Finally, consumers and other agencies want to buy and export agricultural products.

The table above is perfectly balanced, but suppose the 'other demands' from each sector change. How is the total output required recalculated? The answer is to use matrices. One other fundamental assumption is needed before setting up our method, and that is:

In input/output analysis, **total input = total output** for each sector.

With this assumption, the input/output table can be completed and the total input to each sector is written in the last row. You will notice that the sum of individual inputs does not add up to the total input. Therefore a further row is added to the table to allow for other inputs.

		Input to				
		Agric.	Industry	Services	Other demands	Total output
Output from	Agric. →	150	225	125	100	600
	Industry →	210	250	140	300	900
	Services →	170	0	30	100	300
	Other inputs	70	425	5		
	Total input	600	900	300		

Divide the input to each sector by the total input to calculate the fraction (or, if you prefer, multiply the fraction by 100 and quote this as the percentage) of the total input which comes from all sectors:

		Input to				
		Agric.	Industry	Services	Other demands	Total output
Output from	Agric. →	150/600	225/900	125/300	100	600
	Industry →	210/600	250/900	140/300	300	900
	Services →	170/600	0/900	30/300	100	300
	Total input	600/600	900/900	300/300		

The relationship between the total output from all three sectors to the input requirements from other sectors and final other demands may now be described by the matrix equation

$$\begin{pmatrix} \dfrac{150}{600} & \dfrac{225}{900} & \dfrac{125}{300} \\ \dfrac{210}{600} & \dfrac{250}{900} & \dfrac{140}{300} \\ \dfrac{170}{600} & \dfrac{0}{900} & \dfrac{30}{300} \end{pmatrix} \begin{pmatrix} 600 \\ 900 \\ 300 \end{pmatrix} + \begin{pmatrix} 100 \\ 300 \\ 100 \end{pmatrix} = \begin{pmatrix} 600 \\ 900 \\ 300 \end{pmatrix}$$

This equation may be stated in general as

$$AX + d = X$$

where X is the column of total outputs, d is the column of final (other) demands from outside the three sectors. The matrix A is called the matrix of technical coefficients: each column of A gives the fraction of inputs to that sector which comes from each of the three sectors. This equation may be used to solve for the total output (X) required from each sector if the final demands (d) are changed:

$$d = X - AX = (I - A)X \qquad \text{(recall } IX = X\text{)}$$
$$(I - A)^{-1}d = (I - A)^{-1}(I - A)X \qquad \text{premultiply both sides by } (I - A)^{-1}$$
$$(I - A)^{-1}d = X \qquad \text{since } (I - A)^{-1}(I - A) = I$$

For the input/output model $AX + d = X$, the total output required from each sector when final demands d are changed is given by the equation

$$X = (I - A)^{-1}d \qquad\qquad \textbf{(9.37)}$$

WORKED EXAMPLE 9.20
INPUT/OUTPUT ANALYSIS

Given the input/output table for the three-sector economy:

		Input to				
		Agric.	Industry	Services	Other demands	Total output
Output	Agric. →	150	225	125	100	600
from	Industry →	210	250	140	300	900
	Services →	170	0	30	100	300

If the final demands from each sector are changed to 500 from agriculture, 550 from industry, 300 from financial services, calculate the total output from each sector.

SOLUTION

Step 1: Use the underlying assumption total input = total output to complete the input/output table.

		Input to				
		Agric.	Industry	Services	Other demands	Total output
Output	Agric. →	150	225	125	100	600
from	Industry →	210	250	140	300	900
	Services →	170	0	30	100	300
	Other inputs	70	425	5		
	Total input	600	900	300		

Step 2: Calculate the matrix of technical coefficients, A, by dividing each column of inputs by total input.

		Input to				
		Agric.	Industry	Services	Other demands	Total output
Output	Agric. →	150/600	225/900	125/300	100	600
from	Industry →	210/600	250/900	140/300	300	900
	Services →	170/600	0/900	30/300	100	300
	Total input	600/600	900/900	300/300		

Step 3: Get the inverse of the matrix $(I - A)$, since this inverse matrix is required in the equation $X = (I - A)^{-1}d$:

$$(I - A) = \begin{pmatrix} 1 & 0 & 0 \\ 0 & 1 & 0 \\ 0 & 0 & 1 \end{pmatrix} - \begin{pmatrix} \dfrac{150}{600} & \dfrac{225}{900} & \dfrac{125}{300} \\ \dfrac{210}{600} & \dfrac{250}{900} & \dfrac{140}{300} \\ \dfrac{170}{600} & \dfrac{0}{900} & \dfrac{30}{300} \end{pmatrix} = \begin{pmatrix} 0.75 & -0.25 & -0.42 \\ -0.35 & 0.72 & -0.47 \\ -0.28 & 0.00 & 0.90 \end{pmatrix}$$

To calculate the inverse of $(I - A)$, (i) use the elimination method, (ii) use the cofactor method. Set out the table of cofactors:

Elements   Cofactor = minor by (± 1)
of $(I - A)$

0.75 $\begin{pmatrix} \boxed{0.75} & -0.25 & -0.42 \\ -0.35 & 0.72 & -0.47 \\ -0.28 & 0.00 & 0.90 \end{pmatrix} \rightarrow \begin{vmatrix} 0.72 & -0.47 \\ 0.00 & 0.90 \end{vmatrix} (1) = 0.648$

-0.25 $\begin{pmatrix} 0.75 & \boxed{-0.25} & -0.42 \\ -0.35 & 0.72 & -0.47 \\ -0.28 & 0.00 & 0.90 \end{pmatrix} \rightarrow \begin{vmatrix} -0.35 & -0.47 \\ -0.28 & 0.90 \end{vmatrix} (-1) = 0.4466$

-0.42 $\begin{pmatrix} 0.75 & -0.25 & \boxed{-0.42} \\ -0.35 & 0.72 & -0.47 \\ -0.28 & 0.00 & 0.90 \end{pmatrix} \rightarrow \begin{vmatrix} -0.35 & 0.72 \\ -0.28 & 0.90 \end{vmatrix} (1) = 0.2016$

-0.35 $\begin{pmatrix} 0.75 & -0.25 & -0.42 \\ \boxed{-0.35} & 0.72 & -0.47 \\ -0.28 & 0.00 & 0.90 \end{pmatrix} \rightarrow \begin{vmatrix} -0.25 & 0.42 \\ 0.00 & 0.90 \end{vmatrix} (-1) = 0.225$

$\boxed{\begin{aligned} |I - A| &= 0.75(0.648) \\ &+ (-0.25)(0.4466) \\ &+ (-0.42)(0.2016) \\ &= 0.289678 \end{aligned}}$

0.72 $\begin{pmatrix} 0.75 & \boxed{-0.25} & -0.42 \\ -0.35 & 0.72 & -0.47 \\ -0.28 & 0.00 & 0.90 \end{pmatrix} \rightarrow \begin{vmatrix} 0.75 & -0.42 \\ -0.28 & 0.90 \end{vmatrix} (1) = 0.5574$

-0.47 $\begin{pmatrix} 0.75 & -0.25 & \boxed{-0.42} \\ -0.35 & 0.72 & -0.47 \\ -0.28 & 0.00 & 0.90 \end{pmatrix} \rightarrow \begin{vmatrix} 0.75 & -0.25 \\ -0.28 & 0.00 \end{vmatrix} (-1) = 0.070$

$$-0.28 \quad \begin{pmatrix} \boxed{0.75} & -0.25 & -0.42 \\ \boxed{-0.35} & 0.72 & -0.47 \\ \boxed{-0.28} & 0.00 & 0.90 \end{pmatrix} \rightarrow \begin{vmatrix} -0.25 & -0.42 \\ 0.72 & -0.47 \end{vmatrix} (1) = 0.4199$$

$$0.00 \quad \begin{pmatrix} 0.75 & \boxed{-0.25} & -0.42 \\ -0.35 & \boxed{0.72} & -0.47 \\ \boxed{-0.28} & \boxed{0.00} & 0.90 \end{pmatrix} \rightarrow \begin{vmatrix} 0.75 & -0.42 \\ -0.35 & -0.47 \end{vmatrix} (-1) = 0.4995$$

$$0.90 \quad \begin{pmatrix} 0.75 & -0.25 & \boxed{-0.42} \\ -0.35 & 0.72 & \boxed{-0.47} \\ \boxed{-0.28} & 0.00 & 0.90 \end{pmatrix} \rightarrow \begin{vmatrix} 0.75 & -0.25 \\ -0.35 & 0.72 \end{vmatrix} (1) = 0.4525$$

The inverse of $(I - A) = C^T/|I - A|$

$$= \frac{1}{0.289\,678} \begin{pmatrix} 0.648 & 0.4466 & 0.2016 \\ 0.225 & 0.5574 & 0.070 \\ 0.4199 & 0.4995 & 0.4525 \end{pmatrix}^T$$

$$= \frac{1}{0.289\,678} \begin{pmatrix} 0.648 & 0.255 & 0.4199 \\ 0.4466 & 0.5574 & 0.4995 \\ 0.2016 & 0.070 & 0.4525 \end{pmatrix}$$

Step 4: Finally, state the column of new external demands, d, and solve for X, by equation (9.34):

$$X = (I - A)^{-1}d$$

$$\begin{pmatrix} T_{agri.} \\ T_{ind.} \\ T_{serv.} \end{pmatrix} = \frac{1}{0.289\,678} \begin{pmatrix} 0.648 & 0.225 & 0.4199 \\ 0.4466 & 0.5574 & 0.4995 \\ 0.2016 & 0.070 & 0.4525 \end{pmatrix} \begin{pmatrix} 500 \\ 550 \\ 300 \end{pmatrix} = \begin{pmatrix} 1980.5 \\ 2346.5 \\ 949.5 \end{pmatrix}$$

PROGRESS EXERCISES 9.6

The Inverse Matrix and Input/Output

In the following questions, determine the inverse matrices using (i) the elimination method and (ii) the cofactor method.

1. Determine the inverse of the following matrices:

(a) $\begin{pmatrix} 2 & 5 \\ 2 & 6 \end{pmatrix}$ (b) $\begin{pmatrix} 6 & 4 \\ 20 & 10 \end{pmatrix}$ (c) $\begin{pmatrix} -1 & 3 \\ 2 & 5 \end{pmatrix}$ (d) $\begin{pmatrix} 3 & 4 \\ 12 & 0 \end{pmatrix}$ (e) $\begin{pmatrix} 1 & 4 \\ -2 & -8 \end{pmatrix}$

In questions 2, 3, 4 and 5,

(a) Determine the inverse of the matrix.
(b) Use the inverse matrix method to solve the given system of equations.

2. (a) $\begin{pmatrix} 1 & 1 & 0 \\ 2 & 5 & 2 \\ 6 & 3 & 6 \end{pmatrix}$ (b) $\begin{array}{r} x + y = 12 \\ 2x + 5y + 2z = 20 \\ 6x + 3y + 6z = 0 \end{array}$

3. (a) $\begin{pmatrix} 1 & 1 & 0 \\ 2 & -5 & 2 \\ 6 & 3 & 6 \end{pmatrix}$ (b) $\begin{aligned} x + y &= 12 \\ 2x - 5y + 2z &= 20 \\ 6x + 3y + 6z &= 0 \end{aligned}$

4. (a) $\begin{pmatrix} 1 & 1 & 0 \\ 2 & -1 & 2 \\ 2 & 3 & 1 \end{pmatrix}$ (b) $\begin{aligned} x + y &= 2 \\ x - y + 2z &= 20 \\ 2x + 3y + z &= 0 \end{aligned}$

5. (a) $\begin{pmatrix} 1 & 1 & 1 \\ 2 & -5 & 2 \\ 1 & 0 & 1 \end{pmatrix}$ (b) $\begin{aligned} x + y + z &= 12 \\ 2x - 5y + 2z &= 20 \\ x + z &= 0 \end{aligned}$

6. Given the matrix of technical coefficients and the matrix of final demand (other demand),

(i) $A = \begin{pmatrix} 0.5 & 0.4 \\ 0.25 & 0.4 \end{pmatrix}$, $B = \begin{pmatrix} 40 \\ 100 \end{pmatrix}$, (ii) $A = \begin{pmatrix} 0.5 & 0.25 \\ 0.2 & 0.5 \end{pmatrix}$, $B = \begin{pmatrix} 120 \\ 400 \end{pmatrix}$

(a) Determine $(I - A)^{-1}$.
(b) Calculate the total output required from each sector.

7. Given the matrix of technical coefficients, A, and final demand (other demand) B for industries 1, 2, and 3

$$A = \begin{pmatrix} 0.5 & 0.1 & 0 \\ 0.2 & 0.5 & 0.2 \\ 0.1 & 0.4 & 0.6 \end{pmatrix}, \quad B = \begin{pmatrix} 120 \\ 200 \\ 700 \end{pmatrix}$$

(a) Determine $(I - A)^{-1}$.
(b) Calculate the total output required from each sector.

8. Repeat question 7 for the following matrices:

$$A = \begin{pmatrix} 0.4 & 0.4 & 0.2 \\ 0.2 & 0.25 & 0.1 \\ 0.4 & 0.2 & 0.2 \end{pmatrix}, \quad B = \begin{pmatrix} 1020 \\ 1200 \\ 700 \end{pmatrix}$$

9. Given the inter-industrial transaction table for two industries,

	Input to X	Input to Y	Final demand
Output from X	600	400	200
Output from Y	600	200	0

(a) Write down the matrix of technical coefficients.
(b) Calculate the total output required from each industry if the final demands from X and Y change to 500 and 1000 units, respectively.

10. The input/output table for an organic farm is given below:

		Input to				
		Hortic.	Dairy	Poultry	Other demands	Total output
Output	Hortic. →	50	40	200	210	500
from	Dairy →	100	160	0	140	400
	Poultry →	200	0	80	520	800

(a) Write down the matrix of technical coefficients.

(b) Calculate the total output required from each section when other demands change to 200, 800, 1000 from horticulture, dairy and poultry sections, respectively.

11. The input/output table for a three-sector industry is given below:

		A	B	C	Other demands	Total output
					Input to	
Output	A →	200	100	200	500	1000
from	B →	100	0	500	200	800
	C →	100	80	0	620	800

(a) Write down the matrix of technical coefficients.

(b) Calculate the total output required from each industry when other demands change to 400, 400, 1000 from sectors A, B and C, respectively.

12. The input/output table for three interdependent industries is given below:

		Dairy	Beef	Leather	Other demands	Total output
					Input to	
Output	Dairy	400	200	0	400	1000
from	Beef	100	400	200	200	1000
	Leather	200	200	100	600	1000

(a) Write down the matrix of technical coefficients.

(b) Calculate the total output required from each industry when other demands change to 750, 200, 1500 from dairy, beef and leather industries, respectively.

9.6 Excel for Linear Algebra

The elimination methods involve the use of elementary row operations. Examples of elementary row operations include (i) multiplying rows by non-zero constants and (ii) adding multiples of rows to other rows. Excel is ideal for these calculations. The following worked example should demonstrate the usefulness of Excel in carrying out the tedious, error-prone calculations required for elimination methods. Excel is also useful for the evaluation of determinants, hence in solving equations by Cramer's rule.

WORKED EXAMPLE 9.21

USE EXCEL TO SOLVE SYSTEMS OF LINEAR EQUATIONS

(a) Find the inverse of

$$A = \begin{pmatrix} 2 & 1 & 1 \\ 6 & 5 & -3 \\ 4 & -1 & 3 \end{pmatrix}$$

by the elimination method, hence solve the equations

$$2x + y + z = 12$$
$$6x + 5y - 3z = 6$$
$$4x - y + 3z = 5$$

(b) Solve the equations in (a) by Gauss–Jordan elimination.

SOLUTION

(a) Set up the augmented matrix, consisting of the matrix A and the unit matrix of the same dimension. Add multiples of rows to other rows to reduce A to upper triangular form. Enter the Excel formula for each row operation in column 1, then copy the formula across the remaining five columns. See Chapter 1, and remember that formulae start with =.

For example, to divide row 1 by 2, place the cursor in cell B8 and type = B4/2. To copy this formula across, click on the corner of cell B8 until a black cross appears, then drag this across the remaining five columns. This updated row is labelled row 1^1. Similarly, to add row $1^1 \times (-6)$ to row 2 in order to generate the required 0, place the cursor in cell B9 and type in the formula = B5 + B8 * (−6). Copy the formula across the row. In this way carry out all the row operations as indicated.

1	A	B	C	D	E	F	G	
2				Augmented matrix				
3		col 1	col 2	col 3	col 4	col 5	col 6	
4	row 1 →	2	1	1	1	0	0	
5	row 2 →	6	5	-3	0	1	0	
6	row 3 →	4	-1	3	0	0	1	
7								Operation **Formula**
8	row 1^1→	1	0.5	0.5	0.5	0	0	row 1 divided by 2 = **B4/2**
9	row 2^1→	0	2	-6	-3	1	0	row 2 + row $1^1 \times (-6)$ = **B5** + **B8** * (−6)
10	row 3^1→	0	-3	1	-2	0	1	row 3 + row $1^1 \times (-4)$ = **B6** + **B8** * (−4)
11								
12	row 1^1→	1	0.5	0.5	0.5	0	0	
13	row 2^2→	0	1	-3	-1.5	0.5	0	row 2^1 divided by 2 = **B12/2**
14	row 3^2→	0	0	-8	-6.5	1.5	1	row 3^1 + row $2^2 \times (3)$ = **B10** + **B13** * (3)

A is reduced to upper triangular form: row 1^1, row 2^2 and row 3^2. Continue with the row operations to further reduce A to the unit matrix.

15	row $3^3 \rightarrow$	0	0	1	0.81	-0.2	-0.1	row 3^2 divided by -8 = **B14**/(-8)
16	row $1^2 \rightarrow$	1	0.5	0	0.09	0.09	0.06	row 1^1 + row $3^3 \times (-0.5)$ = **B12** + **B15** ∗ (**−0.5**)
17	row $2^3 \rightarrow$	0	1	0	0.94	-0.1	-0.4	row 2^2 + row $3^3 \times (3)$ = **B13** + **B15** ∗ (**3**)
18	row $1^3 \rightarrow$	1	0	0	-0.4	0.13	0.25	row 1^2 + row $2^3 \times (-0.5)$ = **B16** + **B17** ∗ (**−0.5**)

The reduction is now complete. Write out the most recently updated rows in order.

	col 1	col 2	col 3	col 4	col 5	col 6
row $1^2 \rightarrow$	1	0	0	-0.375	0.0125	0.025
row $2^3 \rightarrow$	0	1	0	0.9375	-0.0625	-0.375
row $3^3 \rightarrow$	0	0	1	0.8125	-0.8125	-0.125

Read off the inverse matrix: columns 4, 5 and 6. The inverse of A is

-0.375	0.125	0.25
0.9375	-0.0625	-0.375
0.8125	-0.1875	-0.125

Finally, to multiply A^{-1} by the column of constants from the RHS of the equations, arrange these as shown.

	B	C	D	E	
19		Inverse of A		RHS	
20	-0.375	0.125	0.25	12	
21	0.9375	-0.0625	-0.375	6	
22	0.8125	-0.1875	-0.125	5	
23					
24	x = -2.5	= **B20** ∗ **E20** + **C20** ∗ **E21** + **D20** ∗ **E22**			
25	y = 9	= **B21** ∗ **E20** + **C21** ∗ **E21** + **D21** ∗ **E22**	Formulae used to		
26	z = 8	= **B22** ∗ **E20** + **C22** ∗ **E21** + **D22** ∗ **E22**	calculate x, y and z		

(b) To solve by Gauss–Jordan elimination, go back to the augmented matrix in (a) and replace col. 5 (first column of the unit matrix) by the column of constants from the RHS of each equation. Delete col. 5 and col. 6 (see below). To carry out the Gauss–Jordan elimination, copy across the formulae that were entered earlier to find the inverse of A. Read off the solution.

	Augmented matrix			
row 1	2	1	1	12
row 2	6	5	-3	6
row 3	4	-1	3	5
row 1^1	1	0.5	0.5	6
row 2^1	0	2	-6	-30
row 3^1	0	-3	1	-19
row 2^2	0	1	-3	-15
row 3^2	0	0	-8	-64

row 1 divided by 2
row 2 + row 1^1 × (−6)
row 3 + row 1^1 × (−4)
row 2^1 divided by 2
row 3^1 + row 2^2 × (3)

At this stage it is possible to read off the upper triangular matrix: row 1^1, row 2^2 and row 3^2. Solve by back substitution. For Gauss–Jordan elimination continue the row operations.

row 3^3	0	0	1	8
row 1^2	1	0.5	0	2
row 2^3	0	1	0	9
row 1^3	1	0	0	-2.5

row 3^2 divided by − 8
row 1^1 + row 3^3 × (−0.5)
row 2^2 + row 3^3 × (3)
row 1^2 + row 2^3 × (−0.5)

Read off the solution from the reduced augmented matrix: $x = -2.5$, $y = 9$, $z = 8$.

	col 1	col 2	col 3	col 4
row 1^2 →	1	0	0	-2.5
row 2^3 →	0	1	0	9
row 3^3 →	0	0	1	8

9.7 Summary

Linear programming

Find the optimum value of a linear function (objective function) subject to two or more linear constraints.

Method 1: Graph the inequality constraints and one level of the objective function. Determine graphically the largest value (or smallest value for minimisation) of the objective function which contains a corner point, X, in the feasible region. The coordinates of the point X are the values of the variables for which the objective function is optimised.

Method 2: Calculate the corner points of the feasible region. Evaluate the objective function at each corner point, to obtain the point at which it is optimised.

Matrix algebra

1. A matrix is a rectangular array of numbers (elements).
2. A matrix of order $m \times n$ contains m rows and n columns.
3. Matrices added by adding corresponding elements and subtracted by subtracting corresponding elements, therefore the matrices must be of the same order.
4. A matrix is transposed by writing each row as the corresponding column.
5. Scalar multiply matrices by multiplying each element in the matrix by the scalar (the constant).
6. The unit matrix is a square matrix with zeros everywhere except on the diagonal; the diagonal elements are usually 1.
7. The product AB of matrices A and B is possible only when the number of columns in A is equal to the number of rows in B. In general $AB \neq BA$.
8. Gauss–Jordan elimination is a method for solving a system of equations.

Determinants

1. A determinant is a square array of numbers (elements).
2. The value of a 2×2 determinant is given by

$$\begin{vmatrix} a & b \\ c & d \end{vmatrix} = ad - bc$$

3. The value of a 3×3 determinant is given by

$$\begin{vmatrix} a_{1,1} & a_{1,2} & a_{1,3} \\ a_{2,1} & a_{2,2} & a_{2,3} \\ a_{3,1} & a_{3,2} & a_{3,3} \end{vmatrix} = a_{1,1} \begin{vmatrix} a_{2,2} & a_{2,3} \\ a_{3,2} & a_{3,3} \end{vmatrix} - a_{1,2} \begin{vmatrix} a_{2,1} & a_{2,3} \\ a_{3,1} & a_{3,3} \end{vmatrix} + a_{1,3} \begin{vmatrix} a_{2,1} & a_{2,2} \\ a_{3,1} & a_{3,2} \end{vmatrix}$$

4. Cramer's rule: to solve two simultaneous equations

$$a_1 x + b_1 y = d_1$$
$$a_2 x + b_2 y = d_2$$

use determinants

$$x = \frac{\begin{vmatrix} d_1 & b_1 \\ d_2 & b_2 \end{vmatrix}}{\begin{vmatrix} a_1 & b_1 \\ a_2 & b_2 \end{vmatrix}} = \frac{\Delta_x}{\Delta} \qquad y = \frac{\begin{vmatrix} a_1 & d_1 \\ a_2 & d_2 \end{vmatrix}}{\begin{vmatrix} a_1 & b_1 \\ a_2 & b_2 \end{vmatrix}} = \frac{\Delta_y}{\Delta}$$

Cramer's rule can be extended to the solution of three linear equations in three unknowns (x, y, z):

$$x = \frac{\Delta_x}{\Delta}, \quad y = \frac{\Delta_y}{\Delta}, \quad z = \frac{\Delta_z}{\Delta}$$

The inverse of a square matrix

1. The minor of an element is the determinant of what remains when the row and column containing that element are crossed out.
2. The cofactor of an element is the value of the minor multiplied by ± 1. The ± 1 is given by the determinant of signs.
3. Given a matrix A, the inverse of A is defined as

$$A^{-1} = \frac{C^T}{|A|}$$

where C^T is the matrix in which every element is replaced by its cofactor:

$$A^{-1} = \frac{1}{|A|} \begin{pmatrix} C_{1,1} & C_{1,2} & C_{1,3} \\ C_{2,1} & C_{2,2} & C_{2,3} \\ C_{3,1} & C_{3,2} & C_{3,3} \end{pmatrix}^T$$

To find the inverse by Gauss–Jordan elimination, set up the augmented matrix as $(A|I)$, where A is the square matrix to be inverted and I is the unit matrix of the same dimension. Use Gauss–Jordan elimination to reduce A to the unit matrix, hence $(A|I) \rightarrow (I|A^{-1})$. Read off the inverse of A.

Applications of inverse matrices

1. Simplifying calculations and other operations on large arrays of data.
2. Solving simultaneous equations.

Input/output analysis

Input/output analysis is based on the condition

$$\text{total input} = \text{total output}$$

For the input/output model $AX + d = X$, the total output required from each sector when final demands d are changed to d_{new} is given by the equation

$$X = (I - A)^{-1} d_{new}$$

where

> A = the matrix of technical coefficients (the elements in each column are the proportion of that sector's input derived from each other sector),
> d = the column matrix of final demands from each sector,
> X = the column matrix of total output from each sector.

www.wiley.com/college/bradley

Go to the website for Problems in Context

TEST EXERCISES 9

1. Graph the inequality constraints and shade in the feasible region:
   (a) $10x + 2.5y \leq 50$; $2x + 12y \leq 24$: $x \geq 0; y \geq 0$
   (b) $4x + 5y \geq 50$; $2x + 12y \geq 24$: $x \geq 0; y \geq 0$
2. A small iron works manufactures two types of gate. The requirements for each stage of production, along with the limitations on the available labour-hours, are given in the following table:

Gate type	Stages of production			Selling price	Unit profit
	Welding	Finishing	Sales		
Security	4.5	1.0	1.0	£7200	£180
Ornamental	2.0	2.0	1.0	£665	£315
Max. hours	900	400	250		

   (a) Write down the equations for (i) profit, (ii) total revenue.
   (b) Write and graph the inequality constraints. Hence, determine graphically the number of each type of gate which should be produced and sold to maximise:
       (i) profit and (ii) total revenue.
   (c) Confirm the answers in (b) algebraically.
3. Given the matrices

$$A = \begin{pmatrix} 4 & 2 \\ 1 & 1 \end{pmatrix}, \quad B = \begin{pmatrix} 1 \\ -1 \end{pmatrix}, \quad C = \begin{pmatrix} 1 & 0 \\ 0 & 4 \\ -2 & 2 \end{pmatrix}, \quad D = \begin{pmatrix} -5 & -5 \\ 0 & 10 \end{pmatrix}$$

   Calculate the following, if possible (if not possible, give reasons):
   (a) $A + D$ (b) CD (c) AB (d) $A + C$ (e) AC^T
4. State Cramer's rule. Hence
   (a) Solve the following simultaneous equations:
       (i) $5x - 8y = 55$; $10x + 2y = 20$ (ii) $3Q + 12P + 60 = 0$; $10Q = 4P + 200$.

(b) The equilibrium condition in,
   (i) the goods market, is given as $Y = C + I$, where $C = 330 + 0.2\,Y$; $I = 150 - 0.4r$.
   (ii) the money market, is $M_d = M_s$, where $M_d = 100 + 0.1Y - 0.2r$; $M_s = 156$.
Write each equilibrium condition in the form $a\,Y + br = c$. Hence use Cramer's rule to solve for the equilibrium levels of Y and r.

5. What is the basic assumption underlying the input/output model? The input/output table for a three-sector economy is given in the following table:

		Sector A	Input to sector B	Sector C	Final demand	Total output
Output	Sector A	15	5	0	30	50
from	Sector B	5	15	10	10	40
	Sector C	0	0	10	70	80

(a) Write out the matrix of technical coefficients.
(b) Calculate the total output required from each sector if the final demand changes to 10 from A, 20 from B and 100 from C.

6. The input/output table for a two-sector economy is given as follows:

		Input to sector A	Input to sector B	Final demand
Output from	Sector A	200	50	50
Output from	Sector B	100	400	100

(a) Write out the matrix of technical coefficients.
(b) Calculate the total output required from each sector when final demand increases to 200 from sector A and 250 from sector B.
(c) Confirm your answer by writing out the full input/output table.

DIFFERENCE EQUATIONS

10

10.1   Introduction to Difference Equations
10.2   Solution of Difference Equations (First-order)
10.3   Applications of Difference Equations (First-order)
10.4   Summary

Chapter Objectives

Difference equations are used to model systems where changes occur at discrete points in time. The terminology used to describe and solve difference equations is similar to that used for differential equations. At the end of this chapter you should be able to:

- Write down a difference equation to model simple systems that change at discrete points in time
- Solve first-order difference equations
- Solve lagged income models
- Solve Cobweb models
- Solve Harrod–Domar growth models
- Plot the solution of a difference equation and describe the stability of the solution.

10.1   Introduction to Difference Equations

Difference equations model dynamic systems in which changes occur at fixed, equally spaced, intervals only. There are endless situations in everyday life which are modelled by difference equations. For example:

(a) If income increases by 20% each year, then, for any year, t, income may be written as 120% of the previous year's income,

$$Y_{t+1} = \frac{120}{100}Y_t$$

$$Y_{t+1} - 12Y_t = 0 \tag{10.1}$$

Note: A difference equation describes the relationship between the value of a variable, such as income, in **any** year to its value in the previous year(s). Therefore, an integer constant may be added (or subtracted) to **each occurrence of t** without affecting the relationship expressed by the equation. For example,

$$Y_t - 1.2Y_{t-1} = 0, \quad Y_{t+2} - 1.2Y_{t+1} = 0, \quad \ldots, \quad Y_{t+n} - 1.2Y_{t+n-1} = 0$$

and equation (10.1) all express the same relationship.

(b) If income decreases by 5% each year, but a lump sum of 1000 is given each year, then

$$Y_{t+1} = \frac{95}{100}Y_t + 1000$$

$$Y_{t+1} - 0.95Y_t = 1000 \tag{10.2}$$

(c) The size of the wild duck population in any generation is estimated to be 80% of the population size in the previous generation, plus 20% of the population size two generations before, minus a cull of 2400 per generation:

$$P_n = 0.8P_{n-1} + 0.2P_{n-2} - 2400$$

Note: This equation could also be written as

$$P_{n+2} = 0.8P_{n+1} + 0.2P_n - 2400 \tag{10.3}$$

(d) If I_1 = amount invested in year t, then the difference equation

$$I_{t+2} = 0.6I_{t+1} + 0.4I_t \tag{10.4}$$

states that investment in any year is 60% of the previous year's investment plus 40% of the investment two years ago.

DIFFERENCE EQUATIONS

Table 10.1   Terminology associated with difference equations, (a), (b), (c) and (d)

	Dependent variable	Independent variable	Order of difference equation	Type
(a) $Y_{t+1} - 1.2Y_t = 0$	Y_t	t	1	Homogeneous
(b) $Y_{t+1} - 0.95Y_t = 1000$	Y_t	t	1	Non-homogeneous
(c) $P_{n+2} - 0.8P_{n+1} - 0.2P_n = -2400$	P_n	n	2	Non-homogeneous
(d) $I_{t+2} - 0.6I_{t+1} - 0.4I_t = 0$	I_t	t	2	Homogeneous

Some terminology:

- A difference equation describes the relationship between an **independent variable**, such as time, t (in this chapter we shall deal mostly with time, though occasionally with n, as in example (c)), and a **dependent variable**, such as income, which changes at fixed, equally spaced, intervals in time. See Table 10.1.

  In (a), the dependent variable is income, Y.

  In (b), the dependent variable is income, Y.

  In (c), the dependent variable is population, P.

  In (d), the dependent variable is investment, I.

- Y_t: a discrete function which changes at fixed points in time only, remaining constant in between, while $y(t)$: a continuous function which changes continuously in time.
- The independent variable, t, is used in functional notation for discrete functions, in a similar manner to that for continuous functions. For example, discrete functions:

$$Y_t = 5(t^2) + 10, \quad Y_2 = 5(2)^2 + 10; \quad Y_{t+1} = 5(t+1)^2 + 10$$

continuous functions:

$$f(t) = 5(t^2) + 10, \quad f(2^2) + 10; \quad f(t+1) = 5(t+1)^2 + 10$$

- The **order** of a difference equation is the number of time intervals spanned by the difference equation. See Table 10.1.

$$\text{Example (b):} \quad (t+1) - t = 1: \text{order } 1$$

- When the difference equation is written in the form

$$\text{dependent variables} = \text{RHS}$$

the difference equation is classified as:

Homogeneous when the RHS $= 0$ as in (a) and (d); and

Non-homogeneous when the RHS $\neq 0$ as in (b) and (c).

In this text we shall deal with linear difference equations with constant coefficients, that is, difference equations of the form

$$aY_{t+1} + bY_t = \text{RHS}$$

10.2 Solution of Difference Equations (First-order)

If any value of the dependent variable is given, then the difference equation may be solved by progressively working through from year to year. This method is called iteration.

WORKED EXAMPLE 10.1
SOLVING DIFFERENCE EQUATIONS BY ITERATION

Solve the difference equation (10.1):

$$Y_{t+1} - 1.2Y_t = 0$$

by iteration for years 2, 3, 4 and 5, given that income in year 1 is £18 000.

SOLUTION

Since $Y_{t+1} = 1.2Y_t$ and $Y_1 = 18\,000$, then, similarly,

$$Y_2 = 1.2Y_1$$
$$Y_2 = 1.2(18\,000) = 21\,600$$
$$Y_3 = 1.2(Y_2) = 1.2(21\,600) = 25\,920$$
$$Y_4 = 1.2(Y_3) = 1.2(25\,920) = 31\,104$$
$$Y_5 = 1.2(Y_4) = 1.2(31\,104) = 37\,324.8$$

A general expression for the value of Y_t may be deduced by carefully observing the pattern which evolves, as in Worked Example 10.2.

WORKED EXAMPLE 10.2
GENERAL SOLUTION OF A HOMOGENEOUS
FIRST-ORDER DIFFERENCE EQUATION

(a) Write out the solution of the difference equation (10.1):

$$Y_{t+1} - 1.2Y_t = 0$$

for $t = 1, 2, 3, 4$ and 5 in terms of Y_1.

(b) Hence, deduce a general expression for Y_t in terms of t.

(c) Evaluate Y_{40}, given $Y_1 = £18\,000$.

Comment on the solution.

SOLUTION

(a) Writing out Y_t for $t = 1, 2, 3, 4$ and 5,

$$Y_2 = 1.2Y_1$$

$$Y_3 = 1.2Y_2 = 1.2(1.2Y_1) = (1.2)^2 Y_1$$

$$Y_4 = 1.2Y_3 = 1.2(1.2)^2 Y_1 = (1.2)^3 Y_1$$

$$Y_5 = 1.2Y_4 = 1.2(1.2)^3 Y_1 = (1.2)^4 Y_1$$

(b) In general,

$$Y_t = (1.2)^{t-1} Y_1$$

(c) Substitute $t = 40$ and $Y_1 = £18\,000$ into the expression in (b),

$$Y_{40} = (1.2)^{39}(18\,000)$$

$$= 22\,046\,574$$

Comment: This is a very desirable income in the 40th year and a good example of unlimited growth. **Excel** is particularly useful for repetitive calculations such as those in Worked Examples 10.1 and 10.2.

General and particular solutions of difference equations

From Worked Example 10.2(b) the general expression $Y_t = Y_1(1.2)^{t-1}$ or $Y_t = [(1.2)^{-1} Y_1] \times (1.2)^t = $ constant $\times (1.2)^t$ is called the **general solution** of the homogeneous difference equation.

To generalise: the general solution of a homogeneous first-order difference equation is of the form

$$Y_t = A(a)^t \tag{10.5}$$

where the base, a, is determined from the difference equation (see Worked Example 10.3). The constant, A, may be determined from the general solution, equation (10.5), if conditions are given: for example, the condition $Y_1 = £18\,000$. When the constant, A, is evaluated, we then have a particular solution of the given difference equation. (Similar terminology is used in differential equations, Chapter 8.)

WORKED EXAMPLE 10.3
GENERAL AND PARTICULAR SOLUTIONS OF FIRST-ORDER HOMOGENEOUS DIFFERENCE EQUATIONS

(a) Find the general solution of the difference equation, $Y_{t+1} - 0.8Y_t = 0$.
(b) If $Y_2 = 80$, find the particular solution.
(c) Evaluate Y_1, Y_{20} and Y_{50}.

SOLUTION
(a) The general form of the solution is

$$Y_t = A(a)^t$$

$$\text{therefore } Y_{t+1} = A(a)^{t+1}$$

Substituting these into the given equation, solve for a:

$$Y_{t+1} - 0.8Y_t = 0$$

$$A(a)^{t+1} - 0.8A(a)^t = 0$$

$$Aa^t a - 0.8Aa^t = 0 \quad \text{since } a^{t+1} = a^t a^1$$

$$\underbrace{Aa^t}_{\substack{A \neq 0 \\ a^t \neq 0}} \underbrace{(a - 0.8)}_{\substack{a-0.8=0 \\ a=0.8}} = 0 \quad \text{factoring out } Aa^t$$

Note: If $A = 0$, then $Y_t = A(a)^t = 0(a)^t = 0$. If $a = 0$, then $Y_t = A(a)^t = A(0)^t = 0$. The solution $Y_t = A(a)^t = 0$ is called a trivial solution. This solution is of no interest, therefore we proceed to look for a non-trivial solution, if it exists. In this example the non-trivial solution is $a = 0.8$; hence the general solution is

$$Y_t = A(0.8)^t$$

(b) Given that $Y_2 = 80$, substitute $t = 2$ and $Y_t = 80$ into the general solution to get the particular solution:

$$Y_t = A(0.8)^t$$

$$Y_2 = A(0.8)^2$$

$$80 = A(0.8)^2$$

$$\frac{80}{(0.8)^2} = A$$

$$125 = A$$

Therefore, the particular solution is

$$Y_t = 125(0.8)^t$$

(c) $Y_t = 125(0.8)^t$,

when $t = 1$, $Y_1 = 125(0.8)^1 = 100$

$t = 20$, $Y_{20} = 125(0.8)^{20} = 1.44$

$t = 50$, $Y_{50} = 125(0.8)^{50} = 0.001\,78$

Stability and the time path to stability

From Worked Examples 10.2 and 10.3 the general solution of a difference equation is of the form

$$Y_t = A(a)^t$$

This expression for the dependent variable, Y_t, may then be used to

- Forecast income in future years.
- Determine whether income will stabilise (approach some fixed value) in time.
- Trace how income changes each year until stability is reached, that is, trace the time path to stability.

The stability of the solution as t increases, and the time path to stability, are readily deduced from the exponential term $(a)^t$ as follows:

$(a)^t$ for integer values of $t > 0$

$-\infty < a < -1$ $(a)^t \to \pm\infty$	$-1 < a < 0$ $(a)^t \to 0$	$0 < a < 1$ $(a)^t \to 0$	$1 < a < \infty$ $(a)^t \to \infty$
Solution unstable: time path alternates	Solution stable: time path alternates	Solution stable: time path tends to zero	Solution unstable: time path tends to infinity

For example, in Worked Example 10.3(c), $Y_t = 125(0.8)^t$, as t increased to 20, 50, Y_t decreased to 1.44, 0.001 78, and eventually to zero.
For example, in Worked Example 10.2, $Y_t = 18\,000(1.2)^{t-1}$, Y_t increased from £18 000 to £22 046 574 for $t = 1$ to $t = 40$, eventually going to ∞.
Worked Example 10.4 illustrates the use of these rules.

WORKED EXAMPLE 10.4

STABILITY OF SOLUTIONS OF FIRST-ORDER DIFFERENCE EQUATIONS

(a) Find the general and particular solutions of the difference equations:

(i) $Y_{t+1} - 2Y_t = 0$ (ii) $Y_{t+1} - 0.7Y_t = 0$ (iii) $Y_{t+1} + Y_t = 0$
given $Y_1 = 900$ given $Y_1 = 49$ given $Y_1 = 10$

In each case, state whether the solution is stable, and, if so, state the stable value.

(b) In each case, write out and plot the time path to stability for $t = 0$ to $t = 4$ in steps of one.

SOLUTION

(a) In all three problems the general form of the solution is

$$Y_t = A(a)^t$$

$$\text{therefore} \quad Y_{t+1} = A(a)^{t+1} = A.a^t.a$$

Substituting the general form for Y_t and Y_{t+1}, solve for a, the non-trivial solution.

(i) $Y_{t+1} - 2Y_t = 0$	(ii) $Y_{t+1} + 0.7Y_t = 0$	(iii) $Y_{t+1} + Y_t = 0$
$Aa^{t+1} - 2Aa^t = 0$	$Aa^{t+1} + 0.7Aa^t = 0$	$Aa^{t+1} + Aa^t = 0$
$Aa^t(a - 2) = 0$	$Aa^t(a + 0.7) = 0$	$Aa^t(a + 1) = 0$
$a = 2$	$a = -0.7$	$a = -1$
General solution:	General solution:	General solution:
$Y_t = A(2)^t$	$Y_t = A(-0.7)^t$	$Y_t = A(-1)^t$
Not stable	Not stable	Not stable, but finite

To find the particular solutions, substitute the given conditions:

$Y_t = A(2)^t$	$Y_t = A(-0.7)^t$	$Y_t = A(-1)^t$
$Y_1 = A(2)^t$	$Y_1 = A(-0.7)^1$	$Y_1 = A(-1)^1$
$900 = A2$	$49 = A(-0.7)$	$10 = A(-1)$
$A = 450$	$A = -\dfrac{49}{0.7} = -70$	$A = -10$
Particular solution:	Particular solution:	Particular solution:
$Y_t = 450(2)^t$	$Y_t = -70(-0.7)^t$	$Y_t = -10(-1)^t$

(b) Time path to stability: In each case, set up a table of points, then plot the graph, remembering that, within each time interval, Y_t is constant, so the graph is a horizontal line segment.
(i) Table 10.2 gives the values for $Y_t = 450(2)^t$ over $t = 0$ to $t = 4$. These points are plotted in Figure 10.1.
(ii) Table 10.3 gives the values for $Y_t = -70(-0.7)^t$ over $t = 0$ to $t = 4$. These points are plotted in Figure 10.2.

Table 10.2   Points for $Y_t = 450(2)^t$

t	0	1	2	3	4
Y_t	450	900	1800	3600	7200

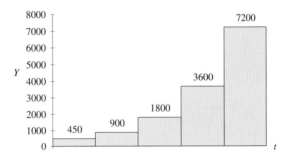

Figure 10.1 $Y_t = 450(2)^t$: Y_t increases without bound

Table 10.3   Points for $Y_t = -70(-0.7)^t$

t	0	1	2	3	4
Y_t	-70	49	-34.3	24.01	-16.81

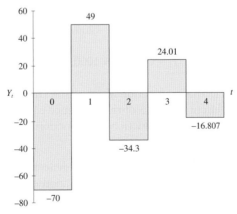

Figure 10.2   Solution oscillates to stability at $Y = 0$

(iii) Table 10.4 gives the values for $Y_t = -10(-1)^t$ over $t = 0$ to $t = 4$. These points are plotted in Figure 10.3.

Table 10.4   Points for $Y_t = -10(-1)^t$

t	0	1	2	3	4
Y_t	−10	10	−10	10	−10

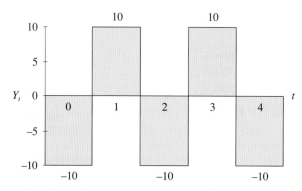

Figure 10.3   Unstable time path: solution oscillates between +10 and −10

Non-homogeneous difference equations (RHS ≠ 0)

The solution of a non-homogeneous difference equation is the sum of two terms: a complementary function and a particular integral: $Y_t = CF + PI$.

- The complementary function (CF): this is the solution of the homogeneous part of the difference equation.
  For example, equation (10.2), $Y_{t+1} - 0.95\, Y_t = 1000$. The complementary function is the solution of $Y_{t+1} - 0.95\, Y_t = 0$.
- The particular integral (PI): this is a function which satisfies the full difference equation. The general form of the PI is deduced from the RHS of the difference equation. In this chapter we shall cover only two types of functions: constant functions and functions of the form constant $\times (b)^t$:

RHS	PI (general form)
Constant	$Y_{t,p} = k$ (another constant, k)
Constant $\times b^t$	$Y_{t,p} = k(b)^t$

Again, the method will be demonstrated by a worked example.

WORKED EXAMPLE 10.5
SOLVE NON-HOMOGENEOUS FIRST-ORDER
DIFFERENCE EQUATIONS 1

(a) Find the general solution of the difference equation $Y_{t+1} - 0.95Y_t = 1000$.
(b) Find the particular solution, given $Y_5 = 20\,950$.
(c) Determine whether the system will stabilise and, if so, what the stable value is. Plot the time path to stability for $t = 0$ to 10 in steps of one.

SOLUTION

(a) The general solution is

$$Y_t = \text{CF} + \text{PI}$$

$$\text{or} \quad Y_t = Y_{t,c} + Y_{t,p}$$

First, find the complementary function by solving the homogeneous equation.

Complementary function (CF): $\quad Y_{t+1} - 0.95Y_t = 0$

General form: $\quad Y_{t,c} = A(a)^t$, therefore $Y_{t+1,c} = A(a)^{t+1} = A.a^t.a$

Substitute the general form for Y_t and Y_{t+1} into the given equation and solve for a,

$$Aa^{t+1} - 0.95\,Aa^t = 0$$

$$Aa^t(a - 0.095) = 0$$

$$a = 0.95$$

Therefore, the complementary function, $Y_{t,c}$, is

$$Y_{t,c} = A(0.95)^t$$

Particular integral (PI): Next, find the particular integral, $Y_{t,p}$. Since the RHS of the full equation is a constant, the general form of the PI is

$$Y_{t,p} = k, \quad \text{therefore} \quad Y_{t+1,p} = k$$

(a constant, k, has the same value at all times: t, $t + 1$, etc.). Substitute the general form into the full difference equation and solve for k:

$$Y_{t+1} - 0.95Y_t = 1000$$
$$k - 0.95k = 1000$$
$$0.05k = 1000$$

$$k = \frac{1000}{0.05} = 20\,000$$

Therefore $Y_{t,p} = 20\,000$.

So the general solution is

$$Y_t = Y_{t,c} + Y_{t,p}$$

$$Y_t = A(0.95)^t + 20\,000$$

(b) Since a condition is given, $Y_5 = 20\,950$, we may substitute this condition into the general equation to find a value for the constant A:

$$Y_5 = A(0.95)^5 + 20\,000$$

$$20\,950 = A(0.95)^5 + 20\,000$$

$$950 = A(0.773\,781)$$

$$1228 = A$$

Therefore the particular solution is

$$Y_t = 1228(0.95)^t + 20\,000$$

(c) Since $a = 0.95 < 1$, the solution will stabilise, that is, $(0.95)^t$ will approach 0 as t increases. Therefore, the stable income is

$$Y_t = 1228(0) + 20\,000$$

$$= 20\,000$$

To plot the time path, evaluate Y_t for several values of t, as in Table 10.5. These points are plotted in Figure 10.4.

Table 10.5   Points for $Y_t = 1228(0.95)^t + 20\,000$

t	0	1	2	3	4	5	6	7	8	9	10
Y_t	21 228	21 167	21 108	21 053	21 000	20 950	20 903	20 858	20 815	20 774	20 735

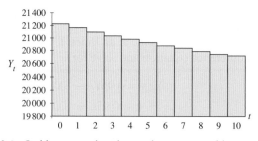

Figure 10.4   Stable time path: solution decreases steadily to 20 000

WORKED EXAMPLE 10.6
SOLVE NON-HOMOGENEOUS FIRST-ORDER
DIFFERENCE EQUATIONS 2

(a) Solve the difference equation $3Y_{t+1} + 2Y_t = 44(0.8)^t$ given $Y_0 = 900$.
(b) Show that the solution stabilises and plot the time path to stability.

SOLUTION
(a) The general solution has the form

$$Y_t = CF + PI$$

$$\text{or} \quad Y_t = Y_{t,c} + Y_{t,p}$$

Complementary function (CF):

General form: $\quad Y_{t,c} = A(a)^t$, therefore $Y_{t+1,c} = A(a)^{t+1} = A.a^t.a$

Substitute into the homogeneous equation:

$$3Y_{t+1} + 2Y_t = 0$$
$$3Aa^t a + 2Aa^t = 0$$
$$Aa^t(3a + 2) = 0$$
$$Aa^t \neq 0 \quad 3a + 2 = 0$$
$$a = -\frac{2}{3}$$

Therefore, $Y_{t,c} = A\left(-\dfrac{2}{3}\right)^t$.

Particular integral (PI):

Since the RHS $= 44(0.8)^t$, the general form is

$$Y_{t,p} = k(0.8)^t, \quad \text{therefore} \quad Y_{t+1,p} = k(0.8)^{t+1} = k(0.8)^t(0.8)^1$$

Substitute into the full difference equation:

$$3k(0.8)^{t+1} + 2k(0.8)^t = 44(0.8)^t$$
$$k(0.8)^t[3(0.8)^1 + 2] = 44(0.8)^t$$
$$k[2.4 + 2] = 44 \qquad \text{divide both sides by } (0.8)^t$$
$$k = \frac{44}{4.4} = 10$$

Therefore $Y_{t,p} = k(0.8)^t = 10(0.8)^t$.

The general solution is

$$Y_t = Y_{t,c} + Y_{t,p}$$

$$= A\left(-\frac{2}{3}\right)^t + 10(0.8)^t$$

Since a condition $Y_0 = 900$ is given, find the particular solution:

$$Y_0 = A\left(-\frac{2}{3}\right)^0 + 10(0.8)^0$$

$$900 = A + 10$$

$$A = 890$$

Therefore, the particular solution is

$$Y_t = 890\left(-\frac{2}{3}\right)^t + 10(0.8)^t$$

(b) **Stability:** Since the base, a, in each of the exponential terms is less than 1, these terms will tend to zero as t increases. So,

$$Y_t = 890(0) + 10(0) = 0$$

Y_t will stabilise at $Y_t = 0$. The time path will fluctuate to stability since the base of one of the exponential terms, $(-2/3)^t$, is negative. To plot the time path, evaluate Y_t for several values of t, as in Table 10.6. These points are plotted in Figure 10.5.

Table 10.6 Points for $Y_t = 890(-2/3)^t + 10(0.8)^t$

t	0	1	2	3	4	5	6	7	8	9	10
Y_t	900	−585	402	−259	180	−114	81	−50	36	−22	17

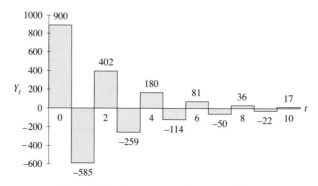

Figure 10.5 Solution oscillates to stability

PROGRESS EXERCISES 10.1

Introduction to Difference Equations

1. For each of the following difference equations, state:
   (i) The order of the equation.
   (ii) Whether the equation is homogeneous or not.
   (a) $P_{t+1} - 0.8P_t = 0$ (b) $Y_{t+2} = 8 - Y_{t+1}$ (c) $Y_{t+2} = 80 + Y_t$

2. Solve each of the following difference equations for the indicated variable by the iteration method.
   (a) $Y_{t+1} - 0.8Y_t = 10$, given $Y_1 = 5$. Find Y_5.
   (b) $P_{t+2} = 4P_{t+1} - 8P_t$, given $P_1 = 20$, $P_2 = 18$. Find P_5.
   (c) $P_t = 0.6P_{t-1} + 80$, given $P_1 = 100$. Find P_5.

3. Find the general and particular solutions of the following. Hence evaluate Y_{20} and Y_{50}.
   (a) $Y_{t+1} - 0.8Y_t = 0$, given $Y_1 = 5$
   (b) $Y_{t+1} - 0.8Y_t = 0$, given $Y_1 = 5$
   (c) $Y_{t+1} - 0.8Y_t = 10$, given $Y_1 = 5$.

4. For each part in question 3, state
   (a) Whether the solution is stable and, if so, state the value of the stable solution.
   (b) Plot the solution for $t = 0, 1, 2, 3, 4$ and 5. Describe the time path to stability.

5. Find the general and particular solutions of the following:
   (a) $Y_{t+1} = 10Y_t + 900$,   given $Y_0 = 20$
   (b) $P_{t+1} = 0.9P_t + 80$,   given $P_1 = 170$.

6. The relationship between P_t, the price of a good in season t, to its price during the previous season is given by the equation $P_t = 1.05P_{t-1}$.
   (a) Outline verbally the relationship defined by the difference equations.
   (b) Solve the difference equation, given that $P_t = 6300$. Use the solution to describe how the price will change in future years.

7. (a) Which of the following difference equations are different from $P_{t+1} = 1.05P_t$? Explain.
       (i) $P_{t+5} - 1.05P_{t+4} = 0$ (ii) $P_t = 1.05P_{t+1}$ (iii) $1.05P_{t+1} - P_{t+2} = 0$
   (b) Which of the following difference equations are different from $P_{n+2} = P_n + (2)^n$? Explain.
       (i) $P_n = P_{n-2} + (2)^{n-2}$ (ii) $P_n = P_{n-2} + (2)^n$
       (iii) $P_{n+1} - P_{n-1} = (2)^n$ (iv) $P_{n+1} = P_{n-1} + (2)^{n-1}$

8. I_t is the amount invested in period t, where

$$I_t = 0.9I_{t-1} + 60$$

   (a) Describe investment I_t in terms of the amount invested in previous periods.
   (b) For each of the initial investments (at $t = 0$): £15m; £180m; £900m;
       (i) Solve the difference equation.
       (ii) Calculate the number of years taken for investment to reach £400m.

9. (a) Solve the difference equation $2Y_t + 10 = Y_{t-1}$.
   (b) If $Y_2 = 80$, find the particular solution.
   (c) Plot and describe the time path for $t = 1, 2, 3, 4$ and 5. Will the system stabilise?

10. A fish population increases by 5% each generation, but a catch of 8000 is lost from each generation. If P_n represents the size of the population for generation n:
    (a) Write the difference equation which expresses P_n in terms of P_{n-1}.
    (b) Solve the difference equation, given $P_0 = 100\,000$. Hence, describe the time path.
11. Solve the difference equations
    (a) $Y_t = 0.6Y_{t-1} + 21(2)^t$ (b) $0.8Y_t = 0.56Y_{t-1} + 12(0.8)^t$, given $Y_4 = 0$
12. Find the general and particular solutions of the following:
    (a) $P_{t+1} = 0.75P_t + 9(3)^t$, given $P_0 = 190$.
    (b) $Y_t = 0.84Y_{t-1} + 5$, given $Y_0 = 40$.
    State whether each solution is stable. Describe and plot the time path for $t = 0$ to $t = 10$.

10.3   Applications of Difference Equations (First-order)

As already mentioned, difference equations may be used to model many situations in everyday life. In this section, difference equations are used in some standard well-known models in both microeconomics and macroeconomics.

The lagged income model

This is a macroeconomic application of difference equations using the standard national income model, in which equilibrium is expressed by the equation $Y = C + I$ with all variables referring to the same time period, that is,

$$Y_t = C_t + I_t$$

(There exist no government or foreign sectors.) However, if consumption depends on the income during the previous period, then

$$C_t = C_0 + bY_{t-1}$$

and

$$I_t = I_0, \text{ a constant}$$

Note: The consumption function, $C_t = C_0 + bY_{t-1}$, is different from the consumption functions that have been used so far. It assumes that planned consumption is dependent on past income, not present income. This is quite logical since present income may not yet be known.
  Substitution of C_t into the equilibrium equation gives a difference equation in Y,

$$Y_t = C_0 + bY_{t-1} + I_0$$
$$Y_t = bY_{t-1} + k_0 \quad \text{where} \quad k_0 = C_0 + I_0$$

This is a linear, first-order difference equation.

WORKED EXAMPLE 10.7
THE LAGGED INCOME MODEL

Assume a simple national income model

$$Y_t = C_t + I_t$$

where $C_t = 5000 + 0.8Y_{t-1}$ and $I_t = 5000$.

(a) Write the national income equation as a difference equation in Y.
(b) Solve the difference equation. Hence, describe the time path. Will the system stabilise?
(c) If $Y_0 = 100\,000$, find the particular solution. Plot the time path for $t = 0$ to $t = 10$.

SOLUTION

(a)

$$Y_t = C_t + I_t$$
$$Y_t = 5000 + 0.8Y_{t-1} + 5000 \quad \text{substituting } C_t \text{ and } I_t \text{ as given}$$
$$Y_t - 0.8Y_{t-1} = 5000 + 5000$$
$$Y_t - 0.8Y_{t-1} = 10\,000$$

(b) The general solution has the form

$$Y_t = CF + PI$$
$$\text{or} \quad Y_t = Y_{t,c} + Y_{t,p}$$

Complementary function (CF):

General form: $Y_{t,c} = A(a)^t$, therefore $Y_{t-1,c} = A(a)^{t-1} = A.a^t.a^{-1}$

Substitute into the homogeneous equation

$$Y_t - 0.8_{t-1} = 0$$
$$Aa^t - 0.8Aa^t a^{-1} = 0$$
$$Aa^t(1 - 0.8a^{-1}) = 0$$
$$Aa^t \neq 0 \quad 1 - \frac{0.8}{a} = 0$$
$$a - 0.8 = 0$$
$$a = 0.8$$

Therefore $Y_{t,c} = A(0.8)^t$.

Particular integral (PI):

The general form of the PI is

$$Y_{t,p} = k, \quad \text{therefore} \quad Y_{t-1,p} = k$$

Substitute the general form of the PI into the full difference equation and solve for k:

$$Y_t - 0.8Y_{t-1} = 10\,000$$
$$k - 0.8k = 10\,000$$
$$0.2k = 10\,000$$
$$k = \frac{10\,000}{0.2} = 50\,000$$

Therefore $Y_{t,p} = 50\,000$.

So the general solution is

$$Y_t = Y_{t,c} + Y_{t,p}$$
$$Y_t = A(0.8)^t + 50\,000$$

As t increases, $(0.8)^t$ will tend to zero. The solution then becomes

$$Y_t = A(0) + 50\,000$$
$$Y_t = 50\,000$$

Since $0 < (a = 0.8) < 1$, Y_t will decline steadily to a stable value of 50 000.

(c) Since a condition $Y_0 = 100\,000$ is given, find the particular solution:

$$Y_0 = A(0.8)^0 + 50\,000$$
$$100\,000 = A + 50\,000$$
$$A = 50\,000$$

Therefore the particular solution is

$$Y_t = 50\,000(0.8)^t + 50\,000$$

The solution stabilises to a value of 50 000.

Table 10.7 gives the values of Y_t for $t = 0$ to $t = 10$. Figure 10.6 plots the time path over the first 10 years. Use Excel, if available.

DIFFERENCE EQUATIONS

Table 10.7 Points for $Y_t = 50(0.8)^t + 50$ (£000s)

t	0	1	2	3	4	5	6	7	8	9	10
Y_t	100	90	82	75.6	70.5	66.4	63.1	60.5	58.4	56.7	55.4

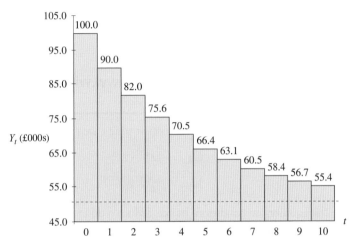

Figure 10.6 Solution decreases steadily to 50 (£000s)

The cobweb model

This is a microeconomic application of difference equations. Consider the demand and supply functions of a good over time given by

$$Q_{d,t} = a - bP_t \tag{10.6}$$

The quantity demanded of the good in time period t is a function of the price in that time period:

$$Q_{s,t} = c + dP_{t-1} \tag{10.7}$$

The quantity supplied of the good in time period t is a function of the price of the good during the previous period. This reflects the fact that a firm's output decision may lag behind the market price. For example, the quantity of agricultural goods supplied frequently depends on the price in the previous period.

The usual condition for market equilibrium is

$$Q_{d,t} = Q_{s,t}$$

Substitution of the RHS of equations (10.6) and (10.7) into the condition for market equilibrium gives rise to a difference equation in P:

$$a - bP_t = c + dP_{t-1}$$

$$-bP_t - dP_{t-1} = (c - a)$$

WORKED EXAMPLE 10.8
THE COBWEB MODEL

 Find animated worked examples at **www.wiley.com/college/bradley**

The demand and supply functions for a good at time t are given as

$$Q_{d,t} = 125 - 2P_t$$

$$Q_{s,t} = -50 + 1.5P_{t-1}$$

(a) State the equilibrium condition, hence deduce a difference equation in P.
(b) Solve the difference equation to find the equilibrium price and quantity.
(c) Given that $P = 60$ at $t = 0$, plot and describe the time path to stability.

SOLUTION

(a) At equilibrium, $Q_{d,t} = Q_{s,t}$, therefore

$$125 - 2P_t = -50 + 1.5P_{t-1}$$

$$-2P_t - 1.5P_{t-1} = -50 - 125$$

$$2P_t + 1.5P_{t-1} = 175$$

(b) The general solution has the form

$$P_t = CE + PI$$

$$\text{or } P_t = P_{t,c} + P_{t,p}$$

Complementary function: This is the solution of the homogeneous equation

$$2P_t + 1.5P_{t-1} = 0$$

Substitute the general form of $P_{t,c} = A(a)^t$ and $P_{t-1,c} = A(a)^{t-1} = A.a^t.a^{-1}$ into the homogeneous equation

$$2P_t + 1.5P_{t-1} = 0$$

$$2Aa^t + 1.5Aa^t a^{-1} = 0$$

$$Aa^t(2 + 1.5a^{-1}) = 0$$

$$Aa^t \neq 0 \quad 2 + \frac{1.5}{a} = 0$$

$$2 = -\frac{1.5}{a}$$

$$2a = -1.5$$

$$a = -0.75$$

Therefore $P_{t,c} = A(-0.75)^t$.

Particular integral: Since the RHS is a constant, the general form of the PI is

$$P_{t,p} = k \quad \text{and} \quad P_{t-1,p} = k$$

Substitute the general form of the PI into the full difference equation and solve for k:

$$2P_t + 1.5P_{t-1} = 175$$

$$2k + 1.5k = 175$$

$$3.5k = 175$$

$$k = \frac{175}{3.5} = 50$$

Therefore $P_{t,p} = 50$.

So the general solution is

$$P_t = P_{t,c} + P_{t,p}$$

$$P_t = A(-0.75)^t + 50$$

(c) If $P_0 = 60$, then

$$P_0 = A(-0.75)^0 + 50$$

$$60 = A + 50$$

$$A = 10$$

Therefore the particular solution is

$$P_t = 10(-0.75)^t + 50$$

As t increases, $10(-0.75)^t$ will oscillate towards 0. Hence P_t will stabilise to $P_t = 50$.

Note: Assuming an equilibrium price, P^*, then $P_t = P^* = P_{t-1} = 50$, and

$$Q_{d,t} = 125 - 2P_t = 125 - 2(50) = 25$$

$$Q_{s,t} = -50 + 1.5P_{t-1} = -50 + 1.5(50) = 25$$

Table 10.8 gives values of P_t for $t = 0$ to $t = 10$. These points are plotted in Figure 10.7.

Table 10.8 $P_t = 10(-0.75)^t + 50$

t	0.00	1.00	2.00	3.00	4.00	5.00	6.00	7.00	8.00	9.00	10.00
P_t	60.00	42.50	55.63	45.78	53.16	47.63	51.78	48.67	51.00	49.25	50.56

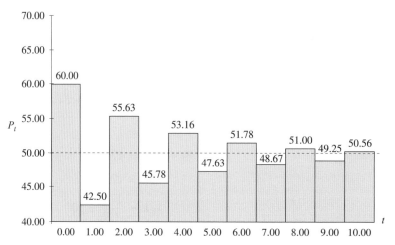

Figure 10.7   The cobweb model: solution oscillates to a stable value of 50

The Harrod–Domar growth model

This is another macroeconomic application of difference equations. The Harrod–Domar growth model is a simple macroeconomic growth model which states that investment is proportional to the rate of change in income:

$$I_t = k(Y_t - Y_{t-1}) \tag{10.8}$$

and savings at time period t depends on income:

$$S_t = sY_t \tag{10.9}$$

where the constant k is the capital–output ratio, usually $k > 1$, and
the constant s is the marginal propensity to save, MPS, where $0 < s < 1$
At equilibrium, $I_t = S_t$, therefore

$$k(Y_t - Y_{t-1}) = sY_t$$
$$(k - s)Y_t - kY_{t-1} = 0$$

thus reducing to a first-order homogeneous difference equation. On dividing by $(k - s)$ this may also be written as

$$Y_t - \left(\frac{k}{k-s}\right)Y_{t-1} = 0$$

which has the general solution

$$Y_t = A\left(\frac{k}{k-s}\right)^t$$

If $\left|\dfrac{k}{k-s}\right| < 1$ the system stabilises.

WORKED EXAMPLE 10.9
THE HARROD–DOMAR GROWTH MODEL

Given the equilibrium condition that

$$I_t = S_t$$

where $I_t = 2.5(Y_t - Y_{t-1})$ and $S_t = 0.1Y_t$:

(a) Write the equilibrium equation as a difference equation in Y_t.
(b) Solve the difference equation, given $Y_0 = £8m$.
(c) State whether the system will stabilise. Plot the time path for $t = 0$ to $t = 7$.

SOLUTION
(a) Substitute the equations for I_t and S_t into the equilibrium equation:

$$I_s = S_t$$
$$2.5(Y_t - Y_{t-1}) = 0.1Y_t$$
$$2.4Y_t - 2.5Y_{t-1} = 0$$
$$Y_t - \frac{2.5}{2.4}Y_{t-1} = 0 \qquad \text{dividing across by 2.4}$$
$$Y_t - 1.04Y_{t-1} = 0$$

(b) Since this is a homogeneous difference equation, the solution has the general form

$$Y_t = A(a)^t, \quad \text{hence} \quad Y_{t-1} = A(a)^{t-1} = A.a^t.a^{-1}$$

Substituting the general form into the given equation

$$Y_t - 1.04Y_{t-1} = 0$$
$$Aa^t - 1.04\,Aa^t a^{-1} = 0$$
$$Aa^t(1 - 1.04a^{-1}) = 0$$

$Aa^t \neq 0$, otherwise we have a trivial solution, hence

$$1 - 1.04a^{-1} = 0$$
$$1 - \frac{1.04}{a} = 0$$
$$1 = \frac{1.04}{a}$$
$$a = 1.04$$

Therefore the general solution is $Y_t = A(1.04)^t$.

Since $Y_0 = £8m$ is given, this condition may be used to find a value for the constant A,

$$Y_0 = A(1.04)^0$$
$$8 = A \quad \text{since } (1.04)^0 = 1$$

Therefore the particular solution is

$$Y_t = 8(1.04)^t$$

(c) Since $a = 1.04 > 1$, the solution is unstable. Y_t will increase steadily without bound as t increases. This is shown in Table 10.9 which is then plotted in Figure 10.8.

Table 10.9   Points for $Y_t = 8(1.04)^t$

t	0	1	2	3	4	5	6	7
Y_t	8	8.32	8.65	9	9.36	9.73	10.12	10.53

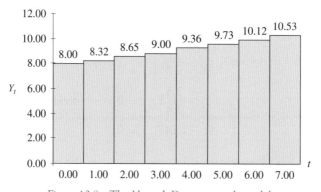

Figure 10.8   The Harrod–Domar growth model

PROGRESS EXERCISES 10.2

First-order Difference Equations: Applications

1. An amount of money, A, is invested at $r\%$ compounded annually.
   (a) Show that the value of the investment at the end of any year, t, is given by the difference equation

   $$P_{t+1} = (1 + r)P_t$$

   (b) Show that the general solution of the equation is $P_t = P_0(1 + r)^t$.
2. Solve each of the following difference equations and plot the time path for $t = 0$ to $t = 20$ (use Excel). State which systems are stable and state the stable value.
   (a) $Y_t = C_t + I_t$ where $C_t = 100 + 0.8Y_{t-1}$ and $I_t = 500$
   (b) $Y_t = C_t + I_t$ where $C_t = 500(1.05)^t + 0.75Y_{t-1}$ and $I_t = 0$.
3. The quantity of cabbage supplied to a market depends on the price during the previous season, while the quantity demanded depends on the present price

   $$Q_{s,t} = -4 + 2P_{t-1}; \quad Q_{d,t} = 80 - 2P_t$$

   (a) Write the equilibrium condition as a difference equation in P and solve it.
   (b) Solve the equation to obtain an expression for price in terms of t, given $P_5 = 10.5$. Hence, determine whether prices will stabilise.
4. The supply and demand functions for a good at time t are given by the equations

   $$Q_{d,t} = 150 - 0.3P_t; \quad Q_{s,t} = 0.1P_t - 30$$

   The price in any season depends on the excess supplied during the previous season as described by the equation

   $$P_{t+1} = P_t - 0.3(Q_{s,t} - Q_{d,t})$$

   (a) Deduce a difference equation which relates the present price to the price in the previous season.
   (b) Solve for an expression for the price at any time t in terms of t. Hence
   (c) Determine whether prices will stabilise.
5. Investment at time t depends on the excess income over the previous year, as given by the difference equation $I_t = 4.5(Y_t - Y_{t-1})$ while dissavings are 12% of income, $S_t = 0.12Y_t$. Given the equilibrium condition $I_t = S_t$:
   (a) Write the equilibrium condition as a difference equation in Y.
   (b) Solve the difference equation, given $Y_0 = £314m$.
   (c) Plot the time path for (i) Y_t, (ii) I_t for $t = 0$ to 4. Will investment stabilise? Explain.

6. Investment at time t depends on the excess income over the previous year, as given by the difference equation $I_t = 0.9(Y_t - Y_{t-1})$, while dissavings are 25.5% of income, $S_t = -0.255Y_t$. Given the equilibrium condition, $I_t = S_t$:
(a) Write the equilibrium condition as a difference equation in Y.
(b) Solve the difference equation, given $Y_0 = £800m$.
(c) Plot the time path for (i) Y_t, (ii) I_t over $t = 0$ to $t = 12$. Will investment stabilise? Explain.

10.4 Summary

1. A difference equation relates the value of a dependent variable which changes at discrete points in time only (such as income) to its values during previous time periods. For example,

$$Y_{t+1} = \frac{120}{100} Y_t$$

In words, the difference equation states 'income in any year is 20% higher than income in the previous year'.

2. The order of a difference equation is the number of time periods spanned by the equation. The order is 1 in the above example.

3. Solution of difference equations is over a reasonably small number of time intervals by iteration, provided conditions are given. Excel is very useful here.

4. Mathematically, the solution of an homogeneous difference equation is of the form: $Y_t = A(a)^t$. To determine the actual solution, substitute the trial solution $Y_t = A(a)^t$ into the difference equation and solve for the base a. The constant A is determined if one condition is given, such as $Y_0 = 30$.

5. The solution of a non-homogeneous difference equation is the sum of two parts: the complementary function (CF) and a particular integral (PI).

   CF is the solution of the corresponding homogeneous equation
   PI is determined by inspection.

Stability of the solution of a difference equation: $Y_t = A(a)^t$ depends on a^t: behaviour of $(a)^t$ as t increases for integer values $t > 0$

$-\infty < a < -1$	$-1 < a < 0$	$0 < a < 1$	$1 < a < \infty$
$(a)^t \to \pm\infty$	$(a)^t \to 0$	$(a)^t \to 0$	$(a)^t \to 0$
Solution unstable: time path alternates	Solution stable: time path alternates	Solution stable: time path steady to zero	Solution unstable: time path steady to infinity

Applications

1. Lagged income model: equilibrium condition: $Y_t = C_t + I_t$
   where $C_t = C_0 + bY_{t-1} : I_t = I_0$
2. The cobweb model: equilibrium condition: $Q_{d,t} = Q_{s,t}$
   where $Q_{d,t} = a - bP_t :$ $Q_{s,t} = c + dP_{t-1}$
   (The quantity of the good supplied in time period t is a function of the price in the previous time period: $Q_{s,t} = c + dP_{t-1}$.)

DIFFERENCE EQUATIONS

3. Harrod–Domar growth model: equilibrium condition: $I_t = S_t$
   where $I_t = k(Y_t - Y_{t-1})$: $S_t = sY_t$

 www.wiley.com/college/bradley

Go to the website for Problems in Context

TEST EXERCISES 10

1. (a) Solve the difference equation by iteration for Y_1 to Y_{10}:

$$Y_{t+1} = 4Y_t + 3, \quad \text{given} \quad Y_1 = 12$$

   (b) Solve the difference equation in (a) by algebraic methods.
2. Determine the general solution of the difference equation

$$Y_{t+1} - 0.4Y_t = 0$$

   State whether the solution is stable or unstable.
3. Given the national income model $Y_t = C_t + I_t$, where $C_t = 220 + 0.4Y_t$, $I_t = 0.35Y_{t-1}$
   (a) Write the equilibrium equation as a difference equation in Y_t.
   (b) Determine the general expression for the equilibrium level of national income.
   (c) Determine the particular solution given $Y_0 = 2200$. Graph the solution for $t = 0$ to 5.
      Comment on its stability.
4. The relationship between the number of whales inhabiting a given area to the numbers inhabiting that area during the previous seasons is given by the difference equation $P_n = 0.8P_{n-1} + 800$.
   (a) Determine the general solution and discuss its stability.
   (b) If $P_0 = 5000$ determine the particular solution.
      Graph the whale population for eight seasons.
5. Given the equation for equilibrium national model $Y_t = C_t + I_t$, where $C_t = 120 + 0.3Y_{t-1}$, $I_t = 5 + 0.2Y_{t-1}$
   (a) Write the equilibrium equation as a first-order difference equation.
   (b) Solve the difference equation, given $Y_0 = 100$. Is the solution stable?
   (c) Calculate the number of years it will take for equilibrium income to reach 450.
6. The price of a good is set each season according to the difference equation $P_t = P_{t-1} + 0.2(Q_{s,t} + Q_{d,t})$, where the quantities demanded and supplied are given by the equations

$$Q_{d,t} = 8 - 0.5P_t, \quad Q_{s,t} = 1.3P_{t-1} - 7$$

   (a) Write the price-setting equation as a second-order difference equation.
   (b) Solve for price; hence determine whether the price will stabilise.

SOLUTIONS TO PROGRESS EXERCISES

Chapter 1

1. $x = 3$

2. $x = 1$

3. $x = 2, x = -4$

4. $x = \sqrt{8}, x = -\sqrt{8}$

5. $x = 0, x = -2$

6. $x = 0, x = 2, x = -4$

7. $x = 0, x = 2$

8. $x = y$: infinitely many solutions

9. $y = -2, x = -2$

10. $x = 0.5, y = -2$

11. $x = 0$

12. $x = \sqrt{-4}, x = -\sqrt{-4}$

13. $x = 0$

14. $x = 0, x = \sqrt{-2}, x = -\sqrt{-2}$

15. $x = -4$

16. $x = 0$

17. $x = 0.5$

18. $x = 0, x = 4, x = -3.8$

19. $P = -20$

20. No solution

21. $x = 5, x = -5$

22. P may be any value, except -2. $Q = 4$

23. $Q = \dfrac{1}{100}$

24. $\dfrac{7}{3}$

25. 0.5

26. 2

27. $0, 4$

28. 4

29. $1, -1$

30. 0

31. 7

32. $\pm\sqrt{\dfrac{63}{20}}$

33. $\dfrac{100}{45}$

34. (a) €303.86 (b) 19 797.02 rupees (c) 1219.90 ringgits

35. (a) €615.24 (b) €186.12

36. 52.98 Brazilian reals

37. Hong Kong dollar exchange rates: $1 =

British pound	US dollar	Japanese yen	Danish krone	Brazil real	Swiss franc	Norwegian krone	Malaysian ringgit
0.0827	0.1289	9.9978	0.7368	0.2272	0.1196	0.7570	0.3978

38. 21 042.89 forints

39. (a) 463.93 Swiss francs (b) A$477.93 (c) S$635.51

40. €9.50

PE 1.3

(d) (i)

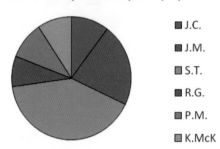

2. (a) $x - 25 > 7$

$\qquad x > 32$ add 25 to both sides

x is greater than 32.

(b) $\quad 5 < 2x + 15$

$\qquad -10 < 2x$ add -15 to both sides

$\qquad -5 < x$ dividing both sides by 2

-5 is less than x or x is greater than -5 ($x > -5$); but $x > 0$, hence solution $x > 0$

(c) $\dfrac{25}{x} < 10$

$\qquad 25 < 10x$ multiply both sides by x (x is positive)

$\qquad \dfrac{25}{10} < x$ dividing both sides by 10

2.5 is less than x or x is greater than 2.5

(d) $\dfrac{x}{2} + \dfrac{x}{3} \geq \dfrac{17}{6}$

$\qquad \dfrac{5x}{6} \geq \dfrac{17}{6}$

$\qquad 5x \geq 17$ multiply both sides by 6

$\qquad x \geq 3.4$ dividing both sides by 5

x is greater than or equal to 3.4

(e) $3x - 29 \leq 7x + 11$

$\qquad -40 \leq 4x$

$\qquad -10 \leq x$ dividing both sides by 4

-10 is less than or equal to x or x is greater than or equal to -10. But x is positive, hence the solution is $x \geq 0$.

3. (a) 651.924   (b) 20.1025   (c) 0.7475   4. (a) £0.791   (b) £6.441

5. £1400   6. 763, 809, 858   7. 320, 256, 205, 164, 131, 105

8. £383.15   **9.** 37.08%   **10.** £458.832   **11.** £642.8   **12.** 12.5%

13. (a) −9% (a decrease in price) (b) Stamp duty is 21 900 in 2003
    (c) 338 793; 345 568.86; 352 480.24

14. (a) The proportion of male students is $\dfrac{676}{1024}$ and female students is $\dfrac{348}{1024}$
    (b) The percentage of male students is 66.02% and female students is 33.98%
    (c) The numbers of male and female students would be 410 and 614, respectively

15. (a) (i) 0.04 (ii) 340
    (b) (i) 0.025 (ii) 310
    (c) (i) 650 (ii) 3.11%

PE 1.4

1. (a) −50.1407   (b) 2.5049   (c) 16.0084
2. (a) 3.061 828   (b) 9.029 86   (c) 44.6069
3. (a) (i) 6.84   (ii) −14.2
    (b) 14p (ii) 44p
    (c) (i) 0.6 (ii) 2.0
    (d) 1.0909
    (e) −2.35
4. (a) (i) $x = 102.4$ (ii) 167.56

(b) (i) $m = \pm\sqrt{\dfrac{v}{0.8} + n^2} = \pm\sqrt{\dfrac{v + 0.8n^2}{0.8}}$ (ii) $n = \pm\sqrt{\dfrac{0.8m^2 - v}{0.8}}$

(c) (i) $Q = \dfrac{AC - 1500}{2.55}$ (ii) 70.5882

(d) (i) $b = \dfrac{\sum y - an}{\sum x}$ (ii) −1.2143

(e) (i) $n = \dfrac{2s_e^2}{s_e^2 + b^2\sigma_x^2 - \sigma_y^2}$ (ii) 3.720 93

For each staff member	J.C.	J.M.	S.T.	R.G.	P.M.	K.McK	Total pay for all staff
(a) Pay	£201.60	£1319.20	£6560.00	£193.75	£503.75	£652.50	£9430.80
(b) Pay − tax	£161.28	£1055.36	£5248.00	£155.00	£403.00	£522.00	£7544.64
(c) Pay − tax + exps.	£281.28	£1110.36	£5248.00	£180.00	£415.00	£677.00	£7911.64

(d) (i)

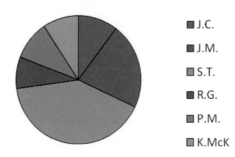

(d) (ii)

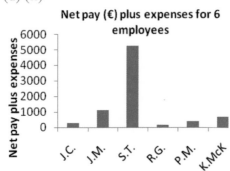

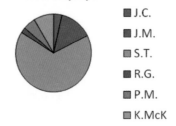

(d) (iii)

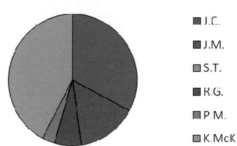

Chapter 2

PE 2.1

1. Points lie on a line: slope = 1; vertical intercept = 2.
2. Points lie on a line: slope = −1; vertical intercept = 6.

SOLUTIONS TO PROGRESS EXERCISES

3.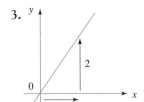

Figure PE 2.1 Q3: cuts the horizontal at $x = 0$

4.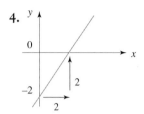

Figure PE 2.1 Q4

5.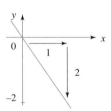

Figure PE 2.1 Q5: cuts the horizontal at $x = 0$

6.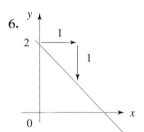

Figure PE 2.1 Q6: cuts the horizontal at $x = 2$

PE 2.2

1. (3) $y = 2x$ (4) $y = x - 2$ (5) $y = -2x$ (6) $y = -x + 2$

2. (a) (i) $m = 1$; $y(0) = 2$; $x(0) = -2$ (ii) 0, 2, 4, 6, 8
   (b) (i) $m = -4$; $y(0) = 3$; $x(0) = 0.75$ (ii) 11, 3, -5, -13, -21
   (c) (i) $m = 0.5$; $y(0) = -2$; $x(0) = 4$ (ii) -3, -2, -1, 0, 1
   (d) Rearrange: $y = 3x + 2$.
       (i) $m = 3$; $y(0) = 2$; $x(0) = -0.67$ (ii) -4, 2, 8, 14, 20

(a) (iii)

(b) (iii)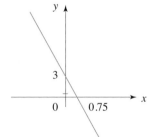

(c) (iii)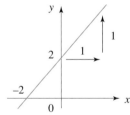

(d) (iii)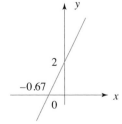

Figures PE 2.2 Q2(iii)

3. (a) $m = 0$; $y(0) = 2$, horizontal line.   (b) $m = \infty$; $x(0) = -2$, vertical line.
(c) $y = -5x - 4 : m = -5$; $y(0) = -4$; $x(0) = -0.8$ (d) $y = x : m = 1$; $y(0) = 0$;
$x(0) = 0$: the $45°$ line. (e) $y = x + 5 : m = 1$; $y(0) = 5$; $x(0) = -5$.

4. (a) (i) $y = 2.5x - 5$ (ii) $y(0) = -5$; $x(0) = 2$ (iii) $m = 2.5$, slope is positive, line is rising or y is
increasing as x increases.
(b) (i) $y = -0.5x + 5$ (ii) $y(0) = 5$; $x(0) = 10$ (iii) $m = -0.5$, slope is negative, line is falling or
y is decreasing as x increases.
(c) (i) $y = -5x + 15$ (ii) $y(0) = 15$; $x(0) = 3$ (iii) $m = -5$, slope is negative; line is falling or y
is decreasing as x increases.

5. (a) (i) $x = 0.4y + 2$ (ii) $x(0) = 2$, this is the vertical intercept; $y(0) = -5$; $m = 0.4$
(b) (i) $x = -2y + 10$ (ii) $x(0) = 10$, this is the vertical intercept; $y(0) = 5$; $m = -2$
(c) (i) $x = -0.2y + 3$ (ii) $x(0) = 3$, this is the vertical intercept; $y(0) = 15$; $m = -0.2$

In question 5, the vertical intercept is the corresponding horizontal intercept in question 4 and vice versa and the slope in question 5 is the slope in question 4 inverted.

6. (a) A, B, C   (b) A, C   (c) A, C
7. (a) $y = 4$ (b) $x = 2$
8.

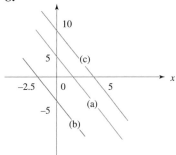

Figure PE 2.2 Q8

All lines have the same slope, $m = -2$, but different intercepts.

PE 2.3

1. (a) (i) $m = -2.5$; $y(0) = 10$; $x(0) = 4$
(ii) $m = 10$; $y(0) = -2.5$; $x(0) = 0.25$
(iii) $m = -0.5$; $y(0) = 15$; $x(0) = 30$

(b) (i) y decreases by 2.5 units
(ii) y increases by 10 units
(iii) y decreases by 0.5 units

2. (a)

Figure PE 2.3 Q2

(b) Number of flights demanded drops by 4.
(c) When $P = 0$, $Q = 64$. Number of flights demanded when $P = £0$.
(d) When $Q = 0$, $P = 16$. When price is £1600, no flights are demanded.

3. (a) The demand function is horizontal. Its slope $= 0$, or no slope, since $P = 65 - 0Q$. The price is not influenced by a change in demand. It remains at $P = 65$.

(b)

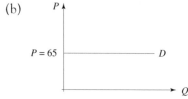

Figure PE 2.3 Q3

4. (a) The vertical intercept is at $P = €500$ when $Q = 0$. Below this price, no wine will be supplied.

(b) When $P = €600$, $Q = 50$ litres of wine.
(c) When $Q = 20$, $P = €540$. When consumers are willing to pay €540, producers supply 20 litres of wine.

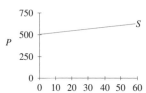

Figure PE 2.3 Q4

5. (a) $Q = -250 + 0.5P$
   (b) $m = 0.5$. When price increases by €1, the producer will supply 0.5 extra litres.

6. (a) The supply function is horizontal. Its slope $= 0$, or no slope, since $P = 50 + 0Q$. The price is not influenced by a change in supply. It remains at $P = 50$.

(b)

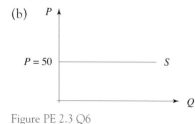

Figure PE 2.3 Q6

7. (a) The supply function is vertical (when plotted as $P = f(Q)$). Its slope $= \infty$. The quantity supplied is fixed and is independent of price.

(b)

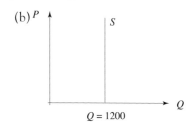

Figure PE 2.3 Q7

8. (a) There is no demand for the wine when the price per bottle is €304 (b) 1520 bottles are demanded when the wine is free (d) The demand drops by 5 bottles when the price increases by €1 (e) 1478 (rounded from 1477.5) bottles are demanded when the price is €8.5.

9. (a) $Q = 12\,800 - 1.6P$
   (b) The slope of the line plotted for $Q = f(P)$ is -1.6. Hence the demand drops by 1.6 when the price increases by £1 – or the demand drops by 16 when the price increases by £10.
   (c) The line cuts the vertical axis at $Q = 12\,800$. Verbally, when the tickets are free the demand will be 12 800. The line cuts the horizontal axis when $P = £8000$. Verbally there will be no demand for tickets when the price is as high as £8000 each.

(d) The demand will be 12 768 when the price is £20 per ticket.
10. (a) (i) $P = 60$. When the price is £60, there is no demand for new potatoes
(ii) $Q = 210$. When potatoes are free, the demand is for 210 kg.
(b) (i) $P = 22.5$. No potatoes will be supplied until the price is £22.5 per kg (ii) $Q = -90$
(iii)The demand and supply functions meet at $P = 40$, $Q = 70$. When the price is £40 per crate, the quantity supplied is the same as the quantity demanded: 70 crates.
11. (b) (i) When $P = 0$, $Q_d = 120$, $Q_s = -100$. When rooms are free 120 will be demanded, none will be supplied.
(ii) When $Q = 0$, $P_d = 96$, $Q_s = 40$. When the price is £96, no rooms will be demanded; no rooms will be supplied until the price reaches £40.
(iii) 45 demanded, 50 supplied.

PE 2.4

1. (a) $FC = 0$

(b) C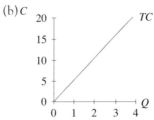

Figure PE 2.4 Q1

(c) $TC = 50$

(d) Each additional unit costs £5

2. (a) $P = 10$

(b) TR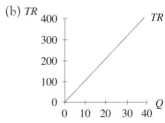

Figure PE 2.4 Q2

3. (a) $TC = 250 + 25Q$

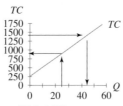

Figure PE 2.4 Q3

(b) When $Q = 28$, $TC = 950$

(c) When $TC = 1400$, $Q = 46$

(d) See arrows on graph

4. (a) $TR = 32Q$ (b) 32 (c) When $Q = 44$, $TR = 1408 > TC = 1350$

5. (a) $TC = 1000 + 15Q$ (c) 1375 (d) 400
6. (a) $TR = 35Q$ (c) 50 (d) The revenue exceeds costs by €600 = profit
7. (a) (i) $FC = 90$ (ii) $VC = 3$ (iii) $TC = 90 + 3Q$ (iv) $TR = 6Q$
(b) $TC = 390$ for 100 watches (c) 55
8. (a) $TC = 1500 + 5Q$ (b) £2000 (c) 160 (d) $TR = 9Q$ (e) 420 (f) £500

SOLUTIONS TO PROGRESS EXERCISES

PE 2.5

1. (a) (i)

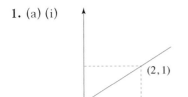

(0,0)

Figure PE 2.5 Q1(a)(i)

(ii)

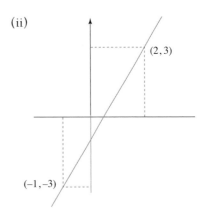

Figure PE 2.5 Q1(a)(ii)

(b) (i) $m = \dfrac{1-0}{2-0} = \dfrac{1}{2}$

(ii) $m = \dfrac{-3-3}{-1-2} = \dfrac{-6}{-3} = 2$

2. (a) $y = x + 2$ (b) $y = 2x$ (c) $y = -0.5x + 4$

3. (a) $TC = 55 + 1.5Q$

(b) When $Q = 8$, $TC = 67$, that is,
$TC = 55 + 1.5(8) = 67$

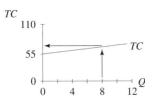

Figure PE 2.5 Q3

4. (a) $Q = 105 - 5P$

(b) (i) Q decreases by 15 units when P increases by £3

(ii) Q increases by 10 units when P decreases by £2

(c) P decreases by £3 when Q increases by 15

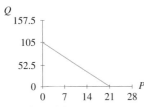

Figure PE. 2.5 Q4

5. (a) $P = 0.125Q - 0.25$
  (b) $Q = 8P + 2$. When P increases by £1, an additional 8 scarves are supplied
  (c) When $P = £8.5$, $Q = 8(8.5) + 2 = 70$
  (d) When $Q = 120$, $P = 0.125(120) - 0.25 = £14.75$
  (e) No scarves are supplied when $P \le -£0.25$ (positive quadrant only)
6. (a) $y = 1.5x + 1$
7. (b) $y = 10 - 2x$

8. (a) 2 (b) $y = 2x + 2$
9. (a) $TC = 64 + 4Q$ (c) 4 (d) 104
    (b)

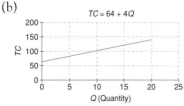

TC = 64 + 4Q

10. (a) $P = \dfrac{1}{5}Q - 5$ (b) 5 (c) €25 (d) 155
11. (a) $Q = 250 - 2.5P$
    (b) (i) When price increases by 10, the demand for tickets drops by 25. (ii) When price increases by 8, the demand for tickets drops by 20.

PE 2.7

1. (a) Slope of demand function is negative, $P > 0, Q > 0$. Therefore

$$\varepsilon_d = \frac{\Delta Q}{\Delta P}\frac{P}{Q} < 0$$

(b) $(\%\Delta Q) = \varepsilon_d(\%\Delta P)$
    (i) $(\% \Delta Q) = (-0.7)\,(5\%) = -3.5\%$ (ii) $(\% \Delta Q) = (-0.7)\,(-8\%) = 5.6\%$
2. $Q = 250 - 5P \rightarrow P = 50 - 0.2Q$

(a) $\varepsilon_d = \dfrac{P}{P - a} = \dfrac{P}{P - 50}$ or $\dfrac{1}{b}\dfrac{P}{Q} = -\dfrac{1}{0.2}\dfrac{P}{250 - 5P} = -\dfrac{P}{50 - P} = \dfrac{P}{P - 50}$

(b) $P = 20, \varepsilon_d = -\dfrac{2}{3}; \quad P = 25, \varepsilon_d = -1; \quad P = 30, \varepsilon_d = -1.5$

3. (a) –3 (b) $-\dfrac{1}{3}$

4. (a) (i) $P = a - bQ \rightarrow \varepsilon_d = 1 - \dfrac{a}{b}\dfrac{1}{Q}$ (ii) $Q = c - dP \rightarrow \varepsilon_d = 1 - \dfrac{c}{Q}$

(b) (i) $\varepsilon_d = 1 - \dfrac{1800}{Q}$. $\varepsilon_d = -17; \; -2.6; \; -1$

(ii) $\varepsilon_d = 1 - \dfrac{120}{Q}$. $\varepsilon_d = -0.2, \; 0.76, \; 0.867$

5. (a) (i) $\varepsilon_d = \dfrac{P}{P - 90}$ (ii) $\varepsilon_d = 1 - \dfrac{1800}{Q}$

(b) $\varepsilon_d = -0.2857, \; -0.5, \; -3.5$ (c) $Q = 900, \; Q = 1800$

6. (a) Slope $= \dfrac{\text{change in quantity}}{\text{change in price}} = \dfrac{\Delta Q}{\Delta P}$.

Same at every point on a linear demand function

$$\varepsilon_d = \frac{\%\text{ change in quantity}}{\%\text{ change in price}} = \frac{\Delta Q}{\Delta P}\frac{P}{Q}. \text{ Different at every point.}$$

(b)

$P =$	20	30	45	70	90
$m =$	-0.05	-0.05	-0.05	-0.05	-0.05
$\varepsilon_d =$	-0.2857	-0.5	-1	-3.5	$-\infty$
$(\%\Delta P) = 10\% \rightarrow (\%\Delta Q) =$	-2.85%	-5%	-10%	-35%	$-\infty$

7. (a) $\dfrac{5}{3}$

  (b) (i) at $Q = 40$, $\varepsilon_d = 2 \rightarrow (\%\Delta Q) = (2)(\%\Delta P) = (2)(10\%) = 20\%$
  (ii) $(P = 40, Q = 40) \rightarrow (P = 44, Q = 48) \rightarrow \%\Delta Q_s = 20\%$

PE 2.9

1. (a) Points (x, y)

  $(0, 2), (1, 3), (2, 4), (3, 5), (4, 6)$

  (c) (i) $y = 2 + x$ (ii) $P = 2 + Q$

(b)
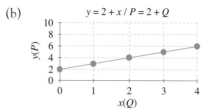
Figure PE 2.9 Q1(b)

2. (a) Demand (Q) decreases by 2.5 for each unit increase in price (b) 42.5, 35, 30

(c)
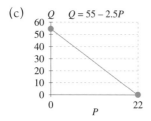
Figure PE 2.9 Q2(c)

(d) $Q = 55 - 2.5P$: $P = 22 - 0.4Q$

3. (a) Points (x, y) $(0, 65), (17, 31), (34, -3)$

  Note: A line may be plotted from two points, use equally spaced x-values to plot from Excel.

  (b) Slope $= -2$: y drops by two units when x increases by one unit. The vertical intercept is 65: $y = 65$ when $x = 0$. The horizontal intercept is $x = 32.5$: $x = 32.5$ when $y = 0$.

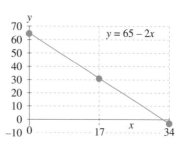
Figure PE 2.9 Q3(a)

4. (a) To plot the graph in the first quadrant, find the horizontal and vertical intercepts. Vertical intercept: $Q = 0$, $P = 80$. Horizontal intercept: $P = 0$, $Q = 20$

(c) Slope $= -4$, vertical intercept $= 80$: horizontal intercept $= 20$

(d) $Q = 20 - 0.25P$

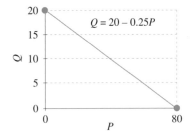

Figure PE 2.9 Q4(d)

(b)

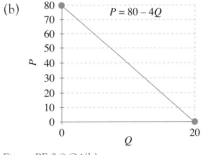

Figure PE 2.9 Q4(b)

5. (a) $Q = 0$, $P = 4$. No positive horizontal intercept, so select any other point to plot the graph, e.g., $Q = 10$, $P = 29$

(c) Slope $= 2.5$: vertical intercept at $P = 4$, horizontal intercept at $Q = -1.6$

(d) $Q = 0.4P - 1.6$

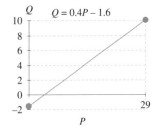

Figure PE 2.9 Q5(d)

(b)

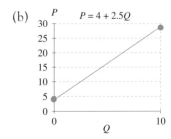

Figure PE 2.9 Q5(b)

6. $P = 60 - 3.5Q$: $Q = 40 - 0.5P$ are demand functions, $P = 5 + 1.2Q$ is a supply function

(a) There are several approaches. For example,

(i) Write each equation in the form $P = a - bQ$: set up a table containing several P-values in the first row, then use the formula $\varepsilon_d = P/(P - a)$ to calculate the corresponding value of elasticity.

(ii) Write each equation in the form $Q = c - dP$; set up a table containing several Q-values in the first row, then use the formula

$$\varepsilon_d = \frac{Q - c}{Q} = 1 - \frac{c}{Q}$$

to calculate the corresponding value of elasticity. Or

(iii) Use the direct method:

$$\varepsilon_d = -\frac{1}{b}\frac{P}{Q}$$

The following tables are examples for the three functions, respectively:

P	2	10	20	30
ε_d	−0.034	−0.2	−0.5	−1
Q	2	10	20	30
ε_d	−19	−3	−1	−0.333
P	2	10	20	30
ε_d	−0.667	2	1.3333	1.2

(b) Set up a table of P-values 0, 5, 10, 15, 20, 25 in the first row. Calculate the corresponding values of Q in the second row. Use the arc elasticity formula for calculations, placing the results in the third row beneath the P-values at the start of the interval:

$P = 60 - 3.5Q$

P	0.00	5.00	10.00	15.00	20.00	25.00
Q	17.14	15.71	14.29	12.86	11.43	10.00
ε_d	−0.04	−0.14	−0.26	−0.41	−0.60	

$Q = 40 - 0.5P$

P	0.00	5.00	10.00	15.00	20.00	25.00
Q	40.00	37.50	35.00	32.50	30.00	27.50
ε_d	−0.03	−0.10	−0.19	−0.28	−0.39	

$P = 5 + 1.2Q$

P	0.00	5.00	10.00	15.00	20.00	25.00
Q	−4.17	0.00	4.17	8.33	12.50	16.67
ε_d	−1.00	3.00	1.67	1.40	1.29	

7. Let x = number of ice-creams; y = number of soft drinks

(a) $12x + 20y = 500 \rightarrow = 25 - 0.6x$

Vertical intercept = 25: horizontal intercept 41.67. See figure PE 2.9 Q7(a).

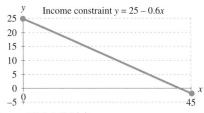

Figure PE 2.9 Q7(a)

Note: The original constraint is shown by the heavy line in this and the following diagrams.

(b) $9x + 20y = 500 \rightarrow y = 25 - 0.45x$

Vertical intercept = 25; horizontal intercept = 55.56 (increased).

See figure PE 2.9 Q7(b).

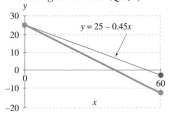

Figure PE 2.9 Q7(b)

(c) $12x + 25y = 500 \rightarrow y = 20 - 0.48x$.
Vertical intercept = 20 (decreased): horizontal intercept = 41.67.
See figure PE 2.9 Q7(c).

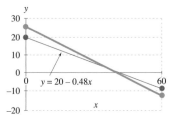

Figure PE 2.9 Q7(c)

(d) $12x + 20y = 750 \rightarrow y = 37.5 - 0.6x$.
Vertical intercept = 37.5 (increased), horizontal intercept = 62.5 (increased).
See figure PE 2.9 Q7(d).

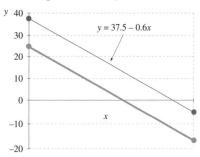

Figure PE 2.9 Q7(d)

8. When $P_X = 25$, $Y = 80$, then $Q_X = 94 + 5P_Y$. If the values of P_Y are entered in row 1 of Table PE 2.9 Q8, then the values of Q_x are calculated in row 2 of the table. In row 3, the values for arc elasticity (called cross-elasticity) are calculated from the data in rows 1 and 2 using the formula

$$\varepsilon_c = \frac{\Delta Q}{\Delta P} \frac{P_1 + P_2}{Q_1 + Q_2}$$

Table PE 2.9 Q8

P_Y	10	20	30	40	50	60	65
Q_X	144	194	244	294	344	394	419
Arc elasticity		0.4438	0.5708	0.6506	0.7053	0.7453	0.7688

SOLUTIONS TO PROGRESS EXERCISES

Chapter 3

PE 3.1

1. $x = 1.5, y = 1.5$ **2.** $x = 7, y = 3$ **3.** $x = 18, y = 1$
4. $x = 1, y = 1$ **5.** $x = 3, y = 2$ **6.** $x = 1, y = 5$
7. $x = 0, y = 4$ **8.** Infinitely many solutions: $y = 5 - 0.5x$
9. $x = 7.04, y = 16.16$

10. No unique solution. Infinitely many: $y = \dfrac{5x - 15}{2}$ **11.** $x = -0.75, y = -1.7$

12. $P = 5.44, Q = 5.92$ **13.** $P = \dfrac{55}{23} = 2.39, \ Q = 1.63$

14. $x = 1, y = 8, z = 7$ **15.** $P_1 = 6, P_2 = 2, P_3 = 0$
16. $y = 5, x = 3$ **17.** $x = 6, y = 4$
18. $p = 8, q = 11$ **19.** $P = -0.4, Q = 2$
20. $x = 28, y = -36$ **21.** $x = 3, y = 4, z = -3$

PE 3.2

1. (a) $y = 1.5x + 6$ (b) $Q = 3P - 45$
2. (a) $P = 632, Q = 84$ (b) 55 (c) 45
3. (a) $P = 165, Q = 47$
4. (a) $P = 104, Q = 90$, (b) 140
5. (a) $Q = 8, P = 26$ (b) $Q_s - Q_d = 16 - 4 = 12$

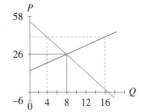

Figure PE 3.2 Q5

6. (a) $Q_d - Q_s = 10 - 4 = 6$
(b) Black market profit $= £72$

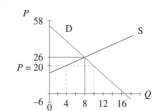

Figure PE 3.2 Q6

7. (a) $L = 10, w = 30$ (b) $L_d - L_s = 12.5 - 5 = 7.5$
(c) $L_s - L_d = 15 - 7.5 = 7.5$

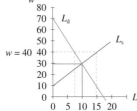

Figure PE 3.2 Q7(c)

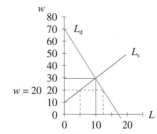

Figure PE 3.2 Q7(b)

PE 3.3

1. (a) $P = 1400, Q = 11$
   (b) $P_d = 1620 - 20Q$
   $P_s = 960 + 40Q$

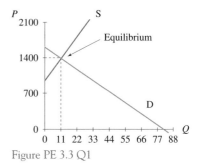

Figure PE 3.3 Q1

2. (a) Break-even when 750 lamps are sold
3. (a) Break-even when 30 bracelets are sold
4. (a) Break-even when 375 meals are sold

5. (a) $(P_s - 120) = 960 + 40Q$
   $P_s = 1080 + 40Q$
   (b) $Q = 9, P = 1440$
   (c) Consumer pays £40, company pays £80

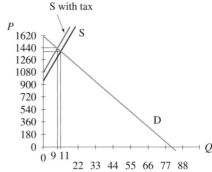

Figure PE 3.3 Q5(a)

6. Pitchers, X: $P_X = 27.5, Q_X = 45$. Putters, Y: $P_Y = 45, Q_Y = 5$
7. (a) $Q = 12, P = £140$
   (b) (i) $(P_s - 9) = 92 + 4Q$
   $P_s = 101 + 4Q$
   (ii) $Q = 11, P = £145$
   (iii) Consumer pays £5, supplier pays £4.

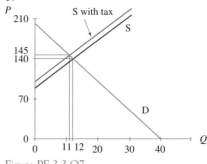

Figure PE 3.3 Q7

8. (a) $TR = 30Q; TC = 200 + 5Q$: therefore $TR = TC \rightarrow 30Q = 200 + 5Q \rightarrow Q = 8$.
   (b) $TR = 30(8) = £240; TC = 200 + 5(8) = £240$.
9. (a) $Q = 75, P = 50$.
   (b) (i) $P_s + 4 = 20 + 0.4Q \rightarrow P_s = 16 + 0.4Q$ (ii) $Q = 80, P = 48$
   (iii) Producer and consumer each receive £2.

10. (a) $TR = 6.6Q; Q = 125$ at break-even
   (b) $P(Q) = 800 + 0.2(Q)$
   $\quad P(160) = 800 + 0.2(160)$
   $\qquad P = 5.2$
   (c)

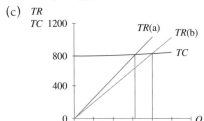

Figure PE 3.3 Q10

11. $P_X = 56$, $Q_X = 6$, $P_Y = 68$, $Q_Y = 14$.

PE 3.4

1. (a) See text.
2. (a) $Q = 180$, $P = 22$
   (b) (i) Consumers pay: $180 \times 22 = 3960$ (ii) Willing to pay: $3960 + 3240 = 7200$
   $\quad$ (iii) $CS = 7200 - 3960 = 3240$
   (c) (i) Producers receive: $180 \times 22 = 3960$
   $\quad$ (ii) Willing to accept: $(4 \times 180) + \dfrac{1}{2}(18 \times 180) = 2340$
   $\quad$ (iii) $PS = 3960 - 2340 = 1620$

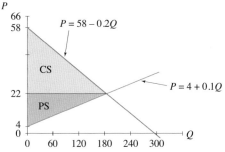

Figure PE 3.4 Q2(a)

3. $\quad P = 500 - 10Q$
   $\qquad P = 100 + 10Q$
   (a) $P = 300$, $Q = 20$
   (b) $CS = 2000$
   (c) $PS = 2000$
   (d) $TS = 4000$

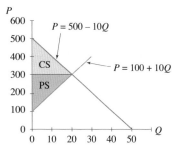

Figure PE 3.4 Q3(a)

4. (a) Price decrease: $P = 250, Q = 25$
(b) $CS = 3125$. Without price decrease, $CS = 2000$
$\therefore$ increase in $CS = 1125$.

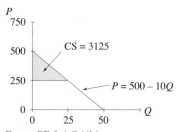

Figure PE 3.4 Q4(b)

5. (a) Price decrease: $P = 250, Q = 15$
(b) $PS = 1125$. Without price decrease, $PS = 2000$
$\therefore$ decrease in $PS = 875$.

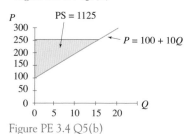

Figure PE 3.4 Q5(b)

6. (a) $P = 80, Q = 30$ (c) $CS = 1800$ (d) $PS = 450$ (e) $TS = 2250$
7. (a) $P = 175, Q = 20$ (b) $CS = 800$ (c) $PS = 1500$
8. (a) $P = 84, Q = 12$ (c) $CS = 216, PS = 360$ (d) $TS = 576$

PE 3.5

1. See text.
2. (a) $E = 360 + 0.6Y$
Intercept $= 360$, slope $= 0.6$
(b) Algebraically, when $E = Y$
$360 + 0.6Y = Y$
$\rightarrow Y_e = 900$
(c) If MPC increases to 0.9, Y_e
increases to $\dfrac{360}{1 - 0.9} = 3600$

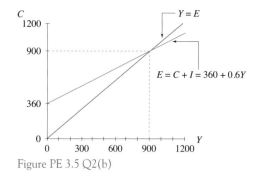

Figure PE 3.5 Q2(b)

3. (a) $E = C + I = 280 + 0.6 Y_d + 80 = 360 + 0.48Y$
Intercept $= 360$, slope $= 0.48$
(b) Algebraically, when $E = Y \rightarrow 360 + 0.48Y = Y \rightarrow Y_e = 692$

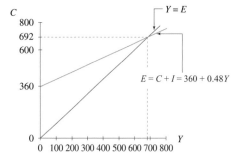

Figure PE 3.5 Q3(b)

(c) If MPC increases to 0.9, then

$$\text{(i)} \quad E = C_0 + b(1 - t)Y + I_0$$
$$= 280 + 0.9(1 - 0.2)Y + 80$$
$$= 360 + 0.72Y$$

Hence (ii) $Y = E \rightarrow Y_e$ increases to $\dfrac{360}{1 - 0.72} = 1287.5$

4. $\Delta Y_e = \frac{1}{1-b}\Delta I_e = £1250\text{m}.$ $Y_e + \Delta Y_e = 800 + 1250 = 2050 = \text{new } Y_e$

PE 3.7

1. (a) $x = 6$, $y = -3$

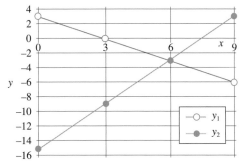

Figure PE 3.7 Q1(a)

(b) $Q = 1.91$, $P = 3.53$

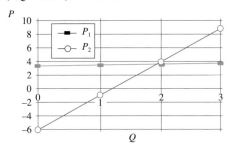

Figure PE 3.7 Q1(b)

2. (a) $P = 200$, $Q = 10$

(b)

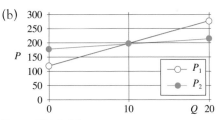

Figure PE 3.7 Q2

3. (a) $Q = 14$, $P = 61$

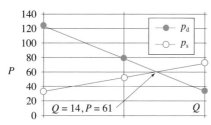

Figure PE 3.7 Q3(a)

(b) (i) $Q = 9.38$, $P = 81.79$ (ii) consumer pays 20.79, producer pays 9.21 (working correct to 2D)

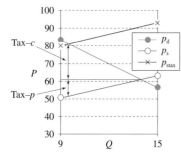

Figure PE 3.7 Q3(b)

4. (a) $Y_e = 414.29$, see figure below.

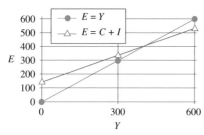

Figure PE 3.7 Q4(a)

(b) Assume $C = 125 + 0.65Y_d$: where $Y_d Y - T$. Then $Y_e = 302.08$ (see figure below).

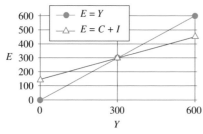

Figure PE 3.7 Q4(b)

SOLUTIONS TO PROGRESS EXERCISES

Chapter 4

PE 4.1

1. $x = 1, x = 5$
2. $Q = 1, Q = 2.5$
3. $Q = 1, Q = 5$
4. $Q = -1, Q = -5$
5. $P = \sqrt{7} = 2.6458,\ P = \sqrt{-7} = -2.6458$
6. $Q = 3$
7. $Q = -1.2426, Q = 7.2426$
8. $Q = 0, Q = 6$
9. $x = -0.7202, x = 9.7202$
10. $P = \pm\sqrt{-9},$ imaginary numbers, $3i$ and $-3i$
11. $P = 1, P = -0.8182$
12. $Q = 0.25$
13. $P = 0, P = 2$
14. $P = -1.5, P = 1.5$
15. $P = -1, P = 0.25$
16. $x = 0, 3, -3$
17. $x = 18.22, -0.22$
18. $Q = 4, 6$
19. $P = -3.61, 3.61$
20. $Q = -0.82, 7.32$
21. (a) TR is zero when $Q = 25$ and again when $Q = 1775$ (b) 900

PE 4.2

1.

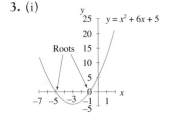

Figure PE 4.2 Q1

2. (a) is a horizontal translation to the left by two units.
   (b) is a horizontal translation to the right by one unit.
   (c) is a vertical translation upwards by one unit.

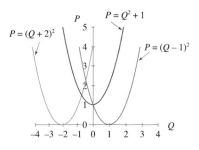

Figure PE 4.2 Q2

3. (i)

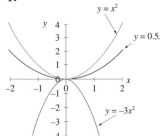

Figure PE 4.2 Q3(a)

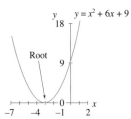

Figure PE 4.2 Q3(b)

(ii) Graph cuts x-axis at:
(a) $x^2 + 6x + 5 = 0$, at $x = -1, x = -5$. Turning point at $x = -3$
(b) $x^2 + 6x + 9 = 0$, at $x = -3$. Turning point at $x = -3$
(c) $x^2 + 6x + 10 = 0$, no real roots, therefore no intersection.
(iii) Graphs are all vertical translations of each other.

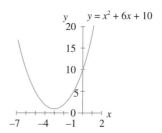

Figure PE 4.2 Q3(c)

4. (a)

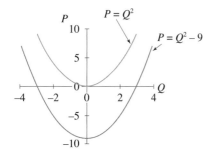

Figure PE 4.2 Q4

(b) $P = Q^2 - 9$. Point of intersection with x-axis: $Q = 3, Q = -3$; y-axis, $P = -9$

5. (a)

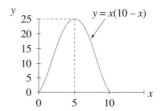

Figure PE 4.2 Q5

(b) Roots: $x = 0, x = 10$
Maximum at $x = 5$
Maximum value, $y = 25$

6. (i)

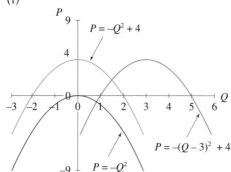

Figure PE 4.2 Q6

(ii)

(a) $P = -Q^2$
(b) $P = -Q^2 + 4$: roots at $Q = -2, Q = 2$
(c) $P = -(Q-3)^2 + 4$: roots at $Q = 1, Q = 5$.

(iii)

Graph (b) is the result of translating (a) vertically upwards by 4 units; (b) is translated horizontally forward by 3 units to give (c).

7. (a)

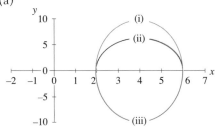

Graphs (i) and (ii) are maximum types, while graph (iii) is a minimum type. Each type may go to any height above or below the axis.

(b) $y = A(x - 2)(x - 6)$

Figure PE 4.2 Q7

PE 4.3

1. (a)

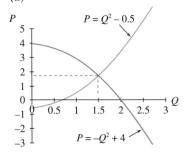

Figure PE 4.3 Q1

(b) Equilibrium price and quantity:
$$P_s = P_d$$
$$Q^2 - 0.5 = -Q^2 + 4$$
$$2Q^2 = 4.5$$
$$Q^2 = 2.25$$

$Q = \pm 1.5$. As $Q = -1.5$ is not economically meaningful, then $Q = 1.5$, $P = 1.75$

2. (a) $TR = (12 - Q)Q = 12Q - Q^2$
(b) $TR = 0$ at $Q = 0$, $Q = 12$
Maximum at $Q = 6$, $TR = 36$

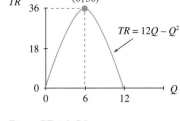

Figure PE 4.3 Q2

3.

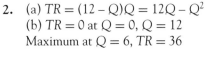

Figure PE 4.3 Q3

$TR = AQ(40 - Q)$
At $Q = 20$, $TR = 1000$, therefore,
$1000 = A(20)(40 - 20) \rightarrow = 2.5$
Therefore, $TR = 2.5Q(40 - Q)$

4. (a) $TR = PQ = (100 - 2Q)Q = 100Q - 2Q^2$:   At $Q = 10$, $TR = 800$

(b) $TR = PQ = P(50 - 0.5P) = 50P - 0.5P^2$:   At $P = 10$, $TR = 450$

5. (a)

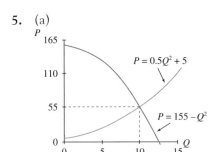

Figure PE 4.3 Q5

(b)

$$P_s = P_d$$
$$0.5Q^2 + 5 = 155 - Q^2$$
$$1.5Q^2 = 150$$
$$Q^2 = 100$$

6.   (a) $TR = (107 - 2Q)Q$

(b) $Q = 26.75$, $TR = 1431$

(c) Break-even: $Q = 2$, $Q = 50$

(d) Between $Q = 2$ to $Q = 50$.

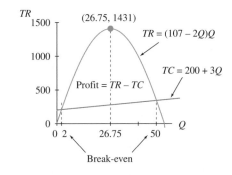

Figure PE 4.3 Q6

7. $\pi = TR - TC = -2Q^2 + 104Q - 200$

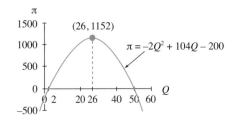

Figure PE 4.3 Q7

(a) $\pi = 0$ at $Q = 2$ and $Q = 50$

(b) Maximum profit at $Q = 26$, $\pi = 1152$

SOLUTIONS TO PROGRESS EXERCISES

8. (a)

(b) Solve
$P_s = P_d \rightarrow Q^2 + 6Q - 40 = 0$: therefore
$Q = 4, Q = -10$. At $Q = 4, P = 36$

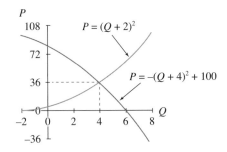

Figure PE 4.3 Q8

9. (a) $TR = PQ = (2400 - 8Q)Q = 2400Q - 8Q^2$
(b) TR is zero when $Q = 0$ and $Q = 300$
(c) $TR = 240Q - 8Q^2$
(d) TR is a maximum when $Q = 150$. The
maximum $TR = 180\,000$

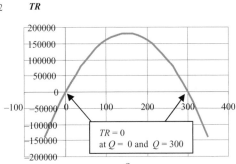

Figure PE 4.3 Q9

10. (a) $AC = 0$ at $Q = 16 \pm 6.63i$
The roots are imaginary; hence the graph
does not cross the axis. Since the graph
crosses the vertical axis at $AC = 1200$, the
graph must be above the horizontal and therefore positive.
(c) Minimum AC at $Q = 16$. The minimum
value of $AC = 176$

(b)

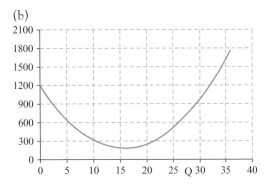

Figure PE 4.3 Q10

11. (a) TC is never zero. The roots of the equation $TC = 800 - 120Q + 5Q^2 = 0$ are $12 \pm 4i$. Complex roots never cross the real axis

(c) The minimum value of TC is 80 at $Q = 12$

(b)

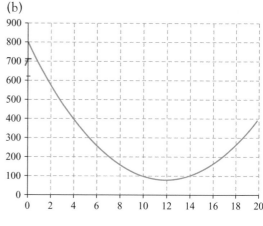

Figure PE 4.3 Q11

12. (a)

Profit $= TR - TC = 2400 - 8Q^2 - (800 - 120Q + 5Q^2) = 120Q + 1600 - 13Q^2$

(c) From the graph, the maximum profit is 1877 at $Q = 4.6$

(b)

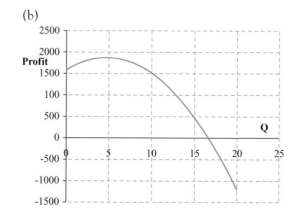

Figure PE 4.3 Q12

PE 4.4

1.

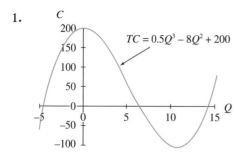

Figure PE 4.4 Q1

Roots at $Q = -4.4, 6.48, 13.94$
max. at $Q = 0, C = 200$
min. at $Q = 10.67, C = -103.4$

SOLUTIONS TO PROGRESS EXERCISES

2.

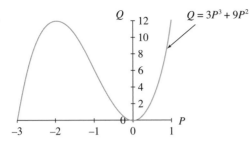

Roots at $Q = -3$, $Q = 0$
max. at $P = -2$, $Q = 12$
min. at $P = 0$, $Q = 0$

Figure PE 4.4 Q2

3.

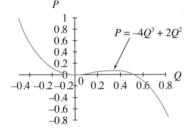

Roots at $Q = 0$, $Q = 0.5$
max. at $Q = \frac{1}{3}$, $P = 0.07$
min. at $Q = 0$, $P = 0$

Figure PE 4.4 Q3

4.

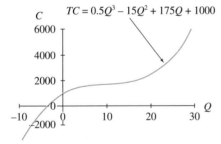

Roots at $Q = -4.09$
no maximum or
minium point

Figure PE 4.4 Q4

PE 4.5

1. (a) 36   (b) 27   (c) 5   (d) 125   (e) 9   (f) 16
   (g) 1   (h) 0.2   (i) 0.0278   (j) 0.008   (k) 1.5811   (l) 0.1317

2. (a) $\dfrac{1}{2}$ (b) $(4)(3^{4.1})$

3. (a) $\dfrac{3^5}{2^3 5^7}$ (b) 2^y

4. (a) a^y (b) $5^x 2^{y-x}$

5. (a) $L^{\alpha-1}K^{\beta}$ (b) $16L^8$

6. (a) $4L^4 K^3$ (b) $2L^2$ (c) $\dfrac{5^{0.5}L^{1.5}}{K^{0.5}}$

7. $\dfrac{0.6K^2}{0.4L^2} = \dfrac{3}{2}\left(\dfrac{K}{L}\right)^2$

8. $\dfrac{L^2}{K^2}$

9. (a) $2Q^{1.5}$ (b) $\dfrac{2Q}{3P^{1.5}}$

10. (a) $\dfrac{K^{0.75}}{2L^{0.75}} = 0.5\left(\dfrac{K}{L}\right)^{0.75}$ (b) $\dfrac{1}{4L^4}$

11. e^{x+2} **12.** $5e^{6x}$ **13.** e^{6-3x} **14.** $\dfrac{120}{e^{1.5t}}$

15. $e^{0.75t} - e^{1.75t}$ **16.** $\dfrac{e^{4t}}{25}$ **17.** $e^{1.25t}$ **18.** $1 + 2e^t + e^{2t}$

19. $5e^{2t} + 5e^{3t}$ **20.** $\dfrac{1}{e^{3t}} + \dfrac{1}{e^{2t}}$

21. $\dfrac{P(1 - PQ^{0.5})}{Q}$ **22.** $\dfrac{1}{2^{0.5}x^{1.5}}$

23. $2xy^2$ **24.** $\dfrac{x(4 - 3xy^4)}{12y}$

25. $x^4 - 2x^2 y + y^2$ **26.** $e^{x-1} + 1$ **27.** e^{2y}

28. pq **29.** $\dfrac{4K^4 - 1}{2K^2} = 2K^2 - \dfrac{1}{2K^2}$ **30.** K^2

PE 4.6

1. $x = -2$ **2.** $Q = 8$ **3.** $x = -12$ **4.** $x = 3$

5. $Q = 0$ **6.** $x = 3$ **7.** $K = 0.125$ **8.** $K = 0.25$

9. $x = 2$ **10.** $t = 32$ **11.** $t = 0$ **12.** $t = 71.5$

13. $t = 3, t = -5$ **14.** $t = -\dfrac{4}{3}$ **15.** $t = \dfrac{2}{3}$ **16.** $x = 0$

17. $t = 7$ **18.** $t = 1$ **19.** $K = 14$ **20.** $L = -2$

21. $x = 7.5$ **22.** $x = \dfrac{5}{6}$ **23.** $t = -0.5$ **24.** $t = -3$

SOLUTIONS TO PROGRESS EXERCISES

PE 4.7

1.

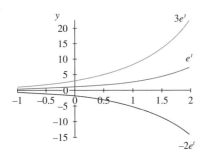

Figure PE 4.7 Q1

Note:

$$(0.5)^t = \left(\frac{1}{2}\right)^t = (2^{-1})^t = 2^{-t}$$

For $a > 1$, $x > 0$, unlimited growth, increasing more rapidly for larger a. For $a = 1$, curve is constant. For $a > 1$, decay curve tends to a limit $y = 0$, i.e., the x-axis.

2.

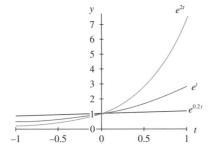

Figure PE 4.7 Q2

For equations of the form $y = ke^t$, unlimited growth when the constant, k, multiplying the exponential is positive; unlimited decay when k is negative. Multiplying the RHS of $y = f(t)$ by a negative number flips the graph through the x-axis.

3.

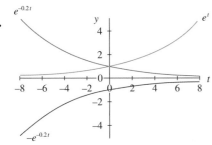

Figure PE 4.7 Q3

Equations of the form $y = e^{\alpha t}$, $\alpha > 0$, increase more rapidly, without limit for larger α.

4.

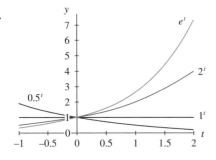

Figure PE 4.7 Q4

The negative exponentials $e^{-0.2t}$ and $-e^{-0.2t}$ are reflections of each other in the x-axis, one growing, the other decaying to the horizontal axis; e^t grows without limit.

5.

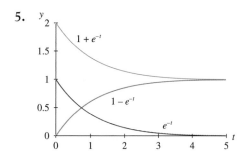

Figure PE 4.7 Q5

All graphs display limited growth or decay: $1 + e^{-t}$ decays to a limit of 1 while $1 - e^{-t}$ grows to a limit of 1; e^{-t} decays to zero as t increases.

6.

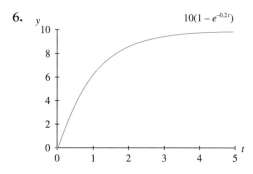

Figure PE 4.7 Q6

Growth curve, tending to a limit of 10.

PE 4.8

1. (a) 125 500

(b)

t	10	30	60	90	100
P	142	180	258	370	417

(c)

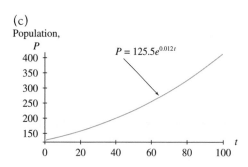

Figure PE 4.8 Q1

2. (a) 9754

(b)

t	5	20	35	45	50	52
S	44	126	165	179	184	185

(c)
Sales (£000)

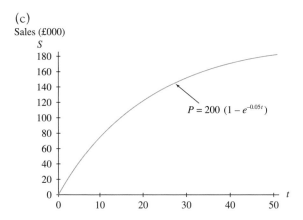

Figure PE 4.8 Q2

3. (a) 1

(b)

t	0	20	40	60	80	100
N	1	7	51	268	631	772

Number of chickens, N

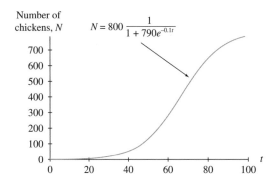

Figure PE 4.8 Q3

(c) As t increases, $e^{-0.1t}$ in the denominator will approach zero, therefore $N = 800$ infected chickens.

PE 4.9

1. (a) 1.3617   (b) 1.7633   (c) –0.6576   (d) 4.0601   (e) 1.3315

(f) –0.5500   (g) 1.3315   (h) 3.0659   (i) 1.2060

2. (a) $\log_a (N) = \text{power}$ (b) $\log_3 10 = 2.0959$ (c) $\log_{10} 3 = 0.4771$

(d) $\log_e 7.389 = 2$ (e) $\log_2 10 = x$ (f) $\log_3 8 = x$

(g) $\log_2 4 = x$ (h) $\log_2 15 = 3x$ (i) $\log_5 4 = Q$

(j) $\log_5 10 = Q + 1$ (k) $\log_{10} 25 = 4x + 3$ (l) $\log_P 12 = 0.3$

(m) $\log_K 5.6 = -0.5$ (n) $\log_L (0.0493) = 0.9$ (o) $\log_{(K+10)} 2 = 0.8$

PE 4.10

1. $x = 1.4472$ **2.** $x = 0.2218$ **3.** $x = 4.6990$

4. $t = 4.8283$ **5.** $t = -0.6098$ **6.** 0.6098

7. $t = -4.8847$ **8.** $t = 0.9316$ **9.** $x = 0.6990$

10.   (a) (i) $t = 0 \to S = 0$ (ii) $t = 1 \to S = 73.93$ (iii) $t = 5 \to S = 275.34$
     (b) Similar to Q2, Progress Exercises 4.8
         When $S = 400\,000$, $t = 10.1$ weeks
         Maximum sales (never reached) $= £500\,000$

11.   (a) $t = 0 \to P = 200$; $t = 4 \to P = 875.27$; $t = 10 \to P = 3918.61$
     (b) Similar to Q3, Progress Exercises 4.8
     (c) $P = 1000$ in 4.39 years; $P = 3000$ in 8.42 years; $P = 4000$ in 10.15 years

12.   (a) $t = 0 \to P = 5.2$m (year 2012); $t = 10 \to P = 5.15$m (year 2022); $t = 50 \to P = 4.95$m
       (year 2062)
     (b) 262.36 years, that is, in AD 2274.36

PE 4.11

1. $(x + 5) = 10^{1.2}$ **2.** $(x + 2) = 10^{3.5/3}$ **3.** $(t - 5) = e^{0.5}$

4. $(x + 1) = 10^{-0.5(2.5)}$ **5.** $2 - 2x = 10^0$ **6.** $x^2 + 1 = 10^{0.3}$

7. $10^{-0.65} = 0.5x$ **8.** $x(x + 1) = 10^1$ **9.** $x - 4.5 = e^2$

10. $\dfrac{x + 1}{x} = e^{3.5}$

11.   (1) 10.8489   (2) 12.6780   (3) 6.6487   (4) –0.9438   (5) 0.5
       (6) ±0.9976   (7) 0.4477   (8) 2.7016, 3.7016   (9) 11.8891   (10) 0.0311

12.   (a) $x = -0.368$ (b) $x = -0.8755$ (c) $x = -17.9435$ (d) $x = -4.4859$
       (e) $t = 0.8473$ (f) $x = 9.8833$ (g) $t = 0.6931$ (h) $x = -5.6711$

13. $\log(12)^2 (20) = 3.4594$

14. $\ln \dfrac{(125)(28)}{(80)} = 3.7785$ **15.** $\ln \dfrac{(138)^3}{(95)^2} = 5.6740$

16. $\ln \dfrac{(2)^8 (8)^3}{(3)^2} = 9.5863$ **17.** $\ln \dfrac{(4)(80)}{(2)^2} = 4.3820$

24.   (a) $\dfrac{\log(27)}{\log(3)} = 3$ (b) $\dfrac{\log(51)}{\log(5)} = 2.4430$ (b) $2\left(\dfrac{\log(30)}{\log(3.5)}\right) = 5.4299$

25.   (a) $x = 25.2982$ (b) $Q = 5.3096$

SOLUTIONS TO PROGRESS EXERCISES

PE 4.12

1.

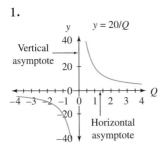

Figure PE 4.12 Q1

2.

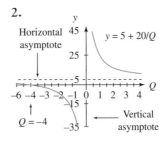

Figure PE 4.12 Q2

3.

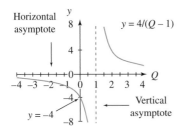

Figure PE 4.12 Q3

4.

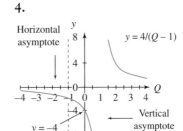

Figure PE 4.12 Q4

5.

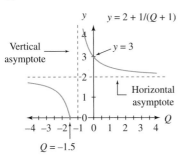

Figure PE 4.12 Q5

6.

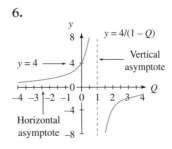

Figure PE 4.12 Q6

7.

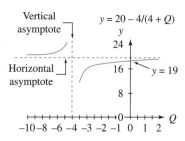

Figure PE 4.12 Q7

8.

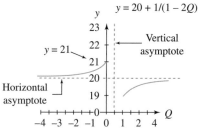

Figure PE 4.12 Q8

9.

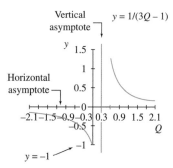

Figure PE 4.12 Q9

PE 4.13

1. $x = -0.5$ 2. $x = -6$ 3. $x = 1, x = 4$ 4. $Q = 4$
5. $Q = 6.5311, Q = -1.5311$ 6. $Q = 0.375$

7. (a)

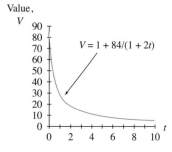

Figure PE 4.13 Q7

(b) 1.71 years.

8. (a) $Q = 11.694, P = 39.388$

(b)

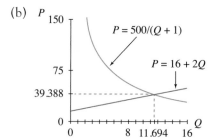

Figure PE 4.13 Q8(b)

9. (a)

t	0	5	10
S_{old}	3000	1125	692
S_{new}	1714	2250	3273

(b) Sales are equal at 1.8 months.

$$S_{old} = S_{new} = 1875$$

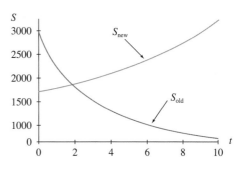

Figure PE 4.13 Q9

10.

(a)

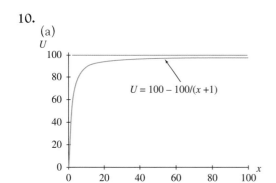

Figure PE 4.13 Q10

(b)

U	10	50	80	100
x	0.11	1	4	∞

Chapter 5

PE 5.1

1. (a) (i) an AP, $a = 4, d = 2$ (ii) a GP, $a = 4, r = 0.5$ (iii) an AP, $a = 5, d = 0.5$

(iv) an AP, $a = 5, d = -0.5$ (v) a GP, $a = 200, r = 1.1$

(b) (i) $T_{10} = 22, S_{20} = 460$, 50th term

(ii) $T_{10} = 0.007\,812\,5, S_{20} = 7.999\,992$, no term exceeds 100

(iii) $T_{10} = 9.5, S_{20} = 195$, 192nd term   (iv) $T_{10} = 0.5, S_{20} = 5$, no term exceeds 100

(v) $T_{10} = 471.5895, S_{20} = 11\,454.999\,89$, first

2. 96   **3.** 25   **4.** 0.5   **5.** $ar^{n-1} = 1.03$

6. $a = 1, d = 1, n = 20$ **7.** $a = 2, d = 2, n = 20$

8. $a = 90, d = -3, n = 5$ or $n = 56$

9. $a = 2, d = 3$: series $2 + 5 + 8 + 11$ **10.** Proof

11. $a = 28\,672, r = \dfrac{1}{8}, S_\infty = 32\,768$

12. (a) $a + a(1.1) + a(1.1)^2 + \ldots; r = \dfrac{110}{100} = 1.1$

(b) $S_8 = 11.43a - 8a = 3.436a$

(c) $a = 800$, $S_8 = 9148.8$, number of tins sold. Profit $= £823.392$. Cost of campaign for eight weeks is £1600. Therefore, advertising campaign not worthwhile.

PE 5.2

1. (a) 594   (b) 6474   (c) $n = 11.47$, the 12th month.
2. (a) (i) $n = 11$, $T_n = a + (n-1)d - 38\,000$. **Note:** $d = -15\,200$ (ii) $1\,064\,000$
(b) $15\,833$ per week
3. (a) 900   (b) 8250

4. (a)

Week	1	5	10	15
Output A	1000	1800	2800	3800
Output B	500	1037	2580	6420

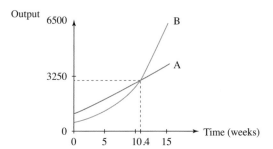

Figure PE 5.2 Q4

Output for both firms is approximately the same in week 11.
(b)   Company A: total production during first 15 weeks: $36\,000$
        Company B: total production during first 15 weeks: $36\,018$
5. $n = 3$ in 2015; $n = 8$ in 2020
(a) Forecast 1:$60\,000$; $85\,000$ Forecast 2:$56\,180$; $75\,182$
(b) Forecast 1:$540\,000$ Forecast 2:$494\,873$
6. (a) (i) 9850 (ii) $230\,250$ (iii) 2 weeks
(b) (i) $22\,639$ (ii) $365\,414$ (iii) Week 14
7. Market 1 (a) 800 (b) 13 600
Market 2 (a) 1199 (b) 18 315

PE 5.3

1. (a) £6381.41   (b) £7012.76   (c) £8052.55
2. 13 400.96
3. (a) 2625   (b) 3113.81   (c) 4072.24   (d) 6633.24
4. £4898   **5.** (a) £5425   (b) £5897.7   **6.** £5572.7
7. £8024.51   **8.** (a) 5.95%   (b) 9.05%   (c) 14.87%   **9.** 4.78 years
10. (a) 4.78 years   (b) 9.43 years   **11.** (a) 11.90 years   (b) 23.45 years
12. $r = 15.83\%$ **13.** $t = 25.89$ years

14.

	$t = 0$	$t = 5$	$t = 10$	$t = 15$	$t = 20$
Simple interest	1	1.5	2.0	2.5	3
Compound interest	1	1.61	2.59	4.18	6.73

15. 7.699%

PE 5.4

1. (a) (i) 52 750, 55 500, 58 250, 61 000, 63 750 (ii) 18.18 years
   (b) (i) 52 750, 55 651.25, 58 712.07, 61 941.23, 65 348 (ii) 12.946 years
   (c) (i) 52 807.24, 55 772.0929, 58 903.406, 62 210.527, 65 703.325 (ii) 12.689 years
   (d) (i) 52 827.03, 55 813.904, 58 969.656, 62 303.837, 65 826.534 (ii) 12.603 years
2. (a) £6420.13 (b) £7095.34 (c) £8243.61
3. (a) 6.136% (b) 6.168% (c) 6.184%
4. (a) £4072.24 (b) £4109.05 (c) £4121.81 (d) £4121.81
5. 6.174% **6.** 8.243% **7.** 8.33% **8.** 3.045%
9. 10.15% **10.** 12.055 years
11. (a) 6567.29 (b) 6581.74 (c) 6584.60 (d) 6584.70
12. Bank A: $APR = 8.99\%$; Bank B: $APR = 9.04\%$; Bank B charges most interest

PE 5.5

1. £65 158.3
2. (a) See text
   (b) Project A: (i) $NPV = £2104$ (ii) $IRR \cong 11.6\%$
       Project B: (i) $NPV = £483$ (ii) $IRR \cong 8.3\%$
3. Project A: at 6%, $NPV = £2331.68$; at 8%, $NPV = -£5964.2$; $IRR = 7.56\%$
   Project B: at 6%, $NPV = £6457.72$; at 8%, $NPV = -£1549.35$; $IRR = 7.6\%$
   At 6% project A is more profitable, at 8% project B: according to the IRR, project B is more profitable.

PE 5.6

1. (a) 2444.49 (b) 409.08
2. (a) 10 172.09 (b) 24 807.07
3. 26 398.97
4. (a) 16 274.54 (b) 62 745.39 (c) interest repaid 10 000 (year 1) 9372.55 (year 2)
5. (a) 3021.07 (b) 10 421.48 (c) 937.5, 898.43, 858.63, 818.09 paid in quarters 1, 2, 3, 4
6. (a) 156 000 (b) 98 677.17 (c) 57 322.8255
7. 11 716.44
8. (a) 69 207.64 (b) 182.89
9. (a) 10 533.25 (b) 109.95 (c) 334.90
10. (a) 165.04 (b) 717.16
11. (a) buy since leasing exceeds the purchase price (b) lease

PE 5.7

1. (a) Annual payment $= r \times$ price of bond $= 0.2 \times 5000 = 1000$.
   (b) and (c)

End of year	Cash flow	0.05	0.10	0.15	0.20	0.25	0.30
1	1000.00	952.38	909.09	869.57	833.33	800.00	769.23
2	1000.00	907.03	826.45	756.14	694.44	640.00	591.72
3	1000.00	863.84	751.31	657.52	578.70	512.00	455.17
4	1000.00	822.70	683.01	571.75	482.25	409.60	350.13
5	1000.00	783.53	620.92	497.18	401.88	327.68	269.33
	NPV	4329.48	3790.79	3352.16	2990.61	2689.28	2435.57
NPV of redeemed bond		3917.63	3104.61	2485.88	2009.39	1638.40	1346.65
Total NPV		8247.11	6895.39	5838.04	5000.00	4327.68	3782.22

2. (a) Annual payment $= r \times$ price of bond $= 0.09 \times 10\,000 = 900$.
   (b)

End of year	Cash flow	Discount rates					
		0.065	0.09	0.12	0.065	0.09	0.12
1	900.00	845.07	825.69	803.57			
2	900.00	793.49	757.51	717.47			
3	900.00	745.06	694.97	640.60			
4	900.00	699.59	637.58	571.97			
5	900.00	656.89	584.94	510.68		(c)	
6	900.00	616.80	536.64	455.97	845.07	825.69	803.57
7	900.00	579.16	492.33	407.11	793.49	757.51	717.47
8	900.00	543.81	451.68	363.49	745.06	694.97	640.60
9	900.00	510.62	414.39	324.55	699.59	637.58	571.97
10	900.00	479.45	380.17	289.78	656.89	584.94	510.68
NPV		6469.95	5775.89	5085.20	3740.11	3500.69	3244.30
NPV of redeemed bond		5327.26	4224.11	3219.73	7298.81	6499.31	5674.27
Total NPV		11\,797.21	10\,000.00	8304.93	11\,038.92	10\,000.00	8918.57

3. (a) £400   (b) (i) £6400 − £5000 = £1400   (ii) £4000 − £5000 = −£1000

4.

End of year	Bond Rate = 0.2			(a) Rate = 0.2		(b) Rate = 0.05		(c) Rate = 0.3	
	Cash flow	PV of cash flow	Balance −1000	PV of cash flow	Balance −1000	PVof cash flow	Balance −1000	PVof cash flow
1	1000.00	833.33	5000	833.33	4250.00	952.38	5500.00	769.23
2	1000.00	694.44	5000	694.44	3462.50	907.03	6150.00	591.72
3	1000.00	578.70	5000	578.70	2635.63	863.84	6995.00	455.17
4	1000.00	482.25	5000	482.25	1767.41	822.70	8093.50	350.13
5	1000.00	401.88	5000	401.88	855.78	783.53	9521.55	269.33

NPV of cash flow	2990.61			2990.61		4329.48	2435.57
PV of redeemed bond	2009.39	PV of final balance	2009.39		670.52		2564.43
NPV cash flow + of final balance	5000.00			5000.00		5000.00	5000.00

Value of redeemed bond is £5000. Final balance in bank is (a) 5000, (b) 855.78, (c) 9521.55. When the interest rate is 30% the final balance is £9521.55 (greater than the value of the bond), but at 5% the final balance is only £855.78.

However, the NPV of the cash flow combined with the PV of the final balance is £5000 in all cases.

Chapter 6

PE 6.1

1. (a) and (b) Draw tangents at $x = -1.5; 0; 2$. Measure slopes. For convenience, $\Delta x = 1$ in the sketch below.

(c) $y = x^2$, $\dfrac{dy}{dx} = 2x = $ slope
at $x = -1.5$, slope $= -3$
at $x = 0$, slope $= 0$
at $x = 2$, slope $= 4$

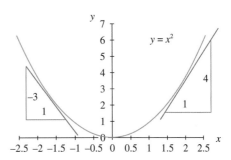

Figure PE 6.1 Q1

2. (a) $5x^4$ (b) $5.5Q^{4.5}$ (c) $80Q^7$ (d) 0   (e) $6Q$ (f) $\dfrac{3.04}{Q^{0.8}}$ (g) $-\dfrac{1.4}{Y^{0.3}}$

3. (a) $-\dfrac{2}{x^3}$ (b) $-\dfrac{10}{x^3}$ (c) $5 - \dfrac{2}{x^3}$ (d) $-\dfrac{12}{x^3}$ (e) $\dfrac{dP}{dQ} = Q^2 + 70 - 30Q$

4. (a) $2x + 7$ (b) $-3 - 1.5x^{-0.5}$ (c) $1 + \dfrac{1}{\sqrt{x}}$ (d) $1 - \dfrac{1}{2x^2} - \dfrac{2}{x^3}$ (e) $1 + \dfrac{2}{x^2} + \dfrac{3}{2x^3}$

(f) $1 + 4\dfrac{1}{x^2} - 8\dfrac{4}{x^3} = 1 + \dfrac{4}{x^2} - \dfrac{32}{x^3}$

5. (a) $\dfrac{5}{2x^{1/2}}$ (b) $-\dfrac{1}{2x^{3/2}}$ (c) $-\dfrac{1}{2Q^{3/2}}$ (d) $-\dfrac{7}{2Q^{3/2}}$

(e) $\dfrac{15}{2Q^{3/2}}$ (f) $\dfrac{1}{9Q^{2/3}}$ (g) 1 (h) $-\dfrac{2}{Q^2}$

6. (a) Q^8 (b) $4Q^{-3/2}$ (c) $x^{-1} + x^{-2}$ (d) $5^{1/2}Q^{3/2}$ (e) $3Q + 8$

7. (a) $8Q^7$ (b) $-\dfrac{6}{Q^{5/2}}$ (c) $-\dfrac{1}{x^2} - \dfrac{2}{x^3}$ (d) $5^{1/2}\left(\dfrac{3}{2}Q^{1/2}\right)$ (e) 3

8. (a) $5x^4$; $20x^3$ (b) $6Q$; 6   (c) $10 + 0.5Q^{-0.5}$; $-0.25Q^{-1.5}$

(d) $\dfrac{1}{3}Q^{-2/3}$; $-\dfrac{2}{9}Q^{-5/3}$ (e) $-\dfrac{1}{x^2}$; $\dfrac{2}{x^3}$ (f) $4Y^3 + \dfrac{4}{Y^5}$; $12Y^2 - \dfrac{20}{Y^6}$

9. $(TC)' = \dfrac{3Q^2}{5} - 16Q + \dfrac{5}{2}$; $(TC)'' = \dfrac{6Q}{5} - 16$; $(TC)''' = \dfrac{6}{5}$

10. $Q = -10P + 27$, $\dfrac{dQ}{dP} = -10$; $P = -\dfrac{1}{10}Q + 2.7$, $\dfrac{dP}{dQ} = -\dfrac{1}{10}$

11. $y = \dfrac{1}{25}x^2$ $\dfrac{dy}{dx} = \dfrac{2}{25}x$; $\dfrac{dx}{dy} = \dfrac{25}{2x}$

12. $\dfrac{dS}{di} = 960i + 200i^4 - 48$

13. $t = \dfrac{18}{6x^3} = \dfrac{3}{x^3} = 3x^{-3}$: $\dfrac{dt}{dx} = -\dfrac{9}{x^4}$; $\dfrac{dx}{dt} = -\dfrac{x^4}{9}$

14. $150 - 4Q$, -4

PE 6.3

1. $Q = 120 - 3P \rightarrow P = 40 - \frac{1}{3}Q$

    (a) $TR = 40Q - \frac{1}{3}Q^2$;

    $MR = 40 - \frac{2}{3}Q$;

    $AR = 40 - \frac{1}{3}Q = P$

    At $Q = 15$, $TR = 525$; total revenue from the sale of 15 units.

    $MR = 30$;

    revenue accrued from sale of 1 extra unit when $Q = 15$.

    $AR = 35$;

    price per unit for each of the first 15 units.
    (b) $AR = 40 - \frac{1}{3}Q$; when $AR = 0$

    $0 = 40 - \frac{1}{3}Q \rightarrow Q = 120$

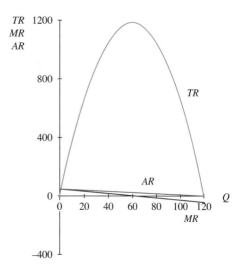

Figure PE 6.3 Q1

    When $Q = 120$, $MR = -40$. Stop selling when $MR = 0$, that is, at $Q = 60$.
    (c) Use Excel, if available.

2. (a) $TR = 125Q - Q^{2.5}$; $MR = 125 - 2.5Q^{1.5}$; $AR = 125 - Q^{1.5}$
    No. Slope $AR = -1.5Q^{0.5}$; slope of $MR = -3.75Q^{0.5}$
    (b) $At Q = 10$, $TR = 933.77$; $MR = 45.94$; $AR = 93.377$
    Comment: AR for first 10 units (per unit) $= 93.377$. $MR = 45.94$, the sale of one extra unit when $Q = 10$ should add £45.94 to TR.

    At $Q = 25$, $TR = 0$; $MR = -187.5$; $AR = 0$

    Comment: $AR > MR$. When $P = 0$, $TR = 0$ as expected. $MR = -187.5$, the sale of one extra unit causes a decrease of £187.5 in TR.
    (c) $MR = 125 - 2.5Q^{1.5}$; when $MR = 0 \rightarrow 0 = 125 - 2.5Q^{1.5} \rightarrow Q = 13.57$

    $AR = 125 - Q^{1.5}$; when $AR = 0 \rightarrow 0 = 125 - Q^{1.5} \rightarrow Q = 25$

    Stop selling when $Q = 13.57$. When $Q > 13.57 \rightarrow MR < 0$.
    (d) Use Excel, if available.
3. (a) $TC = FC + VC \rightarrow TC = 1000 + 3Q$.   When $Q = 20$, $TC = 1060$.
    (b) $MC = d(TC)/dQ = 3$.   When $Q = 20$, $MC = 3$
    MC is the slope of the linear TC function. It is the amount by which TC increases for each extra unit produced. Slope is constant.

4. (a) If AC decreases, then $\dfrac{d(AC)}{dQ} < 0$. $\dfrac{d(AC)}{dQ} = -\dfrac{10}{Q^2} < 0$ for $Q > 0$

The average cost decreases as the number of units produced increases (economies of scale).

(b) Since $AC = \dfrac{TC}{Q} \rightarrow TC = Q(AC)$.

$TC = 50Q + 10$. $TVC = 50Q$; $AVC = 50$; $FC = 10$.

(c) $MC = 50$. MC is the amount by which TC increases for each extra unit produced. This is the same, whether it is for the first unit or the 900th!

5. (a) $TC = Q^3 - 9Q^2 + 150 + 75Q$. When $Q = 15$, $TC = 2625$.

(b) $TVC = Q^3 - 9Q^2 + 75Q$ and $FC = 150$.

(c) $MC = 3Q^3 - 18Q + 75$. $MC = 480$ when $Q = 15$.

MC is a quadratic curve. If MC does not cross the Q-axis, then the curve lies entirely above the Q-axis. Solve

$$MC = 0, \quad 3Q^2 - 18Q + 75 = 0 \rightarrow Q = \frac{18 \pm \sqrt{-576}}{6}$$

These are imaginary values (confirm these findings by plotting MC using Excel, if available).

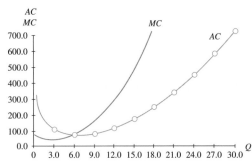

Figure PE 6.3 Q5(c)

6. (a) (i) $TR = Q(AR) = 180Q - 12Q^2$ (ii) $TR = 12Q$

(b) (i) $MR = 180 - 24Q$ (ii) $MR = 12$

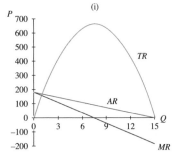

Figure PE 6.3 Q6d(i)

(c) (i) Slope of MR is twice slope of AR (monopolist)

    (ii) Slope of MR is equal to slope of AR (perfectly competitive firm).

(d) Use Excel, if available.

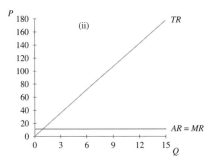

Figure PE 6.3 Q6d(ii)

7. (a) (i) $\dfrac{dy}{dx} = 6x - 4 - \dfrac{2}{x^2}, \dfrac{d^2y}{dx^2} = 6 + \dfrac{4}{x^3}$

(ii) $\dfrac{d(TR)}{dQ} = 200 - 8Q, \dfrac{d^2(TR)}{dQ^2} = -8$

(iii) $\dfrac{d(TC)}{dQ} = Q^2 - 24Q + 164, \dfrac{d^2(TC)}{dQ^2} = 2Q - 24$

(b) (i) $MR = \dfrac{d(TR)}{dQ} = 200 - 8Q$ (ii) $AR = \dfrac{TR}{Q} = \dfrac{200Q - 4Q^2}{Q} = 200 - 4Q$

(iii) $MC = \dfrac{d(TR)}{dQ} = Q^2 - 24Q + 164$

(iv) $AC = \dfrac{TC}{Q} = \dfrac{1}{Q}\left(\dfrac{Q^3}{3} - 12Q^2 + 164Q + 100\right) = \dfrac{Q^2}{3} - 12Q + 164 + \dfrac{100}{Q}$

(v) Profit $= TR - TC = 36Q + 8Q^2 - \dfrac{1}{3}Q^3 - 100$

8. (a) (i) $TR = PQ = (a - bQ)Q = aQ - bQ^2$ (ii) $MR = \dfrac{d(TR)}{dQ} = a - 2bQ$

(iii) $AR = \dfrac{TR}{Q} = \dfrac{aQ - bQ^2}{Q} = a - bQ$

(b) at $Q = 0$, (i) $TR = 0$ (ii) $MR = a$ (iii) $AR = a$

(c) $MR = 0$ at $Q = a/2b$. Substitute $Q = a/2b$ into AR: $AR = a - b(a/2b) = a - a/2 = a/2$.

9. (a) (i) $TVC = aQ^3 - bQ^2 + cQ$ (ii) $FC = d$ (iii) $MC = 3aQ^2 - 2bQ + c$

(iv) $AC = aQ^2 - bQ + c + d/Q$

(b) $MC = Q^2 - 4Q + 3$. Note: this is a quadratic function.

(i) $MC = 0$ at $Q = 1$ and $Q = 3$ (ii) MC is a minimum at $Q = 2$. The minimum is $MC = -1$.

PE 6.4

1. (a) (i) $TLC = wL = 180L$; $ALC = \dfrac{TLC}{L} = 180$; $MLC = \dfrac{d(TLC)}{dL} = 180$

(ii) $TLC = 200L + 5L^2$; $ALC = 200 + 5L$; $MLC = 200 + 10L$

(iii) $ALC = MLC$ (iv) ALC for all values of $L > 0$.

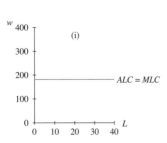

Figure PE 6.4 Q1(i)

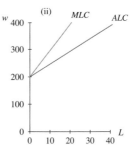

Figure PE 6.4 Q1(ii)

2. (a) (i) $Q = 20L - L^2$, $MP_L = \dfrac{dQ}{dL} = 20 - 2L$.

This is a linear function which decreases (negative slope) as L increases.

(ii) $Q = 225L - \dfrac{1}{3}L^3$, $MP_L = \dfrac{dQ}{dL} = 125 - L^2$

This is a quadratic function which decreases as L increases.

(b) (i) $APL = \dfrac{Q}{L} = \dfrac{20L - L^2}{L} = 20 - L$. Also a linear function with negative slope

(ii) $APL = \dfrac{225L - \frac{1}{3}L^3}{L} = 225 - \dfrac{1}{3}L^2$

(c) (i) Slope of MP_L is double the slope of APL

(ii) $MP_L < APL$

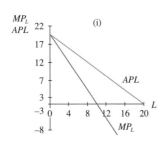

Figure PE 6.4 Q2(i)

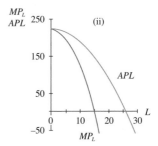

Figure PE 6.4 Q2(ii)

3. (a) $APL = \dfrac{Q}{L} \rightarrow Q = L(APL)$　　(b) $MP_L = \dfrac{dQ}{dL}$

(i) $Q = 150L$ (ii) $Q = 30L - 2L^2$　　　　(i) $MP_L = 150$ (ii) $MP_L = 30 - 4L$

(c) (i) $APL = MP_L$

(ii) Slope of MP_L is double slope of APL, both are linear decreasing functions.

4. (i) $C = 40 + 0.8Y$

(a) $MPC = 0.8$; $APC = \dfrac{40}{Y} + 0.8$

$APC > MPC$

$\dfrac{40}{Y} + 0.8 > 0.8$ True

At $Y = 10$, $APC = 4.8$; $MPC = 0.8$

At $Y = 20$, $APC = 28$; $MPC = 0.8$

(ii) $C = 50 + 0.5Y^{0.8}$

(a) $MPC = \dfrac{0.4}{Y^{0.2}}$; $APC = \dfrac{50}{Y} + \dfrac{0.5}{Y^{0.2}}$

$APC > MPC$

$\dfrac{50}{Y} + \dfrac{0.5}{Y^{0.2}} > \dfrac{0.4}{Y^{0.2}}$

$\dfrac{50}{Y} + \dfrac{0.1}{Y^{0.2}} > 0$ True

At $Y = 10$, $APC = 5.3$; $MPC = 0.25$

At $Y = 20$, $APC = 2.77$; $MPC = 0.22$

(b) $MPS = 1 - MPC = 1 - 0.8 = 0.2$

$APS = 1 - APC = 0.2 - \dfrac{40}{Y}$

$MPS > APS$

$0.2 > 0.2 - \dfrac{40}{Y}$ True

At $Y = 10$, $MPS = 0.2$; $APS = -3.8$

At $Y = 20$, $MPS = 0.2$; $APS = -1.8$

(b) $MPS = 1 - \dfrac{0.4}{Y^{0.2}}$

$APS = 1 - \dfrac{50}{Y} - \dfrac{0.5}{Y^{0.2}}$

$MPS > APS$

$1 - \dfrac{0.4}{Y^{0.2}} > 1 - \dfrac{50}{Y} - \dfrac{0.5}{Y^{0.2}}$

$1 + \dfrac{0.1}{Y^{0.2}} + \dfrac{50}{Y} > 0$ True

At $Y = 10$, $MPS = 0.75$; $APS = -4.3$

At $Y = 20$, $MPS = 0.78$; $APS = -1.8$

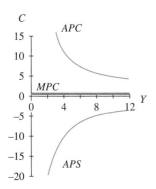

Figure PE 6.4 Q4(i)

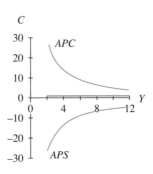

Figure PE 6.4 Q4(ii)

5. (a) $U = 5x^{0.8}$ $MU = \dfrac{d_U}{d_x} = 4x^{-0.2} = \dfrac{4}{x^{0.2}}$

(b) If MU declines, then $\dfrac{d_{(MU)}}{d_x} < 0$. Therefore, $\dfrac{d_{(MU)}}{d_x} = -\dfrac{0.8}{x^{1.2}}$ which is negative for $x > 0$.

(c) $MU \to 0$, but never becomes negative. There is always benefit to be gained from study.

PE 6.5

Possible turning points

1. $\dfrac{dy}{dx} = 2x - 6 = 0$:   turning point at $x = 3$.

2. $\dfrac{dy}{dx} = 6x(x - 1)$:   turning point at $x = 0$, $y = 0$ and $x = 1$, $y = -1$.

3. $\dfrac{dP}{dQ} = -4Q + 8$:   turning point at $Q = 2$, $P = 8$.

4. $\dfrac{dC}{dY} = 1$:   no turning point.

5. $\dfrac{dQ}{dP} = -2.5$:   no turning point.

6. $\dfrac{dy}{dx} = 3x^2 - 6x - 9$:   turning point at $x = 3$, $y = -27$; and $x = -1$, $y = 5$.

7. $\dfrac{d(TC)}{dQ} = 144 - \dfrac{1}{Q^2}$:   turning point at $Q = \pm\dfrac{1}{12}$

8. $\dfrac{dy}{dx} = -2x + 1$:   turning point at $x = \dfrac{1}{2}$, $y = 8.25$.

9. $\dfrac{dU}{dx} = \dfrac{4.8}{x^{0.4}}$:   no turning point.

10. $\dfrac{dy}{dx} = 2x^2(2x - 3)$:   possible turning point at $x = 0$, $y = 0$; $x = 3/2$, $y = -26/16$.

11. Possible turning points at $Q = 1$, $TC = 256/3$ and $Q = 3$, $TC = 84$.

12. Possible turning point at $Q = 50$, $TR = 5000$.

PE 6.6

1. $y' = 2x - 6$, $y'' = 2 > 0$ (a minimum).
2. $y' = 6x^2 - 6x$, $y'' = 12x - 6$.
    At $x = 0$, $y'' = -6 < 0$ (a maximum). At $x = 1$, $y'' = 6 < 0$ (a minimum).
3. $P' = -4Q + 8$, $P'' = -4 < 0$ (a maximum).
4. $C' = 1$, $C'' = 0$,   no turning point.

5. $Q' = -2.5$, $Q'' = 0$, no turning point.

6. $y' = 3x^2 - 6x - 9$, $y'' = 6x - 6 = 0$, at $x = 3$, $y'' = 12 > 0$ (a minimum);
at $x = -1$, $y'' = -12 < 0$ (a maximum).

7. $(TC)' = 144 - \frac{1}{Q^2}$, $(TC)'' = \frac{2}{Q^3}$
< 0 when $Q = -\frac{1}{12}$ (a maximum), > 0 when $Q = \frac{1}{12}$ (a minimum).

8. $y' = -2x + 1$, $y'' = -2 < 0$ (a maximum).

9. No turning point

10. $\frac{dy}{dx} = 4x^2(9 - x)$, $y'' = 72x - 12x^2$. Possible turning point at $x = 0$, $y = 0$; $x = 9$, $y = 2187$.
At $x = 0$, $y'' = 0$: By slope test this is not a turning point.
At $x = 9$, $y'' = -324$: This is a maximum point, by slope and second-derivative tests.

11. Maximum at $Q = 1$ and minimum at $Q = 3$.

12. Maximum TR at $Q = 50$.

13. Min. at $x = 10$, $y = -80$

14. Max. at $t = -1$, $y = 10.67$; min. at $t = 5$, $y = -25.33$

15. Min. at $x = 4$, $y = -10$

16. Max. at $x = 0$, $y = -3$; min. at $x = 2$, $y = -13/3$

17. Max. at $x = 0$, $y = 0$; min. at $x = 1$, $y = -2$

18. Max. at $P = -1/8$, $Q = -16$; min. at $P = 1/8$, $Q = 16$

19. Max. at $Q = 0$, $P = 200$ **20.** Max. at $Q = 4$, $P = 200$

21. Max. at $Q = 2$, $C = 8$ **22.** None

PE 6.7

To find the intervals along which a function is increasing or decreasing, first find the turning point, then find the type of turning point.

1. y increases for $0 < x < \infty$
$y'' = 2 > 0$ (a minimum).

2. $(AC)' = 2Q - 20$,   turning point at $Q = 10$.
$(AC)'' = 2 > 0$,   turning point is a minimum.
AC decreases until $Q = 10$ and increases after $Q = 10$.

3. $(TR)' = 50 - 2Q$,   turning point at $Q = 25$.
$(TR)'' = -2 < 0$,   turning point is a maximum.
TR increases until $Q = 25$ and decreases after $Q = 25$.

4. $(AR)' = -2$,   AR decreases for all Q. $(AR)'' = 0$.

5. $(MC)' = 2Q - 18$,   turning point at $Q = 9$.
$(MC)'' = 2 > 0$,   turning point is a minimum.
MC decreases until $Q = 9$ and increases after $Q = 9$.

6. $P' = \dfrac{-0.8}{Q^{0.6}},$ P' is decreasing, no turning point.

$P'' = \dfrac{0.48}{Q^{1.6}},$ slope is increasing (becoming less negative).

7. $(TC)' = Q^2,$ $(TC)' = 0$ at $Q = 0.$
$(TC)'' = 2Q,$ $(TC)'' = 0$ (no max. or min.).
$(TC)' > 0,$ TC is increasing for all $Q.$

8. C increases for $Y > 0.$

PE 6.8

1.

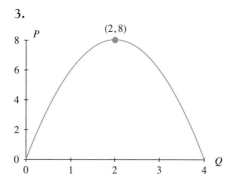

Figure PE 6.8 Q1

2.

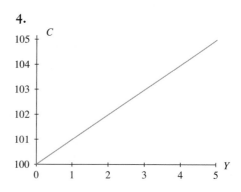

Figure PE 6.8 Q2

3.

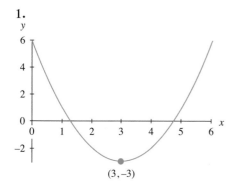

Figure PE 6.8 Q3

4.

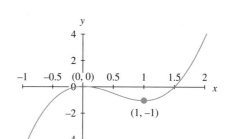

Figure PE 6.8 Q4

SOLUTIONS TO PROGRESS EXERCISES

5.

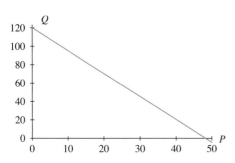

Figure PE 6.8 Q5

6.

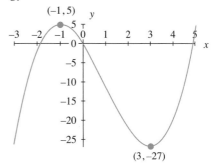

Figure PE 6.8 Q6

7.

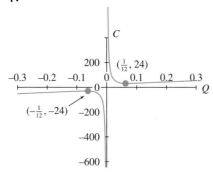

Figure PE 6.8 Q7

8.

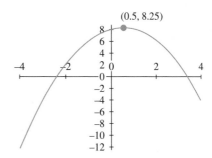

Figure PE 6.8 Q8

9.

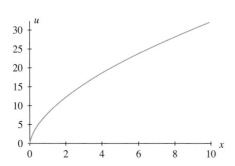

Figure PE 6.8 Q9

10.

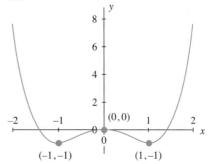

Figure PE 6.8 Q10

PE 6.9

1. (a) TR is a maximum at $Q = 15$ and maximum $TR = £675$.

(b) $AR = 90 - 3Q$, $MR = 90 - 6Q$

For $Q < 15$, AR and MR are positive, but declining. At $Q = 15$, $MR = 0$.

For $Q > 15$, $MR < 0$ while AR remains positive until $Q = 30$.

(c) When $TR = 0$, $AR = 90$ and $MR = 90$.

When TR is a maximum, $AR = 45$ and $MR = 0$.

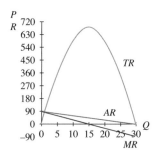

Figure PE 6.9 Q1

2.

	(i) $Q = 150 - 0.5P$	(ii) $P = 80 - 2Q$	(iii) $P = 45$
(a) Total revenue	$300Q - 2Q^2$		$45Q$
(b) TR maximum at Maximum TR	$Q = 75$ $TR = 11\,250$	$80Q - 2Q^2$ $Q = 20$ $TR = 800$	No maximum
(c) $MR = 0$ at $AR = 0$ at	$Q = 75$ $Q = 150$	$Q = 20$ $Q = 40$	MR is never zero AR is never zero

3. (a) $TR = 240Q - 10Q^2$. $\pi = TR - TC = 240Q - 10Q^2 - 120 - 8Q$
$$= 232Q - 10Q^2 - 120$$

(b) (i) $\pi' = -20Q + 232 = 0$. Maximum profit at $Q = 11.6$

(ii) Maximum TR at $Q = 12$

(c) $MR = 240 - 20Q$ and $MC = 8$. $MC = MR$ at $Q = 11.6$

(d) (i) from the graphs and algebraically, break-even at $Q = 0.5$ and $Q = 22.67$

(ii) MC curve intersects MR curve at the point of maximum profit, i.e., $Q = 11.6$.

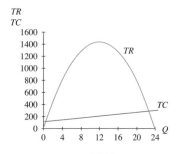

Figure PE 6.9 Q3(d)(i)

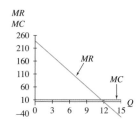

Figure PE 6.9 Q3(d)(ii)

4. (a) $TR = 25Q; TC = 15Q + 8000; MR = 25; MC = 15$

(b) Break-even when $TR = TC : 25Q = 15Q + 8000 \rightarrow Q = 800$

(c) $\pi = TR - TC = 25Q - 15Q - 8000 = 10Q - 8000$.

Maximum profit when $\dfrac{d\pi}{dQ} = 0$, but $\dfrac{d\pi}{dQ} = 10 \neq 0$.

$MR = 25$ and $MC = 15$, therefore $MR \neq MC$, since $MR - MC = 10$.

SOLUTIONS TO PROGRESS EXERCISES

Marginal profit is 10 for each unit produced and sold, etc.

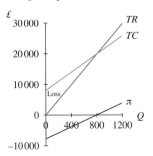

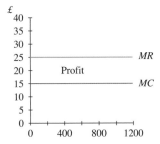

Figure PE 6.9 Q4(d), TR, TC   Figure PE 6.9 Q4(d), MR, MC

5.

Pricing strategy I	Pricing strategy II

Pricing strategy I

(a) $MR = 2374$; $MC = 120$;
$\pi = 2254Q - 608\,580$
(i) No maximum profit since profit function is linear.
(ii) $\pi' = 2254$, therefore can never be zero, so no maximum or minimum.
Comment: since $\pi = 2254Q - 608\,580$ profit increases indefinitely by 2254 for each unit produced.
(b) At $Q = 3365$, $\pi = £6\,976\,130$
(c) Break-even at $Q = 270$

Pricing strategy II

(a) $MR = 5504 - 1.6Q$; $MC = 120$;
$\pi = 5384Q - 0.8Q^2 - 608\,580$
(i) $MR = MC$:
$5504 - 1.6Q = 120 \rightarrow Q = 3365$
At $Q = 3365$,
$P = 5504 - 0.8(3365) = 2812$
$\dfrac{d\pi}{dQ} = 5384 - 1.6Q$
(ii) $= 0 \rightarrow Q = 3365$
$\dfrac{d^2\pi}{dQ^2} = -1.6 < 0$,
thus confirming a maximum.
(b) At $Q = 3365$, $\pi = £8\,450\,000$
(c) Break-even at $Q = 115$ and $Q = 6615$

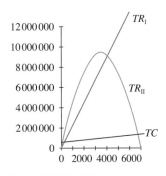

Figure PE 6.9 Q5(c)

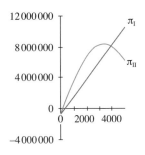

Figure PE 6.9 Q5(d)

Same total revenue occurs at $Q = 0$ and $Q = 3912.5$.
(d) Profits are equal at $Q = 0$ and $Q = 3912.5$.
II is more profitable than I up to the point at which total revenues are equal. After this point, I becomes more profitable than II.
I makes a loss until break-even at $Q = 270$. II makes a loss until break-even at $Q = 115$ and after $Q = 6615$. Choice of pricing strategy depends on the quantity produced.

6. (a) (ii) $y = 5e^x$ increases exponentially with no turning point (since there is no solution to $\frac{dy}{dx} = 0$), crossing the y-axis at $y = 5$. It approaches, but does not touch the negative x-axis.

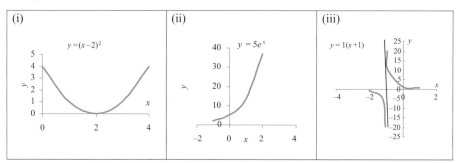

(iii) $y = \frac{1}{x+1}$ has a vertical asymptote at $x = -1$. The graph crosses the y-axis at $y = 1$, it has no turning points (since there is no solution to $\frac{dy}{dx} = 0$), but increases towards the negative x-axis for $x < -1$ and decreasing towards the positive x-axis for $x > -1$.

   (b) (i) $TR = Q(500 - 2.5Q)$ (iii) TR is a max. at $Q = 100$. Maximum $TR = 15\,000$

    (iv) $\varepsilon = -1$

7. (a) (i) $AC = \dfrac{80}{Q} - 20Q + 600$ (ii) $MC = 600 - 40Q$ (iii) $TR = PQ = 110Q - 0.5Q^2$

    (iv) $MR = 110 - Q$ (v) Profit $= 19.5Q^2 - 490Q - 80$

   (b) Maximum cost at $Q = 15$. The maximum cost $= 4580$. When cost is a maximum profit $= -3042.5$ (a loss!).
   (c) Minimum profit at $Q = 12.56$, hence profit is increasing for $Q > 12.5$. The minimum profit $= -3158.3$.

8. (a) See text (b) Market A, $P = 9.5$ at $Q = 16.5$. Market B, $P = 9.2$ at $Q = 34$

9. When AVC is a minimum, $MC = AVC = -b^2/(4a) + c$

10. Profit is maximised at $Q = \dfrac{a - e}{2(b + d)}$

PE 6.10

Point of inflection (PoI) at $y'' = 0$; plus change in sign before and after,

curvature : $y'' > 0$, concave up.
curvature : $y'' < 0$, concave down.

1. $y' = 2x + 2$
       $y'' = 2$ (no point of inflection)
       $y'' > 0$ (concave up)

2. $y' = -3x^2 + 5$
       $y'' = -6x$; $y'' = 0$ at $x = 0$ (PoI)
       $y'' < 0$ for $x > 0$ (concave down)

3. $y' = 12x^3 + 2x$
       $y'' = 36x^2 + 2$ (no point of inflection)
       $y'' > 0$ (concave up)

4. $L' = \dfrac{0.125}{K^{0.75}}$

       $L'' = -\dfrac{0.093\,75}{K^{1.75}} < 0$ for $k > 0$ (no PoI)

       $L'' < 0$ (concave down)

5. $(TC)' = 3Q^2 - 12Q + 3$
$(TC)'' = 6Q - 12$ (PoI at $Q = 2$)
$(TC)'' > 0$ for $Q > 2$ (concave up)
$(TC)'' < 0$ for $Q < 2$ (concave down)

6. $(TC)' = -4Q^3$
$(TC)'' = -12Q^2$ (no PoI)
$(TC)'' < 0$ for all Q (concave down)

7. $(AC)' = 2Q - \dfrac{2}{Q^2}$

$(AC)'' = 2 + \dfrac{4}{Q^3}$ (PoI at $Q = -1.26$)

$(AC)'' > 0$ for all $Q > 0$ (concave up)

8. $(TC)' = 120 - 1.6Q$
$(TC)'' = -1.6$ (no PoI)
$(TC)'' < 0$ (concave down)

9. Simplify: $y = \dfrac{1}{x} + \dfrac{2}{x^2} - \dfrac{2}{x^2} - \dfrac{1}{x}$

$y' = -\dfrac{1}{x^2}$

$y'' = \dfrac{2}{x^3}$ (no point of inflection)

$y'' > 0$ for $x > 0$ (concave up)

10. $(AFC)' = -\dfrac{100}{Q^2}$

$(AFC)'' = \dfrac{200}{Q^3}$ (no PoI)

$(AFC)'' > 0$ for $Q > 0$ (concave up)

PE 6.11

1. (a) $MP_L = \dfrac{dQ}{dL} = -6L^2 + 24L$. $APL = -2L^2 + 12L$

At $L = 1.5$, $MP_L = 22.5 > APL = 13.5$. See text for further description.
(b) MP_L is a maximum at $L = 2$, $MP_L = 24$. APL is maximum at $L = 3$, $APL = 18$. When $L = 3$, $MP_L = 18$, point of intersection.
Graph similar to Figure 6.36(b)
(c) Turning points: $L = 0$ (min.); $L = 4$ (max.). Point of inflection: $L = 2$. MP_L increases when $L < 2$, decreases when $L > 2$, etc.
Graph similar to Figure 6.36(b)
(d) Left to the reader to describe the results in (a), (b) and (c) in terms of quantity of sandwiches and number of labour-hours hired.

2. (a) $MC = Q^2 - 60Q + 2800$. $AC = \dfrac{Q^2}{3} - 30Q + 2800 + \dfrac{900}{Q}$

$AVC = \dfrac{Q^2}{3} - 30Q + 2800$. $AFC = \dfrac{900}{Q}$

(b) $\dfrac{d(MC)}{dQ} = 2Q - 60 = 0 \rightarrow Q = 30$. $\dfrac{d^2(MC)}{dQ^2} = 2 > 0$,

minimum $MC = 1900$ at $Q = 30$.
(c) Graphs similar to those in text (Figure 6.38), MC and AVC intersect at $Q = 45$, the minimum point on AVC.

3. (a) Plot the graph (Excel or otherwise). From the graph, TR curve > TC curve between $Q = 6$ and $Q = 41$ (approx.). Therefore, the company makes a profit when output is between $Q = 6$ and $Q = 41$.
(b) $\pi = -\dfrac{1}{3}Q^3 + 13Q^2 + 56Q - 750$

(i) $\dfrac{d\pi}{dQ} = -Q^2 + 26Q + 56$; $\dfrac{d^2\pi}{dQ^2} = -2Q + 26$

Maximum π at $Q = 28$, $\pi = 3692.67$; minimum π at $Q = -2$, $\pi = -807.33$ (loss).
(ii) At maximum (or minimum) profit $MR = MC$ and

$$(MR)' > (MC)' \text{ for a minimum}$$

$$(MR)' < (MC)' \text{ for a maximum}$$

(c) Marginal profit, $\dfrac{d\pi}{dQ} = -Q^2 + 26Q + 56$. Maximum $\pi = 225$ at $Q = 13$.

Marginal π is a max. at $Q = 13$ (PoI), increasing for $Q < 13$ and decreasing for $Q > 13$.

4. (a) Graph similar to Figure 6.27, hence the firm makes a profit when $TR > TC$ between $Q = 21$ and $Q = 101$ (approx.).

(b) $\pi = -\dfrac{1}{3}Q^3 + 36Q^2 - 140Q - 9900$

(i) Maximum π at $Q = 70$, $\pi = £42\,366.67$;
Minimum π at $Q = 2$, $\pi = -£10\,038.67$

(ii) Optimum profit when $MR = MC$, but use second-order conditions to classify points as max. or min.

PE 6.12

1. $10e^x$

2. $\dfrac{e^x}{10}$

3. $-\dfrac{243}{x^2} + \dfrac{1}{x}$

4. $\dfrac{20}{Q}$

5. $\dfrac{1}{3L^{2/3}}$

6. $-150e^t$

7. $104.4e^t$

8. $\dfrac{1}{Q} - \dfrac{2}{Q^2}$

9. $\dfrac{1}{12P} + \dfrac{3}{P^{3/2}}$

10. $10.85e^t$

11. $-\dfrac{150}{Y^2} + 0.8e^Y$

12. $2x - 5e^x + \dfrac{1}{x}$

13. $e^y - \dfrac{12}{Y^2}$, $e^y + \dfrac{24}{Y^3}$

14. $y = \ln x - 12$, $\dfrac{dy}{dx} = \dfrac{1}{x}$

15. $\dfrac{dt}{dY} = \dfrac{1}{Y} \to \dfrac{dY}{dt} = Y = e^{15+t}$; $\dfrac{d^2Y}{dt^2} = e^{15+t}$

16. $Y = e^{x+5} - 10 \to e^x.e^5 - 10$; $\dfrac{dy}{dx} = e^x.e^5$, $\dfrac{d^2y}{dx^2} = e^x.e^5$

17. $4e^t + 2e^{2t}$

18. $\dfrac{e^x}{2} + \dfrac{1}{2}$

19. $4e^{4x}$

20. $2 - 4x$

21. $2e^{2t} - 3e^{3t}$

22. $1.6e^{-0.4t}$

PE 6.13

1. $14(2x - 5)^6$

2. $-15(4 - 5x)^2$

3. $\dfrac{4}{(1 - 0.8x)^6}$

4. $-\dfrac{5}{2(5x + 12)^{3/2}}$

5. $\dfrac{Q}{\sqrt{Q^2 + 12}}$

6. $\dfrac{(5Q^4 + 6Q)}{2\sqrt{Q^5 + 3Q^2}}$

7. $\dfrac{0.85(2L + 3)}{3(L^2 + 3L)^{2/3}}$

8. $-\dfrac{1200}{(P + 5)^2}$

9. $9.72e^{0.81x}$ **10.** $1333.07e^{1.09t}$ **11.** $-1120e^{-1.4Q}$ **12.** $222e^{-1.2t}$

13. $-2.5e^{2.5t}$ **14.** $-\dfrac{320}{e^{4t}}$ **15.** $-\dfrac{15e^t}{(1+e^t)^2}$ **16.** $26.4e^{0.88Y}$

17. $\dfrac{2Y}{Y^2+4}$ **18.** $\dfrac{500(3K^2+8)}{K^3+8K}$ **19.** $-\dfrac{2}{(x+2)^2}-\dfrac{2}{2x+2}$

20. $P = \ln\left(\dfrac{Q^2-2Q}{Q}\right) = \ln(Q-2),\ \dfrac{dP}{dQ} = \dfrac{1}{Q-2}$ **21.** $-5014.375e^{-0.5t}$

22. $C = 425 + 1.2[\ln Y - \ln(Y+2)],\ \dfrac{dC}{dY} = 1.2\left[\dfrac{1}{Y}-\dfrac{1}{Y+2}\right]$

23. $161.2e^{-0.65Y}$ **24.** $-\dfrac{2}{Q^{3/2}}+\dfrac{5}{Q}$

PE 6.14

1. $1+\ln x$ **2.** $50-3\sqrt{Q}$ **3.** $e^t(2t+t^2)$ **4.** $-\dfrac{1}{Q^2}-1-\ln Q$

5. $\sqrt{L+5}+L\left(\dfrac{1}{2\sqrt{L+5}}\right)$ **6.** Simplify: $y = x^{1.5}9^{1/2}x^{1/2} = 3x^2,\ \dfrac{dy}{dx} = 6x$

7. $-e^Q - Qe^Q$ **8.** $\sqrt{1+2Q}+Q\left(\dfrac{1}{\sqrt{1+2Q}}\right)$ **9.** $100e^{-0.5Y}(1-0.5Y)$

10. $\dfrac{1}{Y}\cdot\dfrac{1}{(Y+8)}-\dfrac{1}{Y^2}\ln(Y+8)$ **11.** $\dfrac{\sqrt{Y}}{Y}+\dfrac{1}{2\sqrt{Y}}\ln(8Y)$ **12.** $\dfrac{3}{20}e^{x-5}(1+x)$

13. $\dfrac{Q^2}{5}+\sqrt{3Q-1}+\dfrac{3Q}{2\sqrt{3Q-1}}-1$ **14.** $-\sqrt{5}\left(\dfrac{\sqrt{Q}}{Q}+\dfrac{1}{2\sqrt{Q}}\ln Q\right)$

15. $(3t-5)^{-4}-12t(3t-5)^{-5}$

16. Simplify: $x = y^3 3\ln y = 3(y^3\ln y),\ \dfrac{dy}{dx} = 3\left(3y^2\ln y+\dfrac{y^3}{y}\right) = 3y^2(3\ln y+1)$

PE 6.15

1. $\dfrac{1-\ln x}{x^2}$ **2.** $\dfrac{-100}{(50+Q)^2}$

3. $\dfrac{2t(t^2-1)-2t(t^2+1)}{(t^2-1)^2} = \dfrac{2t(t^2-1-t^2-1)}{(t^2-1)^2} = \dfrac{-4t}{(t^2-1)^2}$

4. $\dfrac{1}{Q^2}\left(\dfrac{Q}{2\sqrt{Q+1}}-\sqrt{Q+1}\right)$ **5.** $-\left(\dfrac{(Q+1)-Q}{(Q+1)^2}\right) = -\left(\dfrac{1}{(Q+1)^2}\right)$

6. $919e^{-0.8Y}\left(\dfrac{0.8Y+1}{Y^2}\right)$

7. Simplify: $y = 2\ln(x) + \dfrac{1}{x};\ \dfrac{dy}{dx} = \dfrac{2}{x} - \dfrac{1}{x^2}$

8. $205\left(\dfrac{1-\ln Y}{(Y)^2}\right)$

PE 6.16

1. (a) $TR = (24 - 6\ln Q)Q$. TR is maximised at $Q = 20.09$.
   (b) $MR = 18 - 6\ln Q$. Rate of change of $MR = \dfrac{d(MR)}{dQ} = -\dfrac{6}{Q} < 0$.
      $(MR)'$ is negative for $Q > 0$. Therefore MR is decreasing for $Q > 0$.
      $(MR)'' = \dfrac{6}{Q^2} > 0$, therefore concave up.

2. (a) MPC $= 160e^{-0.2Y}$. Decreases as Y increases.
   (b) Maximum consumption at $dC/dY = 0$. That is, at $160e^{-0.2Y} = 0$. However, there is no value of Y to satisfy this condition; therefore, no max. or min.
   (c) Curvature: concave down. $C'' = -32e^{-0.2Y}$.

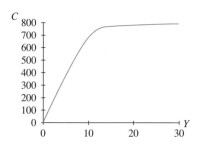

Figure PE 6.16 Q2

3. (a) $U = 20xe^{-0.1x}$. U is a maximum at $x = 10$ and $U = 73.58$.
   (b) $MU_x = 20e^{-0.1x}(1 - 0.1x)$. $(MU)' = 4e^{-0.1x}(0.05x - 1)$.
      When $x > 10$, MU_x is declining; enjoyment from wine is diminishing after 10 glasses.

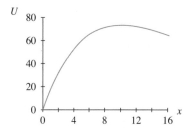

Figure PE 6.16 Q3

SOLUTIONS TO PROGRESS EXERCISES

4. (a) $TR = \dfrac{50Q}{e^{0.05Q}} = 50Qe^{-0.05Q}$. $MR = 50(-0.05Q + 1)e^{-0.05Q}$

(b) Maximum TR when $Q = 20$, $P = 18.39$ and $TR = 367.88$.

5. (a) $U = 95 - \ln(t + 5)$.

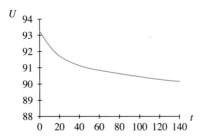

Figure PE 6.16 Q5

U (enjoyment) declines with time.

(b) $MU_x = -\dfrac{1}{(t + 5)} < 0$; $(MU_x)' = \dfrac{1}{(t + 5)^2} > 0$: enjoyment declines at an increasing rate.

6. $AC = \dfrac{100e^{0.05(Q-10)}}{Q + 1}$ (a) Minimum AC at $Q = 19$ and $AC = 7.842$

(b) $TC = AC \times Q = \dfrac{100Qe^{0.05(Q-10)}}{Q + 1}$

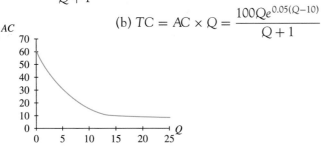

Figure PE 6.16 Q6

7. (a) (i) $4.5c^{3.5} - \dfrac{1}{\sqrt{x}}$ (ii) $10e^{\frac{t}{8}}$ (b) (i) $TR = PQ = (8000 - 20Q)Q$: $MR = 8000 - 40Q$

(b) (ii) Maximum $TR = 800\,000$ at $Q = 200$

8. (a) (i) $\dfrac{dy}{dt} = 40e^{-2t}$ (ii) $\dfrac{dQ}{dL} = 2.4L^{-0.2} = \dfrac{L^{\frac{1}{3}}}{3} = \dfrac{2.4}{L^{0.2}} - \dfrac{1}{3\sqrt{L^3}}$

(b) (i) $TR = PQ = 1600Q - 4Q^2$ and $MR = 1600 - 8Q$ (ii) TR is maximised at $Q = 200$

(c) (i) $TC = 500 + 3Q$ and $MC = 3$. (ii) Profit $= TR - TC = 1597Q - 4Q^2 - 500$

(iii) Profit is maximised when $Q = 199.625$; maximum profit $= £158\,900.5625$. Hence, in practical terms, 200 T-shirts must be produced and sold to give a maximum profit $= £158\,900$.

9. (a) (i) $\dfrac{dy}{dx} = \dfrac{2x + 5}{x^2 + 5x}$. Maximum at $x = -2.5$ (ii) $\dfrac{dy}{dx} = e^x(4 - x)$. Minimum at $x = -4$

9. (b) (i) Minimum AC = 14.75 at $Q = 10$. AC decreasing over the interval $0 < AC < 10$

(ii) $TC = 125 + \dfrac{1}{16}Q^3 - 4Q$. The minimum value of $TC = 112.68$ at $Q = 4.62$

(c) With the subsidy, the total cost function is given as $TC = 125 + \dfrac{1}{16}Q^3 - Q$. The subsidised cost function has a minimum value of 123.46 at $Q = 2.31$.

10. (a) $\dfrac{dS}{dt} = -10\,000e^{-0.05t}$ (b) There is no value of t that satisfies $\dfrac{dS}{dt} = -10\,000e^{-0.05t} = 0$, hence there is no maximum. As t increases, $e^{-0.05t}$ tends to zero, hence S approaches 200 000.

11. (a) $\dfrac{dN}{dt} = \dfrac{63\,200e^{-0.1t}}{(1 + 790e^{-0.1t})^2}$ (b) There is no solution to $\dfrac{dN}{dt} = \dfrac{63\,200e^{-0.1t}}{(1 + 790e^{-0.1t})^2} = 0$, hence no maximum. PoI at $\dfrac{d^2N}{dt^2} = \dfrac{6320e^{-0.1t}(790e^{-0.1t} - 1)}{(1 + 790e^{-0.1t})^3} = 0$, at $t = 66.72$ days. The spread of the virus increases at an increasing rate for the first 66.72 days; thereafter the spread continues to increase but at a decreasing rate to a limiting value of 800.

PE 6.17

1. (a) $\varepsilon_d = \dfrac{\text{\% change in quantity}}{\text{\% change in price}} = \dfrac{(\Delta Q/Q) \times 100}{(\Delta P/P) \times 100} = \dfrac{\Delta Q}{\Delta P}\dfrac{P}{Q} \rightarrow \dfrac{dQ}{dP}\dfrac{P}{Q}$, at a point

(b) $\varepsilon_d = \dfrac{(\text{\% change in } Q)}{(\text{\% change in } P)} \rightarrow -0.8 = \dfrac{(\text{\% change in } Q)}{(5\%)} \rightarrow$ % change in $Q = -4\%$

If ε_d is constant, the demand function is of the form $Q = a/P^{0.8}$

2. (a) (i) $\varepsilon_d = -\dfrac{1}{2}\dfrac{P}{Q}$, $1 - \dfrac{40}{Q}$, $\dfrac{P}{P - 80}$ (ii) $\varepsilon_d = -4\dfrac{P}{Q}$, $1 - \dfrac{120}{Q}$, $\dfrac{P}{P - 30}$

(iii) $\varepsilon_d = \infty$ (iv) $\varepsilon_d = -\dfrac{1}{b}\dfrac{P}{Q}$, $1 - \dfrac{a}{b}\dfrac{1}{Q}$, $\dfrac{P}{P - a}$

(b) (i) -1.67; -0.33 (ii) 2.5; -3 (iii) ∞; ∞ (iv) $\dfrac{50}{50 - a}$; $1 - \dfrac{a}{b}\dfrac{1}{30}$

3. (a) $\varepsilon_d = -\dfrac{100}{Q}$, when $Q = 100$, $\varepsilon_d = -1$: when $Q < 100$, $\varepsilon_d < -1$ so demand is elastic, the % change in Q is greater than the % change in P : when $Q > 100$, $-1 < \varepsilon_d < 0$ so demand is inelastic, the % change in Q is smaller than the % change in P.

(b) When $Q = 150$, $P = 111.57$. Since $\varepsilon_d = -\dfrac{100}{150}$ at $Q = 150$, the response to a 10% increase in P is a 6.67% decrease in Q.

4. (a) $\varepsilon_d = \dfrac{2P^2}{P^2 - 192}$. When $\varepsilon_d = -1$, $P = £8$, $Q = 128$ seats available.

(b) Ten seats is an increase of 7.8125% in Q. P will drop by 7.8125% approx., new price $= £7.348$ approximately.

5. (a) See Worked Example 6.42 (b) (i) $\varepsilon_d = -1.2$, (ii) decrease by 6% approximately, (iii) increase by 6% approximately.

6. (a) $Q = 20\,260$ (b) If P increases by 5%, the new price $= £31.5$ with the corresponding demand, $Q = 19\,108$. Percentage decrease in Q is 5.686%. The approximate % decrease in 5 (b) (ii) is 6%. This is approximate because the elasticity as well as the values of P and Q change over the interval $P = £30$ to £31.5, but the formula uses the slope and the ratio P/Q at the start of the interval to calculate the % change in Q throughout the interval.

7. (a) $TR = 50Q - 0.5Q^2$: $MR = 50 - Q$: $AR = 50 - 0.5Q$. (b) Max. TR at $Q = 50$, $P = 25$
(c) $\varepsilon_d = -2\dfrac{P}{Q}$ (i) $\dfrac{P}{P-50}$ (ii) $1 - \dfrac{100}{Q}$ (d) See text for proof.

8. (a) Graph similar to Figure 6.41. (b) $\varepsilon_d = -1$ at $P = 25$ (see Figure 6.41).

9. (a) $TR = 1500Qe^{-0.025Q}$ $MR = 1500(1 - 0.025Q)e^{-0.025Q}$ $AR = 1500e^{-0.025Q}$
(b) $Q = 40$, $P = €551.82$ (c) $\varepsilon_d = -\dfrac{40}{Q}$: $\varepsilon_d = -\dfrac{1}{7.31 - \ln P}$
(c) Max. $TR = €22\,072.8$ at $\varepsilon_d = -1$, hence $Q = 40$, $P = 551.82$

10. (a) No, $\varepsilon_d = -\dfrac{1}{b}\dfrac{P}{Q} = \dfrac{P}{P-a}$ changes as P changes. (b) (i) $MR = -5044.44$,
(ii) $P = 95\,844.44 - 1056.05Q$ (iii) max. $TR = 2\,174\,710$ at $P = 47\,922.22$, $Q = 45.38$

11. (a) See equation (6.36) with $MR = MC$ for max. profit. (b) Max. profit at $P = 151$, $Q = 149$, $\varepsilon_d = -151/149$; substitute $P = 151$, $MC = 2$, $\varepsilon_d = -(151/149)$ into equation in (a) to confirm its validity: take any other point on the demand function, such as $P = 100, Q = 200$, $\varepsilon_d = -(100/200)$, $MC = 2$, to confirm that the equation in (a) is not true.

Chapter 7

PE 7.1

1. $z_x = 2x + 2y$; $z_y = 2x$
2. $z_x = 10xy^3$; $z_y = 15x^2y^2$
3. $Q_x = \dfrac{1}{x}$; $Q_y = \dfrac{3}{y}$
4. $Q_L = 8L^{-0.2}K^{0.2}$; $Q_K = 2L^{0.8}K^{-0.8}$
5. $P_t = 96e^{0.8t}$
6. $U_x = 8xy^2$; $U_y = 8x^2y$
7. $z_x = 2x(1 + 2y)$; $z_y = 2x^2$
8. $z_x = 5y - \dfrac{y}{x^2}$; $z_y = 5x + \dfrac{1}{x}$
9. $Q_x = \dfrac{1}{x} + 3\ln(y)$; $Q_y = \dfrac{3x}{y}$

10. $Q_L = 8L^{-0.2}K^{0.2} - 6$; $Q_K = 2L^{0.8}K^{-0.8} - 4$ 11. $C_Q = -96e^{-0.8Q}$
12. $Z_L = 0.5L^{-0.5}K^{0.5} + 2\lambda$; $Z_K = 0.5L^{0.5}K^{-0.5} + 3\lambda$; $Z_\lambda = -(50 - 2L - 3K)$
13. Proof:

$$Q = LQ_L + KQ_K$$
$$Q = L(0.5L^{-0.5}K^{0.5}) + K(0.5L^{0.5}K^{-0.5})$$
$$Q = 0.5L^{0.5}K^{0.5} + 0.5L^{0.5}K^{0.5}$$
$$Q = L^{0.5}K^{0.5}$$

14. (a) $K = \dfrac{200}{L}$

    (b) $L = 5$, $K = 40$; $L = 10$, $K = 20$;

    $L = 15$, $K = 13.33$; $L = 20$, $K = 10$.

    e.g., $Q = 2LK = 2(5)(40) = 400$

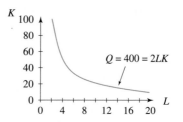

Figure PE 7.1 Q14

15. (a) $K = \dfrac{100}{L^2}$

    (b) $L = 5$, $K = 4$; $L = 10$, $K = 1$;

    $L = 15$, $K = 0.44$; $L = 20$, $K = 0.25$.

    e.g., $Q = 2LK^{0.5} = 2(5)(4)^{0.5} = 20$

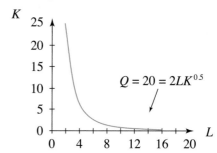

Figure PE 7.1 Q15

PE 7.2

1. $z_{xx} = 2$; $z_{yy} = 0$; $z_{xy} = 2$

2. $z_{xx} = 10y^3$; $z_{yy} = 30x^2y$; $z_{xy} = 30xy^2$

3. $Q_{xx} = -\dfrac{1}{x^2}$; $Q_{yy} = -\dfrac{3}{y^2}$; $Q_{xy} = 0$

4. $Q_{LL} = -1.6L^{-1.2}K^{0.2}$; $Q_{KK} = -1.6L^{0.8}K^{-1.8}$; $Q_{Lk} = 1.6L^{-0.2}K^{-0.8}$

5. $P_{tt} = 76.8e^{0.8t}$

6. $U_{xx} = 8y^2$; $U_{yy} = 8x^2$; $U_{xy} = 16xy$

7. $z_{xx} = 2(1 + 2y)$; $z_{yy} = 0$; $z_{xy} = 4x$

8. $z_{xx} = \dfrac{2y}{x^3}$; $z_{yy} = 0$; $z_{xy} = 5 - \dfrac{1}{x^2}$

9. $Q_{xx} = -\dfrac{1}{x^2}$; $Q_{yy} = -\dfrac{3x}{y^2}$; $Q_{xy} = \dfrac{3}{y}$

10. $Q_{LL} = -1.6L^{-1.2}K^{0.2}$; $Q_{KK} = -1.6L^{0.8}K^{-1.8}$; $Q_{LK} = 1.6L^{-0.2}K^{-0.8}$

11. $C_{QQ} = 76.8e^{-0.8Q}$

12. $Z_{LL} = -0.25L^{-1.5}K^{0.5}$; $Z_{KK} = -0.25L^{0.5}K^{-1.5}$; $Z_{\lambda\lambda} = 0$

    $Z_{LK} = 0.25L^{-0.5}K^{-0.5}$; $Z_{KL} = 0.25L^{-0.5}K^{-0.5}$; $Z_{K\lambda} = 3$ $Z_{L\lambda} = 2$

SOLUTIONS TO PROGRESS EXERCISES

PE 7.3

1. $dz = (2x)dx$ **2.** $dz = (2x)dx + (2y)dy$ **3.** $dz = (3e^x)dx + dy$

4. $dz = (2L^{-0.8}K^{0.6})dL + (6L^{0.2}K^{-0.4})dK$ **5.** $dz = \left(\frac{1}{x}\right)dx + \left(\frac{5}{y}\right)dy$

6. $\Delta Q = 3\%Q$ **7.** $\Delta U = 4\%U$ **8.** $\Delta(TR) = 0.6\%TR$

9. $\Delta(TR) = 1.5\%TR$ **10.** $\Delta(TR) = 4.4\%TR$ **11.** $\Delta Q = 0.06\%Y + 0.9\%P_s$

12. Proof : $z = 0.5(e^{x+y} + e^{x+y}) = z$ **13.** 5%

14. (a) $z_x = 2x + 5y$, $z_{xx} = 2$, $z_y = 5x + 2y$, $z_{yy} = 2$, $z_{xy} = 5$ (b) Profit is reduced by 15.8%

15. (a) $U_x = 75x^{-0.7}y^{0.7}; U_{x,x} = -52.5x^{-1.7}y^{0.7}; U_y = 175x^{0.3}y^{-0.3};$

$U_{y,y} = -52.5x^{0.3}y^{-1.3}; U_{x,y} = 52.5x^{-0.7}y^{-0.3}$

(b) U increases by 0.6%

16. 4.76% increase in U

PE 7.4

1. (a) $MP_L = 5L^{-0.5}K^{0.5}; \quad MP_K = 5L^{0.5}K^{-0.5}$

(b) $Q = L(5L^{-0.5}K^{0.5}) + K(5L^{0.5}K^{-0.5}) = 5L^{0.5}K^{0.5} + 5L^{0.5}K^{0.5} = 10L^{0.5}K^{0.5}$

2. $Q_L = 40L^{-0.5}K^{0.2}; Q_K = 16L^{0.5}K^{-0.8};$

$Q_{LL} = -20L^{-1.5}K^{0.2}; Q_{KK} = -12.8L^{0.5}K^{-1.8}; Q_{LK} = 8L^{-0.5}K^{-0.8}$

(a) See text.

(b) Q_L and Q_K both decrease: their derivatives Q_{LL} and Q_{KK} are negative.

(c) When $Q = 80 : 80 = 80L^{0.5}K^{0.2} \rightarrow K^{0.2} = \dfrac{1}{L^{0.5}} \rightarrow K = \dfrac{1}{L^{2.5}}$

L	2	5	10	
K	0.18	0.018	0.003	
MP_L	20.07	8.01	3.96	decreases as L increases
MP_K	89.2	890	5277	decreases as K increases

3. (a) $MP_L = 40L^{-0.5}K^{0.2}; APL = 80L^{-0.5}K^{0.2} : APL > MP_L$

(b) $MP_K = 16L^{0.5}K^{-0.8}; APK = 80L^{0.5}K^{-0.8} : APK > MP_K$

4. (a) $K = \dfrac{64}{L^2}$

L: 2, 4, 6

K: 16, 4, 1.78

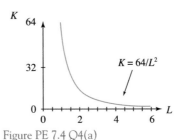
Figure PE 7.4 Q4(a)

(b) $Q_L = MP_L = 10K^{0.5}$

$Q_K = MP_K = 5LK^{-0.5}$

L	2	4	6	
K	16	4	1.78	
MP_L	40	20	13.34	decreases as L increases
MP_K	2.5	10	22.49	increases as L increases (decreases as K increases)

(c) $-MP_L/MP_K$ -16 -2 -0.59 magnitude decreasing confirming law of diminishing MRTS

5. (a) See text: $\dfrac{dK}{dL} = -\dfrac{MP_L}{MP_K} = -2.5\dfrac{K}{L}$

(b) $K = \dfrac{1024}{L^{2.5}}$

L	4	9	16	
K	32	4.21	1	
$\dfrac{dK}{dL} = -2.5\dfrac{K}{L}$	–20	–1.17	–0.16	magnitude decreasing, confirming law of diminishing MRTS

6. MRTS $= \dfrac{dK}{dL} = -\dfrac{Q_L}{Q_K} = -\dfrac{15L^{-0.4}K^{0.5}}{12.5L^{0.6}K^{-0.5}} = -1.2\dfrac{K}{L}$

Given $\alpha = 0,\ \beta = 0.5$: $MRTS = -\dfrac{\alpha K}{\beta L} = -\dfrac{0.6\ K}{0.5\ L} = -1.2\dfrac{K}{L}$, as above.

7. Q decreases by 0.5%.

8.

L	2	4	6	
$APL = 50L^{-0.7}K^{0.5}$	$30.77K^{0.5}$	$18.95K^{0.5}$	$14.26K^{0.5}$	decreases as L increases
$MP_L = 15L^{-0.7}K^{0.5}$	$9.23K^{0.5}$	$5.68K^{0.5}$	$4.28K^{0.5}$	decreases as L increases
				$MP_L < APL$

PE 7.5

1. (a) $U_x = 2x^{-0.8}y^{0.8}$; $U_y = 8x^{0.2}y^{-0.2}$

(b) $U = xU_x + yU_y = 2x^{0.2}y^{0.8} + 8x^{0.2}y^{0.8} = 10x^{0.2}y^{0.8} = U$

2. (a) $y = \dfrac{50}{2 + \ln x}$

(b) $(x = 2,\ y = 18.57)$;
$(x = 10,\ y = 11.62)$
$(x = 25,\ y = 9.58)$

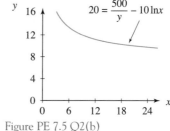

Figure PE 7.5 Q2(b)

(c) Slope $= \dfrac{dy}{dx} = \dfrac{-50}{x(2 + \ln x)^2}$

x	2	10	25
Slope	–3.45	–0.27	–0.07

slope is decreasing in magnitude; diminishing MRS

3. (a) $U_x = -\dfrac{10}{x}$; $U_y = -\dfrac{500}{y^2}$

(b) $\dfrac{dy}{dx} = -\dfrac{U_x}{U_y} = -\dfrac{10}{x} \cdot \dfrac{y^2}{500} = -\dfrac{10}{x} \cdot \dfrac{1}{500}\left(\dfrac{50}{2 + \ln x}\right)^2 = \dfrac{50}{x(2 + \ln x)^2}$, as above.

4. (a) $MU_x = 12.5x^{-0.5}y^{0.2}; \ MU_y = 5x^{0.5}y^{-0.8}$

(b) $MRS = \dfrac{dy}{dx} = -\dfrac{U_x}{U_y} = -2.5\dfrac{y}{x}$

(c) $\alpha = 0.5, \ \beta = 0.2: \ MRS = -\dfrac{\alpha y}{\beta x} = -\dfrac{0.5}{0.2}\dfrac{y}{x} = -2.5\dfrac{y}{x}$

5. Utility decreases by -7.4%.

6. (a) Constant, $\alpha + \beta = I$ (b) Increasing, $\alpha + \beta < 1$ (c) Decreasing, $\alpha + \beta < 1$

7. (a) $Q = 371.45$

(b) $L = 55, K = 220$, hence $Q = 420.44$. A 10% increase gives $Q = 408.6$. Therefore, the production function exhibits increasing returns to scale.

(c) Q increases by 9.5%.

PE 7.6

1. $\varepsilon_d = -0.25; \quad \varepsilon_Y = 0.5243; \quad \varepsilon_c = 0.076$

2. (a) $MP_L = 150L^{-0.4}K^{0.2}; \quad APL = 250L^{-0.4}K^{0.2}$

(b) $\varepsilon_{QL} = 0.6; \quad$ (c) $\varepsilon_{QL} = \dfrac{MP_L}{APL} = \dfrac{150L^{-0.4}K^{0.2}}{250L^{-0.4}K^{0.2}} = 0.6$

3. (a) $Y_e = 800$

(b) (i) investment multiplier $= 1.82$, (ii) government expenditure multiplier $= 1.82$

(iii) income tax rate multiplier $= -872.73$

4. Production function I: $\varepsilon_{QL} = 0.6; \varepsilon_{QK} = 0.4$

(a) $\varepsilon_{QL} = \dfrac{\% \text{ change in } Q}{\% \text{ change in } L} \rightarrow \% \text{ change in } Q \cong 0.6(3) = 1.8\%$

(b) $1.8\% \ (\Delta K = 0)$

Prodution function II: $\varepsilon_{QL} = 0.4; \varepsilon_{QK} = 0.3$

(a) % change in $Q = \varepsilon_{QK}(5) = 1.5\%$

(b) Q increases by 1.5%. Answers (a) and (b) are the same.

5. (a) $MP_K = 50L^{0.6}K^{-0.8}; \quad APK = 250L^{0.6}K^{-0.8}$

(b) $\varepsilon_{QK} = 0.2$ (c) $\varepsilon_{QK} = \dfrac{MP_K}{APK} = \dfrac{50L^{0.6}K^{-0.8}}{250L^{0.6}K^{-0.8}} = 0.2$

PE 7.7

1. Min. $= -61$ at $(x = 5, \ y = 6)$ **2.** Max. $= 116$ at $(x = 10, \ y = 4)$

3. Min. $= -68$ at $(x = 6, \ y = 2)$ **4.** Min. $= -108$ at $(x = 6, \ y = 6)$

5. Max. $= 15.78$ at $(L = -50, \ K = 5)$

6. Min. $= 147$ at $(x = 3, \ y = 2)$;   Max. $= 280$ at $(x = -3, \ y = -2)$;

Saddle points $= 172$ and 248 at $(x = 3, \ y = -2)$ and $(x = -3, \ y = 2)$

PE 7.8

1. Maximum at $(x = 6, y = 23)$

2. (a) $TR = 320x + 106y - 4x^2 - 20y^2 + 4xy$

(b) $x = 43.5$; $P_X = 160$; $y = 7$; $P_Y = 53$; $TR = 7331$

3. (a) $\pi = 280x + 80y - 4x^2 - 20y^2$

(b) Maximum profit $= 4980$ when $x = 35$; $P_X = 184$; $y = 2$; $P_Y = 136$; $TR = 6712$

4. (a) $TR = 190P_1 + 120P_2 - 4P_1^2 - 3P_2^2 - 3P_1P_2$

(b) $P_1 = 20$; $Q_1 = 100$; $P_2 = 10$; $Q_2 = 50$

(c) Maximum revenue $= 2500$

5. Outline proof: $TC_{Q1} = 4Q_2 + 10 = 0$; $TC_{Q2} = 4Q_1 + 15 = 0$.

First-order conditions give negative values of Q.

6. Minimum cost at $(x = 100, y = 7)$.

7. Maximum at $(x = 4, y = 2)$.

8. (a) $TR = 8x + 20y - 2x^2 - 0.5y^2$; $\pi = 4x + 14y - 2x^2 - 0.5y^2 - 10$

(b) Maximum $TR = 208$ at $x = 2$; $P_X = 4$; $y = 20$; $P_Y = 10$

(c) Maximum profit $= 90$ at $x = 1$; $P_X = 6$; $y = 14$; $P_Y = 13$ and $TR = 188$

(d) $\varepsilon_{dX} = \dfrac{P}{P-a} = \dfrac{6}{6-8} = -3$; $\varepsilon_{dY} = \dfrac{P}{P-a} = \dfrac{13}{13-20} = -1.86$

9. (a) Maximum profit $= 722.5$ at $x = 15$; $P_X = 42.5$; $y = 6$; $P_Y = 65$

(b) Maximum profit $= 682.0$ at $P \ 47$, $x = 21$

(c) $\varepsilon_{dX} = -1.13$; $\varepsilon_{dY} = -1.08$

10. Profit $= 8x + 14y - 2x^2 - y^2 - 25$. Profit is a maximum at $x = 2$ and $y = 7$.

11. (a) $TR = 200x - x^2 + 500y - 75y^2 + 10xy$ (b) $x = 175, \ y = 15, \ P_X = 85, \ P_Y = 425$

PE 7.9

1. Maximum utility at $47.5 \ x = 9$; $y = 72$

2. (a) $U_x = 0.2x^{-0.8}y^{0.8}$; $U_y = 0.8x^{0.2}y^{-0.2}$

(b) $\dfrac{U_x}{U_y} = \dfrac{y}{4x} = \dfrac{72}{36} = \dfrac{1}{2} =$ ratio of prices; $P_X = 4, \ P_Y = 2$

3. (a) $MU_x = x^{-0.8}y^{0.8}$: $MU_y = 4x^{0.2}y^{-0.2}$

(b) $\dfrac{MU_x}{MU_y} = \dfrac{P_X}{P_Y} \rightarrow \dfrac{y}{4x} = \dfrac{5}{2} \rightarrow 2y - 20x = 0$ (equation (1))

Second equation is the constraint, $5x + 2y = 200$ (equation (2)). Solution of equations (1) and (2) is $x = 8$, $y = 80$.

(c) Same as (b), $x = 8$, $y = 80$ when utility is maximised; that is, 8 hours skating, 80 hours bridge per month.

4. Maximum $Q = 846.932$ when $L = 60$, $K = 50$.

5. Minimum costs $= 193.62$ when $K = 32.27$, $L = 19.36$

6. Maximum output $= 39.48$ when $L = 1.875$, $K = 12.5$

7. (a) Maximum profit $= 136$ when $x_1 = 4$; $P_1 = 30$; $x_2 = 5$; $P_2 = 20$.

(b) $\varepsilon_{d1} = -1.5$; $\varepsilon_{d2} = -2$

8. Cost constraint: $C = 200_{xA} + 400_{xB}$

Maximum utility $= 26.88$ when $x_A = 6$; $x_B = 6$

9. (a) $P_L = 8$; $P_K = 4$ (b) $L = 30$; $K = 40$

(c) $\dfrac{MP_L}{MP_K} = \dfrac{30K}{20L}$. At maximum this is reduced to $\dfrac{30(40)}{20(30)} = 2$.

Ratio of prices $\dfrac{P_L}{P_K} = \dfrac{8}{4} = 2$

10. (a) $20x + 2y = 1450$

(b) $x = 14.5$ units of strawberries, $y = 580$ units of blackberries.

Maximum $Q = 5546.8$

(c) $\dfrac{MP_X}{MP_Y} = \dfrac{y}{4x}$.

At maximum Q, $\dfrac{MP_X}{MP_Y} = \dfrac{580}{58} = 10$ and the ratio of prices, $\dfrac{P_X}{P_Y} = \dfrac{20}{2} = 10$

11. (a) $Q_1 = 4$, $Q_2 = 8.2$. Maximum profit $= 376.2$
(b) $Q_1 = 6.05$, $Q_2 = 13.95$. Maximum profit $= 232.105$
12. $x = 6$, $y = 28$. Maximum $= 1001.8$

Chapter 8

PE 8.1

Simplify before attempting to integrate

1. $\dfrac{x^2}{2} + \dfrac{x^4}{4} + \dfrac{x^{4.5}}{4.5} + c$

2. Simplify $\dfrac{1}{x^{5/2}} = x^{-5/2}$. Answer : $\dfrac{x^{-3/2}}{-\dfrac{3}{2}} = -\dfrac{2}{3x^{3/2}} + c$

3. Simplify to $x^{-3/2}$. Answer : $\dfrac{x^{-1/2}}{-\dfrac{1}{2}} = -\dfrac{2}{x^{1/2}} + c$

4. Simplify to $1 + x^{-1/2}$. Answer : $x + 2x^{1/2} + c$ **5.** $x^2 + 2\ln(x) + c$

6. $x^2 + \dfrac{1}{2}\ln(x) + c$ **7.** $\dfrac{2}{5}x^{5/2} + 3x + c$ **8.** $\dfrac{5x^2}{2} + \sqrt{5}\dfrac{2}{3}x^{3/2} + c$

9. Simplify to $x^3 - 6x^2 + 9x$. Answer : $\dfrac{x^4}{4} - 2x^3 + \dfrac{9x^2}{2} + c$

10. Simplify to $1 + \dfrac{7}{x}$. Answer : $x + 7\ln(x) + c$

11. Simplify to $x^2 + 1$. Answer : $\dfrac{x^3}{3} + x + c$

12. Simplify to $Q^2 - Q^{3/2}$. Answer : $\dfrac{Q^3}{3} - \dfrac{2}{5}Q^{5/2} + c$

13. Simplify to $t + 2$. Answer : $\dfrac{t^2}{2} + 2t + c$ **14.** $4t + 4\ln(t) + c$

15. $5Q + \ln(Q) + c$ **16.** Simplify to $2 - \dfrac{5}{Q}$. Answer : $2Q - 5\ln(Q) + c$

17. Simplify to $20Q - 0.5Q^2$. Answer : $10Q^2 - \dfrac{0.5Q^3}{3} + c$

18. Simplify to $\dfrac{1}{Q} + 4Q^{-2}$. Answer : $\ln(Q) - \dfrac{4}{Q} + c$.   **19.** $\dfrac{1}{5}\ln(Q) - 5Q + c$

20. $1.25Y^2 + c$

PE 8.2

1. $\dfrac{(x-5)^{1.2}}{1.2} + c$ **2.** $\dfrac{1}{2}\dfrac{(2x-5)^{1.2}}{1.2} + c$ **3.** $\dfrac{(3x-5)^{1.2}}{3(1.2)} + c$

4. $\dfrac{(9x-5)^{1.5}}{9(1.5)} + c$ **5.** $\dfrac{(9x-5)^{4/3}}{9(4/3)} = \dfrac{(9x-5)^{4/3}}{12} + c$

6. $10\ln(x-1) + c$ **7.** $\dfrac{1}{10}\ln(10x-1) + c$

8. Simplify to $10x^{-2}$; answer $-\dfrac{10}{x} + c$ **9.** $-\dfrac{2.5}{(x-1)^4} + c$

SOLUTIONS TO PROGRESS EXERCISES

10. $\dfrac{(20-2Q)^{1.45}}{(-2)(1.45)} = -\dfrac{(20-2Q)^{1.45}}{2.9} + c$

11. $\dfrac{\ln(2+7Q)}{7} + c$

12. $\dfrac{4\ln(2+7Q)}{7} + c$

13. $-\dfrac{\ln(2-7Q)}{7} + c$

14. $-\dfrac{15}{6Q^6} + c$

15. $-\dfrac{5}{2(Q-5)^5} + c$

16. $-\dfrac{5}{4(2Q-5)^6} + c$

17. $\dfrac{2}{3}(Q+1)^{3/2} + 2Q^{1/2} + c$

18. $\dfrac{10(4+Y)^{3/2}}{9} + c$

19. $-\dfrac{10(4-Y)^{3/2}}{9} + c$

20. $-\dfrac{(1+0.8t)^{2.12}}{1.696} + c$

21. $-\dfrac{(1-0.8t)^{2.12}}{1.696} + c$

22. $\dfrac{e^{12x}}{12} + c$

23. $-2e^{-2x} + c$

24. $-58.8e^{-0.85Q} + c$

25. $100(Y + 2e^{-0.5Y}) + c$

26. Simplify to $e^{-2x} + 2e^{-x}$; answer $-0.5e^{-2x} - 2e^{-x} + c$

27. $3e^{1+Y} + c$

28. $-2e^{2-5Q}$

29. $-0.5e^{-2t} + 2e^{t+2} + 3t + c$

PE 8.3

1. 14

2. 80

3. 47.25

4. –6.67

5. $48 + 4\ln 5 = 54.44$

6. $\ln 10 + \dfrac{28}{15} - \ln 3 = 3.07$

7. 102.22

8. 14 666.67

9. 396.75

10. $10(\ln 15 - \ln 11) = 3.1$

11. 14.66

12. $\dfrac{583}{2} + 12\ln 12 = 321.32$

13. $4\ln 5 = 6.44$

14. $50 + \ln 6 = 51.79$

15. 24.67

16. 201.2144

17. 614.7809

18. Simplify to $\displaystyle\int_{Q=0}^{Q=5}(1 + 4e^{-2Q})dQ = 6.9999$

19. 6.4917

20. (a)

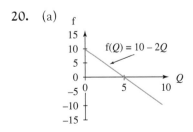

(b) Net area $= \displaystyle\int_{0}^{10} f(Q)dQ = 0$

(c) $\displaystyle\int_{0}^{5} f(Q)dQ = 25$ area above axis

$\displaystyle\int_{0}^{10} f(Q)dQ = -25$ area below axis

Figure PE 8.3 Q20

21. (a)

$f(Q) = Q^2 - 4$

Figure PE 8.3 Q21

(b) Net area $= \int_0^4 f(Q)dQ = \dfrac{16}{3} = 5.33$

(c) $\int_2^4 f(Q)dQ = \dfrac{32}{3} = 10.67$ area above axis

$\int_0^2 f(Q)dQ = -\dfrac{16}{3} = -5.33$ area below axis

22. (a)

$f(Q) = 16 - Q^2$

Figure PE 8.3 Q22

(b) Net area $= \int_0^{10} f(Q)dQ = -173.33$

(c) $\int_0^4 f(Q)dQ = 42.67$ area above axis

$\int_4^{10} f(Q)dQ = -216$ area below axis

23. (a)

$f(Q) = 50 + 5Q - Q^2$

Figure PE 8.3 Q23

24. (a)

$f(Q) = 18 + 8Q - Q^2$

Figure PE 8.3 Q24

(b) Net area $= \displaystyle\int_0^{10} f(Q)dQ = 416.67$
     $=$ area above axis

(b) Net area $= \displaystyle\int_0^5 f(Q)dQ = 148.33$
     $=$ area above axis

25. (a)

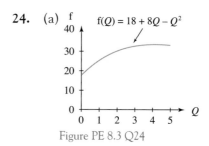

$f(Q) = 4 + 2/Q$

Figure PE 8.3 Q25

(b) Net area $= \displaystyle\int_{0.5}^4 f(Q)dQ = 18.16$
     $=$ area above axis

SOLUTIONS TO PROGRESS EXERCISES

PE 8.4

1. (a)

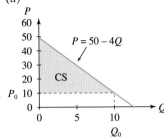

Figure PE 8.4 Q1

(b) $CS = 200$

2. (a)

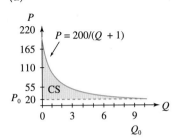

Figure PE 8.4 Q2

(b) $CS = 280.52$

3. (a)

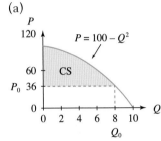

Figure PE 8.4 Q3

(b) $CS = 341$

4. (a)

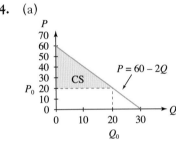

Figure PE 8.4 Q4

(b) $CS = 400$

5. (a)

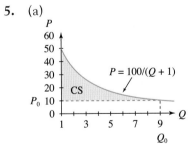

Figure PE 8.4 Q5

(b) $CS = 140.26$

6. (a)

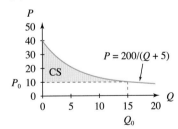

Figure PE 8.4 Q6

(b) $CS = 127.26$

7. (a)

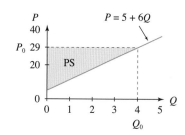

Figure PE 8.4 Q7

(b) $PS = 48$

8. (a)

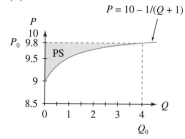

Figure PE 8.4 Q8

(b) $PS = 0.81$

9. (a)

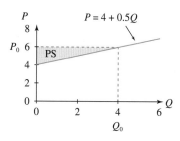

Figure PE 8.4 Q9

(b) $PS = 4$

10. (a)

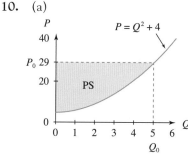

Figure PE 8.4 Q10

(b) $PS = \dfrac{250}{3} = 83.33$

11. (a) $Q = 6$, $P = 38$

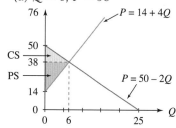

Figure PE 8.4 Q11

(b) $CS = 36$, $PS = 72$

12. (a) $CS = 90$, $PS = 55$

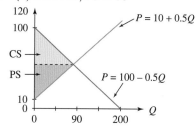

Figure PE 8.4 Q12

(b) $CS = 2025$, $PS = 2025$

13. (a) Equilibrium at $Q = 4$, $P = 34$ and $Q = P = -50$ (not economically meaningful)
     (b) $CS = 42.67$, $PS = 48$

14. (a)

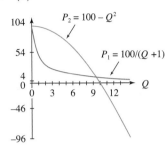

Figure PE 8.4 Q14

(b)

$$P_1 = \frac{100}{Q+1}$$

(i) $TR = 239.79$ for first 10 units

(ii) $TR = 36.77$ for $Q = 8$ to $Q = 12$

$P_2 = 100 - Q^2$

(i) $TR = 666.67$ for first 10 units

(ii) $TR = -5.33$ for $Q = 8$ to $Q = 12$

Revenue from market 2 exceeds that from market 1 until $Q \cong 9.5$. However, revenue from market 2 is zero at $Q = 10$, and negative for $Q > 10$, while revenue from market 1 is always positive but small.

15.

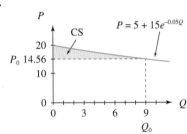

Figure PE 8.4 Q15

$CS = 22.6716$

16.

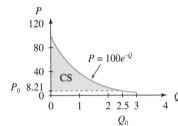

Figure PE 8.4 Q16

$CS = 71.27$

17.

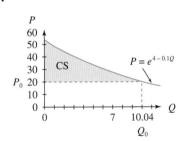

Figure PE 8.4 Q17

$CS = 144.2261$

18.

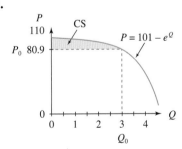

Figure PE 8.4 Q18

(b) $CS = 41.21$

19.

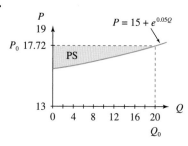

Figure PE 8.4 Q19

$PS = 20$

20.

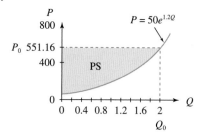

Figure PE 8.4 Q20

(b) $PS = 684.6853$

PE 8.5

1. $y = \dfrac{15x^2}{2} + c$ **2.** $y = 5t^2 - t + c$ **3.** $y = \ln x + c$

4. $C = 5Q^2 + c$ **5.** $P = \dfrac{Q^3}{3} + c$ **6.** $y = 2e^{0.5t} + c$

7. $y = 2.5x^2 - 6$ **8.** $y = 5t^2 + 2t + 15$ **9.** $y = -0.5x + 1.5\ln x + 6.5$

10. $C = 10\ln Q - 0.5Q^2 + 50.5$ **11.** $P = \dfrac{Q^3}{3} + 2.5Q^2 + 10$

12. $y = 20e^{0.6t} + 60$

PE 8.6

1. (a) $MC = \dfrac{d(C)}{dQ}(TC)$ (b) $TC = -\dfrac{Q^3}{3} + 40Q^2 + 500$ (c) $4833 \left(\displaystyle\int_{Q=3}^{Q=12} MCdQ \right)$

2. (a) $MPC = \dfrac{dC}{dY}$ (b) $C = 26.25 - 1.25e^{-0.8Y}$

3. (a) $MR = \dfrac{d(TR)}{dQ}$ (b) $TR = 12Q - Q^2$

4. (a) $TC = \dfrac{Q^3}{3} - 21Q^2 + 200Q + 900$ (b) 852

5. (a) $MLC = \dfrac{d}{dL}(TLC)$ (b) $TLC = 3L + 2L^2$ (c) 114

6. (a) $MP_L = \dfrac{dQ}{dL}$ (b) $Q = 10L - L^2$ (c) 9

7. (a)

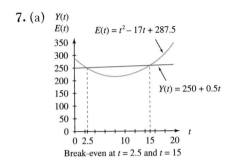

$E(t) = t^2 - 17t + 287.5$

$Y(t) = 250 + 0.5t$

Break-even at $t = 2.5$ and $t = 15$

Figure PE 8.6 Q7(a)

(b) (i) −44.271 (loss); (ii) 325.521

8. (a)

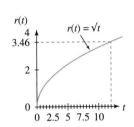

$r(t) = \sqrt{t}$

Figure PE 8.6 Q8(a)

(b) 980 gallons

9. (a)

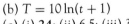

$n = 10/(t + 1)$

Figure PE 8.6 Q9

(b) $T = 10\ln(t + 1)$
(c) (i) 24; (ii) 6.5; (iii) 3.9

10. (a)

$C(t) = 2 + t^{1.5}$

Figure PE 8.6 Q10

(b) (i) 32.36; (ii) 114.13
(c) (i) AC = 6.472; (ii) AC = 22.826

11. (a) $TC = 800 + 80Q$ (b) $TR = 200Q - 10Q^2$ (c) Maximum profit at $Q = 6$
12. $TR = 88Q - 4Q^2$. Maximum TR at $Q = 11$. CS $= 242$
   Profit $= 87Q - 3.5Q^2 + c$. Maximum profit at $Q = 11.6$. CS $= 269.12$

PE 8.7

1. $y(t) = 50e^{2t}$

2. $y = 800e^{-0.1t}$

3. $N = 1807e^{-0.4t}$

4. $y^2 = 2x^2 - 100$

5. $y = 50e^{t^3}$

6. $P = 21 + 20t - \dfrac{t^2}{12}$

7. $y = 20e^{x^2/2}$

8. $I = 900e^{-0.05t}$

9. $Q = \dfrac{1}{P}$

10. $P = 120e^{-Q}$

PE 8.8

1. (a) $A = Ce^{0.02t}$. The amount of ore mined increases at a constant proportional rate
   (b) $t = 0$ at the beginning of 1989, then $A = 400e^{0.02t}$

(c) $t = 1995 - 1989 = 6$ at the beginning of 1995; $t = 2000 - 1989 = 11$ at the beginning of 2000

$$\text{Therefore, ore mined} = \int_{t=6}^{t=11} 400e^{0.02t}\, dt = 2372 \text{ million tons}$$

2. (a) $P = Ce^{0.01t}$ (b) $P = 58.6e^{0.01t}$; $P = 441.75$ million at the beginning of 2200
   (c) $P = 70$ million when $t = 17.78$ years
   $P = 70$ million in $(1998 + 17.78) = 2015.78$, i.e., during 2015

3. (a) $P = 50e^{-0.2t}$
   (b) (i) $P = 22.47 = $ amount present after 4 hours, $P = 42.61$ when $t = 8$
   (ii) 11.5 hours

4. $I = 12e^{-0.05t}$
   (a) $I = 9.11 \times 10^3$ after 5.5 years
   (b) (a) $I = 5 \times 10^3$ when $t = 17.5$ years

5. (a) $\dfrac{dQ}{dP}\dfrac{P}{Q} = -1 \rightarrow \dfrac{dQ}{dP} = -\dfrac{Q}{P}$
   (b) $Q = \dfrac{576}{P}$

6. (a) $F = 80(1 - e^{-0.12t}) \times 10^3$
   (b) Maximum capacity is 80 000
   (c) $F = 40\,000$ at $t = 5.78$ years

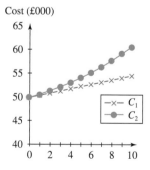

Figure PE 8.8 Q6

7. (a) $C_1 = 50e^{0.01t}$; $C_2 = 50 + 0.8t^{1.25}$
   (b) At $t = 5$, $C_1 = 52.56$, $C_2 = 54.67$
   C_1 is more cost effective than C_2; the
   rate of increase is slower.

Figure PE 8.8 Q7

8. (a) $\dfrac{dQ}{dP}\dfrac{P}{Q} = -\dfrac{10}{Q}$ (b) $P = 20e^{-0.1Q}$

9. Proportional rate of growth at P_0 is $\dfrac{1}{P_t}\left(\dfrac{dP_t}{dt}\right)\Big|_{P_t=P_0}$

$$\frac{1}{P_t}\left(\frac{dP_t}{dt}\right) = \frac{100e^{0.2t}}{500e^{0.2t}} = \frac{1}{5} = 0.2, \text{ a constant rate for any } t$$

Proportional rate $= 0.2$ at $t = 5$
Proportional rate $= 0.2$ at $t = 10$

10. Proportional rate $= \dfrac{-68.5e^{-0.1t}}{68.5e^{-0.1t}} = 0.1$, constant rate for any t

Proportional rate $= -0.1$ at $t = 5$
Proportional rate $= -0.1$ at $t = 10$

11. Proportional rate $= \dfrac{-2/(t+2)^2}{2/(t+2)} = \dfrac{-1}{t+2}$; this depends on t

Proportional rate $= -0.1429$ at $t = 5$
Proportional rate $= -0.0833$ at $t = 10$
Therefore the rate is not constant

12. Proportional rate $= \dfrac{50t^{1.5}}{20t^{2.5}} = \dfrac{2.5}{t}$; this depends on t

Proportional rate $= -0.5$ at $t = 5$
Proportional rate $= -0.25$ at $t = 10$
Therefore the rate is not constant

Chapter 9

PE 9.1

1.

(i)

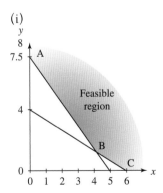

Figure PE 9.1 Q1(i)

A = (0, 7.5), B = (4.2, 1.2), C = (6, 0)

(ii)

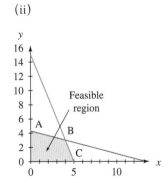

Figure PE 9.1 Q1(ii)

A = (0, 4.33), B = (4, 3), C = (5.0)

2. (a) and (b)

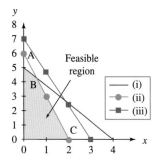

Figure PE 9.1 Q2

(c) A = (0, 5), B = (0.57, 4.29), C = (2, 0). Constraint (iii) is redundant.

3. (a) and (b)

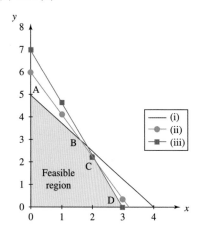

Figure PE 9.1 Q3

(c) A = (0, 5), B = (1.63, 3), C = (2.18, 1.91), D = (3, 0)

4. (a) and (b)

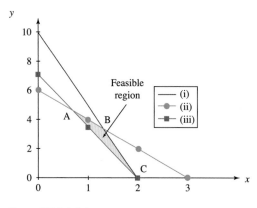

Figure PE 9.1 Q4

(c) A = (0.67, 4.67), B = (1.33, 3.33), C = (2, 0)

SOLUTIONS TO PROGRESS EXERCISES

5.

(a) $y = -0.5x + \dfrac{TC}{24}$

Slope $= -0.5$

Isocost: (i) Let $TC = 24$
          (ii) Let $TC = 48$
or any two values of TC

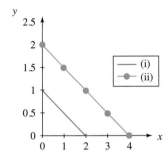

Figure PE 9.1 Q5(a)

(b) $y = -3x + \dfrac{TR}{3}$

Slope $= -3$

Isorevenue: (i) Let $TR = 9$
          (ii) Let $TR = 18$
or any two values of TR

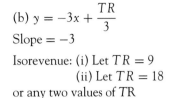

Figure PE 9.1 Q5(b)

(c) $y = -\dfrac{5.5}{2.85} + \dfrac{\pi}{2.85}$

Slope $= -\dfrac{5.5}{2.85}x + \dfrac{\pi}{2.85}$

Isoprofit: (i) Let $\pi = 5.5(2.85)$
        (ii) Let $\pi = 10(2.85)$

or any two values of π

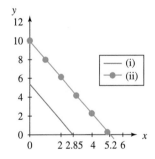

Figure PE 9.1 Q5(c)

6. (a) Algebraically: Corner points of the feasible region were determined in question 3:
A $= (0.5)$, B $= (1.6, 3)$, C $= (2.18, 1.91)$, D $= (3,0)$. Let W $=$ objective function.
W $= 10$ at A;   W $= 10.8$ at B;   W $= 10.36$ at C;   W $= 9$ at D.
Therefore W is a maximum at point B: $x = 1.6$, $y = 3$.
(b) Graphically, the function to be maximised (objective function) is drawn:
$y = -1.5x + \dfrac{W}{2}$. This is a maximum at B.

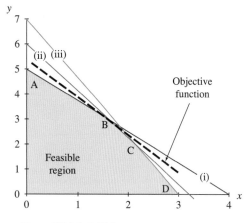

Figure PE 9.1 Q6(b)

7. (a) Algebraically, the minimum value of the objective function is at point D, when $W = 9$.
   (b) Graphically, the objective function is a minimum at D.

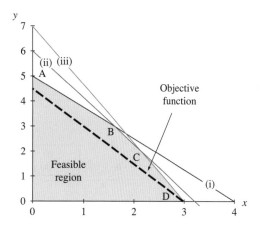

Figure PE 9.1 Q7

8. (a)

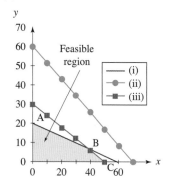

Figure PE 9.1 Q8

9. (a)

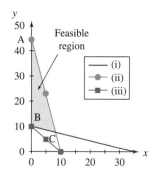

Figure PE 9.1 Q9

(b) Constraint (ii), $60x + 70y \leq 4200$, is not a limitation.
(c) Corner points:
A = (0, 20), B = (37.5, 7.5), C = (50.0)

(b) Constraint (iii), $x + y \leq 10$, is not a limitation.
(c) Corner points:
A = (0, 44), B = (0, 10), C = (8.32, 7.40)

10. (a) (i) $6A + 3B \leq 360$
   (ii) $2A + 4B \leq 280$
   (iii) $2A + 2B \leq 160$
   $\pi = 20A + 30B$

(b)

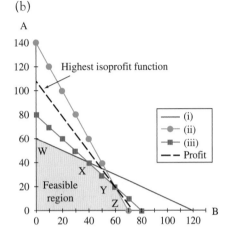

Figure PE 9.1 Q10

$W = (0, 60); \quad X = (40, 40); \quad Y = (60, 20); \quad Z = (70, 0)$

(c) Profit at $W = 20(60) + 30(0) = 1200$
   Profit at $X = 20(40) + 30(40) = 2000$
   Profit at $Y = 20(20) + 30(60) = 2200$
   Profit at $Z = 20(0) + 30(70) = 2100$
   Therefore maximum profit at Y. 20 A, 60 B machines used.

11. (a) (i) $4y + 20x \geq 80$
   (ii) $y + 2x \geq 15$
   (iii) $2y + x \geq 10$
   $X = 18x + 6y$

(b)

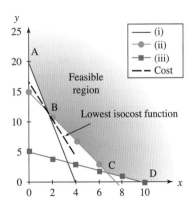

Figure PE 9.1 Q11

$A = (0, 20); \quad B = (1.67, 11.65); \quad C = (6.67, 1.66); \quad D = (10, 0)$

(c) Cost at A $= 18(0) + 6(20) = 120$
Cost at B $= 18(1.67) + 6(11.65) = 100$
Cost at C $= 18(6.67) + 6(1.66) = 130$
Cost at D $= 18(10) + 6(0) = 180$

Therefore, minimum cost at B. The gardener should use $1\frac{2}{3}$ kg of mix X and 11.65 kg of mix Y.

PE 9.2

1. (a) $\begin{pmatrix} 5 & -1 \\ -7 & -9 \end{pmatrix}$ (b) $\begin{pmatrix} -3 & -7 \\ 7 & -9 \end{pmatrix}$ (c) $\begin{pmatrix} 17 & 8 \\ -28 & -9 \end{pmatrix}$ (d) $\begin{pmatrix} 2 & -4 \\ 0 & -8 \end{pmatrix}$

(e) $\begin{pmatrix} 1 & -4 \\ 0 & -9 \end{pmatrix}$ (f) $A + C$ not possible: different dimensions

(g) $\begin{pmatrix} 5 & -11 \\ 3 & -9 \end{pmatrix}$ (h) $\begin{pmatrix} 56 & -4 & 2 \\ -35 & 7 & 7 \end{pmatrix}$ (i) Not possible: $CB = (2 \times 3)(2 \times 2)$

(j) Not possible : $CB^{\mathrm{T}} = (2 \times 3)(2 \times 2)$ (k) $\begin{pmatrix} 32 & 63 \\ 3 & 0 \end{pmatrix}$ (l) Not possible

(m) $\begin{pmatrix} 5 & -128 \\ -1 & 4 \\ -1 & -14 \end{pmatrix}$ (n) $\begin{pmatrix} 56 & 35 \\ -4 & 7 \\ 2 & 7 \end{pmatrix}$ (o) Not possible

2. (a) $A + C$ not possible: different dimensions   (b) $\begin{pmatrix} 2 & 0 \\ 4 & -4 \\ -1 & 0 \end{pmatrix}$

(c) $\begin{pmatrix} 5 & -10 \\ 3 & -6 \end{pmatrix}$ (d) $\begin{pmatrix} 5 & 4 \end{pmatrix}$ (e) Not possible

3. Let quantity sold matrix $= Q$ and profit per unit matrix $= P$.
  (a) π, total profit $= QP^{\mathrm{T}}$ (a 3×3 matrix) where
        $\pi\,11 = $ profit for burgers $= £485$
        $\pi\,22 = $ profit for chips $= £1165$
        $\pi\,33 = $ profit for drinks $= £750$
  (b) Profit for each shop (unit profit $\times$ quantity for all three items)
            $= Q^{\mathrm{T}}P$ (a 3×3 matrix) where
        $\pi\,S_{11} = $ profit for shop A $= £785$
        $\pi\,S_{22} = $ profit for shop B $= £850$
        $\pi\,S_{33} = $ profit for shop C $= £765$

4. Let $R = $ party by area matrix and $V = $ voters in areas 1, 2 and 3.
  (a) Votes per candidate, X
          $X = RTV$ where
        $X_{11} = $ votes for A $= 79\,240$
        $X_{22} = $ votes for B $= 52\,400$
        $X_{33} = $ votes for C $= 51\,360$

SOLUTIONS TO PROGRESS EXERCISES

5. (a) $AB = \begin{pmatrix} 1 & 0 \\ -3 & 10 \end{pmatrix}$ $\qquad BA = \begin{pmatrix} 4 & 3 \\ 6 & 7 \end{pmatrix}$

(b) (i) AC not possible as $(2 \times 2)(3 \times 1)$

(ii) $AD = \begin{pmatrix} 6 & 1 & 3 \\ 12 & 7 & 11 \end{pmatrix}$ $\qquad$ (iii) $DC = \begin{pmatrix} 14 \\ 8 \end{pmatrix}$ $\qquad$ (iv) $\begin{pmatrix} 14 & 70 & 42 \\ 8 & 40 & 24 \end{pmatrix}$

PE 9.3

1. (a, b) $x = -49/4$, $y = 17/4$, $z = 6$

2. (a) $x = 7$, $y = 5$, $z = -9.5$ $\quad$ (b) $x = 15.33$, $y = -3.33$, $z = -13.67$
(c) On completing the elimination, the equation $O = -4$ is obtained. This contradiction indicates no solution.
(d) $x = 26/3$, $y = -6$, $z = -14/3$

3. (a) $x = 1$, $y = 21$, $z = -17$

4. $P_1 = 26$, $P_2 = 38$, $P_3 = 66$

5. $Y = 500$, $C = 100$, $T = 100$

6. $x = 12/5$, $y = 3/5$, $z = 0$

7. $x = 3.8$, $y = 1.53$, $z = 1.53$

8. $P_1 = 7$, $Q_1 = 51$; $P_2 = 9$, $Q_2 = 103$; $P_3 = 10$, $Q_3 = 39$

PE 9.4

1. (a) 3 $\qquad$ (b) -2 $\qquad$ (c) 35 $\qquad$ (d) 0 $\qquad$ (e) 6
(f) 0 $\qquad$ (g) $c^2 - 2c$ $\quad$ (h) a $\qquad$ (i) $1 - \dfrac{a}{1-a}$

2. $x = 0$; $y = 5$ $\qquad$ **3.** $x = 4.6$; $y = -2.8$ $\qquad$ **4.** $P_1 = 0.73$; $P_2 = 3.59$

5. $P = 56.92$; $Q = 5.38$ $\qquad\qquad$ **6.** $P = 4.27$; $Q = -2.96$

7. $Y = 37.6$; $r = 88$ $\qquad$ **8.** $P_1 = 15.09$, $Q_1 = 39.27$; $P_2 = 6.09$, $Q_2 = 69.08$

9. (a) $Y - C = I_0$; $-bY + C = C_0$
(b) $Y_e = \dfrac{I_0 + C_0}{1 - b}$; $\quad C_e = \dfrac{C_0 + bI_0}{1 - b}$

10. (a) $0.8Y + 0.4r = 247.8$; $\quad 0.1Y - 0.3r = 29.225$
(b) $Y = 307.25$; $\quad r = 5$

11. Write equations $Q = f(P)$, hence solve for P_1, P_2 by Cramer's rule.
$P_1 = 0.6Q_1 + 0.8Q_2 - 3$; $\quad P_2 = 0.4Q_1 + 0.2Q_2 - 2$

PE 9.5

1. (a) -8 (b) -78 (c) -234

2. (a) $x = 7; y = 5; z = -9.5$ (b) $x = \dfrac{46}{3}; y = -\dfrac{10}{3}; z = -\dfrac{41}{3}$

     (c) No unique solution, $\Delta = 0$.   (d) $x = \dfrac{26}{3}; y = -6; z = -\dfrac{14}{3}$

3. $x = 1; y = 1; z = 1$ **4.** $P_1 = 2; P_2 = 4; P_3 = 1$

5. $Y = 500; C = 100; T = 100$ **6.** $x = \dfrac{12}{5}; y = \dfrac{3}{5}; z = 0$

7. $x = 3.8; y = 1.53; z = 1.53$

8. $P_1 = 7, Q_1 = 51; P_2 = 9, Q_2 = 103; P_3 = 10, Q_3 = 39$

9. (b) $P_1 = 10.5; P_2 = 3.6; P_3 = 11.8$

PE 9.6

1. (a) $\begin{pmatrix} 3 & -2.5 \\ -1 & 1 \end{pmatrix}$ (b) $\begin{pmatrix} -0.5 & 0.2 \\ 1 & -0.3 \end{pmatrix}$ (c) $\begin{pmatrix} -\dfrac{5}{11} & \dfrac{3}{11} \\ \dfrac{2}{11} & \dfrac{1}{11} \end{pmatrix}$

(d) $\begin{pmatrix} 0 & \dfrac{1}{12} \\ \dfrac{1}{4} & -\dfrac{1}{16} \end{pmatrix}$ (e) No inverse since $|A| = 0$

2. (a) $A^{-1} = \begin{pmatrix} 1 & -\dfrac{1}{4} & \dfrac{1}{12} \\ 0 & \dfrac{1}{4} & -\dfrac{1}{12} \\ -1 & \dfrac{1}{8} & \dfrac{1}{8} \end{pmatrix}$ (b) $x = 7; y = 5; z = -9.5$

3. (a) $A^{-1} = \begin{pmatrix} 1 & \dfrac{1}{6} & -\dfrac{1}{18} \\ 0 & -\dfrac{1}{6} & \dfrac{1}{18} \\ -1 & -\dfrac{1}{12} & \dfrac{7}{36} \end{pmatrix}$ (b) $x = \dfrac{46}{3}; y = -\dfrac{10}{3}; z = -\dfrac{41}{3}$

4. (a) $A^{-1} = \begin{pmatrix} \dfrac{7}{4} & \dfrac{1}{4} & -\dfrac{1}{2} \\ -\dfrac{3}{4} & -\dfrac{1}{4} & \dfrac{1}{2} \\ -\dfrac{5}{4} & \dfrac{1}{4} & \dfrac{1}{2} \end{pmatrix}$ (b) $x = 8.5; y = -6.5; z = 2.5$

5. (a) No inverse since $|A| = 0$ (b) No unique solution.

6. (i) (a) $(I - A)^{-1} = \begin{pmatrix} 3 & 2 \\ 1.25 & 2.5 \end{pmatrix}$ (b) Total output $= \begin{pmatrix} 320 \\ 300 \end{pmatrix}$

(ii) (a) $(I - A)^{-1} = \begin{pmatrix} 2.5 & 1.25 \\ 1.0 & 2.5 \end{pmatrix}$ (b) Total output $= \begin{pmatrix} 800 \\ 1120 \end{pmatrix}$

7. (a) $(I - A)^{-1} = \begin{pmatrix} 2.4 & 0.8 & 0.4 \\ 2.0 & 4.0 & 2.0 \\ 2.6 & 4.2 & 4.6 \end{pmatrix}$ (b) 728; 2440; 4372

8. (a) $(I - A)^{-1} = \begin{pmatrix} 2.9 & 1.8 & 0.95 \\ 1.0 & 2.4 & 0.5 \\ 1.7 & 1.4 & 1.85 \end{pmatrix}$ (b) 5783; 3770; 4709

9. (a) $A = \begin{pmatrix} 0.5 & 0.5 \\ 0.5 & 0.25 \end{pmatrix}$ (b) 7000 units from X; 6000 units from Y

10. (a) $A = \begin{pmatrix} 0.1 & 0.1 & 0.25 \\ 0.2 & 0.4 & 0 \\ 0.4 & 0 & 0.1 \end{pmatrix}$

(b) 808.81 from Horticulture; 1602.9 from Dairy; 1470.6 from Poultry

11. (a) $\begin{pmatrix} 0.2 & 0.125 & 0.25 \\ 0.1 & 0 & 0.625 \\ 0.1 & 0.1 & 0 \end{pmatrix}$

(b) 1086.8; 1281.7; 1236.8 from sectors A, B and C, respectively.

12. (a) $A = \begin{pmatrix} 0.4 & 0.2 & 0 \\ 0.1 & 0.4 & 0.2 \\ 0.2 & 0.2 & 0.1 \end{pmatrix}$

(b) 1719.0 from Dairy; 1406.9 from Beef; 2361.3 from Leather

Chapter 10

PE 10.1

1. (a) Order 1, homogeneous.
(b) Order 1, non-homogeneous.
(c) Order 2, non-homogeneous.

2. (a) 5, 14, 21.2, 26.96, 31.568, therefore $Y_5 = 31.568$
(b) 20, 18, −88, −496, −1280, therefore $P_5 = -1280$
(c) 100, 140, 164, 178, 187, therefore $P_5 = 187$

3. (a) $Y_t = A(0.8)^t$, $Y_t = 6.25(0.8)^t : Y_{20} = 0.072$, $Y_{50} = 0.000\,089$
(b) $Y_t = A(0.8)^t$, $Y_t = -6.25(-0.8)^t : Y_{20} = -0.072$, $Y_{50} = -0.000\,089$.
(c) $Y_t = A(0.8)^t + 50$, $Y_t = -56.25(0.8)^t + 50 : Y_{20} = 49.35$, $Y_{50} = 49.999$

4. (a) Stable. $Y_t = A(0.8)^t$ decreases to 0 as t increases. See Y_{20} and Y_{50}.
   (b) Stable: Y_t oscillates to 0 as t increases.
   (c) Stable: Y_t increases to 50 as t increases.

5. (a) $Y_t = A(10)^t - 100$; $Y_t = 120(10)^t - 100$. Not stable.
   (b) $P_t = A(0.9)^t + 800$; $P_t = -700(0.9)^t + 800$. Stable, P_t increases to 800.

6. (a) Price increases by 5% each season.
   (b) $P_t = 6000(1.05)^t$. Price will increase indefinitely.

7. (a) (ii) is different: $P_{t+1} = \dfrac{1}{1.05} P_t$
   (b) (ii) is different, the index should be $(n - 2)$.
      (iii) is different, the index should be $(n - 1)$.

8. (a) Investment in any year is 90% of that in the previous year, plus £60m (an injection of £60m).
   (b) (i) $I_0 = £15m$, then $I_t = -585(0.9)^t + 600$. Investment increases to £600m. Reaches £400m in 10.1 years.
      (ii) $I_0 = £180m$, then $I_t = -420(0.9)^t + 600$. Investment increases to £600m, more slowly than (i). Reaches £400m in 7 years.
      (iii) $I_0 = £900m$, then $I_t = 300(0.9)^t + 600$. Investment decreases to £600m, therefore never drops to £400m.

9. (a) $Y_t = A(0.5)^t - 10$
   (b) $Y_t = 360(0.5)^t - 10$. Stabilises, decreases steadily to -10. $Y_t = 0$ at $t = 5.17$

10. (a) $P_n = 1.05 P_{n-1} - 8000$
    (b) $P_n = A(1.05)^n + 160\,000$
    (c) If $P_0 = 100\,000$, $P_n = -60\,000(1.05)^n + 160\,000$

11. (a) $Y_t = A(0.6)^t + 30(2)^t$
    (b) $Y_t = A(0.7)^t + 120(0.8)^t$. $Y_t = -204.71(0.7)^t + 120(0.8)^t$

12. (a) $P_t = 186(0.75)^t + 4(3)^t$. Not stable.
    (b) $Y_t = 8.75(0.84)^t + 31.25$. Stable.

PE 10.2

1. (a) $P_{t+1} = P_t + r P_t = (1 + r) P_t$
   (b) Solve $P_{t+1} - (1 + r) P_t = 0$
      $$P_t = A(1 + r)^t$$
      When $t = 0$, $P_t = P_0$, therefore $P_0 = A(1 + r)^0 = A$, therefore $A = P_0$.
      Therefore $P_t = P_0(1 + r)^t$

2. (a) $Y_t = A(0.8)^t + 3000$. Stabilises to 3000.
      If $A > 0$, Y_t decreases to 3000. If $A < 0$, Y_t increases to 3000.
   (b) $Y_t = A(0.75)^t + 1750(1.05)^t$. Not stable, increases indefinitely.

3. (a) $P_t + P_{t-1} = 42$. $P_t = A(-1)^t + 21$. Price oscillates about 21.
   (b) $P_t = 10.5(-1)^t + 21$. Price alternates between 31.5 and 10.5.

SOLUTIONS TO PROGRESS EXERCISES

4. (a) $P_{t+1} - 0.88P_t = 54$
   (b) $P_t = A(0.88)^t + 450$
   (c) Price stabilises to 450.

5. (a) $0.12Y_t = 4.5(Y_t - Y_{t-1}) \rightarrow 4.38Y_t - 4.5Y_{t-1}$
   (b) $Y_t = 314(1.027)^t$
   (c) $I_t = 4.5(Y_t - Y_{t-1}) = 4.5(314)(1.027)^t\{1 - (1.027)^{-1}\} = 37.148(1.027)^t$
   Therefore, investment will continue to grow. All units in £m.

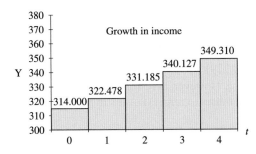

Figure PE 10.2 Q5(c)(i)

Figure PE 10.2 Q5(c)(ii)

6. (a) $-0.255Y_t = 0.9(Y_t - Y_{t-1}) \rightarrow 1.155Y_t - 0.9Y_{t-1} = 0$
   (b) $Y_t = 800(0.779)^t$

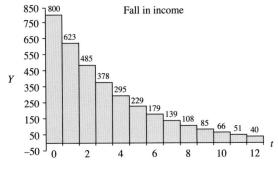

Figure PE 10.2 Q6(b)

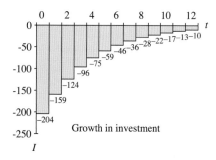

Figure PE 10.2 Q6(c)

(c) $I_t = 0.9(Y_t - Y_{t-1}) = 0.9(800)(0.779)^t\{1 - (0.779)^{-1}\} = -204.26(0.779)^t$

Income is decreasing, but investment is increasing. However, investment is negative (borrowing instead of investing). Investment eventually stabilises to zero. All units in £m.

WORKED EXAMPLES

Items in italics are on the website only, not in the textbook

1	Mathematical Preliminaries	1
1.1	Addition and subtraction	4
1.2	Multiplication and division	4
1.3	Add and subtract fractions	7
1.4	Multiplying fractions	8
1.5	Division with fractions	9
1.6	Solving equations	11
1.7	Solving a variety of simple algebraic equations	13
1.8	Currency conversions	15
1.9	Solving simple inequalities	20
1.10	Calculations with percentages	22
1.11	Evaluation of formulae	25
1.12	Transposition and evaluation of formulae	27
1.13	Using Excel to perform calculations and plot graphs	29
2	The Straight Line and Applications	37
2.1	Plotting lines given slope and intercept	42
2.2	Determine the equation of a line given slope and intercept	44
2.3	Calculation of horizontal and vertical intercepts	47
2.4	To graph a straight line from its equation $y = mx + c$	49
2.5	Plot the line $ax + by + d = 0$	51
2.6	Linear demand function	62
2.7	Analysis of the linear supply function	65
2.8	Linear supply function 2	67
2.9	Linear total cost function	71
2.10a	Linear total revenue function	73
2.10b	Linear profit function	74
2.11	Calculating the slope given two points on the line	77
2.12	Equation of a line given the slope and a point on the line	78
2.13	Equation of a line given two points on the line	79
2.14	*Vertical translations of linear functions*	website only
2.15	*Horizontal translations of linear functions*	website only
2.16	*Horizontal and vertical translations of linear functions*	website only
2.17	*The effect of an excise tax on a supply function*	website only
2.18	*Translating linear cost functions*	website only
2.19	Determining the coefficient of point elasticity of demand	85
2.20	*Calculating the coefficient of point elasticity at different prices*	website only

2.21 *Calculating the coefficient of income elasticity of demand* website only
2.22 *Graphing and interpreting the budget constraint* website only
2.23 *Budget constraint, changes in prices and income* website only
2.24 Use Excel to show the effect of price and income changes on a budget constraint 93

3 Simultaneous Equations 101
3.1 Solving simultaneous equations 1 102
3.2 Solving simultaneous equations 2 104
3.3 Solving simultaneous equations 3 106
3.4 Simultaneous equations with no solution 107
3.5 Simultaneous equations with infinitely many solutions 108
3.6 Solve three equations in three unknowns 110
3.7 Goods market equilibrium 112
3.8 Labour market equilibrium 113
3.9 Goods market equilibrium and price ceilings 115
3.10 Labour market equilibrium and price floors 116
3.11 Equilibrium for two substitute goods 119
3.12 Taxes and their distribution 121
3.13 Subsidies and their distribution 123
3.14 Calculating the break-even point 125
3.15 Consumer and producer surplus at market equilibrium 130
3.16 Equilibrium national income when $E = C + I$ 134
3.17 *Effect of changes in MPC and I_0 on Y_e* website only
3.18 *Equilibrium national income and effect of taxation* website only
3.19 *Expenditure multiplier with imports and trade balance* website only
3.20 *IS-LM analysis* website only
3.21 Cost, revenue, break-even, per unit tax with Excel 137
3.22 Distribution of tax with Excel 138

4 Non-linear Functions and Applications 147
4.1 Solving less general quadratic equations 150
4.2 Solving quadratic equations 151
4.3 Sketching a quadratic function $f(x) = \pm x^2$ 153
4.4 *Comparing graphs of quadratic functions* website only
4.5 Vertical and horizontal translations of quadratic functions 155
4.6 Sketching any quadratic equation 156
4.7 Non-linear demand and supply functions 159
4.8 Non-linear total revenue function 160
4.9 Calculating break-even points 162
4.10a Plotting cubic functions 165
4.10b Graphs of more general cubic functions 166
4.11 TR, TC and profit functions 168
4.12 Graphing exponential functions 171
4.13 Simplifying exponential expressions 174
4.14 Solving exponential equations 178
4.15 Unlimited growth: population growth 180
4.16 Limited growth: consumption and changes in income 182

4.17	Use logs to solve certain equations	187
4.18	Finding the time for the given population to grow to 1750	188
4.19	Graphs of logarithmic functions	189
4.20	Using log rules	191
4.21	Solving certain equations using rule 3 for logs	193
4.22	Solve equations containing logs and exponentials	195
4.23	Sketches of hyperbolic functions	198
4.24	Hyperbolic demand function	200
4.25	Total cost functions with Excel	203

5   Financial Mathematics   209
5.1	Sum of an arithmetic series	211
5.2	Sum of a geometric series	212
5.3	Application of arithmetic and geometric series	214
5.4	*Simple interest calculations*	website only
5.5	Compound interest calculations	220
5.6	Future and present values with compound interest	220
5.7	Calculating the compound interest rate and time period	221
5.8	Compounding daily, monthly and semi-annually	224
5.9	Continuous compounding	225
5.10	Annual percentage rates	227
5.11	Future value of asset and reducing-balance depreciation	229
5.12	Present value of asset and reducing-balance depreciation	230
5.13	Calculating net present value	231
5.14	*IRR determined graphically (Excel) and by calculation*	233
5.15	Compound interest for fixed periodic deposits	238
5.16	Annuities	239
5.17	Present value of annuities	241
5.18	Mortgage repayments	243
5.19	How much of the repayment is interest?	244
5.20	Sinking funds	246
5.21	The interest rate and the price of bonds	249
5.22	Growth of an investment using different methods of compounding (Excel)	252

6   Differentiation and Applications   259
6.1	Equation for the slope of $y = x^2$ from first principles	261
6.2	Using the power rule	264
6.3	More differentiation using the power rule	267
6.4	Calculating higher derivatives	268
6.5	*Determining the rate of change of the slope of a curve:* $y = 3x^2 - 0.1x^3$	website only
6.6	Calculating marginal revenue given the demand function	271
6.7	Calculating marginal revenue over an interval	272
6.8	Derive marginal cost equation from total cost function	274
6.9	*MR, AR* for a perfectly competitive firm and a monopolist	276
6.10	Derive marginal cost from average cost	279
6.11	Deduce the equation for the marginal and average product of labour from a given production function	282

6.12	*Find TLC, MLC, ALC given the labour supply functions*	website only
6.13	*Find TLC, MLC, ALC for a perfectly competitive firm and a monopsonist*	website only
6.14	MPC, MPS, APC, APS	284
6.15	*Calculating the marginal utility*	website only
6.16	Finding turning points	288
6.17	Maximum and minimum turning points	292
6.18	Intervals along which a curve is increasing or decreasing	296
6.19	Derived curves	297
6.20	Sketching functions	302
6.21	Maximum TR and a sketch of the TR function	304
6.22	Break-even, profit, loss and graphs	307
6.23	Maximum and minimum output for a firm over time	309
6.24	Profit maximisation and price discrimination	311
6.25	Profit maximisation for a perfectly competitive firm	313
6.26	Profit maximisation for a monopolist	315
6.27	Curvature of curves: convex or concave towards the origin	323
6.28	Locate the point of inflection, PoI = point at which marginal rate changes	326
6.29	*Relationship between the APL and MP_L functions*	website only
6.30	*Point of inflection on the production function: law of diminishing returns*	website only
6.31	Relationship between TC and MC	327
6.32	Relationship between AC, AVC, AFC and MC functions	329
6.33	Derivatives of exponentials and logs	335
6.34	Using the chain rule for differentiation	336
6.35	Using the product rule for differentiation	339
6.36	Using the quotient rule for differentiation	341
6.37	Find MC given a logarithmic TC function	343
6.38	Demand, TR, MR expressed in terms of exponentials	344
6.39	Expressions for point elasticity of demand in terms of P, Q or both for linear and non-linear demand functions	349
6.40	Point elasticity of demand for non-linear demand functions	351
6.41	Constant elasticity demand function	352
6.42	*Elasticity, marginal revenue and total revenue*	website only
7	**Functions of Several Variables**	**361**
7.1	Plot isoquants for a given production function	364
7.2	Partial differentiation: a first example	367
7.3	Determining first-order partial derivatives	368
7.4	Determining second-order partial derivatives	371
7.5	Differentials for functions of one variable	375
7.6a	Differentials and incremental changes	376
7.6b	Incremental changes	378
7.7	MP_L and MP_K: increasing or decreasing?	382
7.8	Slope of an isoquant in terms of MP_L, MP_K	385
7.9	Constant, increasing and decreasing returns to scale	389
7.10	Indifference curves and slope	392
7.11	Partial elasticities of demand	396
7.12	Partial elasticities of labour and capital	397

7.13	Use partial derivatives to derive expressions for various multipliers	398
7.14	Optimum points for functions of two variables	402
7.15	Monopolist maximising total revenue for two goods	404
7.16	Maximise profit for a multi-product firm	405
7.17	Monopolist: price and non-price discrimination	407
7.18	Maximising total revenue subject to a budget constraint	411
7.19	Lagrange multipliers and utility maximisation	413
7.20	Use Lagrange multipliers to derive the identity $U_x / U_y = P_X / P_Y$	415
7.21	Meaning of λ	416
7.22	Maximise output subject to a cost constraint	417
7.23	Minimise costs subject to a production constraint	419

8 Integration and Applications — 427

8.1	Using the power rule for integration	430
8.2	Integrating sums and differences, constant multiplied by variable term	433
8.3	Integrating more general functions	434
8.4	Integrating functions containing e^x	435
8.5	Integrating functions of linear functions by substitution	437
8.6	Integrating linear functions raised to a power	439
8.7	More examples on integrating functions of linear functions	340
8.8	Evaluating the definite integral	443
8.9	Definite integral and e^x	444
8.10	Definite integration and net area between curve and x-axis	445
8.11	Definite integration and logs	447
8.12	Using the definite integral to calculate consumer surplus	449
8.13	Using the definite integral to calculate producer surplus	451
8.14	Consumer and producer surplus: exponential functions	453
8.15	Solution of differential equations: $dy/dx = f(x)$	458
8.16	Find total cost from marginal cost	459
8.17	Differential equations and rates of change	461
8.18	Solving differential equations of the form $dy/dx = ky$	465
8.19	Solving differential equations of the form $dy/dx = f(x)g(y)$	467
8.20	Limited growth	469
8.21	Determining the proportional rates of growth	470
8.22	*Integration of certain products by substitution*	website only
8.23	*Integration of certain products by substitution and/or quotients by substitution*	website only
8.24	*A further example of substitution to integrate certain products*	website only
8.25	*Integration by parts 1*	website only
8.26	*Integration by parts: choosing u*	website only
8.27	*First-order differential equations: $dy/dx = f(x)g(y)$*	website only

9 Linear Algebra and Applications — 477

9.1	Find the minimum cost subject to constraints	478
9.2	Profit maximisation subject to constraints	483
9.3	Adding and subtracting matrices	490
9.4	Multiplication of a matrix by a scalar	491
9.5	Matrix multiplication	492

9.6   Applications of matrix arithmetic   494
9.7   Solution of a system of equations: Gaussian elimination   499
9.8   More Gaussian elimination   500
9.9   Gauss–Jordan elimination   502
9.10 Using determinants to solve simultaneous equations   505
9.11 Using Cramer's rule to solve simultaneous equations   509
9.12 Find the market equilibrium using Cramer's rule   510
9.13 Use Cramer's rule for the income-determination model   511
9.14 Evaluation of a 3×3 determinant   513
9.15 Solve three simultaneous equations by Cramer's rule   513
9.16 Equilibrium levels in the national income model   515
9.17 The inverse of a matrix: elimination method   518
9.18 The inverse of a 3×3 matrix   521
9.19 Solve a system of equations by the inverse matrix   524
9.20 Input/output analysis   527
9.21 Use Excel to solve systems of linear equations   531

10 Difference Equations   539
10.1 Solving difference equations by iteration   542
10.2 General solution of a homogeneous first-order difference equation   542
10.3 General and particular solutions of first-order homogeneous difference equations 544
10.4 Stability of solutions of first-order difference equations   546
10.5 Solve non-homogeneous first-order difference equations 1   549
10.6 Solve non-homogeneous first-order difference equations 2   551
10.7 The lagged income model   555
10.8 The cobweb model   558
10.9 The Harrod–Domar growth model   561

INDEX

absolute values 84, 287, 431
abstract models 55, 56
accuracy 2–3
addition 3–4
   fractions 6–7
   indices 174
   logs 191
   matrices 489–90
aggregate expenditure 133
algebra 2, 13–14
   see also linear algebra
algebraic substitution 106, 107
   integration by 436–41, 473
amortisation of a loan 242
annual percentage rate (APR) 225–8, 254–5
annuities 238–41, 255
   present value of 241–2
   progress exercises 247–8
annuity factor 241
antilogs 194, 428
arc(-price) elasticity of demand 90, 98
area of triangles 130
area between two curves 455
area under a curve 441–6
arithmetic operations 3–6
arithmetic operators, precedence 5–6
arithmetic sequence 210, 211, 216
arithmetic series/progressions 210–11
   applications of 214–17
   progress exercises 213–14
   sum of terms formula 211, 256
assets, depreciation 229–30
asymptote 173, 198, 199
average cost (AC) 276
   marginal cost from 279
   progress exercises 280–1
   relationship to AVC, AFC and MC 329–33
   total cost from 278
average fixed cost (AFC) 276
   relationship to AC, AVC and MC 329–33

average functions 275–9, 358
   progress exercises 280–1
   relationship to marginal functions 382, 383
average product of capital (APK) 382, 383, 397
average product of labour (APL) 282, 382, 383
   curvature and points of inflection 327
   progress exercises 285–6
average propensity to consume and save 283–5
   progress exercises 285–6
average revenue (AR) 275
   for a perfectly competitive firm and a monopolist 276–8
   progress exercises 280–1
   relationship to price 275
average variable cost (AVC) 276
   relationship to AC, AFC and MC 329–33

back substitution 498, 500, 502
black market profits 115–16
bond prices, link to interest rates 248–51
brackets 2, 5
   'y depends on x' notation 53
break-even point(s) 125–6, 142
   Excel worked example 137–8
   and linear profit function 74–5
   non-linear functions 162–3, 168–9, 203–5, 307–9
   progress exercises 126–7
budget constraints 91–2, 98, 410
   effect of price and income changes 92–6
   maximising total revenue subject to 411–12
   utility maximisation 413–17
budget line 92, 93–5, 98

calculator 24
   evaluation of formulae using 24–6
   exercises using 33–4
   exponentials 170, 171
   logarithms 186, 194
   transposition of formulae 26–8
calculus 261

capital
average product of (APK) 382, 397
marginal product of (MP_K) 381–7
partial elasticity of 396–7
capital recovery factor 243, 244, 255
chain rule for differentiation 336–7, 357
progress exercises 338, 345–7
circular flow model 58
Cobb-Douglas
production function 380, 424
and average products of capital and labour 383
homogenous function 389
MRTS in reduced form for 387
partial derivatives for 381
and partial elasticity of labour and capital 396–7
and returns to scale 388–9
utility function 390
cobweb model 557–60, 564–5
coefficients
income elasticity of demand 90
inverse matrices 524, 526–7
point elasticity of demand 85–8, 89
price elasticity of demand 84, 88–9
price elasticity of supply 90
cofactor method, inverse matrices 520–1
common difference, arithmetic sequence 210, 216, 254
common ratio, geometric sequence 210, 211, 216, 254
complementary function (CF) 548, 549, 551, 555, 558, 564
complementary goods 59
market equilibrium for 118–19
compound interest 219
applications of 221–2
compounded several times a year 223–4
continuous compounding 224
for fixed periodic deposits 236–8
formula 219–20
present value at 220–1
progress exercises 222–3
concave down/up, curvature 320, 321, 322, 358
constant elasticity demand function 352–3
constant proportional rate of growth 470–1
constant returns to scale 388, 389, 424
constant term, integral of 432, 434, 436
constrained optimisation 410–11
Lagrange multipliers 411–13, 424
linear programming 478–88
maximum output subject to production constraints 417–19
maximum utility subject to budget constraint 413–17

minimum cost subject to production constraint 419–20
progress exercises 420–2
consumer surplus (CS) 128–9, 142, 448–9, 474
exponential functions 453
at market equilibrium 130–1
progress exercises 131–2
using definite integral to calculate 449–51
consumption
Cramer's rule for income-determination model 511
and lagged income model 554
and limited growth 181–3
marginal and average propensity to consume 283–5
and national income model 133–5, 142–3, 515–16
rate of change in 461–2
continuous compounding 225
and annual percentage rate (APR) 225–7
Excel worked example 252, 253
conversion periods, compound interest 223–4
cost see average cost; marginal cost; total cost
cost constraints 91–2, 98
maximum output subject to 417–19
cost function 71–2
equation and graph of 480–1
minimising subject to constraints 419–20, 478–82
progress exercises 75–6
translation of linear 83
Cramer's rule 535–6
applications of 515
income-determination model 511, 515–16
market equilibrium for two products 510
solution of 2 equations in 2 unknowns 505–11
solution of 3 equations in 3 unknowns 513–15
cross-price elasticity of demand 395, 396, 424
cubic equations/functions 165
break-even points 168–9, 203–5
general properties of 168
plotting graphs of 165–7
total cost functions 203–5
cumulative present value factor 241
currency conversions 14–16
curvature 358
convex or concave towards origin 323–4
economic applications 322
and second derivative 320–2
for total cost functions 327–8
see also points of inflection (PoI)
curves
derived curves 297–300
indifference curves, utility functions 391–3, 414–15

sketching 300–4
see also graphs

debt repayments 242–3, 255
  by creating sinking fund 245–7
  progress exercises 247–8
  proportion of interest in 244–5
decay *see* growth (and decay)
decimal places, rounding numbers 2–3
decreasing returns to scale 388, 389, 424
definite integration
  area between two curves 455
  area under a curve 441–3
  definite integral, evaluation of 443–4
  and ex 444–5
  and net area between curve and x-axis 445–6
  progress exercises 447–8
  when F(x) = ln |x| 446
demand function 59–60, 98
  calculating marginal revenue from 271–2
  constant elasticity 352–3
  equation of 61
  exponential 344
  hyperbolic shape 200–1, 450
  linear 59–64
    elasticity 83–91
  non-linear 158–60
    point elasticity of demand for 351–2
  partial elasticities 395–6
  progress exercises 68–70
dependent variable 53
  difference equations 541, 564
depreciation 228–30, 255
  progress exercises 236
derivatives 268
  rules for finding 263–6, 334–5, 357
  see also second derivative
derived curves 297–300
determinants 504, 535–6
  Cramer's rule and applications 507–11
  definitions 504–5
  evaluation of 2 x 2 determinants 505
  evaluation of 3 x 3 determinants 513–17
  progress exercises 512, 517
  solution of simultaneous equations 505–6
deterministic model 55
difference equations, first-order 539–41
  applications of 554–64
  homogenous 542–5
  non-homogenous 548–52
  progress exercises 553–4

stability and time path to stability 545–8
  summary 564–5
  terminology 541
  test exercises 565
differential equations 457
  of the form $dy/dx = f(x)$ 458–9
  of the form $dy/dx = f(x)g(y)$ 467
  of the form $dy/dx = ky$ 464–6
  general and particular solutions 457–8
  for limited and unlimited growth 468–73
  limited and unlimited growth/decay 468–73
  and rates of change 460–2
differentials
  for functions of two variables 376
  and small changes 374–80
differentiation 259
  applications 304–20
  chain, product and quotient rules 334–47
  curvature and points of inflection 320–34
  elasticity 347–56
  integration reversing 428–9
  marginal and average functions 270–81
  marginal propensity to consume and save 283–5
  optimisation for functions of one variable 286–304
  production functions 281–2
  and slope of a curve 260–70
  summary 357–9
  test exercises 359–60
  see also partial differentiation
diminishing marginal rate of substitution 391–2
diminishing marginal rate of technical substitution 383–4
diminishing returns 381–2
discount rate, NPV 220–1, 231–2, 234–5
division 5
  fractions 8–9

e^x *see* exponential functions
economic models 57–9
elastic demand 88–9
elasticity (ε) 83
  of demand, supply and income 83–91, 98
  and the derivative 347–56
  see also partial elasticities
elimination methods 498
  to find inverse of a matrix 518–20
  Gauss-Jordan 502–3
  Gaussian 498–502
  progress exercises 503–4
  solving simultaneous equations 102–5
equal matrices 489

equations
  cubic 165–8
  demand function 61
  elimination methods 498–504
  with exponentials, solving 178–9
  horizontal and vertical lines 47–8
  hyperbolic 200–2
  quadratic, solving 149–52
  simple algebraic 2, 11–14
  simultaneous 102–10, 505–10
  for slope of $y = x^2$, deriving from first principles
    261–3
  solving using inverse matrices 524–5
  straight line 43–6, 48–52, 78–81, 97
  supply function 64–5
  using log rules to solve 193–6
  writing in matrix form 523
  see also difference equations, first-order; differential
    equations
equilibrium 111–12
  break-even analysis 125–6
  complementary and substitute goods 118–20, 142
  condition, written as a difference equation 561–2,
    564–5
  goods and labour markets 112–14
  national income 133–7
  in the national income model 398, 515–16
  price controls 114–18
  progress exercises 126–7
  taxes, subsidies and their distribution 120–5
equivalent fractions 9–10
Excel 28–9
  calculations and graph plotting 29–32
  cost, revenue, break-even, per unit tax 137–8
  distribution of tax 138–41
  financial mathematics 251–4
  for linear algebra 531–4
  for linear functions 92–6
  for non-linear functions 202–5
  usefulness of 99
excise tax 83
expenditure multiplier 136
expenditure, national income model 133, 142–3
exponential functions (exponentials) 170–1
  consumer and producer surplus 453–5, 474
  definite integration 435–6
  derivatives of 334–6
  graphing 171–3
  properties of 173
  rules for using 173–4

simplifying 174–6
solving equations
  containing 10^x or e^x 187–8
  containing logs and exponentials 195
using Excel to evaluate 202–3
expression, definition of term 2
extreme point theorem 482

feasible region 480, 481, 482, 484, 485
financial mathematics 209–10
  annuities, debt repayments and sinking funds
    236–48
  arithmetic and geometric sequences and series
    210–17
  depreciation 228–30
  Excel for 251–4
  interest rates and price of bonds 248–51
  net present value and internal rate of return 230–6
  simple and compound interest and APRs 218–28
  summary 254–5
finite series 210
first derivative (y') 268, 317, 358
  and curvature and points of inflection 320–2
first-order difference equations see difference
    equations, first-order
first-order differential equations 456–7, 474
  applications 459–60
  $dy/dx = f(x)$ 458–9
  $dy/dx = f(x)g(y)$ 467
  $dy/dx = ky$ 464–6
  general and particular solutions 457–8
  progress exercises 459, 463–4, 471–3
  rates of change 460–2
  total cost from marginal cost 459–60
first-order partial derivatives 366–9, 378, 423
fixed costs 70, 273
fixed tax per unit of output 120–2, 142
formulae
  compound interest 219, 221
  Cramer's rule 508
  elasticity 84, 85, 90
  evaluation of 24–6, 27–8
  Excel 92–3
  financial mathematics 254–5
  incremental (small) changes 374, 390
  quadratic ('minus b') 149, 150
  series 211, 212, 254, 256
  tax distribution 124–5, 143–4
  transposition of 26–8
  see also equations